Advances in Phytonanotechnology for Treatment of Various Diseases

Several plant bioactives or plant-derived therapeutic molecules have been used against life-threatening diseases and their nanoparticle-mediated delivery greatly improves therapeutic efficacy. *Advances in Phytonanotechnology for Treatment of Various Diseases* aims to describe past and recent advances achieved in the field of phytonanotechnology. The chapters of this book provide thorough knowledge of medicinal plants, plant bioactives, plant extract-based nanoparticle synthesis, delivery of plant bioactives through nanoparticles, and their therapeutic activity against life-threatening diseases. This book focuses on the therapeutic activity of phytocomponents present in plants belonging to diverse families, genera, and species. The book also covers future aspects and challenges present in phytonanotechnology for curing human diseases. This book fulfils the thrust of researchers working in this field, which further encourages them to make new discoveries in this area and support the development of effective nanoparticle-mediated plant bioactive delivery platforms for treating human diseases.

Key Features

- Provides information on medicinal plants and plant bioactives used in the treatment of various life-threatening diseases.
- Covers green synthesis of nanoparticles as well as their toxicological and biological effects.
- Explores various nanocarriers routinely used in the delivery of plant bioactives and their efficacy for future drug delivery.
- Envisages future opportunities, challenges, and implementation made in phytonanotechnology-based treatment of human diseases.

Advances in Phytonanotechnology for Treatment of Various Diseases

Edited by
Prashant Kesharwani and Virendra Gajbhiye

CRC Press
Taylor & Francis Group
Boca Raton London New York

CRC Press is an imprint of the
Taylor & Francis Group, an informa business

Designed cover image: © Shutterstock

First edition published 2024
by CRC Press
6000 Broken Sound Parkway NW, Suite 300, Boca Raton, FL 33487-2742

and by CRC Press
4 Park Square, Milton Park, Abingdon, Oxon, OX14 4RN

CRC Press is an imprint of Taylor & Francis Group, LLC

ISBN: 978-1-032-13746-9 (hbk)
ISBN: 978-1-032-13969-2 (pbk)
ISBN: 978-1-003-23172-1 (ebk)

DOI: 10.1201/9781003231721

Typeset in Times
by Newgen Publishing UK

Contents

Preface

Plants are a potent source of therapeutic molecules, and bioactives are routinely used for the treatment of various diseases since ancient times. Active ingredients present in plants include alkaloids, saponins, polyphenols, terpenes, flavonoids, carnitine, choline, coenzyme Q, dithiolthiones, phytosterols, phytoestrogens, glucosinolates, polyphenols, taurine, lignins, and carotenoids, among others. These active components have been widely investigated for their antiviral, antibacterial, antifungal, anticancer, antidiabetic, anti-inflammatory, cardioprotective, and other effects and are proven effective against life-threatening diseases. However, their direct delivery into the human body has limitations such as low bioavailability, low water solubility, and short lifetime in the body.

Nanoparticle-mediated drug delivery is a promising approach in next-generation therapeutics. The properties of nanoparticles, such as small size, high drug-loading ability, and biocompatibility, attract its use as a potent delivery vehicle among the scientific community. It can overcome the limitations associated with traditional drug delivery and increases the availability of therapeutics at the target site with minimal side effects.

Instead of chemically synthesized nanoparticles, plant-based nanoparticles are safer, less toxic, and can deliver a broad range of therapeutic components. Plant extract-based synthesis of diverse types of metal nanoparticles such as silica, zinc oxide, iron oxide, gold, and silver nanoparticles has been described and used for therapeutic application. It has been proven that the delivery of therapeutic molecules through plant-based nanoparticles has promising outcomes against life-threatening diseases such as cancer, diabetes, neurodegenerative diseases, and cardiovascular diseases. Both plant-based phytonanoparticles and chemically synthesized nanoparticles have been investigated for the delivery of plant bioactives.

This book provides an insight into phytonanotechnology, its approaches, and its application in the treatment of various diseases. This book comprises 15 chapters covering diverse medicinal plants, plant bioactives, plant-based phytonanoparticles, and delivery of plant bioactives through nanoparticles. This book focuses on the synthesis of nanoparticles using plant extracts, isolation of plant bioactives, preparation of nanoparticle-based plant bioactive remedies, and therapeutic application of plant bioactives in the treatment of human diseases. The content of the book will greatly influence new researchers and students to develop a potential therapeutic platform by using plant bioactives and nanoparticles. With its comprehensive coverage of fundamental and applied aspects of the subject, the book will be immensely beneficial for readers and stimulate further interest. Furthermore, we hope that this book will encourage the readers toward newer developments and investigations on this subject. We are grateful to all the authors for their valuable and timely contributions. This book brings pioneers of the field and early career scientists together, demonstrating dedication toward the advancement of phytonanotechnology for the treatment of various diseases, which will help develop a potential therapeutic platform.

Dr. Prashant Kesharwani and Dr. Virendra Gajbhiye

Editor biographies

Dr. Prashant Kesharwani is an assistant professor of pharmaceutics in the School of Pharmaceutical Education and Research at Jamia Hamdard University (New Delhi, India). An overarching goal of his current research is the development of nanoengineered drug delivery systems for various diseases, with more focus on dendrimer-based drug delivery systems. He has authored numerous publications in international journals and books and has presented invited talks and presentations at international conferences. He is a recipient of many research grants from various funding bodies. He has also received several internationally acclaimed awards, including the most prestigious "Ramanujan Fellowship" from the Science and Engineering Research Board (SERB), Government of India. He actively participates in outreach and scientific dissemination for the service of the wider community. He has more than 13,000 citations for his research, with an h-index of 61 (according to Google Scholar).

Dr. Virendra Gajbhiye has been working in the field of nanomedicine since the last 16 years. He has a doctoral degree (Ph.D.) in pharmaceutical science with postdoctoral research experience at the University of Wisconsin–Madison and Oregon State University. Since 2013, he has been working as a scientist in nanomedicine at Agharkar Research Institute, Pune, India. He has worked extensively with polymeric nanoparticles, specifically dendrimers. His research interest lies in nanomedicine, targeted drug and siRNA delivery, biomedical application of dendrimers, biomaterial-based unimolecular micelles and imaging, multifunctional polymeric nanoparticles, and nanoparticles in tissue engineering. He has more than 3500 citations for his research, with an h-index of 31 (according to Google Scholar).

Contributors

A. Kumar Department of Pharmacology and Biotechnology, School of Pharmaceutical Sciences, Delhi Pharmaceutical Sciences and Research University, Mehrauli-Badarpur Road, New Delhi 110017, India

A. Sharma Department of Pharmacognosy & Phytochemistry, School of Pharmaceutical Sciences, Delhi Pharmaceutical Sciences and Research University, Mehrauli-Badarpur Road, New Delhi 110017, India

A. Praveen Department of Pharmacognosy & Phytochemistry, School of Pharmaceutical Sciences, Delhi Pharmaceutical Sciences and Research University, Mehrauli-Badarpur Road, New Delhi 110017, India

Amirhossein Sahebkar Applied Biomedical Research Center, Mashhad University of Medical Sciences, Mashhad, Iran; Biotechnology Research Center, Pharmaceutical Technology Institute, Mashhad University of Medical Sciences, Mashhad, Iran; School of Medicine, The University of Western Australia, Perth, Australia

Antony V. Samrot School of Bioscience, Faculty of Medicine, Bioscience and Nursing, MAHSA University, Jenjarom 42610, Selangor, Malaysia

Asif Mahmood Department of Pharmacy, University of Chakwal, 48800 Chakwal, Pakistan

B. Mishra Department of Pharmaceutics, Indian Institute of Technology, Banaras Hindu University, Varanasi 221005, Uttar Pradesh, India

Bhushan P. Chaudhari Biochemical Sciences Division, CSIR-National Chemical Laboratory, Pashan, Pune 411008, Maharashtra, India, and Academy of Scientific and Innovative Research (AcSIR), Ghaziabad 201002, Uttar Pradesh, India

Çiğdem Kahraman Hacettepe University, Faculty of Pharmacy, Department of Pharmacognosy, 06100, Sıhhiye, Ankara, Turkey

Debalina Das Plant Molecular Cytogenetics Laboratory, Centre of Advanced Study, Department of Botany, Ballygunge Science College, University of Calcutta, Kolkata 700019, West Bengal, India

Dinesh Amalnerkar Nano Particles Technology Laboratory, School of Mechanical Engineering, Sungkyunkwan University, Suwon, Gyeonggi-do, 16419, South Korea; and Department of Technology, Savitribai Phule Pune University, Pune 411007, Maharashtra, India

H. Erdoğan University of Health Sciences, Hakan, Turkey

Hnin Ei Thu Research Institute for Medical and Health Sciences, University of Sharjah, Sharjah 27272, United Arab Emirates; Research and Innovation Department, Lincoln University College, Petaling Jaya 47301, Selangor, Malaysia

Indrani Manna Plant Molecular Cytogenetics Laboratory, Centre of Advanced Study, Department of Botany, Ballygunge Science College, University of Calcutta, Kolkata 700019, West Bengal, India

J.A. Badmus Department of Biochemistry, College of Health Sciences, Ladoke Akintola University of Technology, Ogbomosho, Nigeria

K. Sharma Department of Pharmaceutical Chemistry, School of Pharmaceutical Sciences, Delhi Pharmaceutical Sciences and Research University, Mehrauli-Badarpur Road, New Delhi 110017, India

Kamal Dua Discipline of Pharmacy, Graduate School of Health, University of Technology Sydney, PO Box 123 Broadway NSW 2007, Australia

Kavita Rai Gajbhiye Poona College of Pharmacy, Bharati Vidyapeeth (Deemed to be University) Pune 4110338, Maharashtra, India

Krithika Shree S. Department of Biotechnology, School of Bio and Chemical Engineering Sathyabama Institute of Science and Technology Chennai 600119, Tamil Nadu, India

M. Bacanlı University of Health Sciences, İstanbul, Turkey

M. Gupta Department of Pharmaceutics, School of Pharmaceutical Sciences, Delhi Pharmaceutical Sciences and Research University, Mehrauli-Badarpur Road, New Delhi 110017, India

Madhavan Nampoothiri Department of Pharmacology, Manipal College of Pharmaceutical Sciences, Manipal Academy of Higher Education, Manipal 576104, Karnataka, India

Manali M. Bagade Biochemical Sciences Division, CSIR-National Chemical Laboratory, Pashan, Pune 411008, Maharashtra, India, and Academy of Scientific and Innovative Research (AcSIR), Ghaziabad 201002, Uttar Pradesh, India

Manish D. Shinde Centre for Materials for Electronics Technology, Panchvati, Off Pashan Road, Pune 411008, Maharashtra, India

Maumita Bandyopadhyay Plant Molecular Cytogenetics Laboratory, Centre of Advanced Study, Department of Botany, Ballygunge Science College, University of Calcutta, Kolkata 700019, West Bengal, India

Monali Kadu Biodiversity & Palaeobiology Group, Agharkar Research Institute, Pune 411038, Maharashtra, India; Savitribai Phule Pune University, Pune 411007, Maharashtra, India

N. Savita Department of Pharmacognosy & Phytochemistry, School of Pharmaceutical Sciences, Delhi Pharmaceutical Sciences and Research University, Mehrauli-Badarpur Road, New Delhi 110017, India

Nadire Özenver Hacettepe University, Faculty of Pharmacy, Department of Pharmacognosy, 06100, Sıhhiye, Ankara, Turkey

Neha Kanojia Chitkara University, School of Pharmacy, Chitkara University, Himachal Pradesh 174103, India

Neha Parveen Department of Pharmaceutics, School of Pharmaceutical Education and Research, Jamia Hamdard, New Delhi, 110062, India

Niladri Haldar Nanobioscience Group, Agharkar Research Institute, Pune 411038, Maharashtra, India, Savitribai Phule Pune University, Pune-411007, Maharashtra, India

Nitin Verma Chitkara University, School of Pharmacy, Chitkara University, Himachal Pradesh 174103, India

Nor Amlizan Ramli Department of Pharmaceutics, Faculty of Pharmacy, Universiti Teknologi MARA, Puncak Alam Campus, Bandar Puncak Alam 42300, Selangor, Malaysia

Palanirajan Vijayaraj Kumar UCSI University, Faculty of Pharmaceutical Sciences, Department of Pharmaceutical Technology, Kuala Lumpur, Malaysia

Prasada Chowdari Gurram Department of Pharmacology, Manipal College of Pharmaceutical Sciences, Manipal Academy of Higher Education, Manipal 576104, Karnataka, India

Rai Muhammad Sarfraz College of Pharmacy, University of Sargodha, Sargodha, Punjab, Pakistan

Rajalakshmi D Department of Biotechnology, School of Bio and Chemical Engineering, Sathyabama Institute of Science and Technology Chennai 600119, Tamil Nadu, India

Rajesh Salve Nanobioscience Group, Agharkar Research Institute, Pune 411038, India; Savitribai Phule Pune University, Pune 411007, Maharashtra, India

Ritesh Kumar Choudhary Biodiversity & Palaeobiology Group, Agharkar Research Institute, Pune 411038, Maharashtra, India; Savitribai Phule Pune University, Pune 411007, Maharashtra, India

S. Bhatia Natural & Medical Sciences Research Center, University of Nizwa, Birkat Al Mauz, Nizwa, Oman

S. Goswami Department of Pharmacology and Biotechnology, School of Pharmaceutical Sciences, Delhi Pharmaceutical Sciences and Research University, Mehrauli-Badarpur Road, New Delhi 110017, India

Sairaj Satarker Department of Pharmacology, Manipal College of Pharmaceutical Sciences, Manipal Academy of Higher Education, Manipal 576104, Karnataka, India

Sanjana Varma Biochemical Sciences Division, CSIR-National Chemical Laboratory, Pashan, Pune 411008, Maharashtra, India; Academy of Scientific and Innovative Research (AcSIR), Ghaziabad 201002, Uttar Pradesh, India

Sanjay Preeth R Department of Biotechnology, School of Bio and Chemical Engineering, Sathyabama Institute of Science and Technology, Chennai 600119, Tamil Nadu, India

Sathiyasree M Department of Biotechnology, School of Bio and Chemical Engineering, Sathyabama Institute of Science and Technology, Chennai 600119, Tamil Nadu, India

Sejuty Mondal Plant Molecular Cytogenetics Laboratory, Centre of Advanced Study, Department of Botany, Ballygunge Science College, University of Calcutta, Kolkata 700019, West Berngal, India

Shahzeb Khan Department of Pharmacy, University of Malakand, Chakdara, Dir Lower 18800, Khyber Pakhtunkhwa, Pakistan; Division of Molecular Pharmaceutics and Drug Delivery, College of Pharmacy, The University of Texas, 2409 West University Avenue, PHR 4.116, Austin, TX 78712, USA; Discipline of Pharmaceutical Sciences, School of Health Sciences, University of KwaZulu-Natal, Private Bag X54001, Westville 3631, Durban 4000, South Africa

Shobana N. Department of Biotechnology, School of Bio and Chemical Engineering, Sathyabama Institute of Science and Technology, Chennai 600119, Tamil Nadu, India

Sree Lalitha Bojja Department of Pharmacology, Manipal College of Pharmaceutical Sciences, Manipal Academy of Higher Education, Manipal 576104, Karnataka, India

Tan Kin Fai UCSI University, Kuala Lumpur, Malaysia

TD Alabi Oxidative Stress Research Centre, Cape Peninsula University of Technology, Cape Town, South Africa

Vasudev Viswas Department of Pharmacognosy & Phytochemistry, School of Pharmaceutical Sciences, Delhi Pharmaceutical Sciences and Research University, Mehrauli-Badarpur Road, New Delhi 110017, India

Vidhu Aeri Department of Pharmacognosy and Phytochemistry, School of Pharmaceutical Education and Research, Jamia Hamdard University, Hamdard Nagar, New Delhi 110062, India

Vijender Kumar Department of Pharmacognosy & Phytochemistry, School of Pharmaceutical Sciences, Delhi Pharmaceutical Sciences and Research University, Mehrauli–Badarpur Road, New Delhi 110017, India

Waleed H. Almalki Department of Pharmacology and Toxicology, Faculty of Medicine, Umm Al-Qura University, Makkah, Saudi Arabia

Zahid Hussain Department of Pharmaceutics and Pharmaceutical Technology, College of Pharmacy, University of Sharjah, Sharjah 27272, United Arab Emirates; Research Institute for Medical and Health Sciences, University of Sharjah, Sharjah 27272, United Arab Emirates

1 Historical Perspectives and Advances in Phytonanotechnology

Monali Kadu[a,c], Rajesh Salve[b,c], Kavita Rai Gajbhiye[d], Ritesh Kumar Choudhary[a,c], and Virendra Gajbhiye[b,c*]*
[a]Biodiversity & Palaeobiology Group, Agharkar Research Institute, Pune 411004, Maharashtra, India
[b]Nanobioscience Group, Agharkar Research Institute, Pune 411004, Maharashtra, India
[c]Savitribai Phule Pune University, Pune 411007, Maharashtra, India
[d]Poona College of Pharmacy, Bharati Vidyapeeth (Deemed to be University), Pune 411038, Maharashtra, India

1.1 INTRODUCTION

The properties of a material, such as chemical reactivity, melting point, and optical emission (color changes), change as it transforms from the bulk to the nanoscale. Nanoparticles (NPs) are unique materials obtained by such a transformation process, and their size ranges from 1 to 100 nm (nanometer). Nanotechnology is a field of science that deals with the study and application of NPs. In recent years, nanotechnology has developed into a multidisciplinary area with a broad range of applications. Nanotechnology research focuses on the synthesis and application of physical, chemical, and biological systems at sizes ranging from individual atoms or molecules to submicron components. Nanotechnology is a broad term that encompasses the assimilation of the resultant nanomaterials into bigger systems [1]. Nanotechnology has been used for various applications such as metallurgy and materials, energy storage, electronics, defense and security, optical engineering and communication, agriculture and food, and biomedical and drug delivery, among others. Phytonanotechnology is a branch of nanotechnology that deals with the application of NPs prepared from plant-based components. Research on plant-based NPs has been ongoing for nearly two decades. The potential of phytonanotechnology to address global food security challenges has recently sparked the interest of researchers in the nanotechnology field. Plant-based NPs have received increasing attention, as they are more cost-effective, environment-friendly, less time-consuming during synthesis and purification, and offer long-term safety of the product [2].

Recently, natural resources have been more useful in the treatment of several diseases owing to their anticancer, antimicrobial, and antidiabetic activities, among other functions. Historically, many natural compounds exert potent anticancer activity and have been proven very effective in treating advanced stages of common cancers. The ornamental plant *Catharanthus roseus* (L.) G. Don (Family: Apocynaceae), commonly known as periwinkle, contains vinca alkaloids such as vinorelbine, vinblastine, vincristine, and vindesine. The potential activity of these bioactives is restricted owing to their low bioavailability at the target site, hydrophobicity, and side effects. These issues can be resolved by utilizing nanoparticulate drug carriers, which improve drug stability and solubility, extend drug half-life in the bloodstream, reduce adverse effects in nontarget organs, and concentrate the drugs at the target site [3]. Nanotechnology-based pharmaceuticals and drug delivery systems are being developed and marketed to improve cancer therapy and disease management. The

DOI: 10.1201/9781003231721-1

use of NPs may remarkably improve drug delivery to the target site. Currently available therapeutic regimens require selective drug targeting to demonstrate enhanced activity in cancer cells while minimizing the damage to healthy cells.

The improved biological activity of NPs in *in vitro* and *in vivo* scenarios underscores their advantages over traditional drug delivery strategies. Recently, nanotechnology has been aiding researchers explore the benefits of plant-based compounds. Nanomaterials are being used in our daily activities, with a wide range of applications in healthcare and biomedicine [4]. For the synthesis of herbal medicines, it is essential to develop a consistent biological activity, a uniform chemical profile, or a quality assurance procedure. An overview by Choudhary et al. on the extraction and characterization of herbal medicines as well as the standardization of herbal nanomedicines suggested that these medicines have higher solubility and bioavailability owing to nanotechnological modifications [2]. Bioavailability, *Bhasma* (as a metal nanocarrier), and the potential of metabolomics in the preparation of enhanced phytotherapeutics have been evaluated extensively [5]. Physicians and patients alike have started recognizing and appreciating the benefits of herbal medications. A scientific approach that can aid in delivering the components of phytotherapeutics has become necessary. There is a need for a new drug delivery system (NDDS) to facilitate the delivery of herbal components to target sites and to decrease the frequency of administration. NDDSs also boost the therapeutic value by minimizing toxicities and enhancing drug bioavailability. The treatment of chronic disorders such as asthma and diabetes should be provided by utilizing nanocarriers as an NDDS in the standard treatment system [6]. Nanotechnology aids in disease diagnosis at an earlier level than previously possible. Nanosized herbal medications, including active components from sea sandwort, cassia twig, and licorice root, have been useful in treating cancers of the lungs, liver, bones, and skin [7].

1.2 HISTORICAL OVERVIEW AND CURRENT STATUS OF PHYTONANOTECHNOLOGY

The exact period of NP discovery remains unknown. However, some lines of evidence describe the use of NPs in the era of early human civilization. The Lycurgus cup from the fourth century AD, currently in the British Museum collection, is the best example of NPs' use in ancient times. The cup shows a dichroic property and appears in two colors. When light passes through the cup from the front side, the cup appears green, and when the light passes through the cup from the rear, the cup appears red-purple [8]. This phenomenon is attributed to the coating of silver (Ag) and gold NPs (GNPs; 50–100 nm) with copper (Cu) over the cup surface when analyzed by transmission electron microscopy (TEM). From the ninth through 17th centuries, lustrous ceramic glazes used in Islamic culture, and later in Europe, contained Ag, Cu, or other metal NPs [9]. According to some sources, gold extraction began in the fifth millennium BC near Varna (Bulgaria). It peaked at 10 tons per year in Egypt around 1200–1300 BC, which was during the construction of the great statue of Tutankhamun. It is also possible that "soluble" gold appeared first in Egypt and China around the fifth or fourth century BC, during the construction of the great statue of Tutankhamun [10]. Some of the remnants were also discovered in India. The microstructure of wootz steel, which was made in ancient India around 900 AD and exported worldwide, comprised carbon nanotubes and cementite nanowires. In 2009, Alexander extensively reviewed available literature on the history of silver and its uses. The review showed that silver had many health benefits in ancient times, especially in BCE, CE, and the 18th–20th centuries. Before Lister's time, silver nitrate was used to treat skin ulcers, complex fractures, and other infectious diseases. During 1900–1940, colloidal silver was an efficient treatment material used for puerperal sepsis [11].

Following American physicist Richard Feynman's lecture on the topic "There's Plenty of Room at the Bottom" on December 29, 1959, at a meeting of the American Physical Society at Caltech (primarily regarded as initiating the area of nanotechnology), researchers began to primarily focus on NPs and their properties. Feynman proposed a strategy for honing the ability to manipulate individual atoms and molecules using one set of precise tools to build and operate a correspondingly smaller set and so on down until the desired size was reached [12]. Consequently, in the early 1990s, curiosity about "Plenty of Room" surged among the scientific community. Further, the term "nanotechnology" used in the book *Engines of Creation: The Coming Era of Nanotechnology*, by K. Eric Drexler in 1986, explored Feynman's theory of billion tiny factories [13]. Additionally, the discovery of the scanning tunneling microscope by Gerd Binnig and Heinrich Rohrer in Zurich Research Laboratory at IBM opened a new era in nanotechnology. Since then, surface imaging of nanosurfaces has become possible, and for this concept, Gerd Binnig and Heinrich Rohrer were awarded the Nobel Prize in Physics in 1986 [14,15]. Richard Adolf Zsigmondy, who received the Noble Prize in Chemistry for his work on the heterogeneous nature of colloidal solutions, first used the term "nanometer" after observing NPs from gold colloidal sols as small as 10 nm under an ultramicroscope.

The current era of nanotechnology focuses on the synthesis of a wide range of NPs, their characterization, and their application for human well-being. Researchers in this field have developed new synthesis methods that are scalable. Novel technological advances in NP characterization have greatly helped researchers to understand the properties of novel NPs and apply these NPs in daily life. Nanotechnology has greatly influenced healthcare, and numerous nanomedicines have been developed. The application of NPs in diagnostics and therapeutics has shortened the time taken for disease diagnosis and enhanced the therapeutic potential of drugs. Plant-based NPs and plant bioactives play an important and outstanding role in disease therapeutics, as they have a high therapeutic ability and cause minimal side effects. The extracts of many medicinal plants are being used to synthesize NPs, and medicinal plant bioactives are delivered to the target site through synthetic NPs. Such plant-based NPs have shown promising outcomes against cancer, diabetes, neurological disorders, cardiovascular diseases (CVDs), and other diseases. Thus, phytonanotechnology is expected to improve healthcare and therapeutics in the future.

1.3 PLANT EXTRACTS FOR THE GREEN SYNTHESIS OF NANOPARTICLES

NPs can be synthesized in various ways such as by physical [16], chemical [17], and biological [18] methods. Researchers have resorted to biological methods as a means of improving green synthesis, as this approach uses natural reducing, capping, and stabilizing agents instead of toxic and expensive chemicals and considerable energy consumption for NP synthesis [19–21]. Figure 1.1 shows the green synthesis approach for NP synthesis. Various organic and inorganic NPs have been synthesized from plant-based compounds or plant extracts. Plant extracts are prepared from various plant parts including seeds, tuber, root, stem, leaf, fruit, bark powder, fruit peel, and latex, and NPs of metals and metal oxides are prepared from these extracts. NP extraction entails collection and preparation of the plant tissue, homogenization, and size-based separation of bioactive molecules. The plants used for NPs preparation are either collected from the wild or grown from seedlings in a greenhouse. In the extraction process, the material, i.e., shoot, leaves, or root, is sterilized with 1.25% sodium hypochlorite or 0.05% Tween 20. The sterilized material is then homogenized in a sonic dismembrator, and the extract is collected, filtered, and dialyzed. Finally, NPs are separated based on size by sucrose density gradient centrifugation or size-exclusion chromatography [22].

The concept of green synthesis of NPs may be originated from the term "phytomining." This term is used in the mining industry to describe the use of hyperaccumulating plants to extract metals from soil, and metals extracted from the biomass have a high economic value. A new field, named gold phytomining, which involves the extraction of metals from soils or ores for recovery and sale, has emerged and gained attention [23]. Investigations on the synthesis of plant-based NPs have unraveled diverse forms of NPs existing in plants, and some NPs can be synthesized from plant extracts. Most of the investigations have focused on the extraction of metal NPs. Arya et al. showed that the marine alga *Sargassum wightii* can

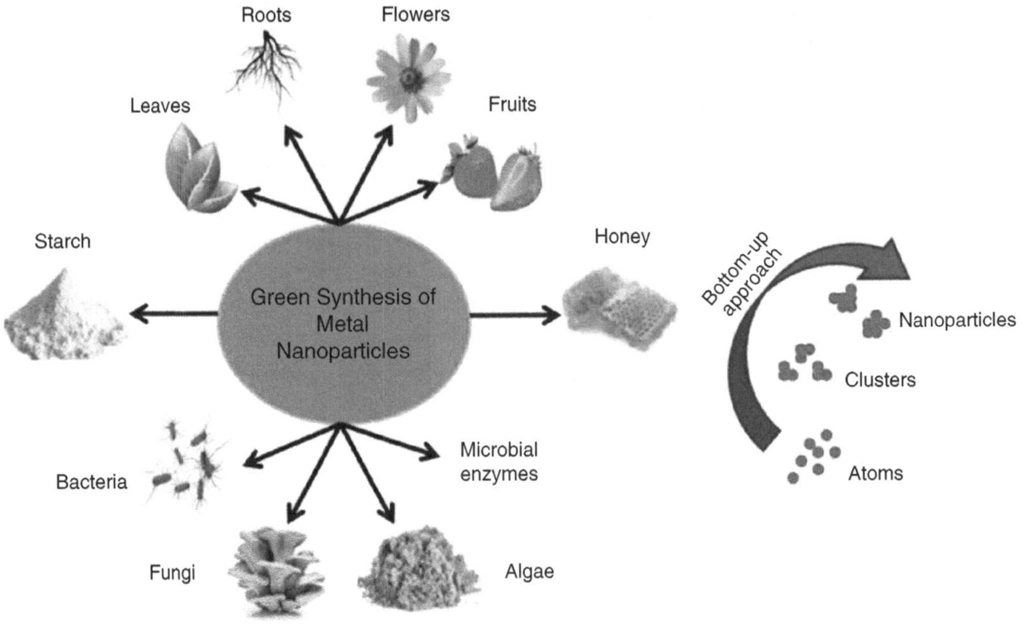

FIGURE 1.1 Different methods of green synthesis used for the preparation of metal nanoparticles. (From Kumar, H., et al., Nanomaterials. 2020 Apr 16;10(4):766. With Permission) [19]. This figure is reproduced under the terms and conditions of the Creative Commons Attribution (CC BY) license 4.0.

be used to rapidly synthesize highly stable extracellular GNPs when compared with other biological techniques. A diverse variety of nanoscale applications such as membranes and controlled-release systems could benefit from the alga's ability to regulate silica NP synthesis and assembly under biological settings [24]. Jha et al. developed a low-cost repeatable green synthesis method for preparing AgNPs from *Cycas* leaf extract. Spherically molded NPs with a diameter of 2–6 nm were obtained. When Rietveld analysis was applied to the X-ray data, AgNPs were found to have a face-centered cubic unit cell structure. The authors also elaborated on the possible formation mechanism of AgNPs and discussed the active involvement of ascorbic/dehydroascorbic acid, amenti/hinoki flavones, and ascorbates/glutathiones/metallothioneins in the synthesis [25].

AgNPs have antibacterial and antiviral properties, which is proven effective against various diseases, which makes them ideal for antiviral and antibacterial drug development. AgNPs were synthesized from *Holarrhena antidysenterica* leaf extract, and their antibacterial activity was determined. The researchers used *Pseudomonas aeruginosa* and *Colletotrichum musae* as test cultures. The most significant inhibitory effect against microorganisms was achieved when 3 mM silver nitrate was used [26]. AgNPs from *Phyllanthus pinnatus* were evaluated for their antibacterial efficacy against *Vibrio cholerae* and *Shigella flexneri*, which displayed a maximum inhibition zone of 1.8 mm diameter. Although the study demonstrated significant antibacterial activity *in vitro*, further research is needed to prove AgNPs' bactericidal potential in animal models for drug development [27]. Furthermore, AgNPs were synthesized from *Andrographis paniculata*, *Phyllanthus niruri*, and *Tinospora cordifolia* extracts and were investigated for their antiviral effects against chikungunya virus (CHIKV). The NPs prepared from *Andrographis paniculata*, *Phyllanthus. niruri*, and *Tinospora cordifolia* extracts showed the highest cytotoxicity at doses of 31.25 μg mL^{-1}, 125 μg mL^{-1}, and 250 μg mL^{-1} on Vero cells, respectively. Additionally, the synthesized AgNPs demonstrated the highest degree of inhibition due to cytotoxic effects *in vitro* and significantly increased the viability of CHIKV-infected cells [28].

Researchers have been scrutinizing the green synthesis of GNPs. Kattie et al. developed a new method for the synthesis of spherical GNPs by combining cumin with sodium tetrachloroaurate (precursor for gold) in an aqueous medium. The synthesis was completed in 30 min, and the phytochemical coating on the NPs enabled nontoxicity of the NPs to normal fibroblast cells. Their research findings reveal a novel "nano-nature" link for the synthesis and application of GNPs [29]. Many lines of evidence have shown that green-synthesized GNPs appeared triangular and hexagonal. Ghodake et al. reported pear fruit-mediated alkaline condition-based GNP synthesis. The authors showed that green synthesis using pear fruit extract and water offers a dependable and economical technique for reducing and growing gold nuclei using an efficient bioresource. The NPs appeared as triangular and hexagonal nanoplates made of hexagonal GNPs, and their size ranged from 200 to 500 nm. The hexagonal nanoplates had a width of 12–20 nm [30]. The hexagonal nanoplate GNPs were also synthesized from the leaf extract of the no-calorie sweetener herb *Stevia rebaudiana*. Small-sized (8–20 nm) well-dispersed octahedral GNPs were observed under high-resolution TEM (HR-TEM) and exhibited Bragg's pattern of reflection confirmed by X-ray diffraction analysis [31].

Some studies have reported the extraction of silica NPs from *Cynodon dactylon* [32,33]. Silica NPs were also extracted from *Azadirachta indica*, *Bambusa vulgaris*, and *Thuja orientalis* and were used for their inhibitory effect on on biofilm formation and for other environmental applications [34–36]. Green-synthesized copper NPs from *Ageratum houstonianum*, *Celastrus paniculatus*, *Hagenia abyssinica*, and *Jatropha curcas* and leaf extract exerted photocatalytic, antifungal, and antibacterial activities [37–40]. Figure 1.2 illustrates the synthesis of copper NPs from leaf extracts. Some research groups synthesized copper NPs from *Abutilon indicum*, *Azadirachta indica*, *Bauhinia tomentosa*, *Camellia sinensis*, *Carica papaya*, *Persea americana*, *Solanum trilobatum*, *Ziziphora tenuior*, pomegranate seeds extract, and green tea leaves. The copper NPs were further investigated for their antibacterial activity, photocatalytic breakdown of remazol yellow RR dye, application toward 1,3-diolein synthesis, and photocatalytic activity for textile dye degradation [41–45]. Karthikeyan et al. reported the synthesis of multi-walled carbon nanotubes (MWNTs) by spraying biodiesel oil from *Jatropha curcas* over a well-distributed ferro-copper-molybdenum

FIGURE 1.2 Scheme of the synthesis of green-synthesized copper nanoparticles. (From Murthy, H.C., et al., J. Nanomater. 2020 Jun 1;2020. With Permission) [37]. The figure has been reproduced under the Creative Commons Attribution license.

catalyst supported on either alumina or silica. These NPs were analyzed by SEM, HR-TEM, and Raman spectroscopy, all of which confirmed the characteristics of MWNTs and the diameter range of 16–24 nm [46].

1.4 ADVANCEMENTS IN PHYTONANOTECHNOLOGY

Plant-based NPs and plant bioactives can be used to treat diverse diseases. This section covers a broad overview of plant-based NPs, their synthesis methods, and delivery of NP-based bioactives for developing effective therapeutics.

1.4.1 Phytonanotechnology for the Delivery of Plant Bioactives

Herbal medicines have been used for centuries in many cultures globally. More than 7500 medicinal plants are being used daily in India to cure a variety of diseases. Kashmir Himalayan medicinal plants account for nearly 40% in terms of their usage in the Indian pharmaceutical industry. Nanotechnology combined with conventional medicine could be a useful tool for developing future medications [47]. Most medicinal plants' active ingredients are water-soluble but poorly absorbed into the body owing to their inability to permeate cellular lipid membranes, chemical instability, low gastrointestinal absorption, poor penetration/bioaccumulation, and scarce biodistribution throughout the body, all of which limit their activity in the treatment of diseases. These characteristics almost probably reduce the rate and amount of oral drug absorption from solid dosage forms and their bioavailability in the body. As conventional therapies fail to satisfy these criteria, NPs extracted from medicinal plants or synthesized from plant extracts have been thoroughly studied for their therapeutic potential against life-threatening diseases. In traditional medicine, *Arrabidaea chica*, an indigenous vine, is used to treat wounds, inflammation, and gastrointestinal colic. Chitosan–sodium tripolyphosphate NPs loaded with the *Arrabidaea chica* standardized extract have been used as an ulcer-healing pharmacological system. This formulation reduced ulcer lesion development at a lower dose than that of the raw extract and could be used to treat ulcers [48]. *Tinospora cordifolia* is a medicinal plant used to treat cancer, either alone or in combination with other herbal therapies. Karuppath et al. evaluated the effects of nanoformulations prepared from *Tinospora cordifolia* stem extract on prostate cancer cells. The nanoformulation exerted antiproliferative activity against prostate cancer cells [49]. Blood–brain barrier (BBB) is a major obstacle for drug delivery to the brain. Herbal medications for neurodegenerative diseases are becoming increasingly popular. Banu et al. developed a novel nano-enabled nasal phytoformulation and reported its application to treat neurodegenerative diseases. Several natural medications, including *Centella asiatica*, have been explored for their therapeutic effects against Alzheimer's disease and dementia [50].

Phytonanotechnology research has also shown that phytonanoparticles can be used to treat coronavirus disease. Bioactive-containing plant extracts have been used to treat severe acute respiratory syndrome coronavirus 2 infection, wherein the bioactives exert antioxidant, anti-inflammatory, and immunomodulatory effects [51]. With recent breakthrough research in nanotechnology, biomedical and pharmacological applications of NPs have recently gained attention. Traditional Chinese herbal medicines have been used as antibiotics and anticancer agents for thousands of years, and the medicinal herbs are obtained from natural resources. Biologically generated AgNPs have challenged physical and chemical techniques to manufacture AgNPs. The physicochemical and biological features of AgNPs synthesized from Chinese herbal medicines were studied in detail for their antibacterial and anticancer activities, wound healing ability, and other biomedical applications. There are several ways by which AgNPs may harm the cell ultrastructure, such as induction of apoptosis, production of reactive oxygen species (ROS), inactivation of proteins, and dysregulation of multiple signaling pathways. Further research into the numerous bioactive components and AgNPs present in

Chinese herbs is needed to develop a synergistic strategy to enhance the therapeutic effectiveness of drug delivery, control antimicrobial resistance, and prevent newly emerging diseases [52].

1.4.2 Phytonanocarriers for Anticancer Activity

The direct administration of anticancer drugs is associated with limitations such as low bioavailability, low water-solubility, short half-life in the body, toxicity, and major side effects. The green synthesis method of anticancer drug delivery can overcome these limitations and has shown successful therapeutic outcomes. Chemically synthesized or plant-based drugs can be delivered to the tumor site through phytonanoparticles. AgNPs synthesized from extracts of *Alternanthera sessilis*, *Allium sativum*, *Azadirachta indica*, *Boerhavia erecta*, *Chlorophytum borivilianum*, *Derris trifoliata*, *Dimocarpus longan*, *Detarium microcarpum*, *Elephantopus scaber*, *Eucalyptus procera*, *Euprenolepis procera*, *Fagonia indica*, *Ganoderma neo-japonicum*, *Ginkgo biloba*, *Jasminum officinale*, *Lavandula dentata*, *Litchi chinensis*, *Mentha arvensis*, *Nepeta deflersiana*, *Noctiluca scintillans*, *Nostoc linckia*, *Solanum trilobatum*, *Olea chrysophylla*, *Perilla frutescens*, *Punica granatum*, *Putranjiva roxburghii*, *Rhizophora apiculata*, *Salvia miltiorrhiza*, *Zanthoxylum rhetsa*, and *Zingiber officinale* have shown anticancer activity when the NPs were delivered to the breast, ovarian, lung, cervical, liver, and prostate cancer cells [53]. In this regard, PEGylated silver-coated graphene nanocomposites were prepared from *Azadirachta indica* extract and delivered doxorubicin (DOX) to the target site. This formulation showed significant cytotoxicity against HeLa cancer cell line and had minimal toxicity on HaCaT normal cell lines [54]. Another study used *Eucalyptus procera* extract to prepare spherical AgNPs (130-nm diameter) that were used to deliver imatinib to MCF-7 breast cancer cells. The NPs showed sustained release of imatinib and dose-dependent cytotoxicity against the MCF-7 cell line [55].

In a similar vein, GNPs exhibit multiple functionalities such as delivery of anticancer agents, contrast agents, and photothermal agents and carrier of drugs. Kamala Priya and Iyer reported that GNPs synthesized with extracts of *Camellia sinensis*, *Mentha arvensis*, *Coriandrum sativum*, *Phyllanthus amarus*, *Mimusops elengi*, *Syzygium aromaticum*, and *Artabotrys hexapetalus* exerted antitumor activity against MCF-7 and that a GNP dose of 2 μg/mL was the minimum concentration for effective cancer therapy [56]. GNPs prepared from *Taxus baccata* demonstrated a caspase-independent cell killing program as an anticancer mechanism, with improved therapeutic efficacy against cancer [57]. Similarly, GNPs prepared from *Dysosma pleiantha* rhizome extract showed antiproliferative and antimetastatic activities against the HT-1080 human fibrosarcoma cell line [58]. To date, chemically synthesized GNPs have mostly been used; only a few studies have documented plant-based green-synthesized GNP toxicity, *in vivo* therapy, and biodistribution. Green-synthesized GNPs prepared from *Peltophorum pterocarpum* leaf extract have been used for DOX administration both *in vitro* and *in vivo* in C57BL6/J female mice. When compared with free DOX and an untreated model, administration of the DOX-loaded drug delivery system inhibited A549 and B16F10 cancer cell growth *in vitro* and tumor formation *in vivo* [59]. Similarly, leaf extracts of Mentha piperita–derived GNPs were examined for their anti-inflammatory and analgesic effects *in vitro* against MDA-MB-231 and A549 cell lines in a Wistar rat model [Figure 1.3]. The GNPs showed substantial *in vitro* antitumor activity. However, an *in vivo* analysis revealed positive findings for both activities, with lower potency than that with regular medications, suggesting that GNPs can be used in combination with standard pharmaceuticals to improve their efficacy [60].

Platinum NPs (PtNPs) have the potential to reduce tumor burden. PtNPs prepared from ethanolic extracts of pomegranate peel have shown apoptosis induction in the MCF-7 cell line in cytotoxicity assays [61]. PtNPs have also been synthesized from *Mentha piperita* (peppermint) aqueous leaf extract. According to the characterization data, green-synthesized PtNPs appeared spherical, with a particle size of 54.3 nm. When these polyphenol-capped PtNPs were tested for cytotoxicity in human colon cancer cell line (HCT 116), the NPs demonstrated reduced cancer cell viability at

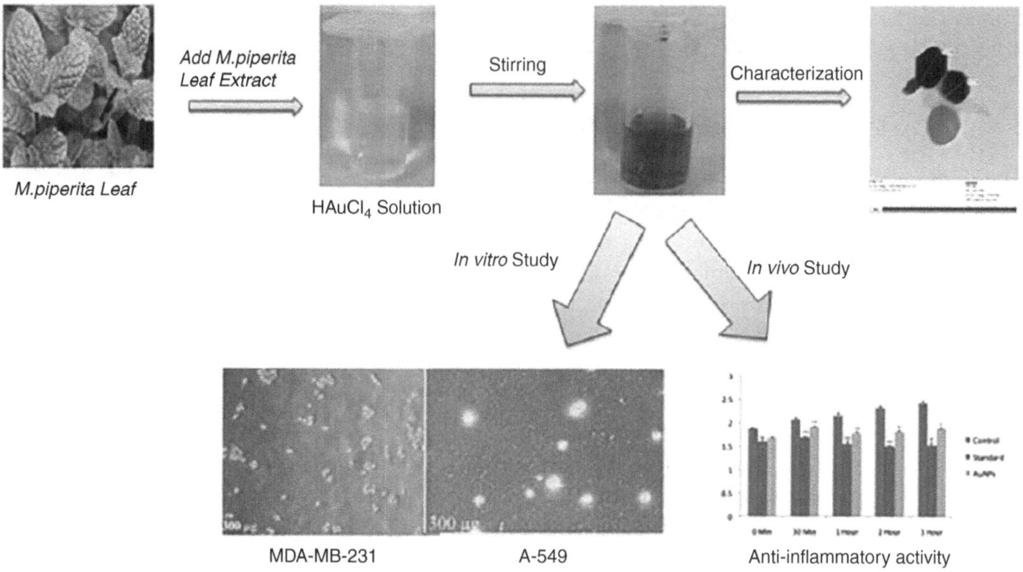

FIGURE 1.3 Schematic illustration of the preparation and application of gold nanoparticles. (From Ahmad, N., et al., Mater. Sci. Eng. C. Mater. Biol. Appl. 2017;78:553–564. With Permission) [60]. Copyright (2017) Elsevier.

low doses, with an IC_{50} value of 20 μg/mL [62]. PtNPs were also synthesized from black cumin seed extract and had an average diameter of 1–6 nm. These NPs showed anticancer effects against HeLa and MDA-MB-231 cell lines, with IC_{50} values of 19.83 μg/mL and 36.86 μg/mL, respectively. They also showed an antibacterial effect against gram-negative and gram-positive bacteria at concentrations of 500 μg/mL and 100 μg/mL NPs, respectively [63]. Platinum-palladium bimetallic NPs (Pt–PdNPs) as well as individual PtNPs and palladium NPs (PdNPs) have been synthesized from *Dioscorea bulbifera* tuber extract. HR-TEM analysis confirmed that monodispersed Pt–PdNPs, PtNPs, and PdNPs of size 2–5 nm, 25 nm, and 10 nm, respectively, were obtained. As for their anticancer activity against HeLa cells, Pt–PdNPs caused 74.25% greater cell cytotoxicity than individual PdNPs (33.15%) or PtNPs (12.6%) and exhibited improved scavenging efficacy against hydroxyl radicals, nitric oxide superoxide, and 2,2-diphenyl-1-picrylhydrazyl [64].

1.4.3 Phytonanocarriers for Antidiabetic Activity

Diabetes is a metabolic disorder involving defective insulin secretion, insulin action, or both, leading to an imbalance in glucose homeostasis. This condition results in higher accumulation of glucose in the blood (hyperglycemia) and low glucose uptake by the body tissues. Diabetes-induced infections have been a major concerning issue worldwide, as they are associated with serious clinical complications. New research outcomes in this field have shown a higher risk of tuberculosis infection, diabetic foot-related complications, dyslipidemia, diabetic peripheral neuropathy, and susceptibility to HIV/AIDS. Insulin administration can gradually cure this disease; thus, investigations have focused on invasive insulin delivery. Other than insulin, extracts from several medicinal plants and various NPs have proven effective against diabetes. Plant-based compounds from *Adansonia digitata*, *Allium sativum*, *Aloe arborescens*, *Aloe vera*, *Brachylaena discolor*, *Cassia fistula*, *Cinnamomum verum*, *Ocimum tenuiflorum*, *Piper nigrum*, *Momordica charantia*, and *Eugenia jambolana* have shown antidiabetic effects [65,66].

Zinc NPs (ZnNPs) are extensively used for diabetes treatment. Zinc is actively involved in glucose metabolism and regulates hepatic glycogenesis, intestinal glucose absorption, and glucagon secretion. It also plays a key role in insulin secretion, synthesis, and storage. Plant-based ZnNPs have shown more prominent antidiabetic activity. Many investigations were conducted as *in vitro* and *in vivo* analysis. The α-amylase and α-glucosidase are two key candidate enzymes involved in diabetes. Inhibition of these enzymes reduces the risk of hyperglycemia in a patient with diabetes. Vinotha et al. have shown that ZnNPs extracted from *Costus igneus* leaf extract have the potential to obstruct α-glucosidase and α-amylase. Another study has shown that the inhibitory rates of α-glucosidase and α-amylase were 82% and 74%, respectively, *in vitro*. These NPs have also exhibited antioxidant, anti-biofilm, and antibacterial activity when tested against the pathogenic bacteria *Lysinibacillus fusiformis*, *Proteus vulgaris*, *Streptococcus mutans*, and *Vibrio parahaemolyticus* [67]. Furthermore, ZnNPs have been synthesized from the fruit extract of *Vaccinium arctostaphylos* L. through a microwave-assisted method and tested for their antidiabetic activity. A study on alloxan-induced diabetes in a rat model showed a reduction in fasting glucose levels and an increase in high-density lipoprotein levels when treated with ZnO NPs synthesized with plant extracts, while such an effect was not observed when treated with insulin, chemically synthesized ZnO, and plant extracts [68].

Hasnain Jan et al. showed that synthesized from *Aquilegia pubiflora* extract had anticancer, antidiabetes, and anti–Alzheimer's disease activity. The 34.23-nm-sized ZnNP particles showed cell viability of 23.68±2.1% on HepG2 cells after 24 hours and excellent inhibitory potential against butyrylcholinesterase (IC_{50}: 125 μg/mL) and acetylcholinesterase (AChE; IC_{50}: 102 μg/mL). They also inhibited the activity of α-glucosidase and α-amylase [69]. ZnNPs synthesized from *Azadirachta indica*, *Hibiscus rosa-sinensis*, and *Moringa oleifera*, *Murraya koenigii*, and *Tamarindus indica* showed higher antioxidant and antidiabetic activities than chemically synthesized ZnNPs *in vitro*. Among these plants, *Tamarindus indica* showed significant α-glucosidase and α-amylase inhibition. In this study, chemically synthesized NPs did not demonstrate free-radical scavenging activity and exhibit lower antidiabetic activity, whereas plant-based NPs exerted potential antidiabetic and free radical scavenging activities [70]. Furthermore, Rajakumar et al. confirmed that ZnNPs generated from *Andrographis paniculata* leaf extract had greater antidiabetic potential in terms of α-amylase inhibitory activity, with an IC_{50} value of 121.42 μg/mL [71]. In another study, ZnNPs at a concentration range of 20–100 μg/mL displayed more effective antidiabetic activity, with inhibition percentages ranging from 20% to 74% for α-amylase and from 36% to 82% for α-glucosidase when prepared with *Costus igneus* leaf extract [67]. Recently, Faisal et al. reported that ZnNPs synthesized from aqueous fruit extracts of *Myristica fragrans* had excellent α-amylase (73.23%) and β-glucosidase (65.21%) inhibitory activity at a concentration of 400 μg/mL and showed that bio-based NPs may have potent antidiabetic abilities, making them a promising therapeutic agent for diabetes treatment as an alternative to expensive and ineffective medications [72].

AgNPs also exert an antidiabetic effect. The study carried out by Agarwal et al. evaluated the microwave irradiation approach versus the conventional heating approach for AgNP synthesis from an extract of lemon grass (*Cymbopogon citratus*). The antidiabetic effect of these NPs was evaluated through the glucose diffusion inhibitory assay and the amylase activity restriction assay, which revealed glucose dispersion retardation and dose-dependent inhibition of α-amylase, respectively [73]. Furthermore, AgNPs synthesized from the medicinal plant *Solanum nigrum* demonstrated antidiabetic activity and were used to treat alloxan-induced diabetic rats. Microscopy techniques confirmed the spherical and monodispersed nature of the NPs, showing an average diameter in the range of 4–25 nm. Rats treated with these NPs showed decreased blood glucose levels, improved dyslipidemia, and body weight gain when compared with control diabetic rats [74]. In a similar approach, *Moringa oleifera* extract was used to synthesize AgNPs, which were tested in diabetic rats. AgNP treatment (0.2 mg/kg) to streptozotocin-induced diabetic rats showed substantial inhibition ($p<0.001$) of blood glucose, triglycerides, low-density lipoprotein, cholesterol, and very low–density lipoprotein levels and body weight gain as compared to the leaf extract–treated group [75].

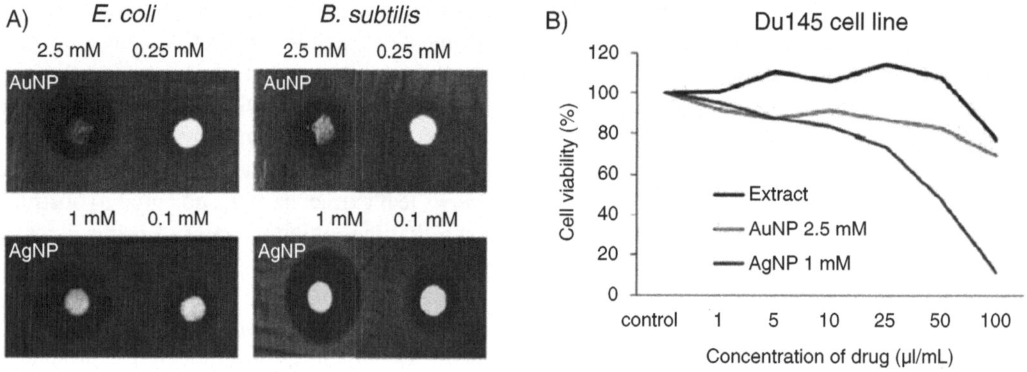

FIGURE 1.4 Screening of the synthesized silver and gold nanoparticles for their (A) antibacterial activity on *Escherichia coli* and *Bacillus subtilis* strains and (B) anticancer activity on prostate cancer cell line (DU145). (From Patra, N., et al., Adv. Nat. Sci.: Nanosci. Nanotechnol. 2018 Jul 25;9(3):035001. With Permission) [78]. This figure has been reproduced under the terms of the Creative Commons Attribution license 3.0.

GNPs have significant antioxidant and antihyperglycemic effect, and hence, they are useful in diabetes treatment. Ponnanikajamideen et al. synthesized GNPs from the insulin plant (*Chamaecostus cuspidatus*) extract that is known for its antidiabetic and wound-healing effects *in vivo*. A 21-day treatment of these NPs showed a dose-dependent free-radical scavenging activity and restored blood glycogen, glucose, and insulin levels in albino rats. NPs also showed cutaneous wound-healing activity together with rapid recovery as compared to the non-treated group [76]. The Procyanidins fraction from *Leucosidea sericea* and the entire extract of the whole plant were used to synthesize GNPs, and the synthesized GNPs were further investigated for their antidiabetic and antioxidant properties. The size of the NPs ranged from 6 to 21 nm and exhibited excellent α-amylase inhibition and strong α-glucosidase inhibition, with IC_{50} values of 1.88 µg/mL and 4.5 µg/mL, respectively, when prepared with the F1 and F2 fractions, respectively. F2- and F1-based NPs showed antioxidant activity of 1521.9±3.0 µM Trolox equivalents per gram and 1834.0±4.7 µM ascorbic acid equivalents per gram, respectively [77]. Furthermore, AgNPs and GNPs prepared from *Saraca asoca* extract displayed antibacterial and anticancer activities [Figure 1.4]. These NPs exerted antibacterial activity against *Escherichia coli* and *Bacillus subtilis*, antidiabetic activity, and prostate cancer cell–killing ability [78]. In another study, GNPs synthesized from *Moricandia nitens* were evaluated for glucosidase inhibitory activity, with an IC_{50} value of 159.3 µg/mL, and anti-*Helicobacter pylori* activity against multidrug-resistant *Helicobacter pylori* strains, with a minimum inhibitory concentration of 31.25 µg/mL. Additionally, these GNPs exhibited anticancer activity against HCT-116 cells (IC_{50}: 36±0.7 µg/mL) and HepG2 cells (IC_{50}: 44.6±0.8 µg/mL) [79].

1.4.4 Neurodegenerative Diseases and Phytonanomedicine

Neurodegenerative diseases are a group of neurological conditions that cause progressive dysfunction of the neuronal structure and function in the brain or spinal cord. Despite breakthroughs in our understanding of the pathophysiology of diseases related to the central nervous system, early detection and treatment of many conditions remain challenging. Biological hurdles such as the BBB contribute to these complications [80]. Neurogenesis and neuroplasticity are compromised in the brain tissues of patients with Alzheimer's disease and Parkinson's disease [81]. Alzheimer's disease involves characteristic alterations in the expression of phosphodiesterase (PDE) 4, PDE8, and PDE7 in the brain. Patients with multiple sclerosis have altered expression of numerous PDE subtypes in the central nervous system. Normalizing cyclic nucleotide signaling through PDE inhibition continues to be an intriguing therapeutic strategy for neurodegenerative diseases [82]. Following are some examples of plants that are useful as phytonanomedicines against neurogenerative diseases. *Centella asiatica* ethanolic extract confers protection to IMR32- and PC12-differentiated cells against Aβ-induced neurotoxicity by enhancing the antioxidative defense system [83]. When the same plant extract was treated to SH-SY5Y cells in the presence of nerve growth factor, it showed neurite outgrowth *in vitro* [84]. In another study of the same plant, the extract showed a neuroprotective effect on the amelioration of decreased AChE activity caused by colchicine induction, neuronal damage caused by asiaticoside induction, and nitric oxide inhibition [85]. *Celastrus paniculatus* water extract enhanced memory and learning by regulating glutamate receptor activation, thereby protecting neurons from glutamate toxicity. The extract also decreased malondialdehyde (MDA) levels in the brain, a critical indicator of oxidative stress [86]. Galantamine, the primary component of *Galanthus nivalis*, acts as a neuroprotective agent in two ways: inhibiting AChE and activating nicotinic receptors [87].

Lycopodium serratum is a fern ally belonging to the family Lycopodiaceae, and the species has unbranched microphylls and vascular threaded leaves. It contains huperzine A, a potential therapeutic option for Alzheimer's disease treatment [88]. An ethanolic extract of *Ocimum sanctum* leaves may help preserve neurons and increase AChE production in the cerebral microvascular endothelial cells of the elderly, which may enhance memory and cognitive function. The study showed that the extract boosts memory and cognitive performance, preserves cellular structure, replenishes and catalyzes enzyme levels and superoxide dismutase (SOD), and minimizes neuronal injury. This activity may be contributed by citric acid, tartaric acid, and maleic acid present in the extract of *Ocimum sanctum*. The *Ocimum sanctum* extract suppresses lipid peroxidation, ROS production, DNA damage, and membrane depolarization. The extract showed antioxidant activity against DPPH (2-diphenyl-1-picrylhydrazyl) and hydroxyl radicals [89]. *Terminalia chebula*, widely known in Tibet as "King of Medicines," is a member of the family Combretaceae. Ayurveda, Siddha, Unani, and Homeopathy use this plant as a traditional medicine. Additionally, *T. chebula* fruit extract protects neuronal cells from ischemia, decreases least formation, and increases the apoptotic rate of microglia cells when lipopolysaccharide is used [90]. In *in vitro* analysis, cis-9-octadecenamide (oleamide) from jujube increases acetylcholine levels and causes progression from mild to moderate cognitive functioning, behavioral difficulties, motor coordination, memory, and learning [91]. Turmeric, a golden spice derived from *Curcuma longa*, belonging to Zingiberaceae, has been used for medicinal purposes for millennia. Curcumin, the primary component of turmeric, exerts neuroprotective effects. Curcumin interacts with Aβ peptides and can be used for treating Alzheimer's disease [92]. Furthermore, *Hypericum perforatum* is a perennial herb that grows as a wild plant in Europe, western Asia, and northern Africa and prevents the increase in AChE and MDA in the brain and enhances SOD, glutathione peroxidase, and catalase levels. Additionally, the extract protects neurons against glutamate-induced cytotoxicity [93].

1.4.5 Wound Healing Activity of Phytonanoparticles

When tissue loss affects the normal architecture of an adult vertebrate, the organ becomes incapable of regeneration. The repair process is initiated by a fibroproliferative response that eventually results in the formation of a fibrotic scar. These events often result in chronic wounds that do not heal or have severe fibrosis [94]. The stages of exudation, proliferation, and extracellular matrix remodeling are sequential events that occur as a result of the integration of dynamic processes. Endothelial cells may develop into mesenchymal components, and this differentiation process seems to be tightly controlled by a series of signaling proteins [95].

Several lines of evidence have listed medicinal plants regularly used in wound healing therapies and how nanoformulations are prepared from these preparations. For instance, *Curcuma longa*, commonly recognized as turmeric rhizome, is an extensively used spice in Indian cuisine, which indicates its usage in foodstuffs. The plant has a diverse array of biological and therapeutic functions including anticancer, antimicrobial, anti-inflammatory, and free-radical scavenging properties [96]. In one nanoformulation study, curcumin-loaded poly(lactic-co-glycolic acid) NPs, curcumin-encapsulated NPs, and curcumin nanocomposite hydrogels were used for wound healing because they contain a high concentration of proteins and vitamin A, which promote collagen synthesis and fibroblast proliferation, thereby promoting fibrogenesis and angiogenesis [97]. *Helianthus annuus* (sunflower) seeds have antibacterial, antiinflammatory, and regenerative properties. Calves' tissues can be repaired by treating them with sunflower seeds [98]. *Moringa oleifera* exerts hepatoprotective, wound healing, and antihypertensive effects [99]. *Aloe vera* maintains a moist environment on the wound surface and boosts the wound's collagen content. Therefore, these results support the development of alternative wound dressings using *Aloe vera*. The combinatorial activity of *Aloe vera* and curcumin loaded onto silver-methacrylic acid nanohydrogels promoted rapid wound healing [Figure 1.5] [100].

Ashwagandha, the root of *Withania somnifera*, has anti-inflammatory effects and promotes wound healing [101]. The leaves, bark, and fruits of neem promote the wound healing mechanism, as they possess antiseptic, insecticidal, astringent, anti-inflammatory, and larvicidal properties [102].

AgNPs can be used as antimicrobial agents, drug delivery vehicles, and optical receptors. These NPs may aid in more rapid wound healing in individuals with diabetes, with less severe scarring. Silver nitrate is frequently used as an antimicrobial agent in the treatment of chronic wounds [103]. Researchers have investigated and assessed the anticarcinogenic and antibacterial effects of AgNPs prepared by the green synthesis method using *Rheum ribes* plant extract. Biologically synthesized AgNPs have a wide range of applications. Literature reports of AgNPs suggest the antibacterial and wound healing properties and sterile band applications of these NPs [104]. Islam et al. investigated

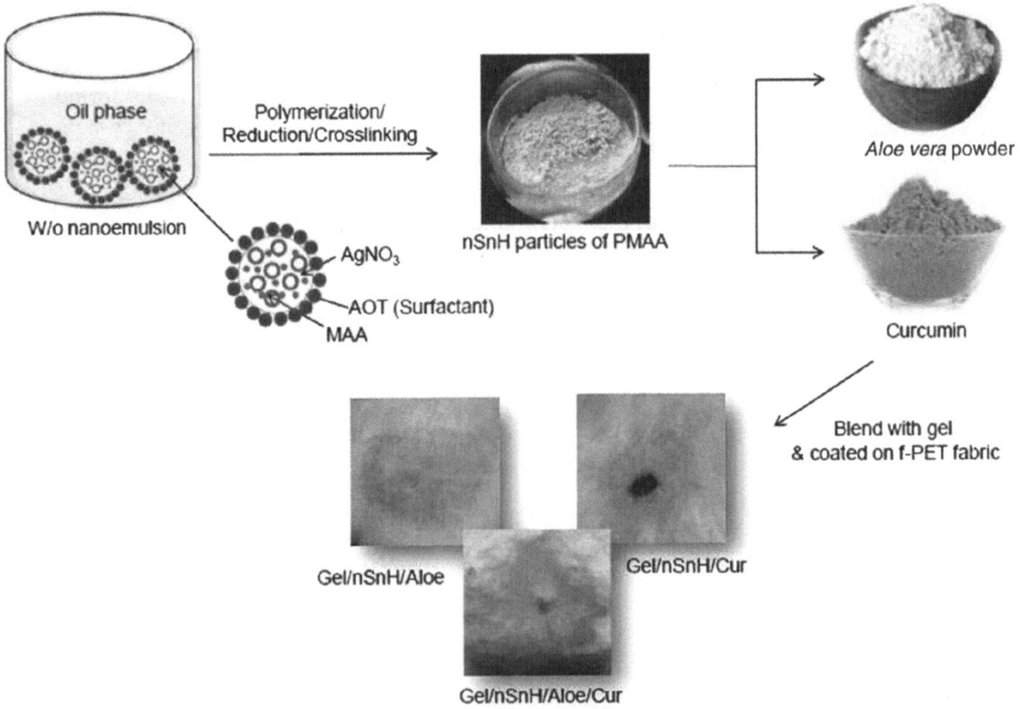

FIGURE 1.5 Schematic representation of the preparation of alternative antimicrobial composite wound dressings. (From Anjum, S., et al., Mater. Sci. Eng. C. Mater. Biol. Appl. 2016 Jul 1;64:157–166. With Permission) [100]. Copyright (2017) Elsevier.

the synergistic or additive effects of AgNP synthesis from Chinese herbs and reported their biological features and superior physicochemical and biomedical applications including antibacterial, wound healing, and anticancer properties [52].

1.4.6 Antifungal and Antibacterial Activities of Phytonanocarriers

Metallic NPs have the potential to restrict the growth of pathogenic bacteria. However, plant-based metallic NPs have more prominent activity than NPs synthesized by other routes. AgNPs possess antibacterial, antifungal, antioxidant, and anticancer effects. Environment-friendly AgNPs synthesized from leaf extracts of the medicinal plant *Tropaeolum majus* showed efficient eradication of *Pseudomonas aeruginosa* when compared with their effects on other microorganisms [105]. AgNPs were also successfully synthesized from *Capparis spinosa* extract, and these NPs were tested for their antibacterial effect on different pathogenic bacteria such as *Salmonella* typhimurium, *Escherichia coli*, and *Staphylococcus aureus*. The AgNPs were synthesized from hot water olive leaf extracts and appeared spherical, and their size ranged from 20 to 25 nm. AgNPs at concentrations of 0.03–0.07 mg/mL effectively suppressed the growth of multidrug-resistant *Pseudomonas aeruginosa*, *Staphylococcus aureus*, and *Escherichia coli* [106]. Iron oxide NPs were synthesized from *Platanus orientalis* leaf extract. The average diameter of these spherical NPs was 38 nm, and they had significant antifungal efficacy against *Aspergillus niger* and *Mucor piriformis* [107]. ZnNPs synthesized from *Solanum nigrum* and *Cinnamomum tamala* leaf extract exhibited antibacterial activity [108,109]. Furthermore, titanium dioxide and iron oxide NPs synthesized from *Azadirachta*

TABLE 1.1
Antibacterial Activity of Plant-Based Nanoparticles

Sr. No	Metal Used in the Nanoparticle	Extract Plant Part	Plant Name	Tested Microorganism (Bacteria)	Reference
1.	Silver	Leaf	*Capparis spinosa*, *Olea europaea*, *Chenopodium murale*, *Camellia sinensis*, *Pedalium murex*, *Ficus benghalensis*, *Skimmia laureola*, *Ceratonia siliqua*, *Ocimum tenuiflorum*, *Mimusops elengi*, and *Tamarix gallica*	*Bacillus pumilus*, *Bacillus subtilis*, *Corynebacterium*, *Escherichia coli*, *Klebsiella pneumoniae*, *Mariniluteicoccus flavus*, *Micrococcus luteus*, *Pseudomonas aeruginosa*, *Salmonella* typhimurium, and *Staphylococcus aureus*	[120]
2.	Copper	Leaf	*Ailanthus altissima*, *Aloe vera*, *Magnolia kobus*, *Malva sylvestris*, *Tabernaemontana divaricata*, and *Tridax procumbens*	*Aeromonas hydrophila*, *Escherichia coli*, *Flavobacterium branchiophilum*, *Pseudomonas fluorescens*, and *Staphylococcus aureus*	[113–116]
3.	Titanium	Leaf	*Azadirachta indica*	*Escherichia coli* and *Salmonella* typhimurium	[110]
4.	Iron oxide	Leaf	*Argemone mexicana*	*Bacillus subtilis* MTCC 441, *Escherichia coli* MTCC 443, and *Proteus mirabilis* MTCC 425	[111]
5.	Zinc	Leaf	*Solanum nigrum* and *Cinnamomum tamala*	*Escherichia coli*, *Salmonella* paratyphi, *Staphylococcus aureus*, and *Vibrio cholerae*	[108,109]

indica and *Argemone mexicana* leaf extract showed significant antibacterial activity [110,111]. Moreover, copper NPs were prepared through biological synthesis methods from *Magnolia kobus*, *Ailanthus altissima*, *Aloe vera*, *Malva sylvestris*, and *Tabernaemontana divaricata* leaf extract and were used as a reducing agent, and they showed antibacterial activity [112–116].

1.4.7 Cardiovascular Activity of Phytonanoparticles

CVDs are a group of disorders affecting the heart and blood vessels. They comprise diseases such as cerebrovascular diseases, coronary heart disease, peripheral arterial disease, rheumatic heart disease, deep vein thrombosis, congenital heart disease, and pulmonary embolism. CVDs are commonly caused by fat deposits on the inner walls of blood vessels. In 2019, the World Health Organization recorded 17.9 million deaths from CVDs, with stroke and heart attacks accounting for 85% of the total deaths [117]. Statins, beta-blockers, aspirin, and angiotensin-converting enzyme inhibitors are some of the useful medications for CVDs. Currently, drug delivery through NPs has attracted the attention of several researchers to develop safe and highly effective therapy against CVDs, which will further reduce surgical dependency. The promising activity of plant-based NPs in the treatment of CVDs has inspired the researchers to explore this area for the development of effective therapeutics. Triterpene glycosides or saponins are plant-derived organic chemicals that reduce low-density lipoprotein and cholesterol levels, thereby exerting a hypolipidemic effect. Saponins are a component of *Panax notoginseng* and possess therapeutic effects such as preventing ischemic brain damage, expanding peripheral vessels, and inhibiting thrombosis, among others [118]. Moreover, phytochemicals such as anolides, berberine, tilianin, baicalin, silymarin, curcumin, cyclovirobuxine D, naringenin, allicin, myricitrin, and resveratrol possess cardioprotective activities [119–122]. The chemical compound allicin, present abundantly in garlic, possesses an antiangiogenic effect and can inhibit clot formation in arteries that may otherwise cause a heart attack or stroke [120]. The medicinal property of curcumin has also been reported for the prevention of stroke, aortic aneurysm, myocardial infarction, and atherosclerosis [121]. *In vitro* experiments conducted by Rachmawati et al. demonstrated that curcumin demonstrated antihypertensive and antihypercholesterolemic effects, analyzed using 3-hydroxy-3-methylglutaryl coenzyme A reductase and AChE assay kits. Nanoemulsified curcumin significantly inhibited AChE and 3-hydroxy-3-methylglutaryl coenzyme A reductase as compared to curcumin only [123]. Another study conducted in a diabetic cardiomyopathy model showed that the multipolymer poly(gamma-benzyl–L-glutamate)-poly(ethylene glycol)-poly (gamma-benzyl-L-glutamate) NPs increased biological efficiency, water solubility, and curcumin dispersion in aqueous solutions. The nanoformulated curcumin enhanced serum Ca^{2+} and H_2S levels in myocardial cells and enhanced the expression of the calcium-sensing receptor cystathionine-γ-lyase and calmodulin. Finally, this study revealed reduced pathological damage to myocardial cells and a decreased incidence of diabetic cardiomyopathy [124]. Myricitrin, a plant glycoside, demonstrated antioxidant, anxiolytic, and anti-inflammatory effects [125]. Quercetin, an antioxidant, prevents the development of CVDs by scavenging free radicals, a crucial event in atherosclerosis. In both anesthetized animals and isolated tissues, phosphatidylcholine liposomes loaded with quercetin conferred protection against peroxynitrite-induced myocardial injuries by acting as a direct scavenger and as a decomposer of endogenously formed peroxynitrite ions, thus restoring normal myocardial contractility [126]. Another study evaluated the cardioprotective effects of polymeric micelles containing a mixture of quercetin and resveratrol against DOX-induced cardiotoxicity. In both *in vitro* and *in vivo* tests, this combination proved effective in reducing cardiotoxicity [127].

Diverse plants such as *Bombax ceiba*, *Centella asiatica*, *Dracocephalum moldavica*, *Ginkgo biloba*, *Nerium oleander*, *Paeonia emodi*, *Panax notoginseng*, *Pinus merkusii*, *Prunus* spp., *Sesbania grandiflora*, *Sonchus asper*, *Syzygium cumini*, *Terminalia arjuna*, *Tinospora cordifolia*, and *Zingiber officinale* from various families showed cardioprotective activity [128]. Panax notoginsenoside (PNS), an active component of *Panax notoginseng*, plays a curative role in CVDs, but their oral administration is restricted owing to its poor absorption in the body. To increase PNS efficiency, methyl ether poly(ethylene glycol)-poly(lactide-co-glycolide)-based NPs were used. This encapsulation delayed the release of PNS components and significantly lowered serum MDA, lactate dehydrogenase, and H_2O_2 levels and increased SOD pituitrin-induced myocardial ischemia in rats [129]. Furthermore, salvianolic acid B and PNS co-delivery in acute myocardial ischemia was achieved using arginyl-glycyl-aspartic acid (RGD)-fabricated lipid polymer hybrid NPs. This nanosystem showed a significant therapeutic effect on infarcts and cardiac distribution *in vivo* when compared with the control group [130]. Berberine is a fluorescent isoquinoline quaternary alkaloid derived from *Berberis* spp. and shows anti-CVD effect by reducing systolic blood pressure and triglyceride levels. Its utility is restricted in medicine owing to its low solubility in aqueous buffers, short half-life in blood circulation, and minimal intestinal absorption rate. Encapsulation of berberine inside liposomes increased its solubility, improved its local delivery, increased accumulation of the payload in the infarct heart tissue, enhanced treatment efficacy, preserved left ventricular ejection fraction, and reduced adverse cardiac remodeling in myocardial infarction, as shown in *in vivo* studies [Figure 1.6] [122].

Tilianin, a natural flavonoid compound, has been used to treat CVDs such as coronary hypertension, heart disease, myocardial ischemia, and atherosclerosis [131]. Breviscapine is a flavone glucuronide isolated from *Erigeron breviscapus*, a traditional Chinese herb used to treat cerebral infarction, angina pectoris, and coronary artery disease. It has shown beneficial effects on reducing

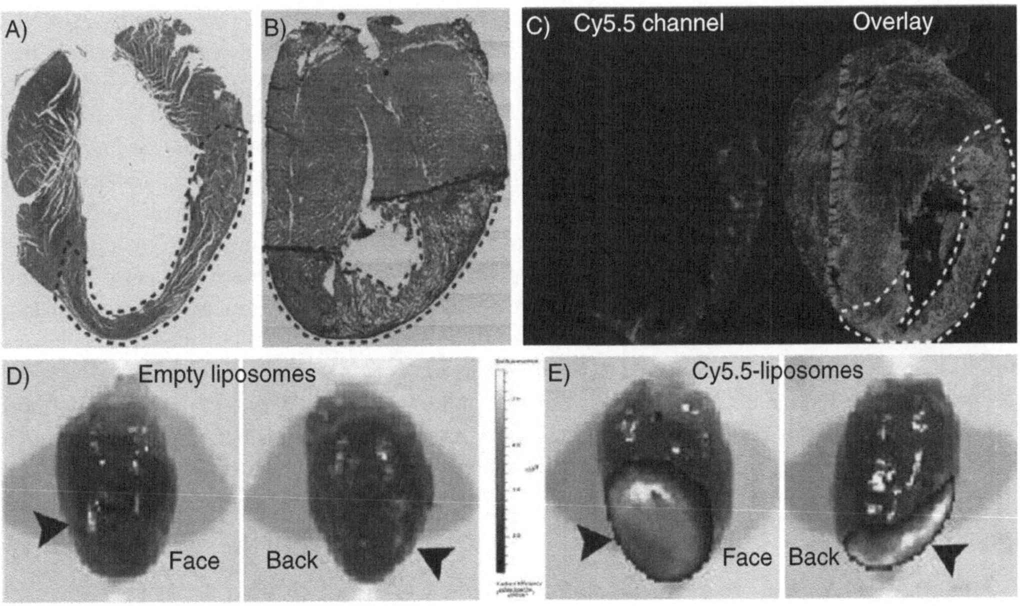

FIGURE 1.6 Infiltration of macrophages and accumulation of liposomes in the infarcted heart tissue. Left ventricles were isolated 3 days after myocardial infarction and analyzed for liposomes and macrophages. A) Masson's trichrome staining indicating the infarcted area. B) Infiltration of macrophages into the infarcted heart tissue. MAC3 staining is shown in brown. C) Liposomes were visualized under a confocal microscope. Cy5.5 is shown in green and merged with the heart tissue obtained under a bright-field microscope. D) Accumulation of control liposomes in the infarcted tissue visualized with IVIS® Spectrum In Vivo Imaging system. E) Accumulation of Cy5.5-liposomes in the infarcted tissue visualized using the IVIS® Spectrum In Vivo imaging system. The infarct region is indicated by dashed lines (A-C) or arrowheads (D-E). (From Allijn, I.E., et al., J. Control. Release. 2017 Feb 10;247:127–133. With Permission) [122]. Copyright (2017) Elsevier.

CVDs by eliminating blood clots and improving blood circulation [132]. Epigallocatechin gallate or epigallocatechin-3-gallate (EGCG), a type of catechin made of the ester of epigallocatechin and gallic acid, can reduce inflammatory reactions and cholesterol accumulation in macrophages, which suppresses atherosclerosis development [133,134]. Tanshinone IIA, a compound derived from the rhizome of *Salvia miltiorrhiza*, has been investigated for its potential to treat CVDs. It possesses a wide range of therapeutic properties including vasodilation, scavenging activity, endothelial dysfunction prevention, macrophage-derived foam cell delay, and inhibition of vascular smooth muscle cell proliferation and migration [135,136]. *Allium sativum*, more generally referred to as garlic, is a member of the Amaryllidaceae family and has been used for medicinal purposes since ancient times. It prevents and cures CVDs and other metabolic diseases such as atherosclerosis, hyperlipidemia, thrombosis, hypertension, dementia, diabetes, and cancer. S-allyl cysteine, a key component of aged garlic extract, has been explored substantially by several researchers. S-allyl cysteine possesses antioxidant activity at both indirect and direct levels. Apart from reducing lipid peroxidation and DNA fragmentation, it also reduces oxidation and nitration [137].

1.5 CONCLUSION AND FUTURE PERSPECTIVES

Plant-based NPs possess unique features such as small size, easy synthesis, ability to deliver payload at the target site, and, most importantly, biocompatibility. Thus, they have the potential to be used as effective therapeutics against diverse severe diseases such as cancer, CVDs, neurodegenerative diseases, diabetes, and bacterial and viral diseases. However, most of the investigations conducted in this field are restricted to metallic NPs such as GNPs, AgNPs, selenium NPs, copper-zinc NPs, and others, necessitating research on various polymeric NPs and some unexplored metallic NPs. Green synthesis of NPs, i.e., plant extract-based NP synthesis, has a few limitations, namely, polydispersity, reproducibility, and low yield, which restricts their use in therapeutic applications. Moreover, other aspects of these NPs, namely, the delivery of synthetic drugs and nucleic acids, should be investigated. A vast number of plants are available in nature; although many of them have been explored for their medicinal use, a remarkable proportion of them are yet to be investigated. Metal NPs such as AgNPs, GNPs, and zinc oxide NPs have intrinsic properties of antibacterial, anti-inflammatory, and antidiabetic activities, but their potential clinical value increases after plant-based synthesis. Therefore, the complete utilization of NPs is contributed by the origin of the NPs and the plant extract components present in them. Therefore, future investigations should also discover the active components present in plant extracts, their characterization, and activity assessment. The pure form of these active components may be effective against life-threatening diseases when delivered through NPs in a targeted manner. As for synthetic NPs, surface modification and targeting has been investigated thoroughly, but plant-based NPs have not been investigated in this regard. We hope that, in the future, researchers from this field of study explore the benefits of plant bioactives and plant-based NPs, thereby contributing to the literature pool.

REFERENCES

[1] Nasrollahzadeh M, Sajjadi M, Issaabadi Z. Interface Sci. Technol. 2019;28: 1–27. Elsevier.
[2] Prakash S, Rajpal VR, Vaishnavi S, Deswal R. ScienceOpen Posters. 2021.
[3] Khan T, Gurav P. Front. Pharmacol. 2018;8:1002.
[4] Monika P, Chandraprabha MN. Functional Bionanomaterials, pp. 111–128. Springer, Cham, 2020.
[5] Choudhary N, Sekhon BS. Indian J. Pharm. Educ. Res. 2011;2(2):55.
[6] Ansari SH, Islam F, Sameem MJ. Adv. Pharm. Technol. 2012;3(3):142.
[7] Sachan AK, Gupta A. Int. J. Pharm. Sci. Res. 2015;6(3):961.
[8] The British Museum. [accessed on 22 July 2019]; Available online: www.britishmuseum.org/research/collection_online/collection_object_details.aspx?objobjec=61219&partId=1.
[9] Pradell T, Climent-Font A, Molera J, Zucchiatti A, Ynsa MD, Roura P, Crespo D. J. Appl. Phys. 2007;101:103518.

[10] Daniel MC, Astruc D. Chem. Rev. 2004;104(1):293–346.
[11] Alexander JW. Surg. Infect. 2009;10(3):289–292.
[12] Gribbin J, Gribbin M. Dutton. p. 170, 1997. ISBN 978-0-452-27631-4.
[13] Drexler KE. Proceedings of the National Academy of Sciences of the United States of America. 1981;78(9):5275–5278.
[14] Binnig G, Rohrer H. Surf. Sci. 1983;126(1–3):236–244.
[15] "Press Release: the 1996 Nobel Prize in Physics." Nobelprize.org. 15 October 1986. Retrieved 12 May 2011.
[16] Kruis F, Fissan H, Rellinghaus B. Mater. Sci. Eng. B. 2000;69:329–334.
[17] Kumar P, Tambe P, Paknikar KM, Gajbhiye V. Colloids Surf B. 2017;156:203–212.
[18] Iravani S. Green. Chem. 2011;13:2638–2650.
[19] Kumar H, Bhardwaj K, Kuča K, Kalia A, Nepovimova E, Verma R, Kumar D. Nanomaterials. 2020;10(4):766.
[20] Korbekandi H, Iravani S, Abbasi S. Crit. Rev. Biotechnol. 2009;29(4):279–306.
[21] Dhillon GS, Brar SK, Kaur S, Verma M. Crit. Rev. Biotechnol. 2012;32(1):49–73.
[22] Lenaghan SC, Zhu Q, Xia L, Zhang M. In: Nanoparticles in Biology and Medicine, 2012 (pp. 381–391). Humana Press, Totowa, NJ.
[23] Lamb AE, Anderson CWN, Haverkamp RG. 2001. Massey University.
[24] Arya V. Dig. J. Nanomater. Bios. 2010;5(1):9–21.
[25] Jha AK, Prasad K. Int. J. Green Nanotechnol. 2010;1(2):110–117.
[26] Paramanantham M, Murugesan A, Muthu MK, Ramamoorthy R. Int. J. Recent Sci. Res. 2014;5(3):677–680.
[27] Balachandar R, Gurumoorthy P, Karmegam N, Barabadi H, Subbaiya R, Anand K, Boomi P, Saravanan MJ. Clust. Sci. 2019;30(6):1481–1488.
[28] Sharma V, Kaushik S, Pandit P, Dhull D, Yadav JP, Kaushik S. Appl. Microbiol. Biotechnol. 2019;103(2):881–891.
[29] Katti K, Chanda N, Shukla R, Zambre A, Suibramanian T, Kulkarni RR, Kannan R, Katti KV. Int. J. Green Nanotechnol. Biomed. 2009;1(1):B39–B52.
[30] Ghodake GS, Deshpande NG, Lee YP, Jin ES. Colloids Surf. B. 2010;75(2):584–589.
[31] Mishra AN, Bhadauria S, Gaur MS, Pasricha R, Kushwah BS. Int. J. Green Nanotechnol. Phys Chem. 2010;1(2):118–124.
[32] Babu RH, Yugandhar P, Savithramma N. Bull. Mater. Sci. 2018;41(3):1–8.
[33] Sethy NK, Arif Z, Mishra PK, Kumar P. J. Polym. Eng. 2019;39(7):679–687.
[34] Durairaj K, Senthilkumar P, Velmurugan P, Dhamodaran K, Kadirvelu K, Kumaran S. J Solgel. Sci. Technol. 2019;90(3):653–664.
[35] Jabeen N, Maqbool Q, Sajjad S, Minhas A, Younas U, Anwaar S, Nazar M, Kausar R, Hussain SZ. IET Nanobiotechnol. 2017;11(5):557–561.
[36] Al-Azawi MT, Hadi SM, Mohammed CH. Iraqi J. Agric. Sci. 2019;50:245–255.
[37] Murthy HCA, Desalegn T, Kassa M, Abebe B, Assefa T. J Nanomater. 2020.
[38] Ghosh MK, Sahu S, Gupta I, Ghorai TK. RSC Adv. 2020;10(37):22027–22035.
[39] Mali SC, Dhaka A, Githala CK, Trivedi R. Biotechnol Rep. 2020;27:e00518.
[40] Chandraker SK, Lal M, Ghosh MK, Tiwari V, Ghorai TK, Shukla R. Nano Express. 2020;1(1):010033.
[41] Bhuiyan MS, Miah MY, Paul SC, Aka TD, Saha O, Rahaman MM, Sharif MJ, Habiba O, Ashaduzzaman M. Heliyon. 2020;6(8):e04603.
[42] Lakshminarayanan S, Shereen MF, Niraimathi KL, Brindha P, Arumugam A. Sci. Rep. 2021;11(1):1–3.
[43] Bibi I, Nazar N, Ata S, Sultan M, Ali A, Abbas A, Jilani K, Kamal S, Sarim FM, Khan MI, Jalal F. J. Mater. Res. Technol. 2019;8(6):6115–6124.
[44] Khalil AT, Ovais M, Ullah I, Ali M, Shinwari ZK, Maaza M. Green. Chem. Lett. Rev. 2017;10(4):186–201.
[45] Arularasu MV, Devakumar J, Rajendran TV. Polyhedron 2018;156 :279–290.
[46] Karthikeyan S, Mahalingam P. Int. J. Green Nanotechnol. Phys. Chem. 2010;2(2):39–46.
[47] Siddique MA, Jeelani SM. In: Biomedical Engineering: Concepts, Methodologies, Tools, and Applications. 2018. pp. 1001–1027. IGI Global.
[48] Servat-Medina L, Gonzalez-Gomez A, Reyes-Ortega F, Sousa IM, Queiroz ND, Zago PM, Jorge MP, Monteiro KM, de Carvalho JE, San Roman J, Foglio MA. Int. J. Nanomed. 2015;10:3897.

[49] Karuppath S, Snima KS, Ravindranath KC, Nair SV, Lakshmanan VK. J Bionanoscience. 2016;10(2):127–133.
[50] Banu HH, Dhrubojyoti M. In: Conference on Drug Design and Discovery Technologies. 2019;355:178. RSC.
[51] Palai S, Patra R, Dehuri M. Ann Phytomed. 2021;S222–S230.
[52] Islam R, Sun L, Zhang L. Nanomaterials. 2021;11(10):2757.
[53] Jain N, Jain P, Rajput D, Patil UK. Micro Nano Syst. Lett. 2021;9(1):1–24.
[54] Jabeen S, Qureshi R, Munazir M, Maqsood M, Munir M, Shah SH, Rahim BZ. Mater Res Express. 2021;8.
[55] Sadat Shandiz SA, Shafiee Ardestani M, Shahbazzadeh D, Assadi A, Ahangari Cohan R, Asgary V, Salehi S. Artif. Cells Nanomed. Biotechnol. 2017;45(6):1082–1091.
[56] Kamala Priya MR, Iyer RP. Appl. Nanosci. 2015;5:443.
[57] Kajani AA, Bordbar A-K, Zarkesh Esfahani SH, Razmjou A. RSC Adv. 2016;6(68):63973–63983.
[58] Karuppaiya P, Satheeshkumar E, Chao WT, Kao LY, Chen ECF, Tsay HS. Colloids Surf. B Biointerfaces 2013;110:163–170.
[59] Mukherjee S, Sau SC, Madhuri D, Bollu VS, Madhusudana K, Sreedhar B, Banerjee R, Patra CR. J. Biomed. Nanotechnol. 2016;121:165–181.
[60] Ahmad N, Bhatnagar S, Saxena R, Iqbal D, Ghosh AK, Dutta R. Mater. Sci. Eng. C Mater. Biol. Appl. 2017;78:553–564.
[61] Şahin B, Aygün A, Gündüz H, Şahin K, Demir E, Akocak S, Şen F. Colloids Surf. B Biointerfaces. 2018;163:119–124.
[62] Yang C, Wang M, Zhou J, Chi Q. Mater. Sci. Eng. C. 2017;77:1012–1016.
[63] Aygun A, Gülbagca F, Ozer LY, Ustaoglu B, Altunoglu YC, Baloglu MC, Atalar MN, Alma MH, Sen F. J. Pharm. Biomed. Anal. 2020;179:112961.
[64] Ghosh S, Nitnavare R, Dewle A, Tomar GB, Chippalkatti R, More P, Kitture R, Kale S, Bellare J, Chopade BA. Int. J. Nanomed. 2015;10:7477.
[65] Premila MS, Conboy L. J. Altern. Complement. Med. 2007;13(8):841–842.
[66] Rahmani AH. Pharmacogn. Res. 2015;7(3):217.
[67] Vinotha V, Iswarya A, Thaya R, Govindarajan M, Alharbi NS, Kadaikunnan S, Khaled JM, Al-Anbr MN, Vaseeharan B. J. Photochem. Photobiol. B: Biol. 2019;197:111541.
[68] Bayrami A, Parvinroo S, Habibi-Yangjeh A, Rahim Pouran S. Artif. Cells Nanomed. Biotechnol. 2018;46(4):730–739.
[69] Jan H, Shah M, Andleeb A, Faisal S, Khattak A, Rizwan M, Drouet S, Hano C, Abbasi BH. Oxid. Med. Cell. Longev. 2021.
[70] Rehana D, Mahendiran D, Kumar RS, Rahiman AK. Bioprocess Biosyst Eng. 2017;40(6):943–957.
[71] Rajakumar G, Thiruvengadam M, Mydhili G, Gomathi T, Chung IM. Bioprocess Biosyst. Eng. 2018;41(1):21–30.
[72] Faisal S, Jan H, Shah SA, Shah S, Khan A, Akbar MT, Rizwan M, Jan F, Wajidullah, Akhtar N, Khattak A. ACS Omega. 2021;6(14):9709–9722.
[73] Agarwal H, Kumar SV, Rajeshkumar S. J. Microbiol. Biotechnol. Food Sci. 2021;2021:371–376.
[74] Sengottaiyan A, Aravinthan A, Sudhakar C, Selvam K, Srinivasan P, Govarthanan M, Manoharan K, Selvankumar T. J Nanostructure Chem. 2016;6(1):41–48.
[75] Kalakotla S, Geetha P, Banu A, Shaik S. 2022, https://doi.org/10.21203/rs.3.rs-1224222/v1
[76] Ponnanikajamideen M, Rajeshkumar S, Vanaja M, Annadurai G. Can. J. Diabetes. 2019;43(2):82–89.
[77] Badeggi UM, Ismail E, Adeloye AO, Botha S, Badmus JA, Marnewick JL, Cupido CN, Hussein AA. Biomolecules. 2020;10(3):452.
[78] Patra N, Kar D, Pal A, Behera A. Adv. Nat. Sci. Nanosci. Nanotechnol. 2018;9(3):035001.
[79] Soliman N, Ismail EH, Abd El-Moaty HI, Sabry DY, Khalil MM. Egypt J. Chem. 2017;61(4):691–703.
[80] Bhattacharya T, Chopra H, Rahman M, Hasan Z, Swain SS, Cavalu S. Materials. 2022;15(3):804.
[81] Hill NL, Kolanowski AM, Gill DJ. Top. Geriatr Rehabil. 2011;27(4):257.
[82] Enciu AM, Nicolescu MI, Manole CG, Mureşanu DF, Popescu LM, Popescu BO. BMC Neurol. 2011;11(1):1–7.
[83] Golla P, Tirupathi H. Int. J. Basic Clin. Pharmacol. 2016;5(5):2017–2020.
[84] Soumyanath A, Zhong YP, Yu X, Bourdette D, Koop DR, Gold SA, Gold BG. J. Pharm. Pharmacol. 2005;57(9):1221–1229.

[85] Kumar A, Seghal N, Padi SV, Naidu PS. Eur. J. Pharmacol. 2006;551(1–3):58–66.
[86] Arora N, Rai SP. Int. J. Pharm. Bio. Sci. 2012;3(3):290–303.
[87] Wattmo C, Jedenius E, Blennow K, Wallin ÅK. 2013;5(1):1–9.
[88] Ferreira A, Rodrigues M, Fortuna A, Falcão A, Alves G. Phytochem. Rev. 2016;15(1):51–85.
[89] Fritea L, Ganea M, Zdrinca M, Dobjanschi L, Antonescu A, Vicas SI, Bodog F, Sindhu RK, Cavalu S. Plants. 2021;10(7):1390.
[90] Lee DY, Kim HW, Yang H, Sung SH. Phytochemistry. 2017;137:109–116.
[91] Heo HJ, Park YJ, Suh YM, Choi SJ, Kim MJ, Cho HY, Chang YJ, Hong B, Kim HK, Kim E, Kim CJ. Biosci. Biotechnol. Biochem. 2003;67(6):1284–1291.
[92] Roy S, Awasthi HI. Int. J. Pharm. Sci. 2017;9(11):1–7.
[93] Mohanasundari M, Sethupathy S, Sabesan M. Indian J. Pharmacol. 2006;38(4):266–270.
[94] Clark RA. Wound Repair. 1988 (pp. 3–50). Springer, Boston, MA.
[95] Gonzalez AC, Costa TF, Andrade ZD, Medrado AR. An. Bras. Dermatol. 2016;91:614–620.
[96] Luthra PM, Singh R, Chandra R. Indian J. Clin. Biochem. 2001;16(2):153–160.
[97] Mandal MK, Chauhan JP. Indian Vet. J. 2000;20(2):90–91.
[98] Chereddy KK, Coco R, Memvanga PB, Ucakar B, des Rieux A, Vandermeulen G, Préat V. J. Control. Release. 2013;171(2):208–215.
[99] Tiloke C, Anand K, Gengan RM, Chuturgoon AA. Biomed. Pharmacother. 2018;108:457–466.
[100] Anjum S, Gupta A, Sharma D, Gautam D, Bhan S, Sharma A, Kapil A, Gupta B. Mater. Sci. Eng. C. Mater. Biol. Appl. 2016;64:157–166.
[101] Sahni YP, Srivastava DN. J. Appl. Anim. Res. 1993;3(2):129–136.
[102] Shukla U, Tewari A, Upadhyay AK. Indian J. Vet. Med. 2002;22(2):109–111.
[103] Arya V. New dimensions in microbiology: Lenin Media Private Limited, Delhi, India. 2015:101–108.
[104] Aygün A, Gülbağça F, Nas MS, Alma MH, Çalımlı MH, Ustaoglu B, Altunoglu YC, Baloğlu MC, Cellat K, Şen F. J. Pharm. Biomed. Anal. 2020;179:113012.
[105] Valsalam S, Agastian P, Arasu MV, Al-Dhabi NA, Ghilan AK, Kaviyarasu K, Ravindran B, Chang SW, Arokiyaraj S. J. Photochem. Photobiol. B: Biol. 2019;191:65–74.
[106] Benakashani F, Allafchian AR, Jalali SA. Karbala Int. J. Mod. Sci. 2016;2(4):251–258.
[107] Devi HS, Boda MA, Shah MA, Parveen S, Wani AH. Green Process. Synth. 2019;8(1):38–45.
[108] Ramesh M, Anbuvannan M, Viruthagiri GJ. Spectrochim. Acta A Mol. Biomol. Spectrosc. 2015;136:864–870.
[109] Agarwal H, Nakara A, Menon S, Shanmugam V. J. Drug Deliv. Sci. Technol. 2019;53:101212.
[110] Thakur BK, Kumar A, Kumar D. S. Afr. J. Bot. 2019;124:223–227.
[111] Arokiyaraj S, Saravanan M, Prakash NU, Arasu MV, Vijayakumar B, Vincent S. Mater. Res. Bull. 2013;48(9):3323–3327.
[112] Lee HJ, Song JY, Kim BS. J. Chem. Technol. Biotechnol. 2013;88(11):1971–1977.
[113] Kumar PP, Shameem U, Kollu P, Kalyani RL, Pammi SV. BioNanoScience. 2015;5(3):135–139.
[114] Awwad AM, Albiss BA, Salem NM. SMU Med J. 2015;2(1):91–101.
[115] Gopalakrishnan K, Ramesh C, Ragunathan V, Thamilselvan M. Dig. J. Nanomater. Biostructures. 2012;7(2):833–839.
[116] Sivaraj R, Rahman PK, Rajiv P, Salam HA, Venckatesh R. Spectrochim. Acta A Mol. Biomol. Spectrosc. 2014;133:178–181.
[117] www.who.int/news-room/fact-sheets/detail/cardiovascular-diseases-(cvds)
[118] Dong TT, Cui XM, Song ZH, Zhao KJ, Ji ZN, Lo CK, Tsim KW. J. Agric. Food Chem. 2003;51(16):4617–4623.
[119] Giannouli M, Karagkiozaki V, Pappa F, Moutsios I, Gravalidis C, Logothetidis S. Mater Today: Proceedings. 2018;5(8):15998–16005.
[120] Li X, Fang Q, Tian X, Wang X, Ao Q, Hou W, Tong H, Fan J, Bai S. Mol. Med. Rep. 2017;16(4):4455–4462.
[121] Campbell MS, Fleenor BS. Crit. Rev. Food. Sci. Nutr. 2018;58(16):2790–2799.
[122] Allijn IE, Czarny BM, Wang X, Chong SY, Weiler M, Da Silva AE, Metselaar JM, Lam CS, Pastorin G, De Kleijn DP, Storm G. J. Control. Release. 2017;247:127–133.
[123] Rachmawati H, Soraya IS, Kurniati NF, Rahma A. Sci. Pharm. 2016;84(1):131–140.

[124] Tong F, Chai R, Jiang H, Dong B. Int. J. Nanomedicine. 2018;13:1945.
[125] Pereira M, Siba IP, Chioca LR, Correia D, Vital MA, Pizzolatti MG, Santos AR, Andreatini R. Prog. Neuropsychopharmacol. Biol. Psychiatry. 2011 Aug 15;35(7):1636–44.
[126] Soloviev A, Stefanov A, Parshikov A, Khromov A, Moibenko A, Kvotchina L, Balavoine G, Geletti Y. Cardiovasc. Toxicol. 2002;2(2):129–139.
[127] Moradi SZ, Momtaz S, Bayrami Z, Farzaei MH, Abdollahi M. Front. Bioeng. Biotechnol. 2020;8:238.
[128] Shaito A, Thuan DT, Phu HT, Nguyen TH, Hasan H, Halabi S, Abdelhady S, Nasrallah GK, Eid AH, Pintus G. Front. Pharmacol. 2020;11:422.
[129] Zhang J, Han X, Li X, Luo Y, Zhao H, Yang M, Ni B, Liao Z. Int. J. Nanomedicine. 2012;7:4299.
[130] Qiu J, Cai G, Liu X, Ma D. Biomed. Pharmacother. 2017;96:1418–1426.
[131] Yuan Y, Cao W, Hong Y, Guo X, Wang Y, Wang Y, Wang X, Hu P. Phytomedicine. 2017;34:106–114.
[132] Fei X, Chen X, Liang G, Yue-Jian C, Hao W, Ning G, Jia-Bi Z. J. Drug. Target. 2009;17(5):408–414.
[133] Hong MH, Kim MH, Chang HJ, Kim NH, Shin BA, Ahn BW, Jung YD. Life Sci. 2007;80(21):1957–1965.
[134] Hong J, Lu H, Meng X, Ryu JH, Hara Y, Yang CS. Cancer Res. 2002;62(24):7241–7246.
[135] Feng J, Li S, Chen H. Eur. J. Pharmacol. 2016;791:632–639.
[136] Gao S, Liu Z, Li H, Little PJ, Liu P, Xu S. Atherosclerosis. 2012;220(1):3–10.
[137] Powell CR, Dillon KM, Matson JB. Biochem. Pharmacol. 2018;149:110–123.

2 Usefulness of Medicinal Plants in Treating Various Diseases

*Asma Praveen[1], Vasudev Biswas[1], Shikha Goswami[2], Arya Lakshmi Marisetti[1], Abhishek Tiwari[4], Varsha Tiwari[4], Nikita Savita[1], Vijender Kumar[1], Madhu Gupta[3], Saurabh Bhatia[5], and Ajay Sharma[*1]*

[1]Department of Pharmacognosy & Phytochemistry, School of Pharmaceutical Sciences, Delhi Pharmaceutical Sciences and Research University, Mehrauli Badarpur Road, New Delhi 110017, India

[2]Department of Pharmacology and Biotechnology, School of Pharmaceutical Sciences, Delhi Pharmaceutical Sciences and Research University, Mehrauli Badarpur Road, New Delhi 110017, India

[3]Department of Pharmaceutics, School of Pharmaceutical Sciences, Delhi Pharmaceutical Sciences and Research University, Mehrauli Badarpur Road, New Delhi 110017, India

[4]Natural & Medical Sciences Research Center, University of Nizwa, Birkat Al Mauz, Nizwa, P.O. Box 33, Birkat Al Mauz, Nizwa 616, Oman

[5]Pharmacy Academy, IFTM University, Lodhipur-Rajput, Moradabad 244102, Uttar Pradesh, India

2.1 INTRODUCTION

Plant-based medicines have been used since ancient times (Seifu et al. 2004). Medicinal plants are used for the treatment of various diseases, and plant parts such as leaf, stem, root, bark, and latex, among others, contain various therapeutic substances that can be used for medicinal purposes and may be precursors for further synthesis of drug molecules (Sofowora et al. 2013). Phytomedicines or herbal medicines are more affordable (low cost), cause less severe side effects, and have better accessibility than synthetic medicines (Asase et al. 2008). The majority of the world's population prefer herbal medicines for supporting primary health care. According to the World Health Organization (WHO), approximately 80% of the world's population in developing countries prefer plant-based medicines. Various synthetic medicines have been developed with plant-based isolated moieties/compounds, and approximately 80% of the active compounds have shown a positive response, such as quinine, baicalein, aspirin, and curcumin, and a correlation between modern medicine and traditional medicine (Sarkar et al. 2015a).

Many plants and plant parts have been used in traditional medicine. Pharmacists and pharmacologists use the key word "crude drug of natural origin or plant origin." This term indicates the medicinal properties of plants or their parts (Sofowora et al. 2013).

DOI: 10.1201/9781003231721-2

According to Sofowora (2008) and Evans (2009), some of the medicinal plants and plant parts used are as follows:

- Cascara bark is used in medicinal compound preparation (galenical preparation), e.g., extraction and decoction.
- Some plant parts are used for the extraction and isolation of pure compounds, and they may be used directly or after hemi-synthesis process; e.g., diosgenin was obtained from *Dioscorea* through hemi-synthesis for the preparation of sex hormones.
- Some microscopic plants such as *Ergot* infected by the fungus *Claviceps purpurea* grow on rye plants, and alkaloidal compounds are isolated for use as antibiotics.
- Cotton, jute, and flax are fibers plants used for the preparation of surgical products.

2.1.1 Strategies for Planning the Development of Traditional Medicine

- A new phytoformulation prepared from medicinal plants should follow the WHO guidelines in terms of safety, efficacy, and quality.
- For enhanced safety and quality, proper guidelines should be followed by evaluating quality assurance and regulatory standard parameters.
- The accessibility and affordability of traditional medicine should be increased.
- The traditional medicine should be appropriately employed for therapeutic use.

Several studies have indicated the use of various medicinal plants in the treatment of various diseases such as diabetes, malaria, Alzheimer's disease (AD), and cardiovascular diseases. Recently, during the coronavirus disease (COVID-19) pandemic, many medicinal plants were used for the treatment of severe acute respiratory syndrome coronavirus 2 (SARS-CoV-2) infection and showed promising effects. This chapter summarizes information on the use of medicinal plants in the treatment of diabetes, malaria, AD, COVID-19, cancer, and cardiovascular disease.

2.2 EFFECTIVENESS OF MEDICINAL PLANTS AGAINST DIABETES

Diabetes mellitus has become a major problem in the health care system. According to tenth edition of the International Diabetes Federation, 536.6 million individuals had diabetes in the year 2021, and this number is expected to increase to approximately 783.2 million by 2045 (Ogurtsova et al. 2022). Diabetes is a serious complex metabolic disorder involving various pathological conditions and demonstrate both acute and chronic symptoms (Soumya & Srilatha 2011). Both genetic and environmental factors influence diabetes development (Murea et al. 2012). During diabetes, body cells cannot metabolize blood glucose in a regular manner, leading to the development of hyperglycemia (Buowari 2013). Diabetes is of various types: type 1 diabetes mellitus (T1DM) and type 2 diabetes mellitus (T2DM).

- T1DM is an insulin-dependent diabetes, and it occurs owing to insulin deficiency, wherein the pancreas does not release sufficient insulin properly because of the destruction of pancreatic β-cells, associated with autoimmune factors. T1DM is an autoimmune condition (Chhabra et al. 2013).
- T2DM is an insulin-resistant diabetes in the peripheral system and improper functioning of β-cells in the Islets of Langerhans. Insulin resistance causes a disturbance in the metabolism of carbohydrates, proteins, and fats (Pathak et al. 2012).

Different modern antidiabetic medicines are available in the market, such as α-glucosidase inhibitors, biguanides (metformin), sulfonylureas (gliclazide) (Adhikari 2021), but these drugs have increased toxicity; for instance, both short-term and long-term metformin exposure can cause

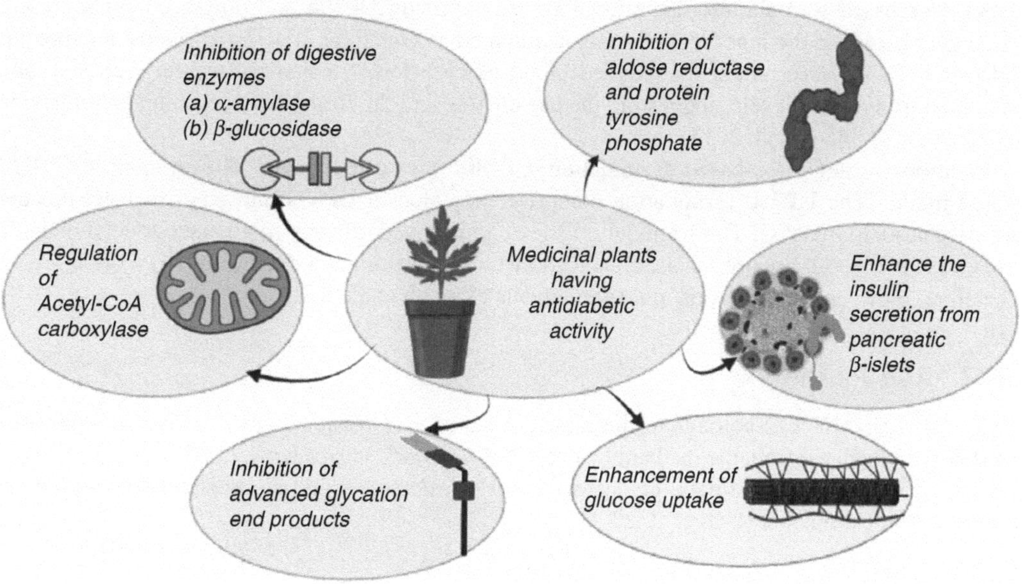

FIGURE 2.1 Possible mechanism underlying the antidiabetic activity of medicinal plants (created in BioRender).

metformin-associated lactic acidosis (MALA). The patient may have alterations in the central nervous system (CNS), which lead to cardiovascular collapse, renal failure, and, finally, death owing to MALA. However, acarbose and thiazolidinediones produce hepatic injury with long-term treatment (Spiller et al. 2006), and synthetic medicines have shown saturation and are associated with different comorbidities; thus, the majority of the world's population prefer traditional herbal medicine (see Figure 2.1). The following sections provide a list of antidiabetic medicinal plants along with their mechanism of action (see Table 2.1).

2.2.1 *Aloe vera*

Aloe barbadensis, which belongs to the family Liliaceae, possesses different pharmacological activities including antidiabetic activity and are used for perfumery. *Aloe vera* contains several phytoconstituents that show antidiabetic activity against T2DM, such as phytosterol compounds such as lophenol, 24-methyl-lophenol, 24-ethyl-lophenol, and 24-methylene-cycloartanol, and cycloartanol; these constituents reduced blood glucose levels and showed antihyperglycemic activity, but β-sitosterol did not show antihyperglycemic effects (Tanaka et al. 2006). When *Aloe vera* extract was treated at a dose of 130 mg/kg/day for 4 weeks in streptozotocin (STZ)-induced diabetic mice, the treatment reduced the levels of blood glucose, triglycerides, low-density lipoprotein (LDL), and cholesterol (Kumar et al. 2011). When *Aloe vera* extract was administered to STZ-induced diabetic rats for 3 weeks, the treatment reduced hyperglycemia by enhancing insulin secretion and regulating β-cell function (Noor et al. 2017). In a study conducted by Noor et al., when *Aloe vera* extract was administered to STZ-induced diabetic rats, the treatment achieved normal fasting plasma glucose levels and conferred protection to the pancreas, liver, and small intestine (Noor et al. 2017). In an another interesting experiment, *Aloe vera* extract administered to STZ-induced diabetic rats reduced blood glucose levels, conferred protection to the pancreas, enhanced β-cell proliferation in the Islets of Langerhans, and improved insulin secretion (Okyar et al. 2001; Salem et al. 2014). In a recent study, *Aloe vera* extract administered orally to alloxan-induced diabetic rats reduced the levels of serum glucose, triacylglycerol, and cholesterol (Mohamed 2011).

An *Aloe*-based formulation, namely, UP780 (comprising UP394 and polysaccharides present in the gel powder of the inner leaf of *Aloe*, formulated as Qmatrix), was tested against an alloxan-induced T1DM mouse model for reduced blood glucose levels, enhanced insulin secretion, and decreased triglyceride levels, suggesting the use of *Aloe* and the formulation as natural antidiabetic supplements (Yimam et al. 2014).

Additionally, the *Aloe*-based formulation UP780 was tested in an alloxan-induced T1DM mouse model. The UP780 formulation comprised the aloesin UP394 and inner leaf gel powder polysaccharide (Qmatrix). The formulation showed improved efficacy in alloxan-induced diabetic mice, reduced blood glucose levels, enhanced insulin secretion, and decreased triglyceride levels. Therefore, they can be used as natural supplements (Yimam et al. 2014).

2.2.2 *Adhatoda vasica*

Adhatoda vasica (Nees.) belongs to the family Acanthaceae. Vasicine and vasicinol (quinazoline alkaloids) are isolated from the methanolic extract of *Adhatoda vasica* leaves. Vasicine and vasicinol have been tested on rat intestinal α-glucosidase, and they demonstrated sucrase inhibitory activity in a dose-dependent manner (Gao et al. 2008).

2.2.3 *Acacia arabica*

Acacia arabica, commonly called babool, belongs to the family Fabaceae. The extract of *Acacia arabica* was tested on BRIN-BD11 cell line, and it demonstrated enhanced secretion of insulin and release of dehydrogenase enzyme, whose activity depended on the dose used.

- *Acacia arabica* extract was tested on 3T3L1 cell line with glucose analogs, which demonstrated enhanced glucose metabolism and insulin secretion.
- *Acacia arabica* contains kaempferol, quercetin, and catechin. Quercetin and kaempferol isolated from *Acacia arabica* were tested on BRIN-BD11 cell line, and they showed enhanced insulin secretion and reduced glucose levels in a dose-dependent manner (Ansari et al. 2021).
- In another study, *Acacia arabica* extract administered orally to STZ-induced diabetic rats showed reduced the levels of serum glucose, cholesterol, triglycerides, LDL, and malondialdehyde (MDA), and the efficacy dose dependently enhanced high-density lipoprotein (HDL) levels (Hegazy et al. 2013).

2.2.4 Cinnamon Bark

Cinnamon belongs to the family Lauraceae. Different species of cinnamon are used for their antidiabetic activity, namely, *Cinnamomum zeylanicum*, *Cinnamomum cassia*, *Cinnamomum burmanni*, and *Camellia japonica*. *Cinnamomum cassia* contains cinnamaldehyde (approximately 5%), polyphenols, proanthocyanidins, procyanidine, catechin, cinnamtannin, and cinnamic acid (Kim et al. 2004; Cheng et al. 2012). Kim et al. administered *Cinnamomum cassia* extract into T2DM animals at different doses, and the treatment was provided for nearly 6 weeks, which reduced blood glucose levels, enhanced insulin secretion and HDL-cholesterol levels, decreased the levels of triglycerides and total cholesterol, and inhibited intestinal α-glucosidase activity. Hence, this extract can reduce blood glucose and lipid levels by stimulating insulin regulation and inhibiting α-glucosidase enzyme activity (Kim et al. 2006).

The ethanolic extract of *Cinnamomum burmanni* was administered orally to diabetic animals (animals who had induced obesity, insulin resistance, and hyperglycemia after treatment with a very high-fat diet), and the extract could remove polysaccharides, reduce blood glucose levels, block hepatic glucose production, and regulate hepatic gluconeogenesis (Cheng et al. 2012). Anand et al.

observed the regulation of pyruvate kinase enzyme activity, increase in insulin production from β-cells, and regulation of glucose absorption by the skeletal muscle tissue (Anand et al. 2010). When T2DM diabetic mice were treated with the extract of *Cinnamomum Cassia* and *Camellia japonica*, the extract decreased blood glucose levels and enhanced extracellular glucose absorption in HePG2 cells (Lu et al. 2011).

2.2.5 *Momordica charantia*

Momordica charantia, commonly called "bitter gourd" and "bitter melon," belongs to the family Cucurbitaceae. It has potent antidiabetic property, and its main constituents are charantin (steroidal glycoside) and vicine (glycoalkaloid). Fernandes et al. administered the extract of *Momordica charantia* to alloxan-induced diabetic rats and observed stimulated insulin production from β-cells of Islets of Langerhans, decreased glycogenesis in the liver, enhanced glucose absorption in the peripheral tissue, reduced total cholesterol and triglyceride levels, and increased HDL-cholesterol levels (Fernandes et al. 2007). In another study, high-fat diet- and STZ-induced diabetic rats treated with *Momordica charantia* extract demonstrated reduced body weight, enhanced insulin secretion, and regulated GLUT4 function, and improved glucose absorption in the peripheral system. In T2DM, *Momordica charantia* extract regulates interferon (IL)-6 and tumor necrosis factor (TNF)-α; protein expression levels of suppressor of cytokine signaling-3 (SOCS-3) and C-Jun N-terminal kinase (JNK); and protein and mRNA expression levels in the liver by Akt stimulation (Ma et al. 2017).

2.2.6 *Zingiber officinale*

Zingiber officinale, commonly called as "ginger," mainly existing as dried roots and rhizomes, belongs to the family Zingiberaceae. Its main component is gingerol, and it possesses many pharmacological activities. The aqueous extract of *Zingiber officinale*, when tested to STZ-induced diabetic rats (500 mg/kg) for 7 weeks, reduced the levels of fasting blood glucose (FBG), total cholesterol, and triglycerides. This treatment also decreased urinary protein levels (Al-Amin et al. 2006). In a clinical trial, 70 patients with T2DM who were treated with *Zingiber officinale* at a dose of 1600 mg/day (800 mg per capsule) before meal had decreased levels of blood glucose level, total cholesterol, and triglycerides after week 12 when compared with those who received placebo (Arablou et al. 2014). In another clinical trial, 64 patients with T2DM who were treated with a tablet of ginger powder had enhanced insulin levels but no change in blood glucose levels, and continuous intake of the ginger powder reduced the levels of triglycerides and LDL–cholesterol after 2 months (Mahluj et al. 2013).

2.2.7 *Enicostemma littorale*

Enicostemma littorale, commonly called mamejava, belongs to the family Gentianaceae. It contains many phytoconstituents such as catechins, sterols, saponin, glycoside, alkaloids, and volatile oils. The main constituents of this plant are the triterpene sapogenin betulin, the glycoside swertiamarin, and the monoterpenes enicoflavine and gentiocurcine (Vishwakarma et al. 2010). The methanolic extract of *Enicostemma littorale* can reduce blood glucose levels and enhance insulin levels in alloxan-induced diabetic rats. The *Enicostemma littorale* extract enhances glutathione concentration and reduces the stimulation of catalase and lipid peroxidase owing to their antioxidant activity (Maroo et al. 2003). Additionally, STZ-induced T2DM rats treated with the aqueous extract of *Enicostemma littorale* showed reduced levels of total cholesterol, triglycerides, and creatinine. *Enicostemma littorale* is a potent antidiabetic plant and exerts nephrotoxicity (Murali et al. 2002). When the aqueous extract of *Enicostemma littorale* was administered to T1DM rats in a dose-dependent manner (0.5, 1, and 2 g/kg), the high-dose concentrations (1 and 2 g/kg) showed better efficacy, reduced FBG and triglyceride

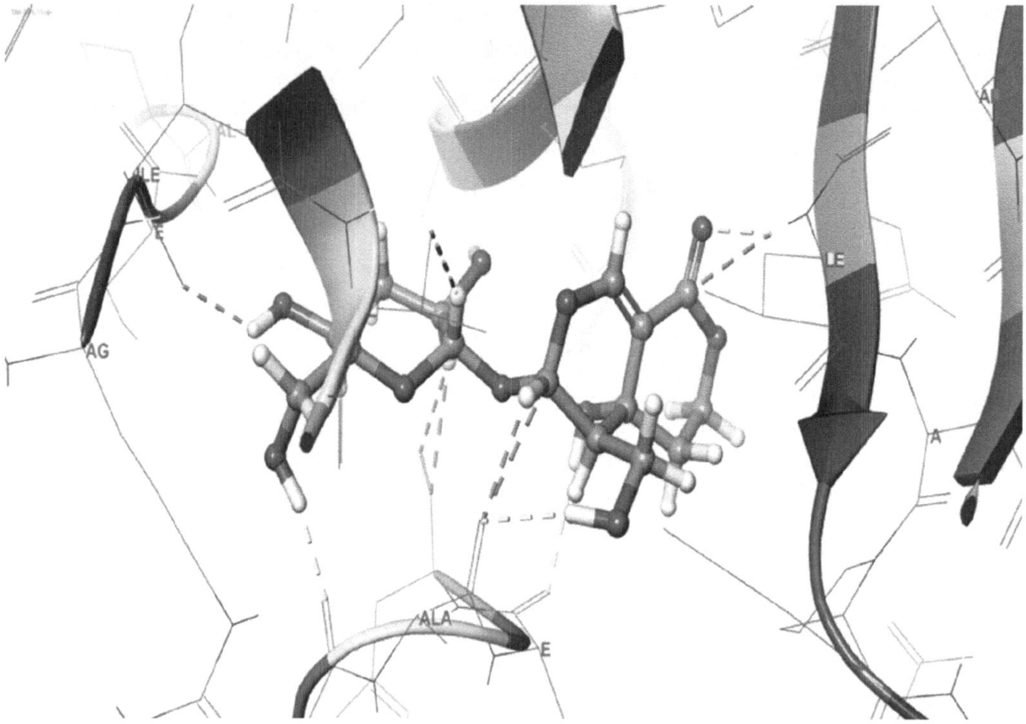

FIGURE 2.2 Three-dimensional diagram showing the interaction of swertiamarin with acetyl-CoA carboxylase receptors.

levels, and enhanced insulin secretion. Swertiamarin is the main compound in *Enicostemma littorale* and regulates insulin secretion through the regulation of acetyl CoA carboxylase (see Figure 2.2) and possesses antidiabetic activity (Vishwakarma et al. 2010).

2.2.8 *Coptis japonica*

Coptis japonica is a Korean medicinal herb belonging to the family Ranunculaceae. It contains isoquinoline alkaloids such as berberine, coptisine, and epiberberine. Berberine shows antidiabetic effects. Berberine stimulates the glutathione and PI3k-Akt pathways and enhances glucose–lipid metabolism (Zhang et al. 2010; Lee et al. 2017). *Coptis japonica* contains berberine sulfate, berberine iodide, and palmatine chloride (isoquinoline alkaloids) and quercetin; when these constituents were treated with rat lens aldose reductase, they exerted potent inhibition on aldose reductase. Quercetin showed better efficacy than berberine and palmatine (Lee 2002).

2.2.9 *Vinca rosea*

Catharanthus roseus, formerly known as *Vinca rosea* and commonly called "vinca," is an alkaloidal medicinal plant belonging to the family Apocynaceae. The leaf and stem of vinca contain alkaloids such as vincristine and vinblastine, which are potent anticancer compounds, and the root contains ajmalicine and serpentine, which are antihypertensive compounds (Kulkarni et al. 1999). In a study of vinca plants, alloxan-induced diabetic rats treated with the methanolic extract of *Catharanthus roseus* had reduced blood glucose levels, enhanced secretion of insulin, increased number of β-cells in the Islets of Langerhans, stimulated lipid metabolism, and reduced body weight (Ahmed et al. 2010).

2.2.10 *JUGLANS REGIA*

Juglans regia (synonym: *Juglans kamaonia*), belonging to the family Juglandaceae. It contains various phytoconstituents such as gallic acid and amino acids such as aspartic acid, threonine, serine, glutamic acid, proline, glycine, alanine, valine, methionine, leucine, isoleucine, tyrosine, histidine, and arginine, among others (Al-Snafi 2018). In a clinical study of 61 patients with non-insulin-dependent diabetes

TABLE 2.1
List of Medicinal Plants Used in Diabetes Treatment

Medicinal Plant	Extract	Chemical Name	Diabetes Type	Activity	References
Aloe vera (*Aloe barbadensis* and *Aloe arborescens*)	Aqueous extract of *Aloe vera*	Cycloartanol (1), lophenol	T1DM and T2DM	Enhances insulin secretion and glucose regulation	(Tanaka et al. 2006)
Adhatoda vasica	Methanolic extract of *Adhatoda vasica*	Vasicine (2) and vascinol (3)	T2DM	α-Glucosidase inhibitory activity	(Gao et al. 2008)
Acacia arabica	Aqueous extract of *Acacia arabica*	Kaempferol (4), quercetin (5), and catechin (6)	T1DM and T2DM	Enhances insulin secretion and glucose uptake	(Ansari et al. 2021)
Cinnamon bark	Aqueous and ethanolic extract of cinnamon	Cinnamaldehyde (7) and cinnamic acid (8)	T1DM and T2DM	Inhibits digestive enzyme activity, enhances glucose absorption and insulin secretion	(Kim et al. 2004, 2006; Cheng et al. 2012)
Momordica charantia	Aqueous extract of *Momordica charantia*	Charantin and vicine (9)	T1DM and T2DM	Inhibits digestive enzyme activity, enhances glucose absorption by the peripheral system, and improves insulin secretion	(Fernandes et al. 2007; Ma et al. 2017)
Zingiber officinale	Tablet of ginger powder	Gingerol (10)	T1DM and T2DM	Enhances glucose absorption and insulin secretion and inhibits AGEs	(Al-Amin et al. 2006; Ma et al. 2017)
Enicostemma littorale	Methanolic extract of *Enicostemma littorale*	Swertiamarine (11)	T1DM and T2DM	Inhibition of AGEs promotes antioxidant activity. Enhances insulin secretion and regulation of glucose uptake	(Vishwakarma et al. 2010; Maroo et al. 2003)
Coptis japonica	Activity evaluation of isolated compounds	Berberine (12), quercetin, and palmitine	T2DM	Inhibition of aldose reductase enzymes	(Zhang et al. 2010; Lee et al. 2017; Lee. 2002)
Vinca rosea	Methanolic extract of *Catharanthus roseus*	Vincristine (17), vinblastine (18)	T1DM	Reduces blood glucose level and enhances insulin secretion	(Kulkarni et al. 1999; Ahmed et al. 2010)
Juglans regia	Capsule of *Juglans regia* powder	Gallic acid (13), aspartic acid (14), arginine (15), leucine (16)	T2DM	Reduces glucose level and total cholesterol level	(Al-Snafi 2018; Hosseini et al. 2014)

AGE: advanced glycation end product; T1DM: type 1 diabetes mellitus; T2DM: type 2 diabetes mellitus

mellitus, continuous treatment with a *Juglans regia*-powdered capsule (100 mg/kg; three capsules per day) for 12 weeks achieved normal levels of blood glucose (FBG), total cholesterol, and triglycerides and conferred protection to the liver and kidney (Hosseini et al. 2014).

Chemical structure of antidiabetic phytoconstituents

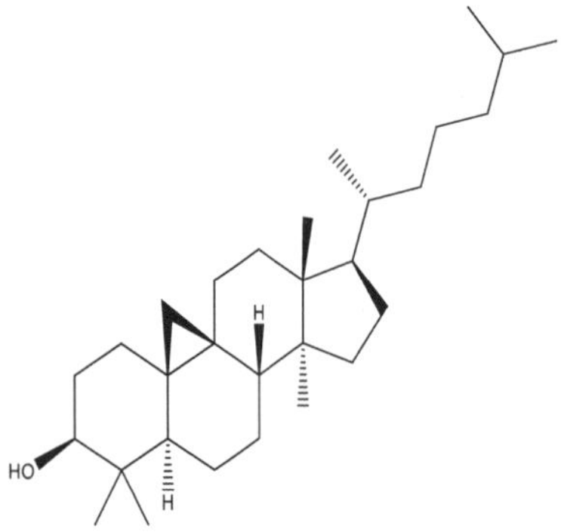

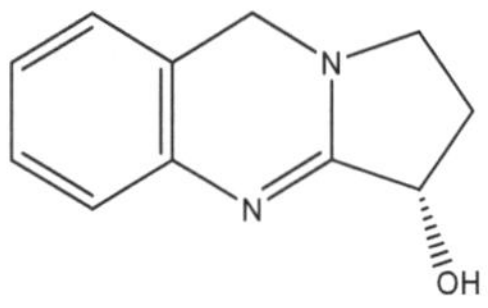

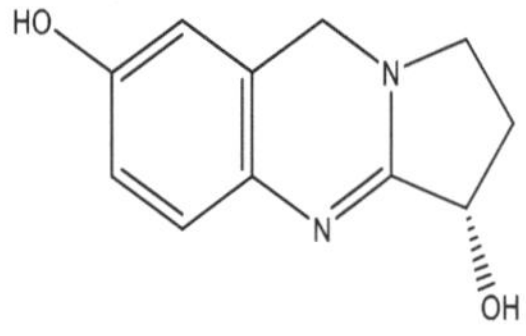

(1) (2) (3)

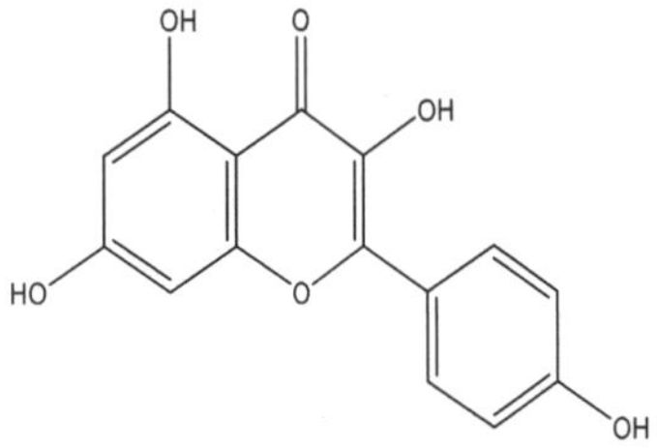

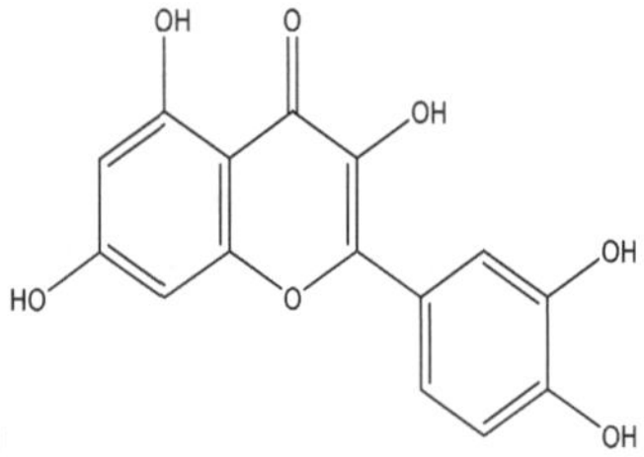

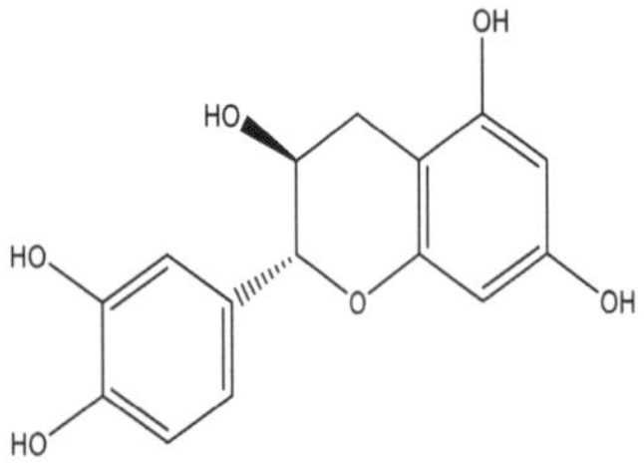

(4) (5) (6)

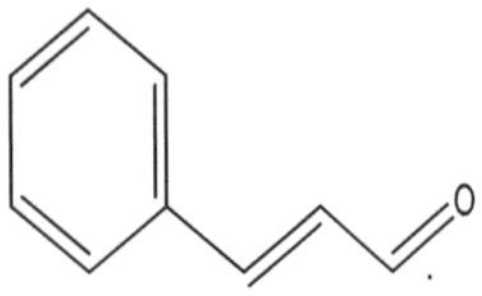

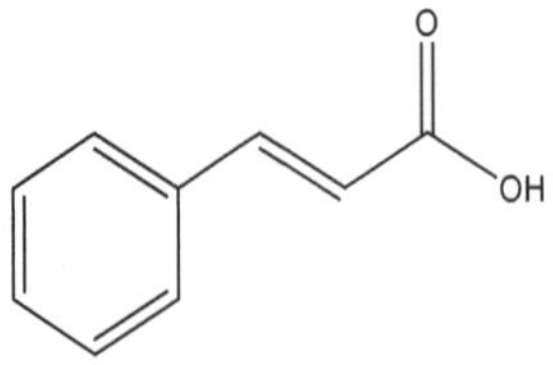

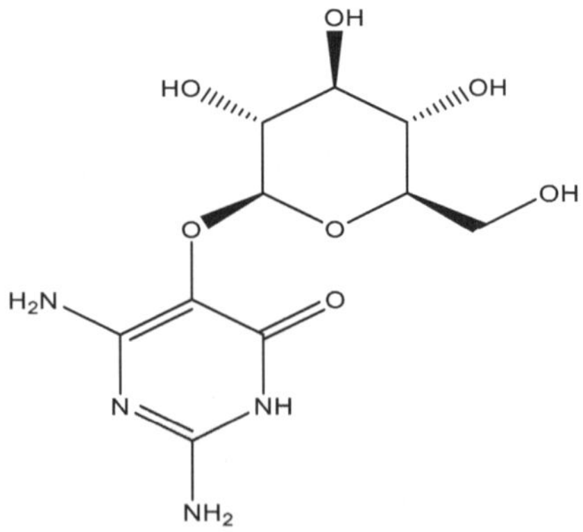

(7) (8) (9)

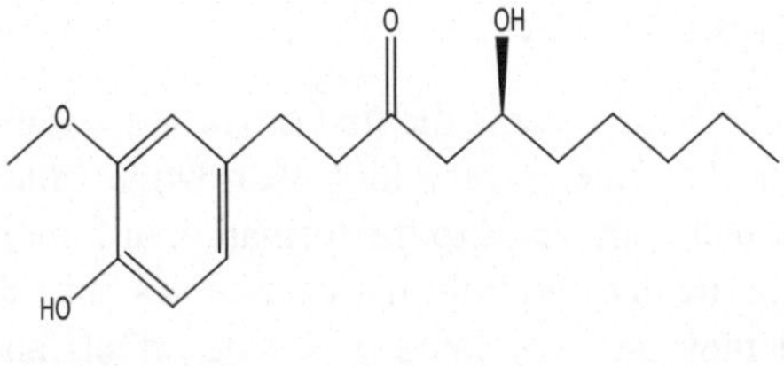

(10)

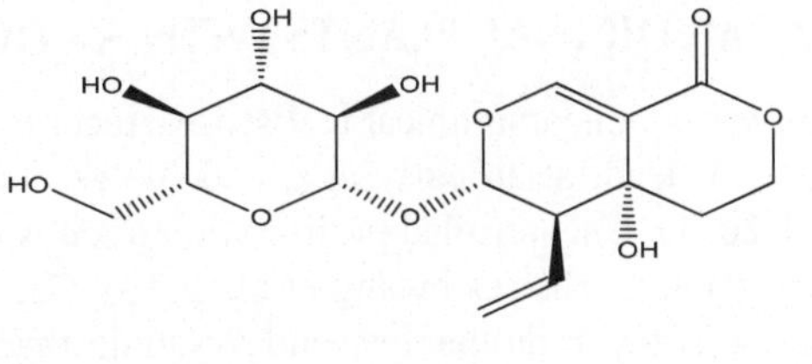

(11)

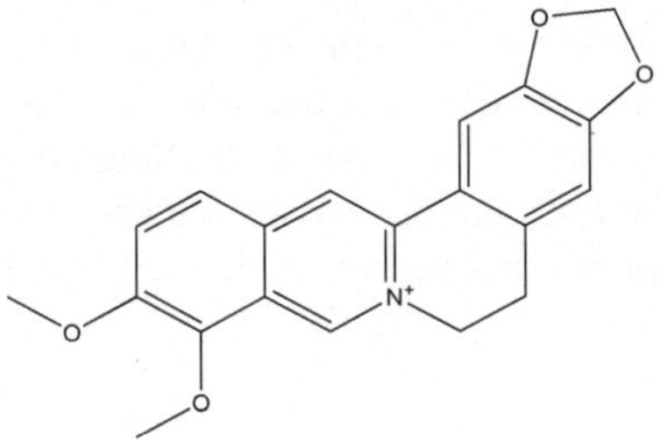

(12)

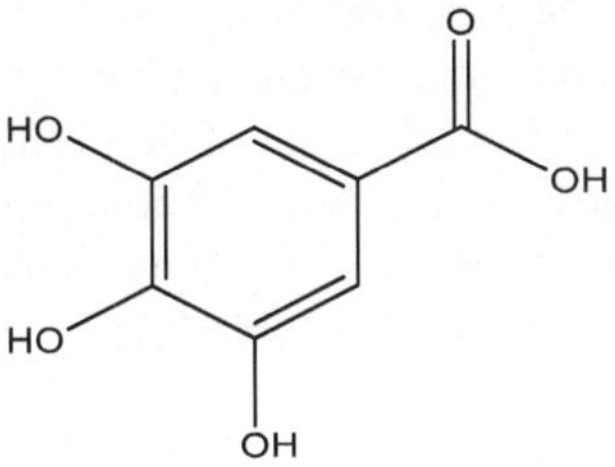

(13)

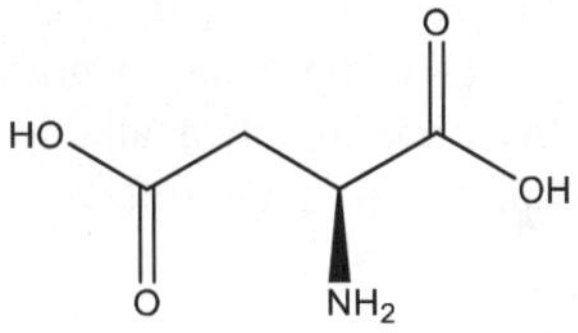

(14)

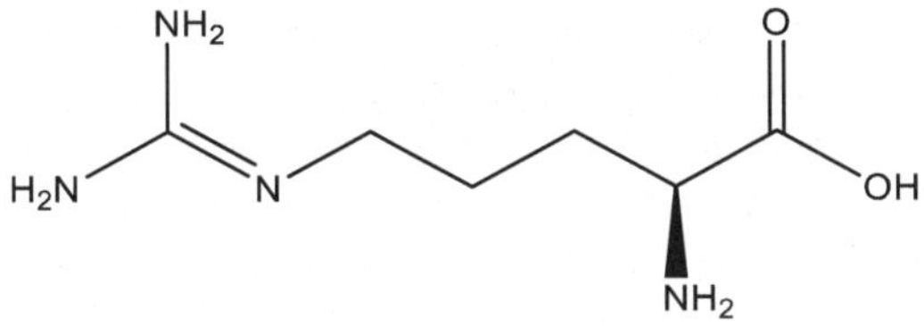

(15)

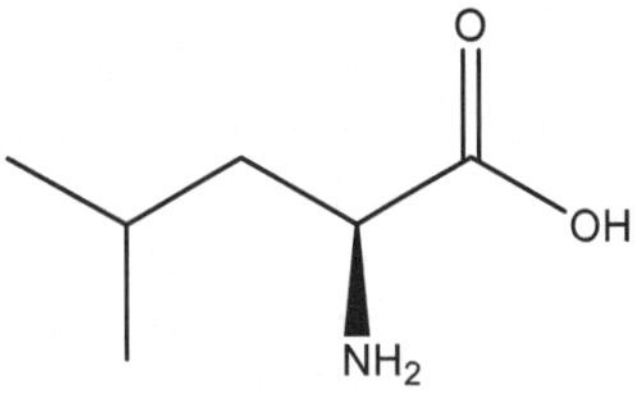

(16)

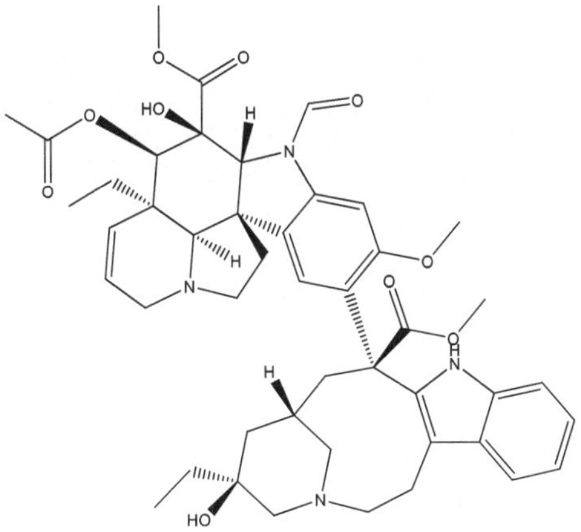

(17)

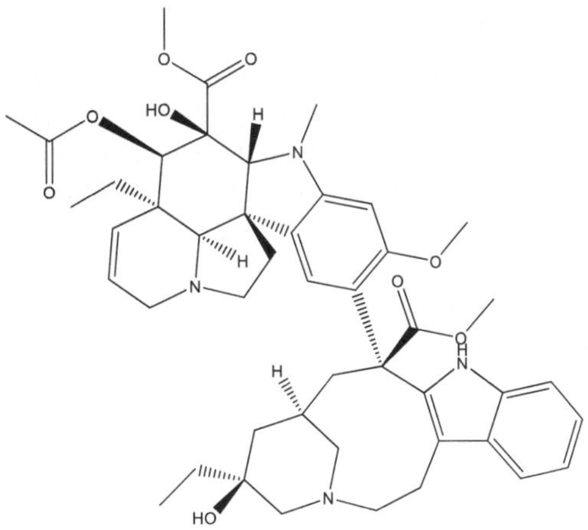

(18)

2.3 MEDICINAL PLANTS WITH ANTICANCER ACTIVITY

Cancer is a noncommunicable disease. According to the WHO, cancer is the first or second leading cause of death at an early age, and the associated mortality rate is very high worldwide (Sung et al. 2021). Uncontrolled proliferation of cells in the human body leads to the formation of tumors by unwanted cells (Ochwang et al. 2014). Cancer can be treated through many strategies such as chemotherapy, radiotherapy, and systemic therapy, but they are associated with side effects and high-level toxicity. Chemotherapy has shown severe treatment-related adverse reactions including alopecia, weakness, and strain (Cancer Research 2015). Because of the higher toxicity level of currently available therapies, the majority of the world's population have moved toward herbal medicine (plant-based medicines), and various ongoing studies are assessing the development of phytomedicines to gain less severe toxicity and better efficacy to kill cancerous cells (Singh et al. 2016) (see Figure 2.4). Currently, many phytoconstituent-based formulations possessing anticancer activity are already available in the market. Some medicinal plants that possess anticancer activity are discussed in the following sections (see Table 2.2):

2.3.1 *Catharanthus roseus* (Vinca)

It is a small shrub that belongs to the family Apocynaceae. This plant contains alkaloids with medicinal properties. It also contains many phytoconstituents such as vincristine, vinblastine, vindesine, and vinorelbine (Singh et al. 2013). Vincristine and vinblastine inhibit tubulin heterodimers, interrupt the function of microtubules, and block cell division at the mitotic phase (Moudi et al. 2013). Semisynthetic derivatives of vinca alkaloids are available in the market, such as vinorelbine, vindesine, vinfosiltine, and vinorelbine. These derivatives can be used either alone or together with other synthetic molecules to block cell division (Almagro et al. 2015). Huang et al. developed a novel controlled-release vinblastine formulation and observed better efficacy and enhanced drug delivery (Huang et al. 2019).

2.3.2 *Taxanes*

These compounds are found in a variety of plants such as *Taxus baccata* and *Taxus canadensis*, belonging to the family Taxaceae, and *Corylus avellana*, belonging to the family Betulaceae. The main taxane constituents are docetaxel and paclitaxel. Taxanes inhibit cell division by blocking the formation of microtubules (β-tubulin), which play a major role in cell division (Xie et al. 2017). They inhibit the mitotic stage of cell division and apoptotic cell death. Taxanes are used against various cancers such as ovarian, breast, and lung cancers (Schutz et al. 2011). A recent study on prostate cancer cell line showed that taxanes affect signals from the androgen receptor (genetic signals). The androgen receptor transfers the signal from the cytoplasm to the cell nucleus, and taxanes block these signals and inhibit gene transcriptional activity, finally blocking cell division (Darshan et al. 2011).

2.3.3 *Camptotheca acuminata*

Camptotheca acuminata belongs to the family Nyssaceae. The main constituent is camptothecin, which is a potent anticancer compound and inhibits topoisomerase I enzyme (Kim et al. 2015). Topoisomerase promotes DNA replication in cell division, and camptothecin derivatives inhibit topoisomerase I enzyme and blocks cell proliferation. There are two derivatives of camptothecin: topotecan (hycantin) and irinotecan. Irinotecan is used in the treatment of colorectal cancer, and topotecan is used in the treatment of ovarian and lung cancers (Rahier et al. 2005).

2.3.4 *Bleekeria vitiensis* and *Ochrosia elliptica*

Bleekeria vitiensis (synonym: *Ochrosia vitiensis*) and *Ochrosia elliptica* belong to the family Apocyanaceae, and their main constituent is ellipticine, which is a potent anticancer compound. Ellipticine, an alkaloidal drug, is isolated from the whole plant of *Bleekeria vitiensis* and *Ochrosia elliptica*. It possesses topoisomerase II inhibitory activity, destroys the double-stranded DNA helix, and blocks cell division (Stiborová et al. 2014). Ellipticine derivatives are available in the market, such as N-2-(diethylaminoethyl)-9-hydroxyellipticinum chloride and 2-N-methyl-9-hydroxyellipticine, which have shown more efficacy; these derivatives are used against various cancers such as ependymoblastoma, leukemia, myeloma, melanoma, breast cancer, and colon cancer (Kizek et al. 2012; Isah et al. 2016).

2.3.5 *Cephalotaxus harringtonia*

Cephalotaxus harringtonia is a cephalotaxus alkaloid belonging to the family Taxaceae, and their components are harringtonine and isoharringtonine, which have potent anticancer activity (Feldman et al. 1996). As shown in studies of the cancer cell lines A549 lung cancer cell line, HeLa cell line, and SGC-7901 gastric cancerous cell line, cephalotaxus alkaloids can inhibit protein synthesis, promote polyribosome degradation but have no effect on new peptide chains, and inhibit cell division by blocking protein synthesis (Heibliga et al. 2014).

2.3.6 Guduchi (*Tinospora cordifolia*), Barberry (*Berberis vulgaris*), *Berberis aquifolium*, and *Rhizoma coptidis*

Guduchi belongs to the family Menispermaceae, Barberry belongs to the family Berberidaceae, and *Rhizoma coptidis* belongs to the family Ranunculaceae. All these medicinal plants contain berberine alkaloid as the main constituent. Berberine possesses potent anticancer activity (Mantena et al. 2006). It enhances cell apoptosis and blocks mitotic cell division. It is used in the treatment of breast, colorectal, and liver cancers (Xu et al. 2013).

2.3.7 Red Pepper (*Capsicum annum*)

Capsicum annum is a dried ripe fruit belonging to the family Solanaceae. The main constituent is capsaicin, which is used against various cancers owing to its antimutagenic, antiangiogenic, and chemopreventive activities (Venier et al. 2015; Clark et al. 2016). It enhances cell apoptosis through the p53 protein (p53 is a cancer-suppressive protein) to achieve normal cell division (Sarkar et al. 2015). A study on the MCF10A h-Ras cell line treated with *Capsicum annum* showed cell apoptosis through an increase in the reactive oxygen species and Rac1 pathways (Kim et al. 2004).

2.3.8 Resveratrol

Resveratrol is a polyphenolic compound obtained from medicinal plants such as mulberry, peanuts, grapes, and blueberries. It is an anticancer agent and is used in the treatment of various cancers such as breast, lung, colorectal, liver, pancreatic, and prostate cancers. It regulates p53 protein together with Bcl-2 and Bax transcriptional proteins and reduces the stimulation of nuclear factor (NF)-κB, AP-1, cyclic enzyme, cyclin-dependent kinase enzyme, and OX-2 protein (Patel et al. 2010; Varoni et al. 2016). It blocks the growth of blood vessels and inhibits angiogenesis, thus controlling cell proliferation (Brakenhielm et al. 2001).

2.3.9 *Curcuma longa* (Turmeric)

Curcuma longa is a dried root and rhizome belonging to the family Zingiberaceae. The main constituent is curcumin. In a study wherein curcumin was tested in glioblastoma human cell line, the treatment regulated gene and protein expression of Bcl-2 and Bax, Erk enzyme, JNK-1, NF-κB, and cyclin D K2 proteins (Vallianou et al. 2015).

2.3.10 Kaempferol

Kaempferol is obtained from various plants such as propolis, black tea, grape fruit, and broccoli. When kaempferol was tested in colorectal and HT-29 cancer cell lines, it could block G1 and G2 mitotic cell division through the inhibition of CDk2, CDC2, CDK4, and CDk2 (cyclin-dependent kinase 2) and inhibits the cell proliferation (Lee et al. 2017b).

2.3.11 Crocus

Colchicum autumnale, which exists as seed and bulb, belongs to the family Iridaceae. The main constituent is colchicine, which inhibits the α- and β-dimers of tubulin (α-β tubulin interface) and disables the microtubules, which is generally important for cell division; colchicine also suppresses microtubule formation (Figure 2.3) (Slobodnick et al. 2015).

2.3.12 Saffron

Crocus sativus, well known for their flower petals, belongs to the family Iridaceae. It contains carotenoids, crocin, crocetin, and safranal (Hoshyar et al. 2017). Crocin is used as an anticancer agent, as it enhances cell apoptosis; crocin and crocetin inhibit gene expression, block DNA and RNA syntheses, suppress RNA polymerase II enzyme, and block RNA pathways, finally inhibiting cell proliferation. Safranal is also used in cancer treatment and has shown low toxicity in other cells (Figure 2.3) (Milajerdi et al. 2016).

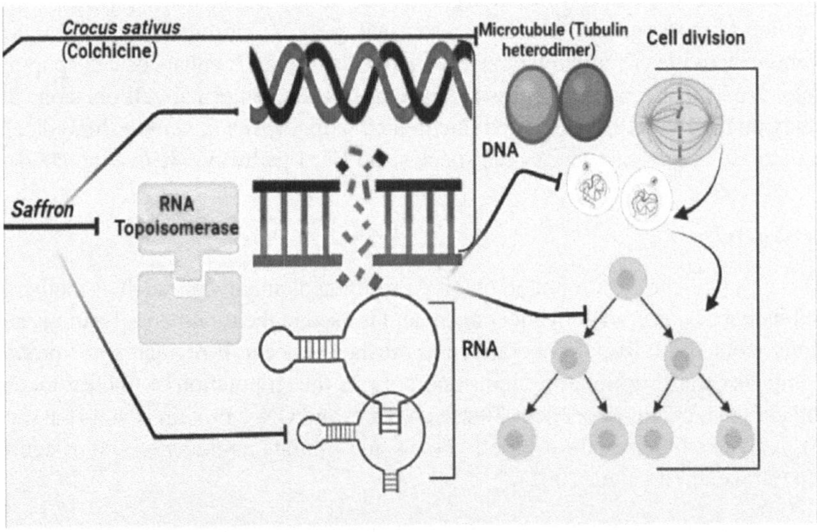

FIGURE 2.3 Possible molecular targets in the anticancer mechanism of *Crocus sativus* (Saffron).

Chemical composition of *Crocus sativus* (Saffron)

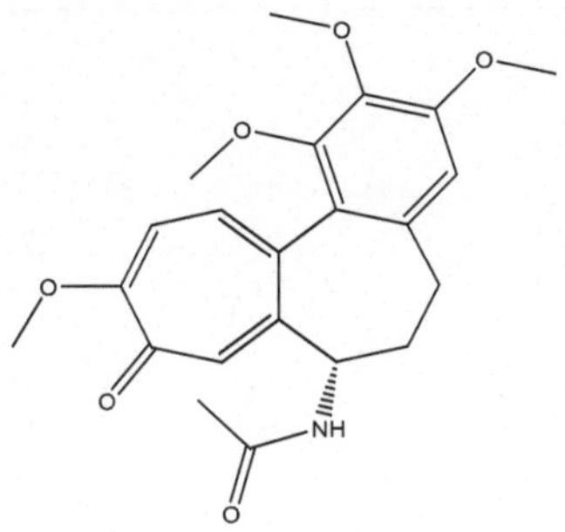

(27)

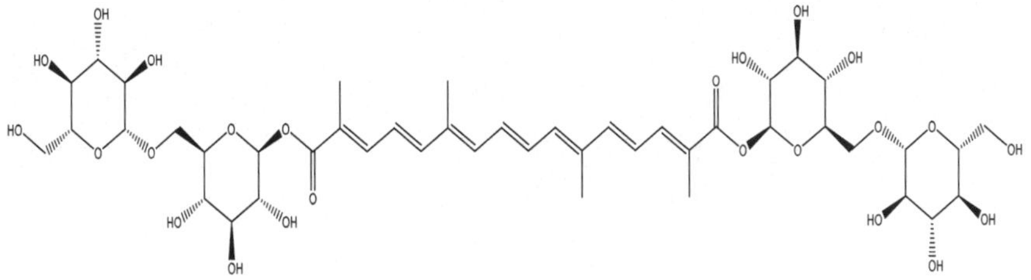

(28)

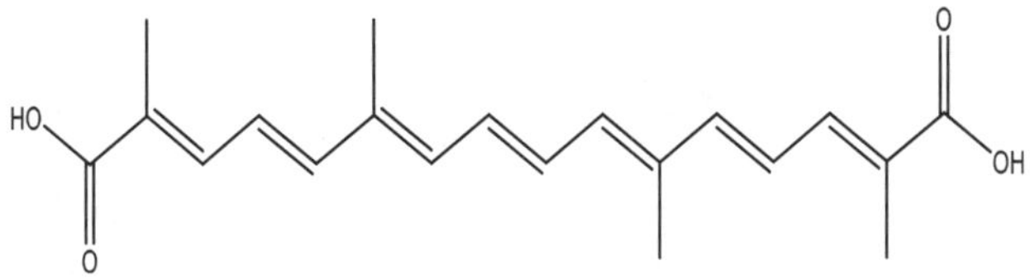

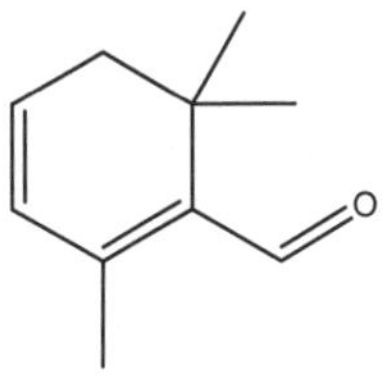

(29) (30)

Crocus sativus flowers and buds (saffron)

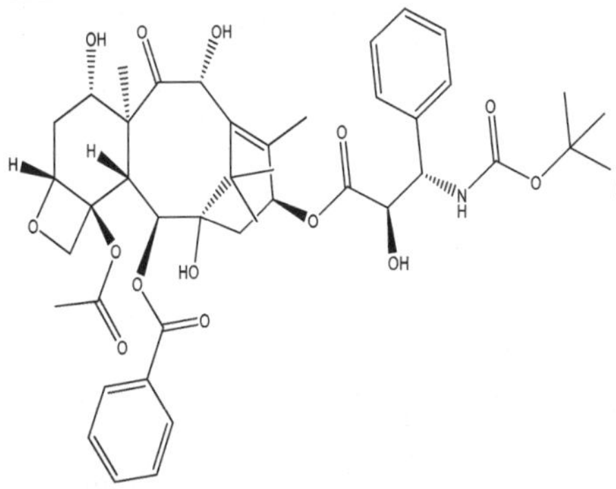

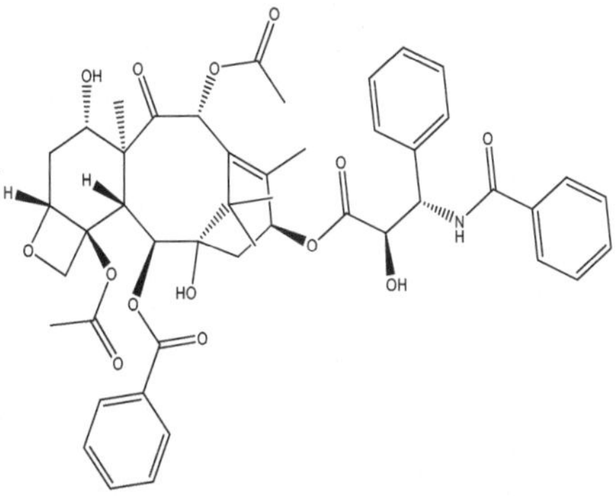

(19) (20)

Chemical structure of anticancer phyto-constituents

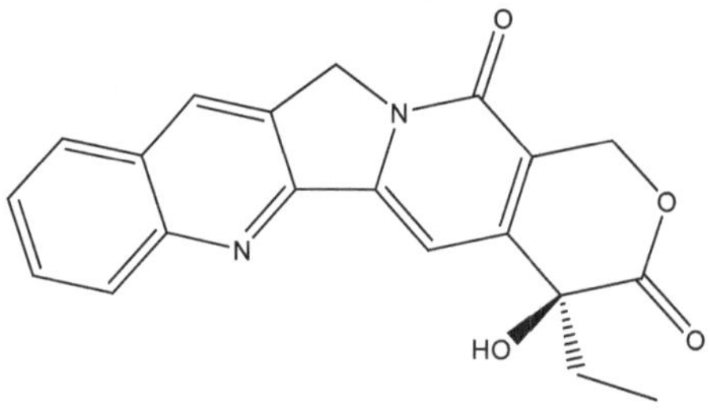

(21)

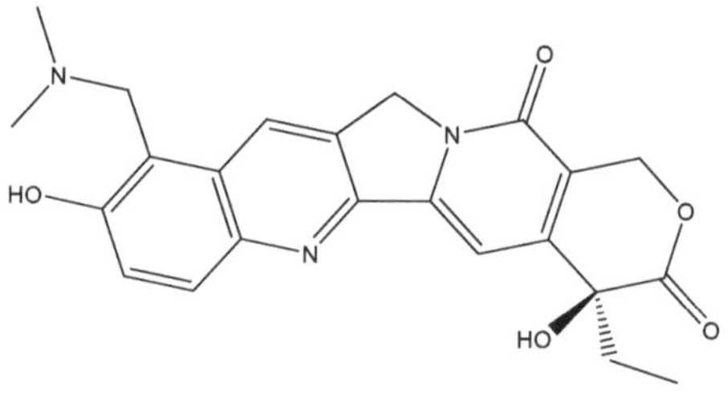

(22)

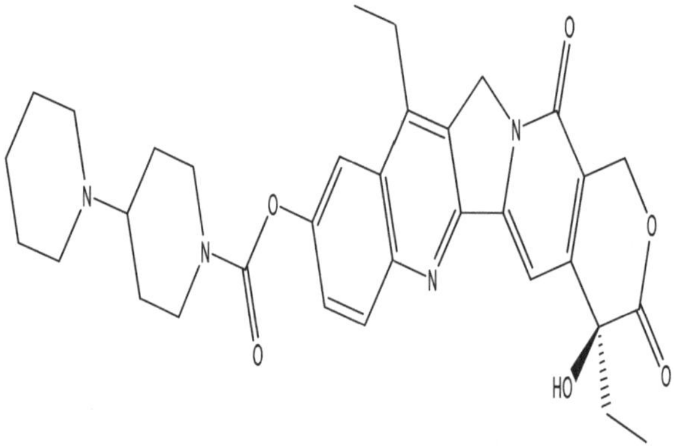

(23)

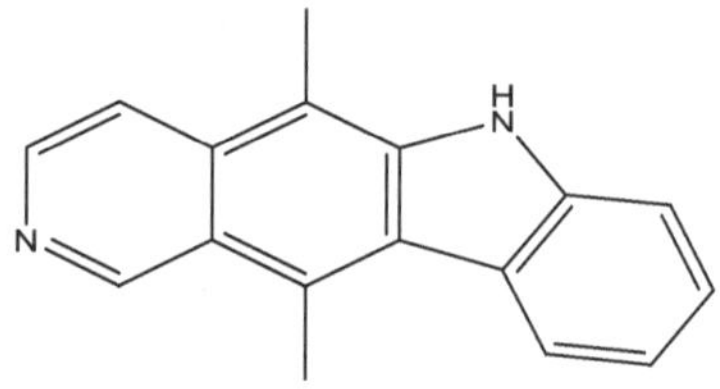

(24)

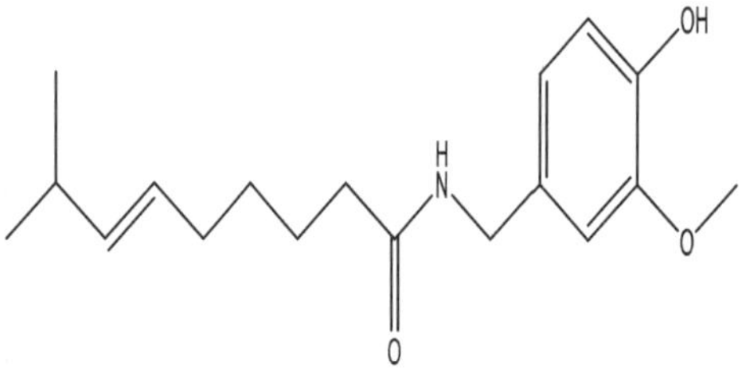

(25)

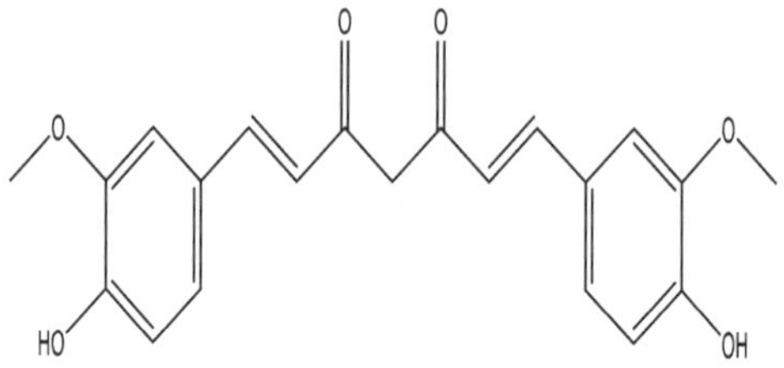

(26)

2.4 MEDICINAL PLANTS WITH CARDIOVASCULAR ACTIVITY

Cardiovascular disease is a rapidly emerging clinical condition worldwide, associated with high mortality and morbidity rates (Irfan et al. 2020). Cardiovascular conditions such as coronary heart failure, myocardial infarction, hypertension, atherosclerosis, and heart attack/stroke. Heart attack/stroke are the main causes of death (80%) among patients with heart ailments, as this condition causes deficiency of oxygen in the brain and heart. Moreover, fatty material accumulates inside the blood vessels and blocks cerebral arteries and coronary arteries (Roth et al. 2017). Cardiovascular diseases are associated with diabetes, obesity, and dyslipidemia (imbalance in the levels of LDL and HDL) (Smith et al. 2015). "Hypertension," which involves an imbalance in blood pressure (systolic

TABLE 2.2
List of Anticancer Medicinal Plants

Medicinal Plant Name	Chemical Compound	Mechanism of Action	References
Catharanthus roseus (Vinca)	Vincristine (17) and vinblastine (18)	Inhibits the tubulin of hetero dimers	(Moudi et al., 2013)
Taxanes (*Taxus baccata*, *Taxus canadensis*, and *Corylus avellana*)	Docetaxel (19) and paclitaxel (20)	Inhibits β-tubulin	(Xie et al. 2017)
Camptotheca acuminata	Camptothecin (21) and the derivatives topotecan (22) (hycantin) and irinotecan (23)	Inhibits topoisomerase I enzymes	(Kim et al. 2015; Rahier et al. 2005)
Bleekeria vitiensis and *Ochrosia elliptica*	Ellipticine (24)	Inhibits topoisomerase II	(Stiborova et al. 2014)
Cephalotaxus harringtonia (alkaloid)	Harringtonine and isoharringtonine	Inhibits protein synthesis	(Feldman et al. 1996; Heibliga et al. 2014)
Tinospora cordifolia (Guduchi), *Berberis vulgaris* (Barberry), *Berberis aquifolium*, and Rhizoma Coptidis	Berberine	Cell apoptosis and blocks mitotic cell division	(Mantena et al. 2006; Xu et al. 2013)
Capsicum annum (red pepper)	Capsaicin (25)	Regulates p53 and enhances cell proliferation	(Venier et al. 2015; Clark et al. 2016; Sarkar et al. 2015)
Mulberry, peanuts, grapes, blueberries	Resveratrol	Regulate p53 protein and inhibits angiogenesis	(Patel et al. 2010; Varoni et al. 2016; Brakenhielm et al. 2001)
Curcuma longa (turmeric)	Curcumin (26)	Regulates gene expression	(Vallianou et al. 2015)
Propolis, black tea, grape fruit, broccoli	Kaempferol	Inhibits G1 and G2 mitotic cell division	(Lee et al. 2017b)
Crocus (*Colchicum autumnale*)	Colchicine (27)	Inhibits α- & β-dimers of tubulin	(Slobodnick et al. 2015)
Saffron (*Crocus sativus*)	Crocin (28), crocetin (29), and safranal (30)	Inhibits RNA polymerase II enzyme	(Hoshyar et al. 2017; Milajerdi et al. 2016)

and diastolic blood pressure; Figure 2.5) (Verma et al. 2021). Currently, various treatments for cardiovascular diseases are available, such as antihypertensives, antiplatelets, and antithrombotic drugs, among others, but they may be associated with complications along with therapeutic profiles (e.g., aspirin causes gastric ulcers) (Smith et al. 2015). Clopidogrel causes aplastic anemia and thrombocytopenia (Barrett et al. 2008; Mackman 2008). Several studies have indicated the use of various medicinal plants in the treatment of cardiovascular diseases (Table 2.3).

2.4.1 *Panax ginseng* (Asian Ginseng)

Panax ginseng belongs to the family Araliaceae. It possesses various pharmacological activities such as antidiabetic, anticancer, and immunomodulatory activities. The main constituents are ginsenoside, ginsenoside, and saponin, classified as panaxadiol (PPD) and panaxatriol (PPT). PPD contain Rb1, Rb2, Rb3, Rc1, Rg3, Rh2, and Rs. PPT contain Re, Rf, Rg1, Rg2, and Rh (Wong et al., 2015). PPD contains Rb1 and has shown better cardiovascular activity, wherein Rb1 regulates nitric oxide (NO) synthase activity, regulates the phosphorylation of SER1177, and enhances NO production in endothelial cells (Yu et al. 2007). PPD contains Rg3 and stimulates endothelial NO synthase (eNOS) activity and Erk/Akt signals and activates Peroxisome proliferator-activated receptor-γ and induced the differentiation of human umbilical

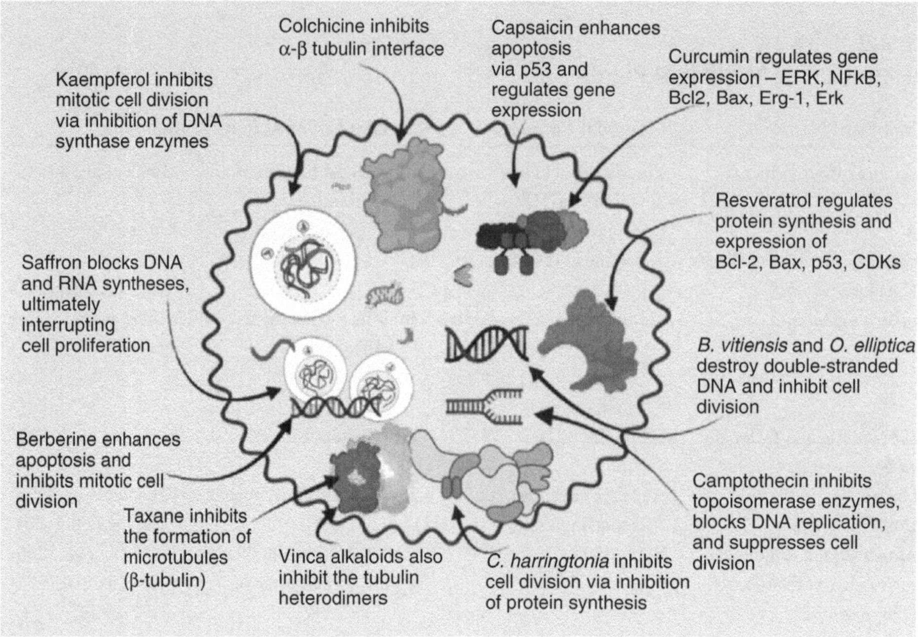

FIGURE 2.4 Diagrammatic representation showing medicinal plants with anticancer activity (created in BioRender).

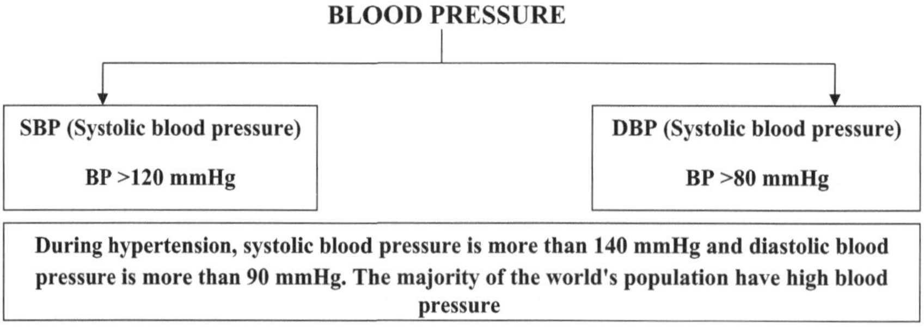

FIGURE 2.5 Classification of blood pressure.

vein endothelial cells (Kwok et al. 2012). In another study, the PPT saponin Rg1 reduced Ca^{2+} channel activation (mechanism is shown in Figure 2.5), blocked Ca^{2+} inflow into cardiomyocytes (cardiomyocytes generate the contraction force and are responsible for the rhythmic beating of the heart), and stimulated antioxidant enzymes in cardiomyocytes during myocardial ischemia (Zhu et al. 2009).

2.4.2 Garlic (*Allium sativum*)

Allium sativum is a perennial herb belonging to the family Liliaceae; it exerts different therapeutic activities such as antidiabetic, antiatherosclerotic, anticancer, antimicrobial, antifungal, and cardiovascular activities (Mohamed et al. 2020). Its main constituents are sulfur-containing compounds such as alliin, allicin (thiosulfinates), vinyldithiines, ajoenes, and flavonoids such as quercetin (Al-Snafi 2013). It also contains S-allyl cysteine and diallyl sulfide. It reduces the levels of cholesterol, LDL, and triglycerides (Mahady & Gail 2009). The chloroform extract of *Allium cepa* can stimulate eNOS and NO production in endothelial cells (Nyeem et al. 2017).

2.4.3 Artichoke (*Cynara scolymus*)

Cynara scolymus Linn., well known for the dried leaves, belongs to the family Asteraceae. It contains various phytoconstituents such as 3-O-caffeoquinic acid (chlorogenic acid), caffeic acid, and 4-O-caffeoquinic acid; sesquiterpenes lactones such as dehydrocynaropicrin and grosheimin; and flavonoids such as scolymoside, cynaroside, and cyanotrioside. In a study, the aqueous extract of *Cynara scolymus* decreased the levels of total cholesterol, triglycerides, and LDL (Mahady & Gail 2009). In a clinical study, 107 patients with mild hypertension who received concentrated leaf juice of *Cynara scolymus* for 12 weeks had improvement in hypertension in a dose-dependent manner, compared with placebo, as the extract enhanced antioxidant activity and regulated eNOS activation. It possesses antithrombotic, antiatherosclerotic, and antihypertensive activities and enhances NO production in endothelial cells (Roghani et al. 2009).

2.4.4 Ginkgo (*Ginkgo biloba*)

Ginkgo biloba belongs to the family Ginkgoaceae. It contains ginkgolides, bilobalides, flavonoids such as quercetin, and other oligomeric compounds such as proanthocyanidins (procyanidin and procyanidin B-2) and hyperosides (Zhou et al. 2004). The extract of *Ginkgo biloba* was tested against mercury II–induced cardiovascular damage in albino rats, and it reduced proinflammatory mediators such as TNF-α, inhibited mercury-induced toxicity, and regulated free radicals. Hg showed a high toxicity level because of the thiol groups, and the *Ginkgo biloba* extract inhibited the toxicity by blocking protein expression of thiol enzyme (Tunali-Akbay et al. 2007). In a recent study, *Ginkgo biloba* extract (EGb761) was administered orally to hypertensive rats in a dose-dependent manner (60/90/180 mg/kg/day), and it reduced hypertension and protein expression of thiol enzyme, regulated glutathione peroxidase activity, inhibited angiotensin-converting enzyme (ACE), and stimulated NO production in endothelial cells (Mansour et al. 2011).

2.4.5 Hawthorn (*Crataegus oxyacantha*)

Crataegus oxyacantha, including *Crataegus monogyna* and *Crataegus laevigata*, as well as other main species of Hawthorn, belongs to the family Rosaceae. It contains vitexin; hyperoxide; rutin; catechin (flavonoids); triterpenoids such as ursolic acid, oleanolic acid, and crataegolic acids; chlorogenic acid; and caffeic acid (Rayyan et al. 2005). The dried extract of hawthorn leaf has shown a positive ionotropic effect and stimulates the sodium-potassium ion channels and increases Ca^{2+} ion flow into cardiomyocytes (Rodriguez et al. 2008). It possesses antioxidant activity and reduces the levels of inflammatory cytokines (Li et al. 2011). It can also reduce the antiarrhythmic effect (Müller et al. 1999). Additionally, it enhances vasodilation in the coronary artery and peripheral blood vessels, stimulates eNOS activity, and inhibits ACE and NO production in endothelial cells (Brixius et al. 2006).

2.4.6 Flaxseed Oil (*Linum usitatissimum*)

Linum usitatissimum belongs to the family Linaceae. It contains linolenic acid and n-3 polyunsaturated fatty acid as the main constituents and has shown potent cardiovascular effects and anti-inflammatory activity (Fleming et al. 2014). It also contains lignans (Anderson et al. 2009). Treatment with flaxseed-derived products at different concentrations in 1004 patients for 3–48 weeks could reduce blood pressure (Khalesi et al. 2015). Flaxseed can reduce LDL-total cholesterol levels, and lignans have shown better effects against hypertension by reducing cholesterol and LDL levels; this lipid-reducing ability is more commonly found in women than in men (Pan et al. 2009).

2.4.7 *Hibiscus sabdariffa* (Rosella)

Hibiscus sabdariffa belongs to the family Malvaceae. Its main phytoconstituents are delphinidin-3-sambubioside and cyanidine-3-sambubioside, both of which possess anticholesteremic, antihypertensive, and antioxidant activities (Abubakar et al. 2015). It possesses various pharmacological activities such as antimicrobial, antioxidant, anticholesterolemic, and antihypertensive activities (Herrera-Arellano et al. 2004; Herrera-Arellano et al. 2007; McKay et al. 2010; Gurrola-Diaz et al. 2010). *Hibiscus sabdariffa* can reduce blood pressure (systolic and diastolic blood pressure) in a dose-dependent manner (Mojiminiyi et al. 2007) and can inhibit ACE activity similar to captopril (Herrera-Arellano et al. 2004). In a clinical trial, treatment of *Hibiscus sabdariffa* extract to patients demonstrated antihypertensive and vasodilatory effects. *Hibiscus sabdariffa* stimulates of NO/cyclic guanosine monophosphate pathways and causes relaxation to endothelium cells and blocks Ca^{2+} channels, thus enhancing vasodilation (Ajay et al. 2007).

2.4.8 Green Tea (*Camellia sinensis*)

Camellia sinensis belongs to the family Theaceae. The main constituents are catechin and its derivatives epicatechin-3-gallate, epigallatocatechin-3-gallate, and epicatechin (Deka et al. 2011). α-Methylated epigallatocatechin-3-gallate blocks ACE, and regular intake of green tea can reduce the risk of systolic blood pressure (San Cheang et al. 2015).

2.4.9 *Rauwolfia serpentina*

Rauwolfia serpentina exists as dried roots and rhizomes and belongs to the family Apocyanaceae. It contains various phytoconstituents such as alkaloids, phenols, tannins, and flavonoids. Reserpine is the main alkaloidal compound and exhibits cardiovascular activity. Other constituents are ajmaline, and serpentine. Ajmaline is effective in blocking the sodium channels (Bunkar et al. 2017). A study reported that the methanolic extract of *Rauwolfia serpentina* tested in a rat model could control hypertension and hypercholesterolemia (Shah et al. 2020).

2.4.10 Gokhru (*Tribulus terrestris*)

Tribulus terrestris is a spiny herb and belongs to the family Zygophyllaceae (Ahmed et al. 2020). It contains various phytocompounds such as aloe emodin, chrysophanol, emodin, isoernodin, and rhein (Sun & Yanfei 2017). In a previous study, rhubarb extract tested in an animal model demonstrated protective effects against chronic renal failure (Wang et al. 2009). Rhubarb blocks sodium and potassium ATPase channels, enhances the osmotic pressure in the blood, increases the water level in blood vessels, and maintains blood volume and microcirculation (Cao et al. 2017).

TABLE 2.3
List of Cardiovascular Agents Derived From Medicinal Plants

Medicinal Plant Name	Chemical Compound	Plant Extract	Mechanism of Action	References
Panax ginseng	Ginsenoside Rb1, Rg1 (31), and Rg3	Compound use of Rb1, Rg1, and Rg3	Regulates eNOS and reduces Ca^{2+} channel activation	(Yu et al. 2007)
Garlic (*Allium sativum*)	Alliin, allicin (32), S-allyl cysteine	Aqueous extract of *Allium sativum*	Stimulates eNOS and vasodilation, reduced LDL	(Al-Snafi. 2013; Nyeem et al. 2017; Mahady & Gail 2009)
Artichoke (*Cynara scolymus*)	3-O-caffeoquinic acid, caffeic acid (33), 4-O-caffeoquinic acid, cynaroside (34), and cyanotrioside	Aqueous extract of artichoke	Antithrombotic, antiatherosclerotic, and antihypertensive activities	(Mahady & Gail 2009)

TABLE 2.3 (Continued)
List of Cardiovascular Agents Derived From Medicinal Plants

Medicinal Plant Name	Chemical Compound	Plant Extract	Mechanism of Action	References
Ginkgo (*Ginkgo biloba*)	Procyanidin (35), and procyanidin B-2, and hyperoside	Aqueous extract of *Ginkgo biloba*	Reduces thiol enzyme, production of endothelial cells, and vasodilation, and anti-inflammatory activity	(Zhou et al. 2004; Tunali-Akbay et al. 2007)
Hawthorn (*Crataegus oxyacantha*)	Ursolic acid (36), oleanolic acid (37), and crategolic acid	Aqueous extract of Hawthorn	Enhances vasodilation and inhibits ACE	(Rayyan et al. 2005; Brixius et al. 2006)
Flaxseed oil (*Linum usitatissimum*)	Linolenic acid (38) and lignans	Flaxseed-derived product	Reduces LDL-total cholesterol level and hypertension	(Fleming et al. 2014; Anderson et al. 2009; Pan et al. 2009)
Hibiscus sabdariffa (Rosella)	Delphinidine-3-sam bubioside, cyanidine-3-sambubioside	Extract of *Hibiscus sabdariffa*	Reduces SBP and DBP, stimulation eNOS, ACE inhibition and relaxation in endothelial cells	(Abubakar et al. 2015; Mojiminiyi et al. 2007; Herrera-Arellano et al. 2004; Ajay et al. 2007)
Green tea (*Camellia sinensis*)	Epicatechin-3-gallate (39), epigallactocatechin-3-gallate, epicatechin	Extract of green tea	Blocks ACE and reduces the risk of SBP	(Deka et al. 2011; San Cheang et al. 2015)
Rauwolfia serpentina	Reserpine (40)	Rauwolfia constituents	Reduces SBP and DBP	(Bunkar et al. 2017; Shah et al. 2020)
Gokhru (*Tribulus terrestris*)	Emodin, rhein aloe-emodin	Aqueous extract of Gokhru	Inhibits the Na^+ and K^+ ATPase channel	(Sun & Yanfei 2017; Cao et al. 2017)

ACE: angiotensin-converting enzyme; DBP: diastolic blood pressure; eNOS: endothelial nitric oxide synthase; LDL: low-density lipoprotein; SBP: systolic blood pressure

Chemical structure of the cardiovascular agents

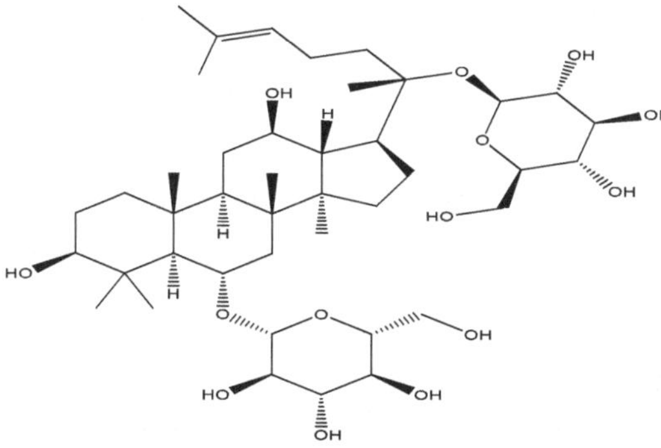

(31)

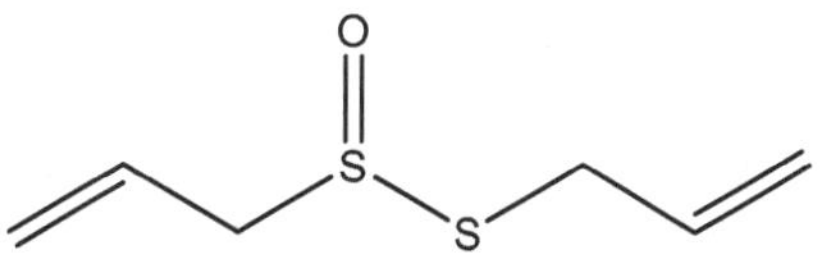

(32)

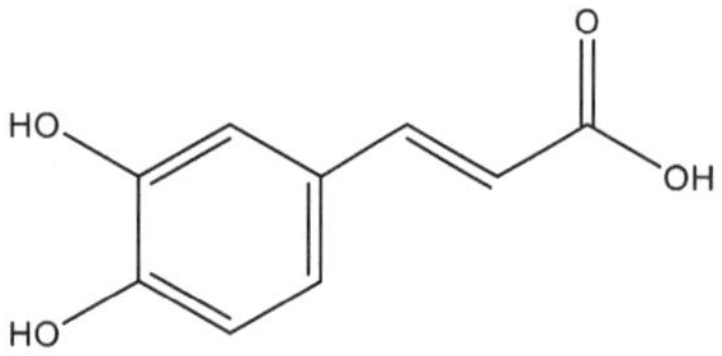

(33)

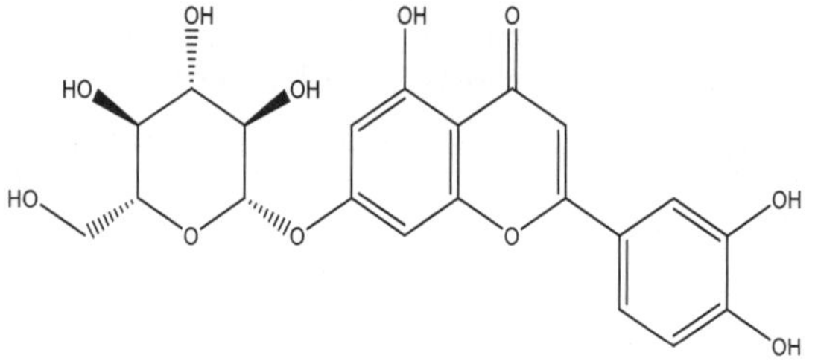

(34)

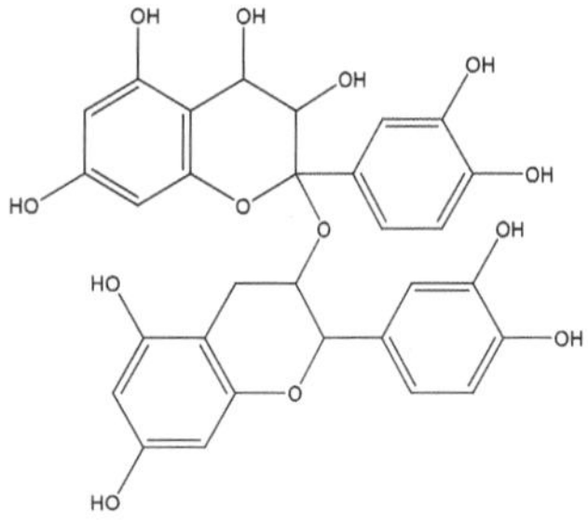

(35)

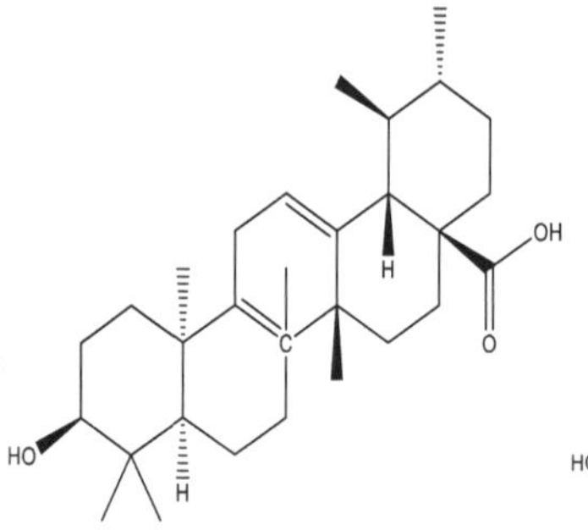

(36)

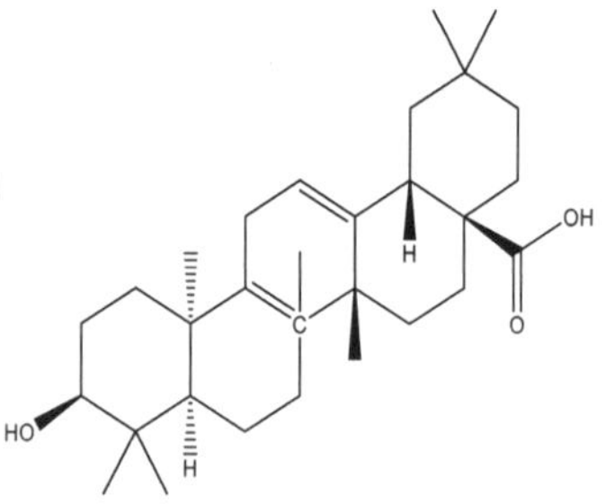

(37)

(38)

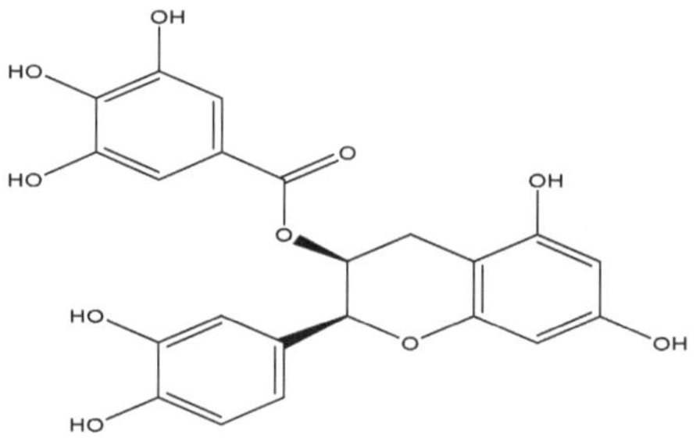

(39)

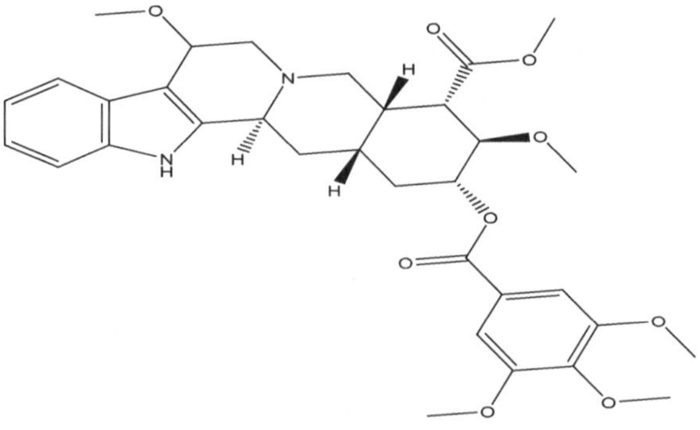

(40)

2.5 MEDICINAL PLANTS EFFECTIVE AGAINST SARS-COV-2 INFECTION

The recent COVID-19 pandemic affected the population globally with its disastrous effects, and it persists to date. It was first detected in Wuhan (China). According to the WHO, COVID-19 is caused by SARS-CoV-2 and mainly affects the respiratory system, hence named as "coronavirus disease." Several studies have focused on both newly developed and existing drugs with better efficacy against this virus. Treatment for COVID-19 (SARS-CoV-2) targets ACE-2. Additionally, many inflammatory cytokines such as IL-1, IL-6, IL-12, interferon-γ, NF-κB, and TNF-α can affect multiple organs (Figure 2.6). Available drugs including antimalarials such as chloroquine and hydroxychloroquine have shown efficacy against SARS-CoV-2 (Gautret et al. 2020; Keyaerts et al. 2007; Zhengli & Shi 2020). Other medications used against COVID-19 are nucleoside analogs, remdesivir, and the anti-inflammatory drug lopinavir/ritonavir (Li et al. 2020). Symptoms of infection are fever, cough, loss of sense, fatigue, and breathing problems (Xu et al. 2020). The pathology of COVID-19 involves diffuse alveolar injury together with cellular fibrous mucus exudates, lymphocyte infiltration into pulmonary interstitium, reduced number of lymphocytes in the circulatory system, immunological disorders, and mucus accumulation into the tracheal cavity and alveoli (Xu et al. 2020b). Medicinal plants are a very important source of medicine, and some of them are used against SARS-CoV-2 infection, showing improvement in the patient's heath (Hong-Zhi et al. 2020). A group of experts recommended different methods for the use of medicinal plants or a mixture of plants for COVID-19 treatment (Jin et al. 2020).

Various medicinal plants have been used for the treatment of cough, general weakness, digestive system disorders, and anxiety. Medicinal plants and herbal preparation can be used against COVID-19.

Lianhua Quingwen (LQ) is a Chinese herbal preparation used against SARS-CoV-2. These medicines have been used since thousands of years, especially against influenza (H1N1) and SARS-CoV-2 (Lu et al. 2020). LQ is also used to treat fever, cough, fatigue, influenza, pneumonia, bronchitis, and early-stage measles (Ding et al. 2017). It possesses better broad-spectrum effects, particularly antivirus, antibacterial, antipyretic, anti-inflammatory, anti-tussive, and immunomodulatory effects.

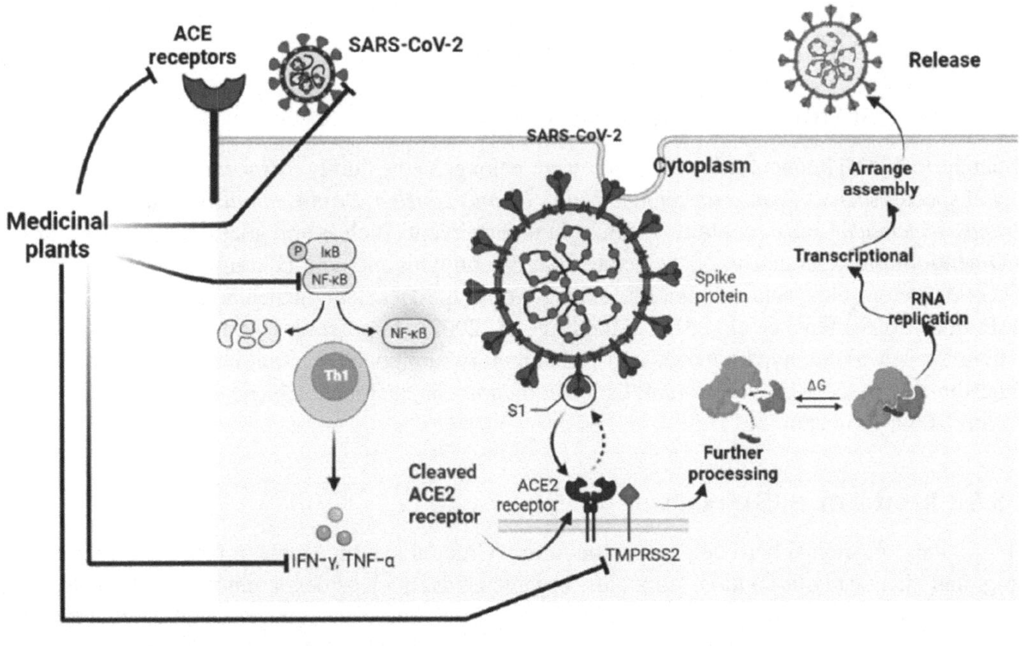

FIGURE 2.6 Possible mechanism of action of medicinal plants against COVID-19 (created in BioRender).

This herbal formulation demonstrated anti-SARS-CoV-2 activity by blocking SARS-CoV-2 replication in a dose-dependent manner and reducing the secretion of proinflammatory cytokines such as TNF-α, IL-6, CCL-2/MCP-1, and CXCL-10/IP-10, and its activity depends on the drug concentration (Runfeng et al. 2020). LQ is a mixture of various medicinal plants (Liu et al. 2010).

2.5.1 *Forsythia suspense* (Thunb.)

Forsythia suspense, well known for its dry fruits, belongs to the family Oleaceae. It is used in the preparation of the Chinese herbal medicine Lian Qiao. It exerts anti-inflammatory activity, has a bitter taste, and is slightly cold (Lee et al. 2018). *Forsythia fructus* reduces the amount of proinflammatory cytokines released, such as TNF-α, IL-1β, NF-κB, and inducible NOS. It possesses anti-inflammatory and antioxidant activities (Qu et al. 2021). The main chemical constituents are forsythoside A and B (phenylethanolamine), phillyrin, and phillygenin (lignans) (Zhang et al. 2020b; Hu et al. 2020; Jiang et al. 2020; Wang et al. 2021).

2.5.2 *Lonicera japonica* (Thunb.)

Lonicera japonica, well known for its dried flower, belongs to the family Caprifoliaceae, and it is used in the traditional medicine Jin Yin Hua. It contains various phytoconstituents such as chlorogenic acid (Yip et al. 2006), hyperoside, aromadendrene, linalool, hexadecanoic acid, and other aldehydes and ketones. It possesses various therapeutic activities such as antibacterial, antioxidant, hepatoprotective, anti-inflammatory, and antibacterial activities (Shang et al. 2011).

2.5.3 *Ephedra* (Ma Huang) (*Ephedra sinica, Ephedra intermedia,* and *Ephedra equisetina*)

Ephedra herba and Ephedra radix are Chinese herbal medicines belonging to the family Ephedraceae. This plant possesses various pharmacological activities such as lowering respiratory complications, antitussive, antipyretic, and anti-inflammatory activities (Pharmacopoeia Chinese 2015). Chemically, these plants contain various alkaloidal compounds such as ephedrine, pseudoephedrine, nor-pseudoephedrine, and methyl ephedrine (Chen et al. 2019).

2.5.4 Apricot Seed (Semen Armeniacae Amarum)

Bitter apricot, well known for its dried ripe seed, belongs to the family *Rosaceae* and includes a variety of species such as *Prunus armeniaca* Linn., *Prunus sibirica*, *Prunus mandshurica* (maxim), and *Prunus armeniaca* Linn. It contains various phytoconstituents such as amygdalin (D-mandelonitrile-β-D-gentiobioside), prunasin, chlorogenic acid, 3-feruloylquinic acid (Zhongzhen & Zhao 2004), glyceride sterols, oleic acid, and linoleic acid (Kwon et al. 2010; Zhongzhen & Zhao 2004; Hempen & Fischer 2009; Tang et al. 2011; Fukuda et al. 2003). It possesses various pharmacological activities such as antihypertensive, anti-inflammatory, anticoagulant, antineoplastic, antitussive, antiasthmatic, mucolytic, expectorant, laxative, menostasia, and menorrhagic activities (Hempen & Fischer 2009; Tang et al. 2011).

2.5.5 Radix isatidis (Banlangen)

This Chinese medicinal herb belongs to the family Cruciferae (Rhoeadales). It is a potent antiviral drug, and chemically, it contains alkaloids, organic acids, nucleosides, amino acids, lignanosides, flavonoids, sulfur-containing organic compounds (goitrin, epigoitrin, sinigrin, and glucobrassicin), and amino acids such as proline. It possesses various pharmacological activities such as anticancer, anti-inflammatory, and antiviral activities. Its therapeutic profile depends on its large chemical profile. This is mainly attributed to its water-soluble portion (sulfur-containing chemoprofile) (Zhou et al. 2013).

2.5.6 Rhizoma of *Dryopteris crassirhizoma*

Dryopteris crassirhizoma belongs to the family Aspiadeaceae. It is used in the preparation of the Chinese herbal medicine Mianma Guanzhong (Zhu et al. 2017). Chemically, it contains ursolic acid, dryopteric acid A and B, kaempferol, rhamnoside, and matteucinol. Dryocrassin ABBA is a lead constituent that has demonstrated antiviral activity against influenza virus H5N1 infection. Dryopteric acid shows potent anti-HIV activity (Lee et al. 2008).

2.5.7 Herba Houttuyniae (*Houttuynia cordata*)

Houttuynia cordata is a flowering perennial herb that belongs to the family Saururaceae. It is used in the Chinese herbal formulation Yu-Xing-Cao. The main component of *Houttuynia cordata* is decanoyl acetaldehyde (Ma et al. 2017b). It is an antibacterial agent and converted to methyl-n-nonane at a high temperature (Chen et al. 2014). Other constituents are essential oils containing ketones, bornyl acetate, and β-myrcene (Lu et al. 2006). Flavonoids such as rutin, hyperoside, and quercetine are also present in this plant (Chen et al. 2012; Xu et al. 2006). Other phenolic compounds are linolenic acid, linoleic acid, oleic acid, palmitic acid, stearic acid, quinic acid, caffeic acid, neochlorogenic acid, chlorogenic acid, and cryptochlorogenic acid (Nuengchamnong 2009). It regulates immunity and improves barriers in the vaginal cavity and oral and intestinal routes. It is also a potent antiviral and anti-inflammatory drug (Wu et al. 2021).

2.5.8 Herba Pogostemonis (*Pogostemon cablin*)

Pogostemon cablin (Blanco), well known for their dried aerial parts, belongs to the family Lamiaceae (Labiateae). Herba pogostemonis is used in the preparation of the Chinese herbal medicine Guang-thio-xing. Pogostone contains β-patchoulene, patchouli alcohol, patchoulene epoxide, pogostone, and pachypodol. It exerts various therapeutic activities such as antimicrobial, anti–peptic ulcer, anti-oxidant, anti-inflammatory, anti-ischemic, analgesic, antitumor, and antidiabetic activities (Junren et al. 2021). It is used for the treatment of gastric ulcers (Chen et al. 2016).

2.5.9 Rhodiola (Hong Jing Tian)

Rhodiola belongs to the family Crassulaceae and includes various species such as *Rhodiola rosea*, *Rhodiola alterna*, *Rhodiola crenulata*, *Rhodiola kirilowii*, *Rhodiola quadrifida*, and *Rhodiola sachalinensis*. The root and rhizomes of *Rhodiola* (Bawa & Khanum 2009). *Rhodiola crenulata* is used as traditional medicines. It contains rhodiosin, which is a herbacetin derivative, and other constituents are the phenylpropanoids cinnamyl alcohol (Peschel et al. 2016), rosiridin (Van Diermen et al. 2009), epigallocatechin-3-gallate (Chen et al. 2015), lignans (Zhou et al. 2015), and many flavonoids (Péter Zomborszki et al. 2019).

2.5.10 Rhei Radix et Rhizoma (Rhubarb/Dahuang)

Root and rhizomes of *Rheum palmatum* Linn., *Rheum tanguticum*, and *Rheum officinale* belong to the family Polygonaceae. It contains anthraquinone, rhein, emodin, aloe-emodin, chrysophanol, and physcion (Agarwal et al. 2000; Zhou et al. 2006; Xie et al. 2008; Wang et al. 2011). It has a laxative property and is used for the treatment of gastric carcinoma (Liu et al. 2015).

2.5.11 Glycyrrhizae Radix et Rhizoma (Ganco)

Glycyrrhiza glabra (licorice), well known for its roots and rhizomes, belongs to the family Leguminosae. It contains glycyrrhizic acid (Lin et al. 2009) and glycyrrhizinic acid, which are metabolized to glycyrrhetinic acid (Ploeger et al. 2001) and are present as two isomers: trans-isomers (18-α-glycyrrhetinic acid) and cis-isomers (18-β-glycyrrhetinic acid) (Borisenko et al. 2016). Other flavonoids present in this plant are liquiritigenin and isoliquiritigenin (Peter et al. 2013). Additionally, other compounds such as glabridin, glycyrol, among others are present. It is used therapeutically for its anti-hepatitis, antitussive, antidemulcent, and anti-peptic ulcer effects (Iino et al. 2001; Van et al. 2001).

LQ has shown better efficacy against SARS-CoV-2 by improving symptoms and the clinical condition. In a clinical study, after the completion of phase 3, 24 patients with COVID-19 had improvement in health and clinical symptoms, both severe and mild conditions (Zhuang et al. 2021).

SARS-CoV-2 enters host cells by binding with the ACE-2 protein. Various medicinal plants that inhibit ACE-2, such as *Berberis integerrima* Bunge, belong to the family Berberidaceae; *Crataegus laevigata* (Poir) is a hawthorn species belonging to the family Rosaceae; *Onopordum acanthium* belongs to the family Asteraceae, and *Quercus infectoria* G. Oliver belongs to the family Fugaceae. *Onopordum infectoris* showed better efficacy by approximately 94% against ACE-2. *Berberis integerrima*, *Crataegus microphylla*, and *Onopordum acanthium* showed better efficacy as an ACE inhibitor and has antioxidant ability (Sharifi et al. 2013).

2.6 MEDICINAL PLANTS WITH ANTI-ALZHEIMER'S ACTIVITY

AD is a dynamic neurodegenerative disorder of the brain associated with aging and accounts for 60% of all mental disorders occurring in the elderly; AD is characterized by loss of memory and behavioral abnormalities. AD is caused by accumulation of misfolded proteins. The β-amyloid proteins aggregate in different regions of the CNS. AD also involves reduced thinking ability and changes in personality. It adversely interferes with activities of the daily living. Approximately 70% of cases are due to genetic reasons and 21% of cases due to environmental factors (Singhal 2012; Roy 2018).

In AD, the cholinesterase enzyme is inhibited, leading to an inactive acetylcholinesterase (AChE) and the consequent increase in acetylcholine levels in the brain. Various AChE-blocking compounds are available, such as rivastigmine, donepezil, tacrine, and galantamine. In modern therapy, there is no proper treatment for AD, and currently available therapies can reduce only the symptoms (Orhan et al. 2009). Several medicinal plants possess potent activity against AD (see Table 2.4), and some of them are listed in the following sections (see Figure 2.7):

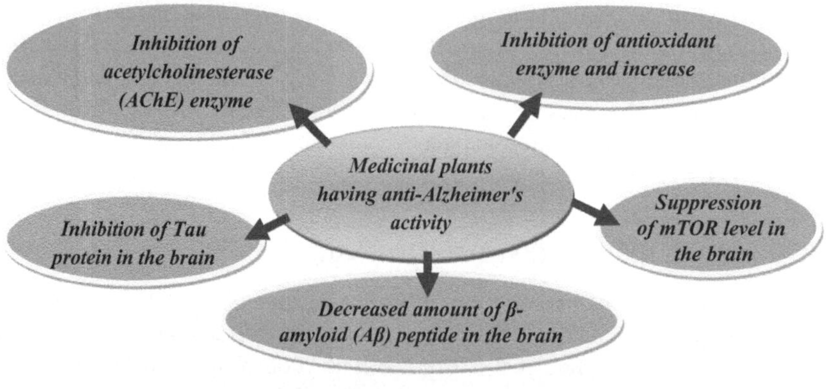

FIGURE 2.7 Possible mechanism of anti-Alzheimer's activity of medicinal plants.

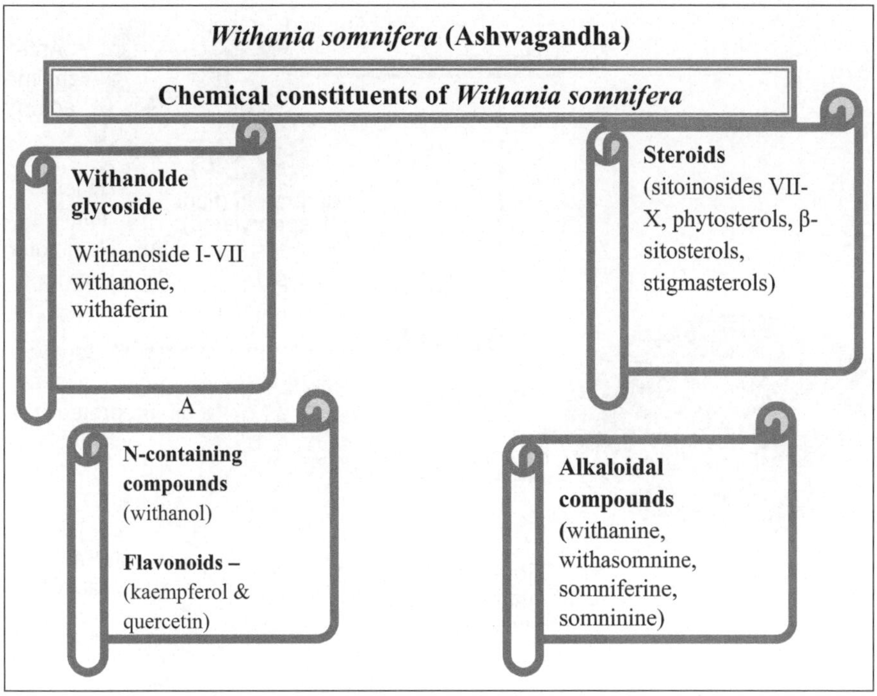

FIGURE 2.8 Chemical constituents of *Withania somnifera* (ashwagandha).

2.6.1 *Withania somnifera* (Ashwagandha)

Withania somnifera, which belongs to the family Solanaceae, is extensively used in Ayurvedic formulations as a nerve tonic. This plant possesses antioxidant property and free radical-scavenging ability (Russo et al. 2001). This plant decreases stress levels, increases memory and concentration power, and reverses forgetfulness, and it causes no adverse drug effects (Abedon et al. 2008). Ashwagandha is used for the treatment of neurodegenerative ailments such as Huntington's disease, Parkinson's disease, and anti-AD. *Withania* is also used as a brain tonic, antistress agent, memory booster, sleep induction, and gouty arthritis treatment, and it is effective against SARS-CoV-2 infection (Dar 2020; Shree et al. 2020). This plant contains various phytoconstituents (see Figure 2.8) such as withanolides, glycosides, withanosides I to VII, withasomidienone, withanine, withasomnine somniferine A, somninine, dehydro withanolide R, withaferin A, withasomniferols A to C, withanone, phytosterols sitoindosides VII to X, beta-sitosterol, alkaloids, amino acids, and high amounts of iron (Matsuda et al. 2001). Withanamides shows free radical-scavenging activity during the initial phase of AD and its progression. It also blocks the neurons triggered by amyloid plaques (Jayaprakasam et al. 2010). In an *in silico* (docking) study, active moiety-like withanamides A and C, which binds to the α-amyloid (Aα 25–35) active motif inhibits fibril development (Figure 2.9) (Kumar et al. 2012).

In an in silico and dynamics study of ashwagandha constituents, withanoside V and somniferine exhibited anti-COVID-19 activity against SARS-CoV-2 M^{pro} (Shree et al. 2020). In an another molecular study, Dharmendra and Deepak explored other constituents of ashwagandha and reported that withaferin A, withanolide B, tinosporide, withanolide, orientin, flavonol glucoside, apigenin, kaempferol, and withanone inhibited SARS-CoV-2 activity (Figures 2.10 and 2.11) (Maurya 2020). *Withania somnifera* is used for treating neurodegenerative ailments such as Parkinson's and Huntington's diseases. The root extract of ashwagandha exhibits anti-Parkinsonism activity through

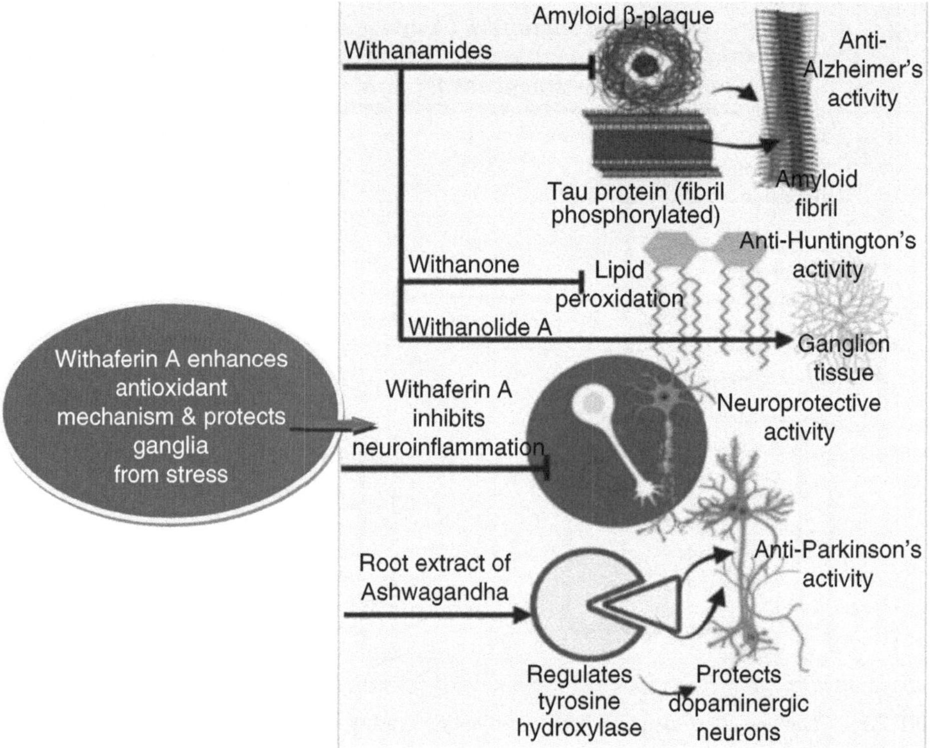

FIGURE 2.9 Possible molecular targets of *Withania somnifera* and its metabolites against different diseases (created in BioRender).

the regulation of tyrosine hydroxylase enzyme and provides safety to dopaminergic neurons through diminished apoptosis (programmed cell death) (Figure 2.9) (Prakash et al. 2014). It also possesses anti-Huntington's disease activity, and withanolide A (metabolite of ashwagandha) enhances the antioxidative activity and preserves ganglionic neurons in case of oxidative stress. In an animal study, this extract upgrades cognitive disfigurement and reimposes glutathione and reduced AChE levels in the brain, cortex, and hippocampus (Figure 2.9) (Kumar & Kumar 2008). In a previous study, the component withaferin A has shown neuroprotective activity and decreases neuronal inflammation (Figure 2.9) (Patel et al. 2015).

Structure of chemical constituents of *W. somnifera*

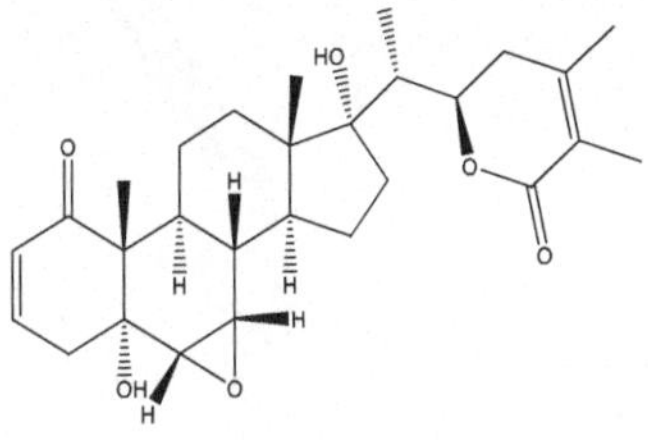

Withanone

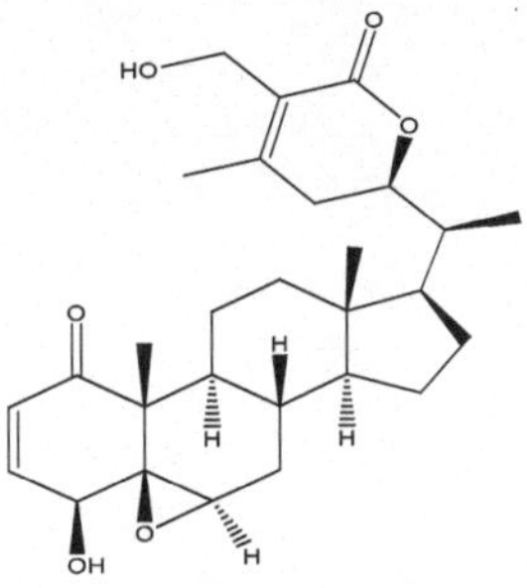

Withaferin A

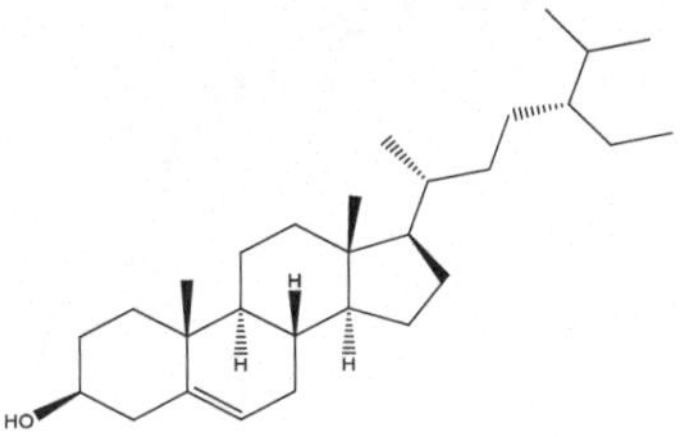

β-Sitosterol

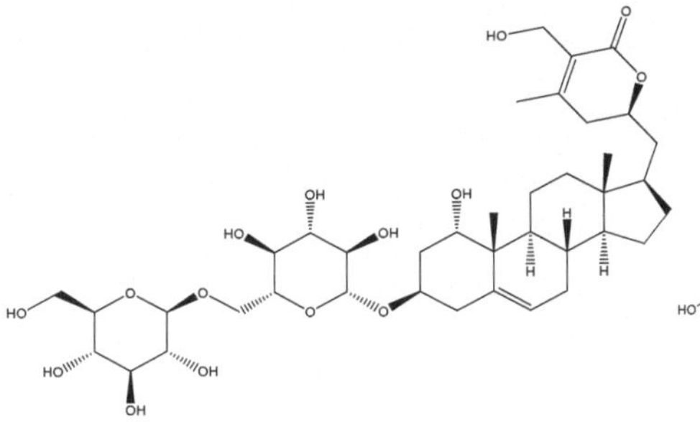

Withanoside I

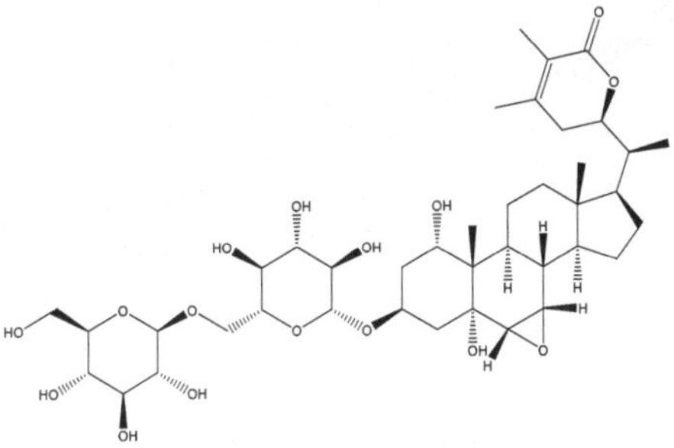

Withanoside II

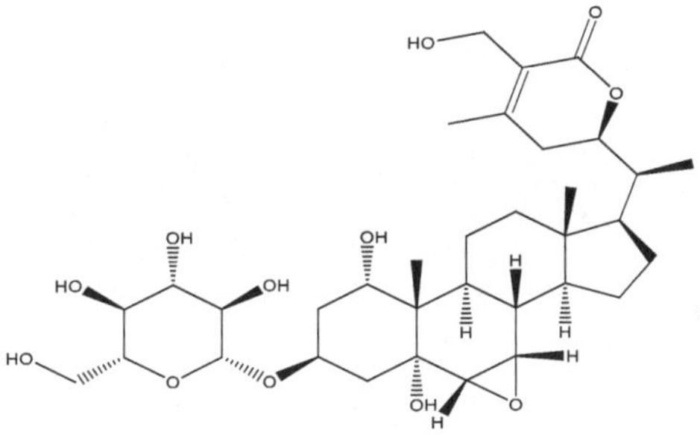

Withanoside III

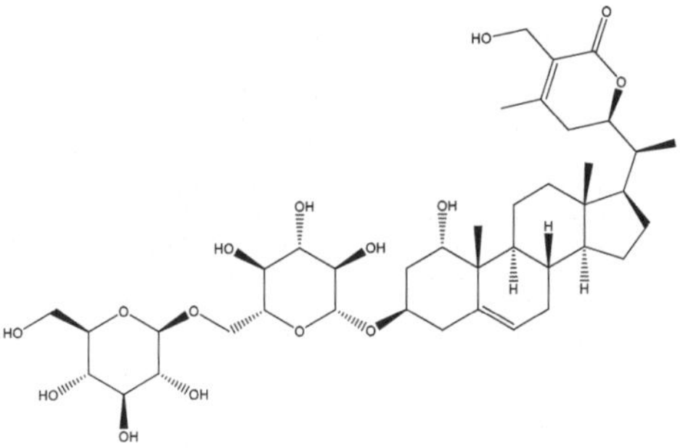

Withanoside IV

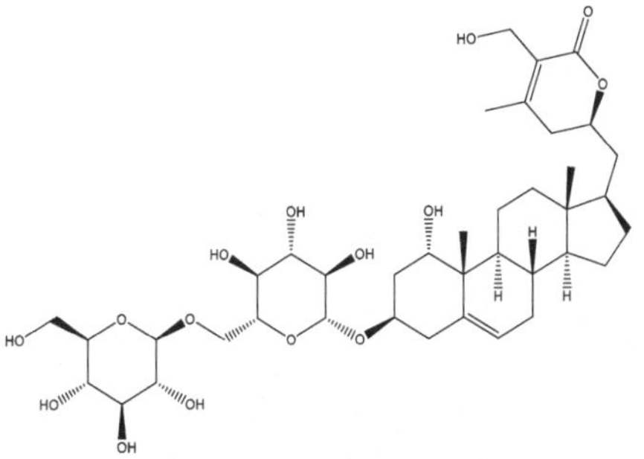

Withanoside V

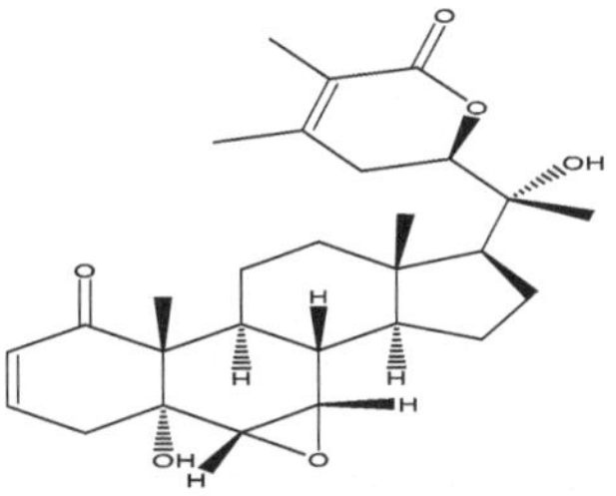

Withanolide A

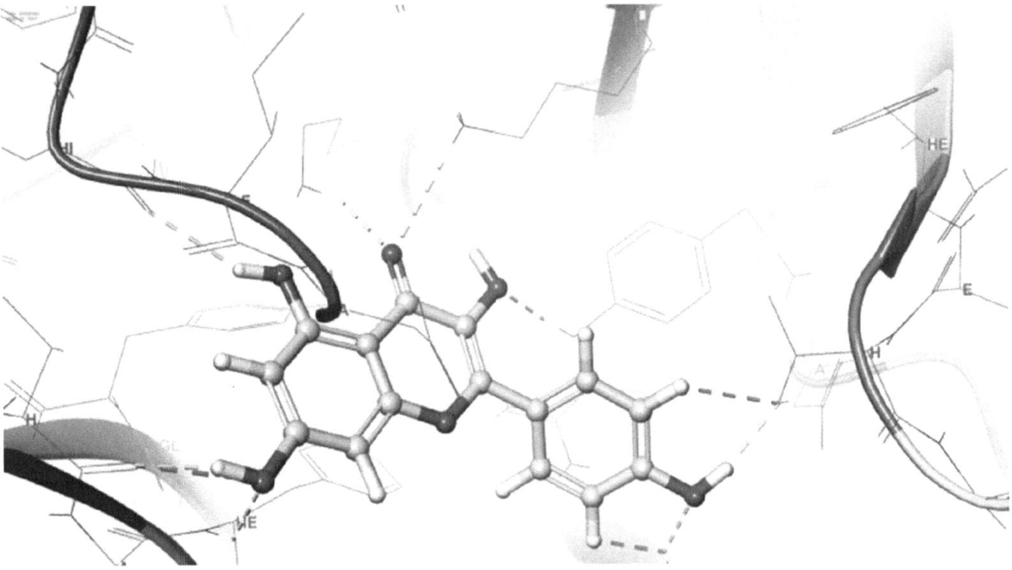

FIGURE 2.10 Three-dimensional diagram showing the interaction of *Withania somnifera* metabolites (kaempferol) with severe acute respiratory syndrome coronavirus 2.

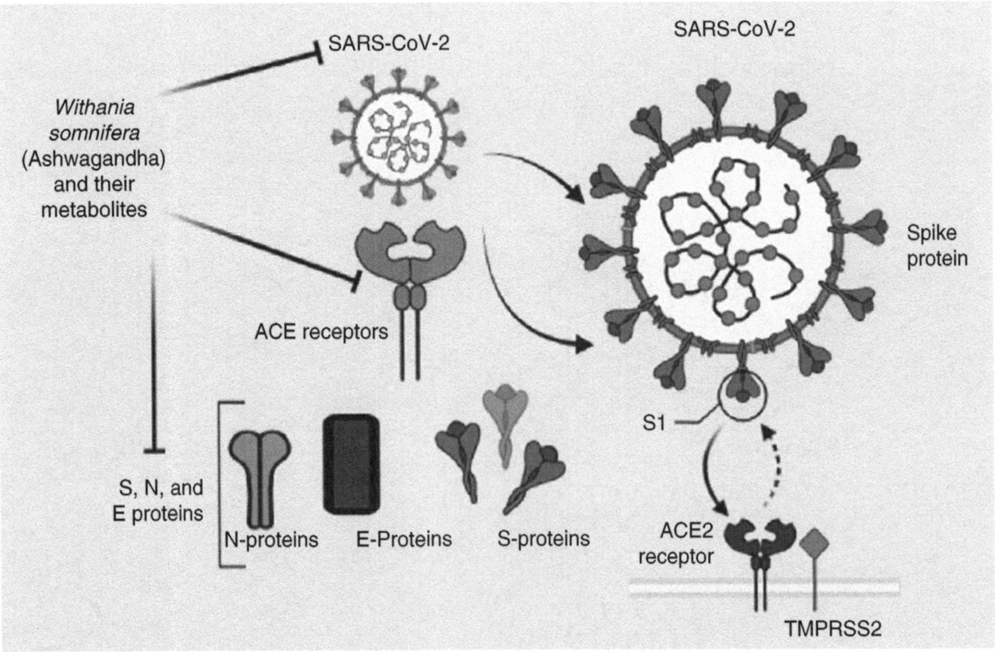

FIGURE 2.11 Possible mechanism of action of *Withania somnifera* and their metabolites against COVID-19 receptors (created in BioRender).

2.6.2 *Bacopa monnieri* (Brahmi)

Bacopa monnieri has been widely helpful in the traditional system of medicines, and it belongs to the family of Plantaginaceae. Chemically, Brahmi contains bacoside A, which consists of bacopaside III, bacopaside X, and bacoside A3. These compounds exhibit neurological activity; other compounds present in Brahmi are cucurbitacin, brahmine, monnierin, hersaponin, mannitol, and nicotine (Abdul Manap et al. 2019). This plant possesses active constituents that increase memory and cognitive function in the brain. Various extracts of this plant have been widely tested to investigate the neurological and nootropic effects (Shinomol & Bharath 2011). The ethanolic extract of *Bacopa monnieri* inhibits the levels of acetylcholine by reducing acetylcholine transferase activity and blocking the binding to muscarinic cholinergic receptors in the frontal cortex of the hippocampus (Bhattacharya et al. 2000). The Brahmi extract interrupts the accumulation of Aβ and Tau proteins and reduces Aβ toxicity (Dubey et al. 2019).

2.6.3 *Centella asiatica* (Gotu kola)

Centella asiatica, a well-known plant, belongs to the family Apiaceae (Roy et al. 2017). It contains phytoconstituents such as triterpenes, asiatic acid, asiaticoside, madecassoside, sapogenins, glycosides, madecassic acid, vellarin, and centelloside (Roy et al. 2017a). Asiatic acid and asiaticoside mainly decrease hydrogen peroxide levels and enhance cell death, thereby decreasing the concentration of free radicals and Aβ cell (Dhanasekaran et al. 2009; Roy et al. 2017b; Roy et al. 2016). *Centella asiatica* is the most useful plant that confers protection to the brain and nerve cells, and it enhances memory, intellect, and longevity (Shinomol & Bharath 2011).

2.6.4 *Rhus mysorensis* (*Searsia mysorensis*)

Rhus mysorensis, commonly known as Dasni, belongs to the family Anacardiaceae and is geographically distributed in India, Africa, and North America (Swathi et al. 2016). As for its therapeutic properties, it has various activities such as anti-inflammatory (Rani et al. 2017), antidiabetic (Penumala et al. 2018), hypolipidemic, antioxidant, and hepatoprotective activities (Reddy et al. 2010). *Rhus mysorensis* contains phytoconstituents such as glycosides, flavonoids, amino acids, carbohydrates, alkaloids, and coumarin derivatives. Flavonoids such as baicalein, quercetin, and fisetin, among others, are a few examples which have shown activity (Rani et al. 2017). The methanolic extract and chloroform extract of this plant show high inhibitory activity against AChE, butyrylcholinesterase, α- and β-Glc enzymes. The chloroform extract of *Rhus mysorensis* acts against hydrogen peroxide–induced oxidative stress in neuronal cells (Penumala et al. 2018).

2.6.5 *Ginkgo biloba*

Ginkgo biloba, commonly known as ginkgo or gingko, belongs to the family Ginkgoaceae has been used traditionally to treat various mental health problems. Various extracts of *Ginkgo biloba* were used to treat improper blood circulation, loss of consciousness, despondency, and headaches (Luo 2006). The *Ginkgo Biloba* (Egb761) leaf extract consists of various phytoconstituents such as flavonoids including quercetin and kaempferol. Terpenoids include ginkgolides A, B, C, and J as well as bilobalide, and other organic acids compounds (EGb761), all of which show better efficacy to the improvement of AD and other mental disorders (Shi et al. 2010). In a study, the *Ginkgo biloba* extract possesses different mechanisms such as inhibition of lipid peroxidation and reduced accumulation of β-amyloid peptides (Luo 2006).

2.6.6 *Magnolia officinalis*

Magnolia officinalis belongs to the family Magnoliaceae, and its major constituent is 4-O-methylhonokiol. The ethanolic extract of *Magnolia officinalis*, and its constituent, namely, 4-O-methylhonkiol, were administered orally, and the treatment was continued for 5 weeks, which inhibited amyloid-β protein (Aβ-protein consists Aβ1–40 and Aβ1–42 species), enhanced apoptotic cell death, decreased the stimulation of β-secretase enzyme, and prevented toxicity due to hydrogen peroxide and Aβ-proteins (mainly Aβ1–42) in the brain and suppressed AChE activity (Lee et al. 2010). In another study, phytoconstituents of *Magnolia officinalis* (honkiol) show potent anti-inflammatory activity through the interruption of the synthesis of reaction oxygen species (Dikalov et al. 2008). *Magnolia officinalis* possesses anti-inflammatory and antioxidant activities, which play an important role for the treatment of AD (Liou et al. 2003).

2.6.7 *Convolvulus pluricaulis* (Shankhpushpi)

Convolvulus pluricaulis belongs to the family Convolvulaceae and is used as a memory enhancer. It contains kaempferol derivatives such as β-sitosterols, ceryl alcohol, and scopoletin, and the chloroform extract contains 20-oxodotriacontanol, tetratriacontanoic acid, and 29-oxodotriacontanol. The aqueous extract and ethyl acetate extract of *Convolvulus pluricaulis* inhibit AChE, decrease the expression of glutathione reductase and superoxide dismutase, and induce the hippocampus by scopolamine (Bihaqi et al. 2011). It also exerts a calming effects on the nerves by regulating the

TABLE 2.4
List of Medicinal Plants Used in the Treatment of Alzheimer's Disease

Medicinal Plant	Extract	Chemical Compound	Activity	References
Withania somnifera	Aqueous extract of *Withania somnifera*	Withanolide A, withanamides, withanone	Decreases the level of stress	(Matsuda et al. 2001; Abedonet al. 2008)
Bacopa monnieri	Ethanolic extract of *Bacopa monnieri*	Colchicine and ibotenic acid	Interrupts the accumulation of Aβ and Tau proteins	(Abdul Manap et al. 2019; Dubey et al. 2019)
Centella asiatica	Ethanolic extract of *Centella asiatica*	Asiatic acid (41), asiaticoside, madecassoside, madecassic acid, vellarin, and centelloside	Reduces the level of H_2O_2, enhances cell death, and decreases the extent of free radicals and β-amyloid cell	(Roy et al. 2017; Dhanasekaran et al., 2009; Roy et al. 2017b; Roy et al. 2016)
Rhus mysorensis	Methanolic extract of *Rhus mysorensis*	Baicalein, quercetin, and fisetin (42)	Inhibitory activity against AChE, BuChE, and α- and β-Glc enzymes	(Rani et al. 2017; Penumala et al. 2018)
Ginkgo biloba	Ethanolic extract of *Ginkgo biloba*	EGb761, quercetin, kaempferol, ginkgolides, and bilobalide	Induces apoptosis, blocks lipid peroxidation in the membrane, and interrupts the accumulation of β-amyloid	(Shi et al. 2010; Luo 2006)
Magnolia officinalis	Methanolic extract of *Magnolia officinalis*	4-O-methylhonokiol	Inhibits acetyl cholinesterase activity	(Lee et al. 2010)
Convolvulus pluricaulis	Methanolic extract of *Convolvulus pluricaulis*	Convolamine (43) and convolvine	Inhibits AChE and decreases the expression of glutathione reductase	(Bihaqi et al. 2011; Dubey et al. 1994)

AChE: acetyl cholinesterase; BuChE: butyrylcholinesterase

synthesis of cortisol and adrenaline (stress hormones) (Sethiya et al. 2009). The ethanolic extract of shankhpushpi was used to enhance learning and memory retention power (Nahata et al. 2008). In hippocampal CA1 and CA3 regions, this plant increases AChE activity and learning ability (Dubey et al. 1994).

Chemical structure of anti-Alzheimer's phytoconstituents

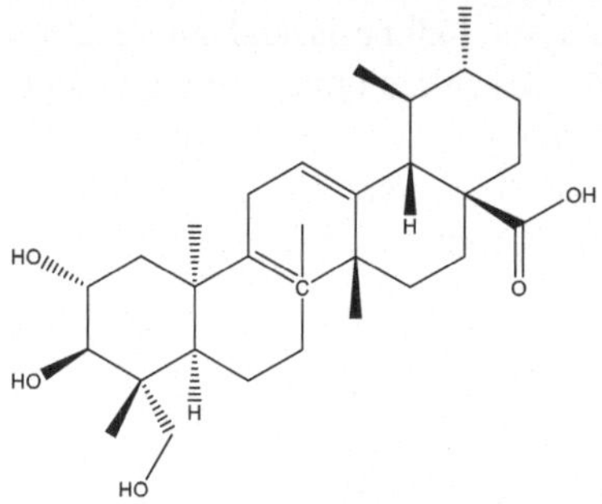

(41)

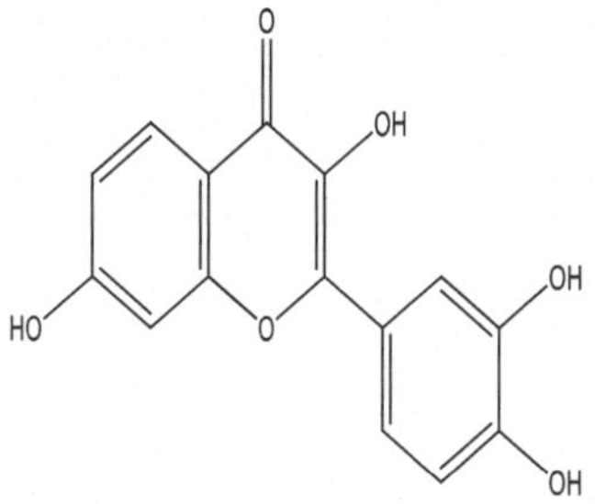

(42)

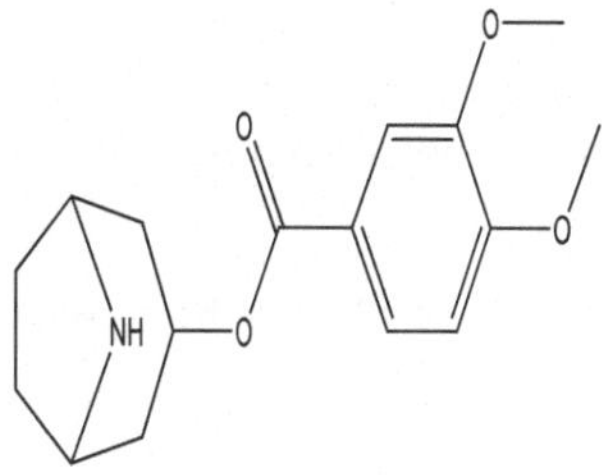

(43)

2.7 MEDICINAL PLANTS WITH ANTIMALARIAL ACTIVITY

Malaria is a parasitic disease, and currently, it is one of the world's most leading causes of death (Fischer et al. 2002). It is a major health issue in the world, especially in the Sub-Saharan Africa. In an estimated that approximately 300 million cases of malaria are diagnosed annually, resulting in 1 million deaths worldwide. In Sub-Saharan Africa, deaths due to malaria disease are common now and it severely affects toddlers up to 5 years of age and pregnant women (World Health Organization 2005). Malaria-related deaths occur because people are unable to buy expensive medicines (Zirihi et al. 2005). As in earlier times, the compounds quinine and artemisinin, which were isolated from plants, have been used as antimalarial agents, and similar active compounds are needed (Basco et al. 1994; Chiyaka 2009) (see Table 2.5).

2.7.1 *Cryptolepis sanguinolenta*

Cryptolepis sanguinolenta (shrubs) belongs to the family Apocynaceae. It mainly contains alkaloidal compounds such as cryptolepinoic acid, methyl cryptole-pionate, cryptolepine, hydroxycryptolepine, and quindoline. The aqueous and ethanolic extracts of the root and leaf of this plant prevent *Plasmodium falciparum* (K1) infection (Paulo et al. 2000). In one study, a patient with malaria was treated with a tea bag formulation of the root extract of *Cryptolepis sanguinolenta*, and the overall cure rate was 93.5% (Bugye et al. 2010).

2.7.2 *Terminalia* (*Terminalia albida* and *Terminalia catappa*)

It belongs to the family Combretaceae and has been traditionally used as an antimalarial drug. In a study, the methanolic extract of the root and stem show potent antiplasmodial activity. It is also used in hemorrhoids, skin diseases, jaundice, yellow fever, pile, ulcer, wounds, and other infections (Balde et al. 2020; Agyare et al. 2009). *Terminalia ivorensis* is popularly known as "black afara." A study revealed that aqueous leaf extract of *Terminalia ivorensis* has shown *in vitro* antiplasmodial activity against *Plasmodium falciparum* (Komlaga et al. 2016). Another study showed that the ethanolic extract of the stem and bark of *Terminalia ivorensis* shows an *in vitro* antimalarial activity against chloroquine-resistant strains of *Plasmodium falciparum* (Annan et al. 2012). Chemically, *Terminalia* species contain ellagic acid, catechin, gallocatechin, epigallocatechin, and

23-galloylarjunolic acid (Camaral et al. 2019). 23-Galloylarjunolic acid possesses antiplasmodial efficacy against chloroquine-sensitive and chloroquine-resistant strains (Machumi et al. 2013).

2.7.3 *Elaeis guineensis*

Elaeis guineensis belongs to the family Arecaceae and is also known as African oil palm tree. This plant grows generally up to 15–20 m in height with a 100-year lifespan and is distributed throughout the tropical region of West Africa as well as in Central Africa, Malaysia, Indonesia, and other regions (Henson 2012). Chemically, it contains α-tocopherol, squalene, β-sitosterol, stigma sterol, arundoin. *Elaeis guineensis* has a variety of activities and can be used for treating malaria, headache, wounds, and gonorrhea (Mshana 2000). In a study, the ethanolic extract of the leaves of *E. guineensis* has shown potent antimalarial activity, against *Plasmodium falciparum* (Annan et al. 2012).

2.7.4 *Phyllanthus emblica*

Phyllanthus emblica belongs to the family Euphorbiaceae and is distributed in many tropical and semitropical regions. It is rich in vitamin C, minerals, and amino acids, which helps regain lost energy and strength. Chemically, it contains essential oils, tannins, amino acids, glycosides, and polyphenols such as flavonoids, corilagin, gallic acid, and ellagic acid (Calixto et al. 1998; Gaire et al. 2014). It is therapeutically used for jaundice, cough, asthma, peptic ulcer, skin diseases, leprosy, intermittent fever, headache, anemia, dizziness, snakebite, and scorpion sting (Mirunalini & Krishnaveni 2010). The methanolic and ethyl acetate extracts of the leaf have shown potent antimalarial activity against *Plasmodium falciparum* strains 3D7, Dd2, and INDO (Bagavan et al. 2011).

2.7.5 *Syzygium aromaticum* (Clove)

Syzygium aromaticum belongs to the family Myrtaceae. Chemically, it contains hydrocarbons, phenolic compounds, terpenes, eugenol, and β-caryophyllene (El-Saber Batiha et al. 2020). It is a well-known spice and has various medicinal uses, being used for the treatment of allergic disorders and toothache; it exerts several activities such as anti-inflammatory, antioxidant, antiviral, and antiparasitic properties (Mittal et al. 2014). In a study, the methanolic extract of *Syzygium aromaticum* showed potent inhibition of *Plasmodium falciparum* chloroquine-sensitive "3D7" and chloroquine-resistant (Dd2 and INDO) strains (Bagavan et al. 2011).

2.7.6 *Casearia sylvestris*

Casearia sylvestris belongs to the family Salicaceae; it is an evergreen herb and has multiple uses. Among folklore, this plant is used for treating wound healing, inflammation, snake bites, fever, gastric ulcers, and diarrhea (Sleumer 1980). Chemically, it contains casearins U and V, sylvestris side A and B, blumenol B, cinnamic acid, soyacerebroside I, methypheophorbides, pheophorbide A, chlorophyll, β-sitosterol, stigmastane-3,6-diol, daucosterol, and β-D-glucopyranoside (Wang et al. 2008). In an interesting study, the ethanolic extract of *Syzygium aromaticum* treated with L-6 cell lines of rats and MRC-5 of humans demonstrated antiplasmodial activity in a dose-dependent manner against chloroquine-resistant *Plasmodium falciparum* strain chloroquine-sensitive "3D7" and chloroquine-resistant "Dd2 and INDO" (De et al. 2007).

2.7.7 *Cupania vernalis*

Cupania vernalis is a 10–22-meter tall semi-deciduous tree belonging to the family Sapindaceae. Various scientific investigations revealed that various types of extracts have bioactive potential against numerous diseases, and the plant serves as a diuretic stimulant, cough expectorant, and sedative and a cure for stomach ache and dermatitis (Lorenzi 2002). In a study, the ethanolic and hexane extracts of *Cupania vernalis* possess antimalarial activity against *Plasmodium falciparum* (De et al. 2007).

2.7.8 *Xylopia emarginata*

Xylopia emarginata is native to Brazil and belongs to the family Annonaceae. It can grow up to 10–20 m tall and have a columnar crown. Therapeutically, it is used as a carminative and aphrodisiac in traditional medicine (Luchi 2016). In a study, hexane extract of the bark, stem, and root of *Xylopia emarginata* has shown activity against *Plasmodium falciparum* (chloroquine-resistant FcB1/Colombia strains) (De et al. 2007).

2.7.9 *Xylopia*

Various species of *Xylopia*, such as *Xylopia aromatic*, *Xylopia aethiopica*, and *Xylopia sericea*, belong to the family Annonaceae. The dichloromethane extract of *Xylopia sericea* tested on HepG2 cell line exhibited better efficacy against *Plasmodium falciparum* W2 strain (Gontijo et al. 2020). The hexane extract of the root shows *in vitro* antimalarial activity in CQ-resistant (FcB1/Colombia) strains of *Plasmodium falciparum* (De et al. 2007).

2.7.10 *Aspidosperma*

Several species of *Aspidosperma* (*Aspidosperma nitidum*, *Aspidosperma pyrifolium*, *Aspidosperma macrocarpon*, and *Aspidosperma ramiflorum*) belong to the family Apocynaceae (Ceravolo et al. 2018), which is geographically distributed in southern regions of Brazil. *Aspidosperma* species is used for the treatment of various ailments. Some of the isolated compounds present in the extract of *Aspidosperma ramiflorum* are 10-methoxy-geissoschizol, β-yohimbine, isositsirikine, ramiflorine A, and ramiflorine B. Total extract fraction, alkaloidal fraction, and isolated compounds have been tested against *Plasmodium falciparum* in HepG2, BGM (monkey kidney) cell line, and human monocytes separated from connective tissue, and the extracts showed slight toxicity and better efficacy, and the compound isositsirikine showed better efficacy and lower toxicity than the other compounds. *In vivo* testing of the extract in a mice model revealed 53% inhibition of *Plasmodium berghei* parasites (Aguiar et al. 2015). The *in vitro* antiplasmodial study of the ethanolic extract revealed an effective activity against *Plasmodium falciparum* (chloroquine-resistant FcB1/Colombia) with an IC_{50} value of 4.9±1.1 µg/mL (De et al. 2007). In a cell line (Vero and HepG2), the ethanolic extract of *Aspidosperma pyrifolium* bark and its isolated constituent "uleine" showed strong affinity against chloroquine-resistant falciparum strain W2. It is an indole alkaloid and has a structure similar to that of quinine. Uleine and the fraction containing other alkaloidal compounds showed better efficacy against falciparum strains, but uleine showed better efficacy with lower toxicity than the whole ethanolic extract (Dolabela et al. 2015).

2.7.11 *Azadirachta indica* (Neem)

Azadirachta indica is an evergreen tree and belongs to the family Meliaceae. It grows up to a height of 15–20 m, and some plants grow up to a height of 35–40 m. It grows in tropical and semi-tropical regions. Neem has numerous health benefits; it is effective against skin diseases, eczema, and psoriasis (Porter 2006). It is equally effective as quinine, and the ethanolic extract of leaves and seeds is effective against malaria-infected cell cultures. The methanolic leaf extract showed *in vitro* antiplasmodial activity against chloroquine- and pyrimethamine-sensitive 3D7 strain and chloroquine-resistant and pyrimethamine-sensitive Dd2 strain, with IC_{50} values of 5.8 and 1.7 µg/mL, respectively (El Tahir et al. 1999).

2.7.12 *Harrisonia abyssinica*

Harrisonia abyssinica belongs to the family Rutaceae, and chemically, it contains harronin I and harronin II (Mayaka et al. 2012). It is used for the treatment of malaria, tuberculosis, gonorrhea, and dysentery. The methanolic stem bark extract inhibited chloroquine-resistant *Plasmodium falciparum* strain Dd2, with an IC_{50} value of 4.7±0.113, while in chloroquine-sensitive *Plasmodium falciparum* strain 3D7, the IC_{50} value was 10±0.114 µg/mL (El Tahir et al. 1999).

2.7.13 *Maytenus senegalensis*

Maytenus senegalensis is an African shrub or tree belonging to the family Celastraceae. It is widely distributed throughout Central and South America, Southeast Asia, Micronesia, Australasia, the Indian Ocean, and Africa; it grows up to a height of 15 m, with its spines growing up to 7 cm long. Folklore use this plant for its anti-inflammatory effects and for treating toothache (Da et al. 2011). The methanolic extract of the stem bark showed antiplasmodial activity, with IC_{50} values of 3.9 and 10 µg/mL when treated *in vitro* on chloroquine-sensitive 3D7 strains and chloroquine-resistant Dd2 strains, respectively (El Tahir et al. 1999). In a recent study, purified ethanolic fraction of *Maytenus senegalensis* (3-hydroxy-20(29)-lupen-28-ol, 20α)-3-hydroxy-2-oxo-24-nor-friedela-1(10), 3, 5, 7-tetraen-carboxylic acid-(29)-methyl ester, 2(4H)-Benzofuranone, 5,6,7,7a-tetrahydro, and phytol) showed strong antimalarial activity (Jigam et al. 2020).

TABLE 2.5
List of Antimalarial Medicinal Plants

Medicinal Plant	Extract	Chemical Compound	Activity
Cryptolepis sanguinolenta	Aqueous and ethanolic extract	Cryptolepinoic acid (44), methyl cryptole-pionate, and cryptolepine	Potent inhibition of the *Plasmodium falciparum* (K1) strain
Terminalia ivorensis	Aqueous extract	Ellagic acid (45), 23-galloyl arjunolic acid	Inhibits *Plasmodium falciparum* CQ-sensitive (3D7) and CQ-resistant (W2) strains
Elaeis guineensis (palm oil)	Ethanolic extract	α-Tocopherol (46), squalene (48), and β-sitosterol (49)	Potent antimalarial activity against CQ-resistant *Plasmodium falciparum*
Phyllanthus emblica	Methanolic extract	Corilagin (50) and gallic acid	Inhibits infection due to CQ-sensitive 3D7 and CQ-resist Dd2 and INDO strains of *Plasmodium falciparum*
Syzygium aromaticum	Methanolic extract	Eugenol (51) and β-caryophyllene	Shows potent activity against *Plasmodium falciparum* strains CQ-sensitive 3D7 and CQ-resistant Dd2 and INDO
Casearia sylvestris	Aqueous extract	Sylvestris side A and B	Active against the CQ-resistant FcB1/Colombia *Plasmodium falciparum* strain

TABLE 2.5 (Continued)
List of Antimalarial Medicinal Plants

Medicinal Plant	Extract	Chemical Compound	Activity
Cupania vernalis	Ethanolic extract	Vernanolide	Activity against the *Plasmodium falciparum* strain and CQ-resistant FcB1/Colombia strain
Xylopia emarginat	Ethanolic extract	Spathulenol (52)	Shows potent inhibition against *Plasmodium falciparum* CQ-resistant FcB1/Colombia strains
Xylopia aromatic	Ethanolic extract	Verbenone (53)	Activity against *Plasmodium falciparum* strains CQ-resistant FcB1/Colombia strains
Aspidosperma macrocarpon/ parvifolium	Ethanolic extract	β-yohimbine, isositsirikine uleine (54)	Active against *Plasmodium falciparum* (CQ-resistant W2), *Plasmodium berghei*
Azadirachta indica	Ethanolic extract	Azadiradione (55)	Activity against CQ-and pyrimethamine-sensitive 3D7 and CQ-resistant and pyrimethamine-sensitive Dd2 strains
Harrisonia abyssinica	Methanolic extract	Harronin I and harronin II	Activity against the *Plasmodium falciparum* Dd2 strain and CQ-sensitive 3D7 *Plasmodium falciparum* strain
Maytenus senegalensis	Methanolic extract	Senegalensis, 5,6,7,7a-tetrahydro- and 3-hydroxy-20(29)-lupen-28-ol, 2(4H)-benzofuranone, phytol (47)	Activity against CQ-sensitive 3D7 and CQ-resistant Dd2 strains

CQ: chloroquine

Chemical structure of antimalarial phytoconstituents

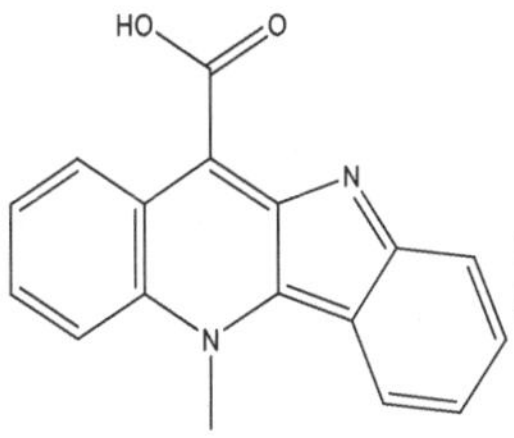

(44)

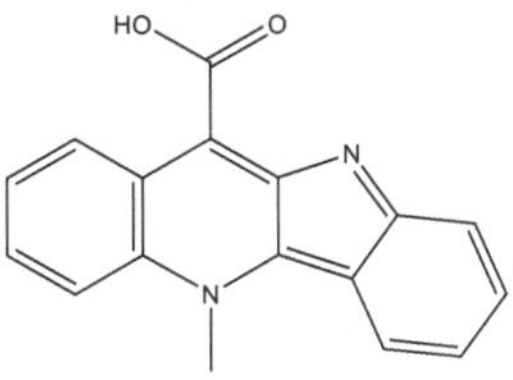

(45)

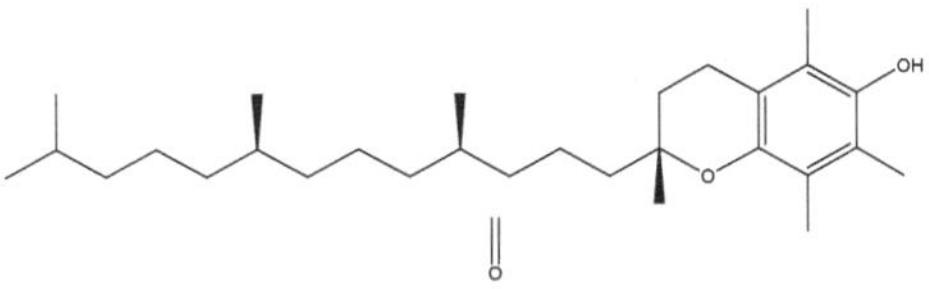

(46)

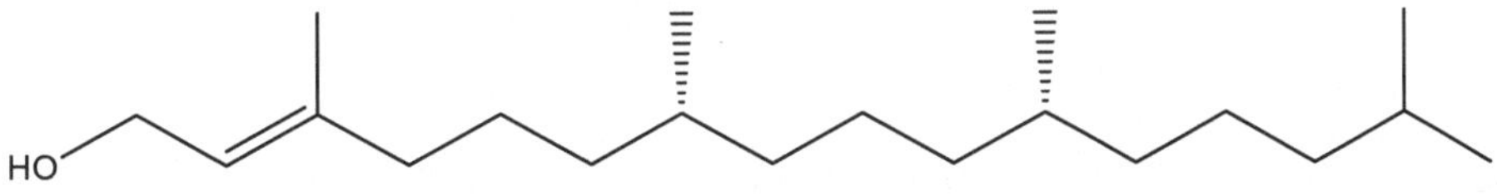

(47)

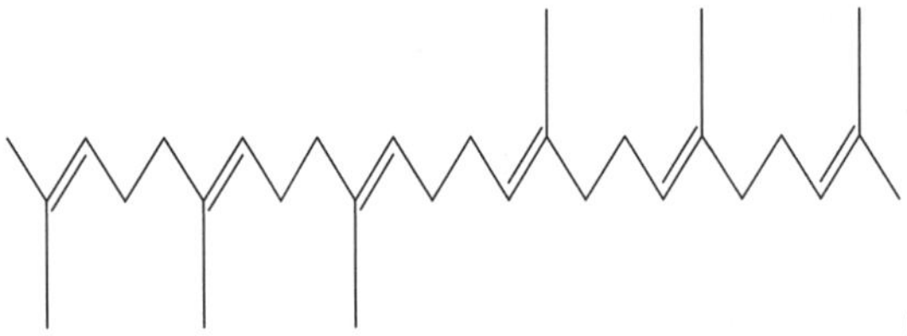

(47)

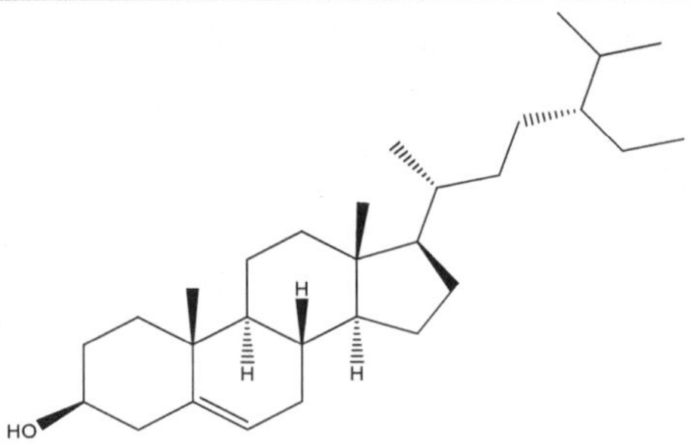

(48)

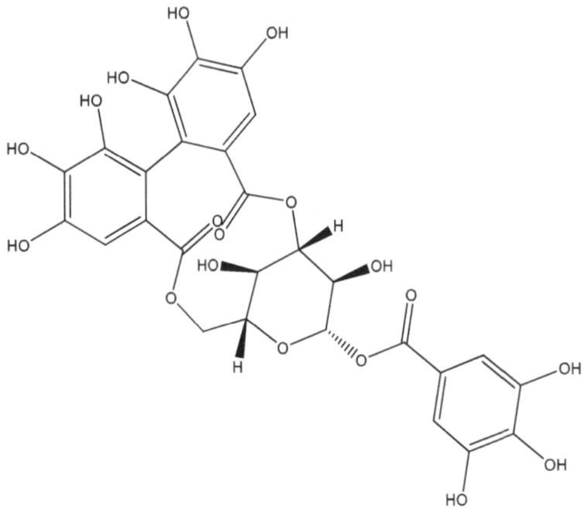

(49)

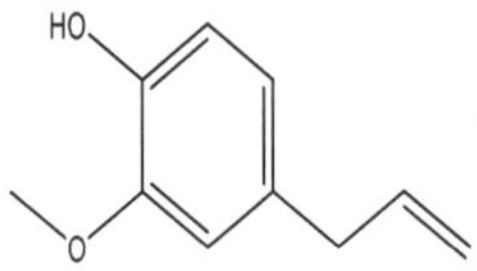

(50)

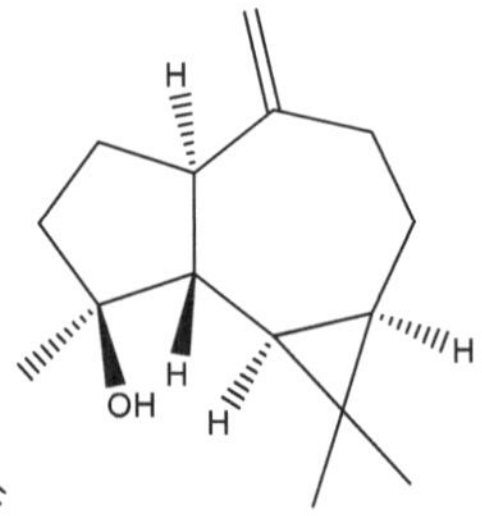

(51)

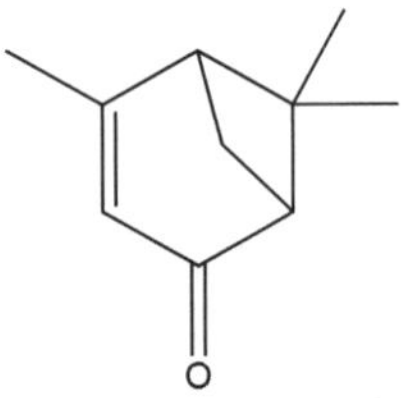

(52)

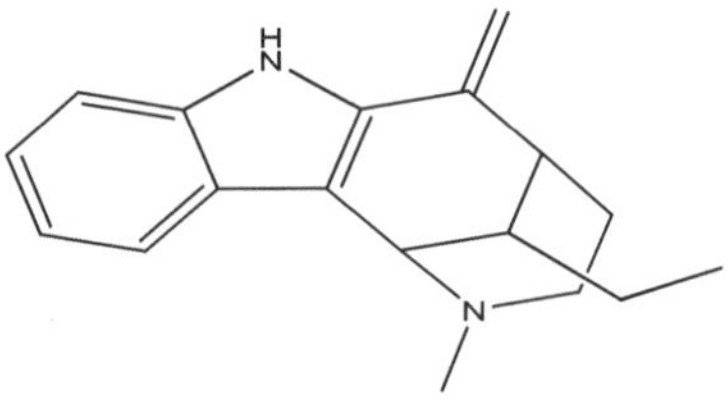

(53)

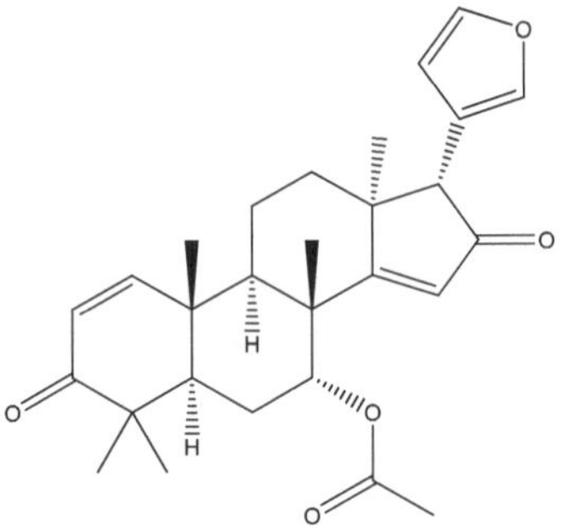

(54)

2.8 CONCLUSION

Medicinal plants provide the foremost dynamic and multivalent therapeutic approach for the management of complicated complex physiological imbalance. Traditional medicine has long been used as the primary health care system worldwide. Mostly, conventional medicines are directly or indirectly derived from natural sources. Plants' secondary metabolites may be directly used for therapeutic purposes and synthesis of novel compounds or preparation of formulations. All plants reviewed in this chapter demonstrate strong efficacy, confirming their different traditional uses and their ability to treat ordinary diseases.

REFERENCES

Abdul Manap, Aimi Syamima, Shantini Vijayabalan, Priya Madhavan, Yoke Yin Chia, Aditya Arya, Eng Hwa Wong, Farzana Rizwan, Umesh Bindal, and Shajan Koshy. Drug Target Insights 13 (2019): 1177392819866412.

Abedon, Bruce, B. Auddy, J. Hazra, A. Mitra, and S. Ghosal. JANA 11 (2008): 50–56.

Abubakar M.G., A.N. Ukwuani, and U.U. Mande. Sky Journal of Biochemistry Research 4, no. 3 (2015): 16–20.

Adhikari, Bikash. Journal of Chemistry 2021 (2021).

Agarwal, S. K., Sudhir S. Singh, Sushma Verma, and Sushil Kumar. Journal of Ethnopharmacology 72, no. 1–2 (2000): 43–46.

Aguiar, Anna CC, Ananda C. Cunha, Isabela Penna Ceravolo, Regina A. Correia Gonçalves, Arildo JB Oliveira, and Antoniana Ursine Krettli. Memorias do Instituto Oswaldo Cruz 110 (2015): 906–913.

Agyare, Christian, Alex Asase, Matthias Lechtenberg, Michael Niehues, Alexandra Deters, and Andreas Hensel. Journal of Ethnopharmacology 125, no. 3 (2009): 393–403.

Ahmed, Mohammed Fazil, Syed Mohammed Kazim, Syed Safiullah Ghori, Syeda Sughra Mehjabeen, Shaik Rasheed Ahmed, Shaik Mehboob Ali, and Mohammed Ibrahim. International Journal of Endocrinology 2010 (2010).

Ahmed, Saad, Asim Ali Khan, Pawan Yadav, Jamal Akhtar, Usama Akram, and Lubna Fatimaand Shamim. International Journal of Herbal Medicine 2020 (2020).

Ajay, Machha, H. J. Chai, A. M. Mustafa, Anwar Hassan Gilani, and Mohammad Rais Mustafa. Journal of Ethnopharmacology 109, no. 3 (2007): 388–393.

Al-Amin, Zainab M., Martha Thomson, Khaled K. Al-Qattan, Riitta Peltonen-Shalaby, and Muslim Ali. British Journal of Nutrition 96, no. 4 (2006): 660–666.

Almagro, Lorena, Francisco Fernández-Pérez, and Maria Angeles Pedreño. Molecules 20, no. 2 (2015): 2973–3000.

Al-Snafi, Ali Esmail. IOSR Journal of Pharmacy 8, no. 11 (2018): 1–21.

Al-Snafi, Ali Esmail. International Journal of Pharmaceutical and Health Care Research 1, no. 4 (2013): 132–147.

Anand, Prachi, K.Y. Murali, Vibha Tandon, P.S. Murthy, and Ramesh Chandra. Chemico-Biological Interactions 186, no. 1 (2010): 72–81.

Anderson, James W., Pat Baird, Richard H. Davis, Stefanie Ferreri, Mary Knudtson, Ashraf Koraym, Valerie Waters, and Christine L. Williams. Nutrition Reviews 67, no. 4 (2009): 188–205.

Annan, Kofi, K. Sarpong, C. Asare, R. Dickson, K.I. Amponsah, B. Gyan, M. Ofori, and S.Y. Gbedema. Pharmacognosy Research 4, no. 4 (2012): 225.

Ansari, Prawej, Peter R. Flatt, Patrick Harriott, J.M.A. Hannan, and Yasser H.A. Abdel-Wahab. Plants 10, no. 6 (2021): 1190.

Arablou, Tahereh, Naheed Aryaeian, Majid Valizadeh, Faranak Sharifi, AghaFatemeh Hosseini, and Mahmoud Djalali. International Journal of Food Sciences and Nutrition 65, no. 4 (2014): 515–520.

Asase, Alex, Tetsuo Kokubun, Renée J. Grayer, Geoffrey Kite, Monique SJ Simmonds, Alfred A. Oteng-Yeboah, and George T. Odamtten. Phytotherapy Research 22, no. 8 (2008): 1013–1016.

Bagavan, Asokan, Abdul Abdul Rahuman, Naveen Kumar Kaushik, and Dinkar Sahal. Parasitology Research 108, no. 1 (2011): 15–22.

Balde A.O., E.S. Balde, F. Bah, A. Camara, M.A. Balde, A. Drame, F. Dembele, H. Barry, M.S. Traore, and A.M. Balde. South African Journal of Botany 131 (2020): 443–447.

Barrett N.E., L. Holbrook, S. Jones, W.J. Kaiser, L.A. Moraes, R. Rana, et al. British Journal of Pharmacology 154, no. 5 (2008): 918–939.

Basco, Leonardo K., Sofia Mitaku, Alexios-Leandros Skaltsounis, Nicole Ravelomanantsoa, Francois Tillequin, Michel Koch, and J. Le Bras. Antimicrobial Agents and Chemotherapy 38, no. 5 (1994): 1169–1171.

Batiha, Gaber El-Saber, Luay M. Alkazmi, Lamiaa G. Wasef, Amany Magdy. Beshbishy, Eman H. Nadwa, and Eman K. Rashwan. Biomolecules 10, no. 2 (2020): 202.
Bawa A.S. and F. Khanum. Phytotherapy Research: An International Journal Devoted to Pharmacological and Toxicological Evaluation of Natural Product Derivatives 23, no. 8 (2009): 1099–1102.
Bhattacharya S.K., A. Bhattacharya, A. Kumar, and S. Ghosal. Phytotherapy Research 14, no. 3 (2000): 174–179.
Bihaqi, Syed Waseem, Avninder Pal Singh, and Manisha Tiwari. Indian Journal of Pharmacology 43, no. 5 (2011): 520.
Borisenko S.N., Lekar A.V., Vetrova E.V., et al. Russian Journal of Bioorganic Chemistry, 42, no. 7 (2016), 716–720.
Brâkenhielm, Ebba, Renhai Cao, and Yihai Cao. The FASEB Journal 15, no. 10 (2001): 1798–1800.
Brixius, Klara, Sonja Willms, Andreas Napp, Paschalios Tossios, Dennis Ladage, Wilhelm Bloch, Uwe Mehlhorn, and Robert H.G. Schwinger. Cardiovascular Drugs and Therapy 20, no. 3 (2006): 177–184.
Bugyei K.A., G.L. Boye, and M.E. Addy. Ghana Medical Journal 44, no. 1 (2010).
Bunkar, Alka Rani. International Journal of Advanced Science and Research 2, no. 2 (2017): 23–26.
Buowari, Omiepirisa Yvonne. Rijeka, Croatia: InTechOpen (2013): 131.
Calixto, João B., Adair R.S. Santos, Valdir Cechinel Filho, and Rosendo A. Yunes. Medicinal Research Reviews 18, no. 4 (1998): 225–258.
Camara, Aissata, Mohamed Haddad, Karine Reybier, Mohamed Sahar Traoré, Mamadou Aliou Baldé, Jade Royo, Philippe Batigne, et al. Malaria Journal 18, no. 1 (2019): 1–15.
Cancer Research UK. [Accessed 23 January 2015] What is cancer? Available at: http://www.cancerresearchuk.org/about-cancer/what-is-cance.Characterization of Lignans in Forsythiae Fructus and Their Metabolites in Rats. 2014.
Cao, Yu-Jie, Zong-Jin Pu, Yu-Ping Tang, Juan Shen, Yan-Yan Chen, An Kang, Gui-Sheng Zhou, and Jin-Ao Duan. Chinese Medicine 12, no. 1 (2017): 1–12.
Ceravolo, Isabela P., Carlos L. Zani, Flávio J.B. Figueiredo, Markus Kohlhoff, Antônio E.G. Santana, and Antoniana U. Krettli. Malaria Journal 17, no. 1 (2018): 1–11.
Chen, Jing, Wenqing Wang, Chunyang Shi, and Jianguo Fang. International Journal of Molecular Sciences 15, no. 12 (2014): 22978–22994.
Chen, Lei, Bo Yu, Yaofang Zhang, Xin Gao, Liang Zhu, Tonghui Ma, and Hong Yang. PLoS One 10, no. 3 (2015): e0119122.
Chen, Shao-Dan, Hao Gao, Qin-Chang Zhu, Ya-Qi Wang, Ting Li, Zhen-Qiang Mu, Hong-Ling Wu, Tao Peng, and Xin-Sheng Yao. Organic Letters 14, no. 7 (2012): 1772–1775.
Chen, Xiao-Ying, Hai-Ming Chen, Yu-Hong Liu, Zhen-Biao Zhang, Yi-Feng Zheng, Zu-Qing Su, Xie Zhang, et al. Experimental Biology and Medicine 241, no. 2 (2016): 193–204.
Chen, Xinlian, Yingxian Cui, Liping Nie, Haoyu Hu, Zhichao Xu, Wei Sun, Ting Gao, Jingyuan Song, and Hui Yao. BioMed Research International 2019 (2019).
Cheng, Diana M., Peter Kuhn, Alexander Poulev, Leonel E. Rojo, Mary Ann Lila, and Ilya Raskin. Food Chemistry 135, no. 4 (2012): 2994–3002.
Chhabra, Preeti, and Kenneth L. Brayman. Stem Cells Translational Medicine 2, no. 5 (2013): 328–336.
Chiyaka, Christinah, Winston Garira, and Shadreck Dube. Theoretical Population Biology 75, no. 1 (2009): 14–29.
Clark, Ruth, and Seong-Ho Lee. Anticancer Research 36, no. 3 (2016): 837–843.
Da Silva G., R. Serrano, and O. Silva. Journal of Natural Science, Biology, and Medicine 2, no. 1 (2011): 59.
Dar, Nawab John. Journal of Ethnopharmacology 256 (2020): 112769.
Darshan, Medha S., Matthew S. Loftus, Maria Thadani-Mulero, Benjamin P. Levy, Daniel Escuin, Xi Kathy Zhou, Ada Gjyrezi, et al. Cancer Research 71, no. 18 (2011): 6019–6029.
De Mesquita, M.L., P. Grellier, L. Mambu, J. E. De Paula, and L. S. Espindola. Journal of Ethnopharmacology 110, no. 1 (2007): 165–170.
Deka, Apranta, and Joseph A. Vita. Pharmacological Research 64, no. 2 (2011): 136–145.
Dhanasekaran, Muralikrishnan, Leigh A. Holcomb, Angie R. Hitt, Binu Tharakan, Jami W. Porter, Keith A. Young, and Bala V. Manyam. Phytotherapy Research: An International Journal Devoted to Pharmacological and Toxicological Evaluation of Natural Product Derivatives 23, no. 1 (2009): 14–19.
Dikalov, Sergey, Tanya Losik, and Jack L. Arbiser. Biochemical Pharmacology 76, no. 5 (2008): 589–596.

Ding, Yuewen, Lijuan Zeng, Runfeng Li, Qiaoyan Chen, Beixian Zhou, Qiaolian Chen, Wang Yutao, Jingping Zheng, Zifeng Yang, and Fengxue Zhang. BMC Complementary and Alternative Medicine 17, no. 1 (2017): 1–11.

Dolabela, Maria Fâni, Marinete Marins Póvoa, Geraldo Célio Brandão, Fabíola Dutra Rocha, Luciana Ferreira Soares, Renata Cristina de Paula, and Alaíde Braga de Oliveira. Malaria Journal 14, no. 1 (2015): 1–12.

Dubey, G.P., S.R. Pathak, and B.S. Gupta. Pharmacopsychoecol 7, no. 3 (1994): 249–51.

Dubey, Tushar, and Subashchandrabose Chinnathambi. Archives of Biochemistry and Biophysics 676 (2019): 108153.

El Tahir, Ahmed, Gwiria M.H. Satti, and Sami A. Khalid. Journal of Ethnopharmacology 64, no. 3 (1999): 227–233.

Evans, William Charles. Trease and Evans' Pharmacognosy. Elsevier Health Sciences, 2009.

Feldman, E.J., K.P. Seiter, T. Ahmed, P. Baskind, and Z.A. Arlin. Leukemia 10, no. 1 (1996): 40–42.

Fernandes, Nafisa PC, Chakradhar V. Lagishetty, Vandana S. Panda, and Suresh R. Naik. BMC Complementary and Alternative Medicine 7, no. 1 (2007): 1–8.

Fischer, Philip R., and Ralf Bialek. Clinical Infectious Diseases 34, no. 4 (2002): 493–498.

Fleming, Jennifer A., and Penny M. Kris-Etherton. Advances in Nutrition 5, no. 6 (2014): 863S–876S.

Fukuda, Toshiyuki, Hideyuki Ito, Teruo Mukainaka, Harukuni Tokuda, Hoyoku Nishino, and Takashi Yoshida. Biological and Pharmaceutical Bulletin 26, no. 2 (2003): 271–273.

Gaire, Bhakta Prasad, and Lalita Subedi. Chinese Journal of Integrative Medicine (2014): 1–8.

Gao, Hong, Yi-Na Huang, Bo Gao, Peng Li, Chika Inagaki, and Jun Kawabata. Food Chemistry 108, no. 3 (2008): 965–972.

Gautret, Philippe, Jean-Christophe Lagier, Philippe Parola, Line Meddeb, Morgane Mailhe, Barbara Doudier, Johan Courjon et al. International Journal of Antimicrobial Agents 56, no. 1 (2020): 105949.

Gontijo, Douglas Costa, Maria Fernanda Alves do Nascimento, Geraldo Célio Brandão, and Alaíde Braga de Oliveira. Natural Product Research 34, no. 24 (2020): 3526–3530.

Gurrola-Díaz, Carmen M., Pedro Macedonio García-López, Sergio Sánchez-Enríquez, Rogelio Troyo-Sanromán, I. Andrade-González, and J.F. Gómez-Leyva. Phytomedicine 17, no. 7 (2010): 500–505.

Hegazy, Gehan A., Amina M. Alnoury, and Hoda G. Gad. Saudi Medical Journal 34, no. 7 (2013): 727–733.

Heiblig, M., M. Sobh, and F.E. Nicolini. Leukemia Research 38, no. 10 (2014): 1145–1153.

Hempen, C.H. and T. Fischer. A materia medica for Chinese medicine, English edition (2009).

Henson, Ian E. A Brief History of the Oil Palm. In: Palm Oil, pp. 1–29. AOCS Press (2012).

Herrera-Arellano, A., S. Flores-Romero, M.A. Chavez-Soto, and J. Tortoriello. Phytomedicine 11, no. 5 (2004): 375–382.

Herrera-Arellano, Armando, Judith Miranda-Sánchez, Pedro Ávila-Castro, Sara Herrera-Álvarez, Jesús Enrique Jiménez-Ferrer, Alejandro Zamilpa, Rubén Román-Ramos, Héctor Ponce-Monter, and Jaime Tortoriello. Planta Medica 73, no. 01 (2007): 6–12.

Hong-Zhi, D.U., H.O.U. Xiao-Ying, M.I.A.O. Yu-Huan, B.-S. Huang, and L.I.U. Da-Hui. Chinese Journal of Natural Medicines 18, no. 3 (2020): 206–210.

Hoshyar, Reyhane, and Homa Mollaei. Journal of Pharmacy and Pharmacology 69, no. 11 (2017): 1419–1427.

Hosseini, Saeed, Leila Jamshidi, Saeed Mehrzadi, Kazem Mohammad, Ali Reza Najmizadeh, Houman Alimoradi, and Hasan Fallah Huseini. Journal of Ethnopharmacology 152, no. 3 (2014): 451–456.

Hu, Naihua, Cheng Wang, Xuyang Dai, Mengting Zhou, Lihong Gong, Lingyuan Yu, Cheng Peng, and Yunxia Li. Journal of Ethnopharmacology 248 (2020): 112361.

Huang, Keng-Shiang, Chih-Hui Yang, Ya-Chin Wang, Wei-Ting Wang, and Yen-Yi Lu. Pharmaceutics 11, no. 5 (2019): 212.

Iino, Shiro, Toshiro Tango, Takashi Matsushima, Gotaro Toda, Kazuhiko Miyake, Kunihiko Hino, Hiromitsu Kumada, et al. Hepatology Research 19, no. 1 (2001): 31–40.

Irfan, Muhammad, Yi-Seong Kwak, Chang-Kyun Han, Sun Hee Hyun, and Man Hee Rhee. Journal of Ginseng Research 44, no. 4 (2020): 538–543.

Isah, Tasiu. Pharmacognosy Reviews 10, no. 20 (2016): 90.

Jayaprakasam, Bolleddula, Kaillathe Padmanabhan, and Muraleedharan G. Nair. Phytotherapy Research 24, no. 6 (2010): 859–863.

Jiang, Qian, Jun Chen, Xiaobing Long, Xiaolong Yao, Xin Zou, Yiping Yang, Guangying Huang, and Huaqiu Zhang. International Immunopharmacology 79 (2020): 106083.

Jigam, Ali A., Rachael Musa, Abdulkadir Abdullahi, and Bashir Lawal. Clinical Phytoscience 6, no. 1 (2020): 1–8.

Jin, Ying-Hui, Lin Cai, Zhen-Shun Cheng, Hong Cheng, Tong Deng, Yi-Pin Fan, Cheng Fang, et al. Military Medical Research 7, no. 1 (2020): 1–23.

Junren, Chen, Xie Xiaofang, Li Mengting, Xiong Qiuyun, Li Gangmin, Zhang Huiqiong, Chen Guanru, et al. Chinese Medicine 16, no. 1 (2021): 1–20.

Keyaerts, Els, Leen Vijgen, Christophe Pannecouque, Els Van Damme, Willy Peumans, Herman Egberink, Jan Balzarini, and Marc Van Ranst. Antiviral Research 75, no. 3 (2007): 179–187.

Khalesi, Saman, Christopher Irwin, and Matt Schubert. The Journal of Nutrition 145, no. 4 (2015): 758–765.

Kim, Se Hye, Jonah A. Kaplan, Yuan Sun, Aileen Shieh, Hui-Lung Sun, Carlo M. Croce, Mark W. Grinstaff, and Jon R. Parquette. Chemistry–A European Journal 21, no. 1 (2015): 101–105.

Kim, Seonhoe, and Aree Moon. Archives of Pharmacal Research 27, no. 8 (2004): 845–849.

Kim, Sung Hee, Sun Hee Hyun, and Se Young Choung. Journal of Ethnopharmacology 104, no. 1–2 (2006): 119–123.

Kizek, Rene, Vojtech Adam, Jan Hrabeta, Tomas Eckschlager, Svatopluk Smutny, Jaroslav V. Burda, Eva Frei, and Marie Stiborova. Pharmacology & Therapeutics 133, no. 1 (2012): 26–39.

Komlaga, Gustav, Sandrine Cojean, Rita A. Dickson, Mehdi A. Beniddir, Soulaf Suyyagh-Albouz, Merlin L.K. Mensah, Christian Agyare, Pierre Champy, and Philippe M. Loiseau. Parasitology Research 115, no. 8 (2016): 3185–3195.

Kulkarni, R.N., K. Baskaran, R.S. Chandrashekara, and S. Kumar. Plant Breeding 118, no. 1 (1999): 71–74.

Kumar, P., and A. Kumar. International Journal of Health Research 1, no. 3 (2008).

Kumar, Rajiv, Bhavna Sharma, Neha R. Tomar, Partha Roy, Atul K. Gupta, and Anil Kumar. Applied Biochemistry and Biotechnology 164, no. 8 (2011): 1246–1256.

Kumar, Suresh, Robin J. Harris, Christopher J. Seal, and Edward J. Okello. Phytotherapy Research 26, no. 1 (2012): 113–117.

Kwok, Hoi-Hin, Guan-Lun Guo, Justin Kai-Chi Lau, Yuen-Kit Cheng, Jiang-Rong Wang, Zhi-Hong Jiang, Man-Hong Keung, Nai-Ki Mak, Patrick Ying-Kit Yue, and Ricky Ngok-Shun Wong. Biochemical Pharmacology 83, no. 7 (2012): 893–902.

Kwon, Ha-Jeong, Je-Hyun Lee, and Seon-Pyo Hong. Archives of Pharmacal Research 33, no. 1 (2010): 81–86.

Lee, Geum-A., Kyung-Chul Choi, and Kyung-A. Hwang. Environmental Toxicology and Pharmacology 49 (2017b): 48–57.

Lee, Heyjin, Ji Hye Jeong, Sang-Jin Lee, Gyu-Un Bae, and Jae-Ha Ryu. Bioorganic & Medicinal Chemistry Letters 27, no. 6 (2017): 1401–1404.

Lee, Hoi-Seon. Journal of Agricultural and Food Chemistry 50, no. 24 (2002): 7013–7016.

Lee, Jae Woong, Yong Kyung Lee, Beom Jun Lee, Sang-Yon Nam, Sang Il Lee, Young Heui Kim, Ki Ho Kim, Ki-Wan Oh, and Jin Tae Hong. Pharmacology Biochemistry and Behavior 95, no. 1 (2010): 31–40.

Lee, Ji Suk, Hirotsugu Miyashiro, Norio Nakamura, and Masao Hattori. Chemical and Pharmaceutical Bulletin 56, no. 5 (2008): 711–714.

Lee, Jung Ju, Kyun Ha Kim, Eun Jeong Kim, Jun-Yong Choi, Sang-Jun Kim, Seung-Il Jeong, Jong-In Kim, and Myungsoo Joo. Journal of Ethnopharmacology 227 (2018): 97–104.

Li, Chunmei, and Myeong-Hyeon Wang. Nutrition Research and Practice 5, no. 2 (2011): 101–106.

Li, H., Y. M. Wang, J. Y. Xu, and B. Cao. Zhonghua jie he he hu xi za zhi= Zhonghua jiehe he huxi zazhi/ Chinese Journal of Tuberculosis and Respiratory Diseases 43 (2020): E002–E002.

Lin, Shiuan-Pey, Shang-Yuan Tsai, Yu-Chi Hou, and Pei-Dawn Lee Chao. Journal of Agricultural and Food Chemistry 57, no. 5 (2009): 1854–1859.

Liou, Kuo-Tong, Yuh-Chiang Shen, Chieh-Fu Chen, Cheng-Ming Tsao, and Shen-Kou Tsai. European Journal of Pharmacology 475, no. 1–3 (2003): 19–27.

Liu, C.Y., X.Q. Li, and S.Q. Cai. Pharmacology and Clinics of Chinese Materia Medica 26, no. 6 (2010): 84–85.

Liu, Xiao, Huan Li, Li Wu, Jiangwa Xing, Yanhong Poh, Hao Cai, and Bao Chang Cai. Journal of Chromatography B 980 (2015): 88–93.

Lorenzi, Harri. Vol. 2. Instituto plantarum de estudos da flora, 2002.

Lu, Hongmei, Xianjin Wu, Yizeng Liang, and Jian Zhang. Chemical and Pharmaceutical Bulletin 54, no. 7 (2006): 936–940.

Lu, Hongzhou. Bioscience Trends 14, no. 1 (2020): 69–71.

Lu, Zhaolian, Qi Jia, Rui Wang, Ximin Wu, Yingchun Wu, Caiguo Huang, and Yiming Li. Phytomedicine 18, no. 4 (2011): 298–302.

Luchi, Agnes E. Hoehnea 43, no. 3 (2016): 517–520.

Luo, Yuan. Life Sciences 78, no. 18 (2006): 2066–2072.

Ma, Chunyu, Hongyu Yu, Ying Xiao, and Huijiao Wang. Pharmaceutical Biology 55, no. 1 (2017): 2170–2177.

Ma, Qinge, Rongrui Wei, Zhiqiang Wang, Wenmin Liu, Zhipei Sang, Yaping Li, and Hongchun Huang. Journal of Ethnopharmacology 195 (2017b): 166–172.

Machumi, Francis, Jacob O. Midiwo, Melissa R. Jacob, Shabana I. Khan, Babu L. Tekwani, Jin Zhang, Larry A. Walker, and Ilias Muhammad. Natural Product Communications 8, no. 6 (2013): 1934578X1300800619.

Mackman, Nigel. Nature 451, no. 7181 (2008): 914–918.

Mahady, Gail B. Ethnopharmacology 2 (2009): 75–99.

Mahluji, Sepide, Vahide Ebrahimzade Attari, Majid Mobasseri, Laleh Payahoo, Alireza Ostadrahimi, and Samad EJ Golzari. International Journal of Food Sciences and Nutrition 64, no. 6 (2013): 682–686.

Mansour, Suzan M., Ashraf K. Bahgat, Aiman S. El-Khatib, and Mohamed T. Khayyal. Phytomedicine 18, no. 8–9 (2011): 641–647.

Mantena, Sudheer K., Som D. Sharma, and Santosh K. Katiyar. Molecular Cancer Therapeutics 5, no. 2 (2006): 296–308.

Maroo, Jyoti, Arna Ghosh, Rajni Mathur, Vihas T. Vasu, and Sarita Gupta. Pharmaceutical Biology 41, no. 5 (2003): 388–391.

Matsuda, Hisashi, Toshiyuki Murakami, Akinobu Kishi, and Masayuki Yoshikawa. Bioorganic & Medicinal Chemistry 9, no. 6 (2001): 1499–1507.

Maurya, Dharmendra Kumar, and Deepak Sharma. 2020. https://doi.org/10.26434/chemrxiv.12110214.v2

Mayaka, Regina K., Moses K. Langat, Josiah O. Omolo, and Peter K. Cheplogoi. Planta Medica 78, no. 04 (2012): 383–386.

McKay, Diane L., C.Y. Oliver Chen, Edward Saltzman, and Jeffrey B. Blumberg. The Journal of Nutrition 140, no. 2 (2010): 298–303.

Milajerdi, Alireza, Kurosh Djafarian, and Banafshe Hosseini. Journal of Nutrition & Intermediary Metabolism 3 (2016): 23–32.

Mirunalini, S., and M. Krishnaveni. Journal of Basic and Clinical Physiology and Pharmacology 21, no. 1 (2010): 93–105.

Mittal, Monika, Nomita Gupta, Palak Parashar, Varsha Mehra, and Manisha Khatri. International Journal of Pharmacy and Pharmaceutical Sciences 6, no. 8 (2014): 67–72.

Mohamed, E., A.E. Taha, Y. Md Abd-Elhakim, H. Prasad Devkota. Nutrients 12 (2020): 872.

Mohamed, Enas Ali Kamel. Australian Journal of Basic and Applied Sciences 5, no. 11 (2011): 1321–1327.

Mojiminiyi, F.B.O., M. Dikko, B.Y. Muhammad, P.D. Ojobor, O.P. Ajagbonna, R.U. Okolo, U.V. Igbokwe, et al. Fitoterapia 78, no. 4 (2007): 292–297.

Moudi, Maryam, Rusea Go, Christina Yong Seok Yien, and Mohd Nazre. International Journal of Preventive Medicine 4, no. 11 (2013): 1231.

Mshana, N. Robert. Organization of African Unity/Scientific, Technical & Research Commission, 2000.

Müller, Andreas, Werner Linke, and Wolfgang Klaus. Planta Medica 65, no. 04 (1999): 335–339.

Murali, B., U.M. Upadhyaya, and R.K. Goyal. Journal of Ethnopharmacology 81, no. 2 (2002): 199–204.

Murea, Mariana, Lijun Ma, and Barry I. Freedman. Review of Disability Studies 9, no. 1 (2012): 6.

Nahata, Alok, U.K. Patil, and V.K. Dixit. Natural Product Research 22, no. 16 (2008): 1472–1482.

Noor, Ayesha, S. Gunasekaran, and M.A. Vijayalakshmi. Pharmacognosy Research 9, no. Suppl 1 (2017): S99.

Nuengchamnong, Nitra, Kamlai Krittasilp, and Kornkanok Ingkaninan. Food Chemistry 117, no. 4 (2009): 750–756.

Nyeem, Mohammad Abu Bin, Md Abdul Mannan, Mohammad Nuruzzaman, K.M. Kamrujjaman, and Samir Kumar Das. Journal of Medicinal Plants Studies 5, no. 2 (2017): 318–324.

Ochwang'i, Dominic O., Charles N. Kimwele, Jemimah A. Oduma, Peter K. Gathumbi, James M. Mbaria, and Stephen G. Kiama. Journal of Ethnopharmacology 151, no. 3 (2014): 1040–1055.

Ogurtsova, Katherine, Leonor Guariguata, Noël C. Barengo, Paz Lopez-Doriga Ruiz, Julian W. Sacre, Suvi Karuranga, Hong Sun, Edward J. Boyko, and Dianna J. Magliano. Diabetes Research and Clinical Practice 183 (2022): 109118.

Okyar, Alper, Ayşe Can, Nuriye Akev, Gül Baktir, and Nurhayat Sütlüpinar. Phytotherapy Research 15, no. 2 (2001): 157–161.
Orhan, Ilkay, and Mustafa Aslan. Journal of Ethnopharmacology 122, no. 2 (2009): 327–332.
Pan, An, Danxia Yu, Wendy Demark-Wahnefried, Oscar H. Franco, and Xu Lin. The American Journal of Clinical Nutrition 90, no. 2 (2009): 288–297.
Patel, Ketan R., Victoria A. Brown, Donald J.L. Jones, Robert G. Britton, David Hemingway, Andrew S. Miller, Kevin P. West, et al. Cancer Research 70, no. 19 (2010): 7392–7399.
Patel, Priyanka, Jean-Pierre Julien, and Jasna Kriz. Neurotherapeutics 12, no. 1 (2015): 217–233.
Pathak, Rajiv, and Ashima Pathak. International Journal of Applied and Basic Medical Research 2, no. 2 (2012): 92.
Paulo, Alexandra, Elsa T. Gomes, Jonathan Steele, Dave C. Warhurst, and Peter J. Houghton. Planta Medica 66, no. 01 (2000): 30–34.
Penumala, Mohan, Raveendra Babu Zinka, Jeelan Basha Shaik, Suresh Kumar Reddy Mallepalli, Ramakrishna Vadde, and Damu Gangaiah Amooru. BMC Complementary and Alternative Medicine 18, no. 1 (2018): 1–13.
Peschel, Wieland, Alfred Kump, Attila Horváth, and Dezső Csupor. Industrial Crops and Products 83 (2016): 787–802.
Péter Zomborszki, Zoltán, Norbert Kúsz, Dezső Csupor, and Wieland Peschel. Pharmaceutical Biology 57, no. 1 (2019): 295–305.
Peter, Karoline, Johann Schinnerl, Susanne Felsinger, Lothar Brecker, Rudolf Bauer, Heimo Breiteneder, Riurong Xu, and Yan Ma. Journal of Ethnopharmacology 149, no. 2 (2013): 562–569.
Pharmacopoeia, Chinese. Commission. Pharmacopoeia of the People's Republic of China. (2015): 191.
Ploeger, Bart, Tjeert Mensinga, Adriënne Sips, Willem Seinen, Jan Meulenbelt, and Joost DeJongh. Drug Metabolism Reviews 33, no. 2 (2001): 125–147.
Porter, Anna Horsbrugh. BBC News 17 (2006).
Prakash, Jay, Shikha Chouhan, Satyndra Kumar Yadav, Susan Westfall, Sachchida Nand Rai, and Surya Pratap Singh. Neurochemical Research 39, no. 12 (2014): 2527–2536.
Qu, Qian, Yuefei Li, Qi Dong, Shupeng Li, Hongliang Du, Zhihua Wang, Xiaopei Gong et al. Frontiers in Pharmacology 12 (2021).
Rahier, Nicolas J., Craig J. Thomas, and Sidney M. Hecht. Anticancer Agents from Natural Products 5 (2005): 22.
Rani, G, Singara Charya Ma, M. Viswanadham and Murali Krishna Thupurani. IOSR Journal of Pharmacy and Biological Sciences 12 (2017): 37–40.
Rayyan, Saleh, Torgils Fossen, H. Solheim Nateland, and Ø. M. Andersen. Phytochemical Analysis: An International Journal of Plant Chemical and Biochemical Techniques 16, no. 5 (2005): 334–341.
Reddy, G.D., G.S. Reddy, A.S. Reddy, and P.V. Reddy. International Journal of Pharmaceutical Sciences Review and Research 4 (2010): 46–48.
Rodriguez, Michelle E., Brian J. Poindexter, Roger J. Bick, and Amitava Dasgupta. Journal of Medicinal Food 11, no. 4 (2008): 680–686.
Roghani-Dehkordi, Farshad, and Amir-Farhad Kamkhah. Journal of Dietary Supplements 6, no. 4 (2009): 328–341.
Roth, Gregory A., Catherine Johnson, Amanuel Abajobir, Foad Abd-Allah, Semaw Ferede Abera, Gebre Abyu, Muktar Ahmed et al. Journal of the American College of Cardiology 70, no. 1 (2017): 1–25.
Roy, A., K. Kundu, G. Saxena, and N. Bharadvaja. Journal of Plant Biochemistry & Physiology 5 (2017b): 198.
Roy, A., K. Kundu, G. Saxena, L. Kumar, and N. Bharadvaja. Advanced Techniques in Biology & Medicine 4 (2016): 172.
Roy, Arpita, and Navneeta Bharadvaja. Current Trends in Biomedical Engineering & Biosciences 5, no. 3 (2017a): 1–5.
Roy, Arpita, and Navneeta Bharadvaja. International Journal of Ayurvedic and Herbal Medicine 7 (2017): 2643–2650.
Roy, Arpita. International Journal of Complementary and Alternative Medicine 11, no. 4 (2018): 205–208.
Runfeng, Li, Hou Yunlong, Huang Jicheng, Pan Weiqi, Ma Qinhai, Shi Yongxia, Li Chufang et al. Pharmacological Research 156 (2020): 104761.
Russo, A., A.A. Izzo, V. Cardile, F. Borrelli, and A. Vanella. Phytomedicine 8, no. 2 (2001): 125–132.
Salem, Mohamed Y., and Nahla El-Eraky El-Azab. Egyptian Journal of Histology 37, no. 3 (2014): 571–578.

San Cheang, Wai, Ching Yuen Ngai, Ye Yen Tam, Xiao Yu Tian, Wing Tak Wong, Yang Zhang, Chi Wai Lau, et al. Scientific Reports 5, no. 1 (2015): 1–10.

Sarkar, Arnab, Shamee Bhattacharjee, and Deba Prasad Mandal. Asian Pacific Journal of Cancer Prevention 16, no. 15 (2015): 6753–6759.

Sarkar, Saikat, Sadaf Zaidi, Amit K. Chaturvedi, R. Srivastava, P.K. Dwivedi, and R. Shukla. Journal of Pharmacognosy and Phytochemistry 3 (2015a): 59–72.

Schutz, Fabio AB, Joaquim Bellmunt, Jonathan E. Rosenberg, and Toni K. Choueiri. Expert Opinion on Drug Safety 10, no. 4 (2011): 645–653.

Seifu, Tesfaye, Kaleab Asres, and Tsige Gebre-Mariam. Ethiopian Pharmaceutical Journal 24, no. 1 (2004): 41–58.

Sethiya, Neeraj Kumar, Alok Nahata, Shri Hari Mishra, and Vinod Kumar Dixit. Zhong Xi Yi Jie He Xue Bao. Journal of Chinese Integrative Medicine 7, no. 11 (2009): 1001–1022.

Shah, Syed Muhammad Ali, Syed Ali Raza Naqvi, Naveed Munir, Sohaib Zafar, Muhammad Akram, and Jaweria Nisar. Dose-Response 18, no. 3 (2020): 1559325820942077.

Shang, Xiaofei, Hu Pan, Maoxing Li, Xiaolou Miao, and Hong Ding. Journal of Ethnopharmacology 138, no. 1 (2011): 1–21.

Sharifi, Niusha, Effat Souri, Seyed Ali Ziai, Gholamreza Amin, and Massoud Amanlou. DARU Journal of Pharmaceutical Sciences 21, no. 1 (2013): 1–8.

Shi, Chun, Jun Liu, Fengming Wu, and David T. Yew. International Journal of Molecular Sciences 11, no. 1 (2010): 107–123.

Shinomol, G. K., and M. M. Bharath. Recent Patents on Endocrine, Metabolic & Immune Drug Discovery 5, no. 1 (2011): 33–49.

Shree, Priya, Priyanka Mishra, Chandrabose Selvaraj, Sanjeev Kumar Singh, Radha Chaube, Neha Garg, and Yamini Bhusan Tripathi. Journal of Biomolecular Structure and Dynamics 40, no. 1 (2022): 190–203.

Singh, S., R. Jarial, and S. S. Kanwar. Wudpecker J. Med. Plants 2 (2013): 53–65.

Singh, Sukhdev, Bhupender Sharma, Shamsher S. Kanwar, and Ashok Kumar. Frontiers in Plant Science 7 (2016): 1667.

Singhal, Anil Kumar, Vijay Naithani, and Om Prakash Bangar. International Journal of Nutrition, Pharmacology, Neurological Diseases 2, no. 2 (2012): 84.

Sleumer, H. O. Flacourtiaceae. New York New York Bot Gard Organ Flora Neotrop 499p Illus, maps, keys. Icones, Maps Geog 4 (1980).

Slobodnick, Anastasia, Binita Shah, Michael H. Pillinger, and Svetlana Krasnokutsky. The American Journal of Medicine 128, no. 5 (2015): 461–470.

Smith, Jennifer N., Jenna M. Negrelli, Megha B. Manek, Emily M. Hawes, and Anthony J. Viera. The Journal of the American Board of Family Medicine 28, no. 2 (2015): 283–293.

Sofowora, A. Ibadan, Nigeria: Spectrum Books Limited (2008): 22–30.

Sofowora, Abayomi, Eyitope Ogunbodede, and Adedeji Onayade. African Journal of Traditional, Complementary and Alternative Medicines 10, no. 5 (2013): 210–229.

Soumya, D., and B. Srilatha. Journal of Diabetes & Metabolism 2, no. 9 (2011): 1000167.

Spiller, Henry A., and Tama S. Sawyer. American Journal of Health-System Pharmacy 63, no. 10 (2006): 929–938.

Stiborová, Marie, Věra Černá, Michaela Moserová, Iveta Mrízová, Volker M. Arlt, and Eva Frei. International Journal of Molecular Sciences 16, no. 1 (2014): 284–306.

Sun, Yanfei. In: 2017 2nd International Conference on Materials Science, Machinery and Energy Engineering (MSMEE 2017), pp. 14–17. Atlantis Press, 2017.

Sung, Hyuna, Jacques Ferlay, Rebecca L. Siegel, Mathieu Laversanne, Isabelle Soerjomataram, Ahmedin Jemal, and Freddie Bray. CA: A Cancer Journal for Clinicians 71, no. 3 (2021): 209–249.

Swathi, H., and T. Shekshavali. Research Journal of Pharmacology and Pharmacodynamics 8, no. 3 (2016): 115.

Tanaka, Miyuki, Eriko Misawa, Yousuke Ito, Noriko Habara, Kouji Nomaguchi, Muneo Yamada, Tomohiro Toida et al. Biological and Pharmaceutical Bulletin 29, no. 7 (2006): 1418–1422.

Tang, W. and Eisenbrand, G., Handbook of Chinese Medicinal Plants – Chemistry, Pharmacology, Toxicology, Vol. 2, Wiley Verlag GmbH & Co. KgaA, Weinheim, 2011.

Tunali-Akbay, Tugba, Goksel Sener, Hanife Salvarli, Ozer Sehirli, and Aysen Yarat. Phytotherapy Research: An International Journal Devoted to Pharmacological and Toxicological Evaluation of Natural Product Derivatives 21, no. 1 (2007): 26–31.

Vallianou, Natalia G., Angelos Evangelopoulos, Nikos Schizas, and Christos Kazazis. Anticancer Research 35, no. 2 (2015): 645–651.
Van Diermen, Daphné, Andrew Marston, Juan Bravo, Marianne Reist, Pierre-Alain Carrupt, and Kurt Hostettmann. Journal of Ethnopharmacology 122, no. 2 (2009): 397–401.
Van Rossum, Tekla G.J., Arnold G. Vulto, Wim CJ Hop, and Solko W. Schalm. The American Journal of Gastroenterology 96, no. 8 (2001): 2432–2437.
Varoni, Elena M., Alfredo Fabrizio Lo Faro, Javad Sharifi-Rad, and Marcello Iriti. Frontiers in Nutrition 3 (2016): 8.
Venier, Natalie A., Toshihiro Yamamoto, Linda M. Sugar, Hans Adomat, Neil E. Fleshner, Laurence H. Klotz, and Vasundara Venkateswaran. The Prostate 75, no. 12 (2015): 1300–1311.
Verma, Tarawanti, Manish Sinha, Nitin Bansal, Shyam Raj Yadav, Kamal Shah, and Nagendra Singh Chauhan. Natural Products and Bioprospecting 11, no. 2 (2021): 155–184.
Vishwakarma, Santosh L., Rakesh D Sonawane, M. Rajani, and Ramesh K. Goyal. Indian Journal of Experimental Biology 48, no. 1 (2010): 26–30.
Wang, Jiabo, Yanling Zhao, Xiaohe Xiao, Huifang Li, Haiping Zhao, Ping Zhang, and Cheng Jin. Journal of Ethnopharmacology 124, no. 1 (2009): 18–25.
Wang, Jia-bo, Yi Qin, Wei-jun Kong, Zhe-wei Wang, Ling-na Zeng, Fang Fang, Cheng Jin, Yan-ling Zhao, and Xiao-he Xiao. Food Chemistry 129, no. 4 (2011): 1737–1743.
Wang, W., J. Zhao, Y.H. Wang, T. A. Smillie, X.C. Li, I. A. Khan. Planta Med, 75, no. 13 (2009): 1436–1441.
Wang, Xuebing, Xiao Li, Xuefang Wang, Ling Chen, Erjuan Ning, Yi Fan, Hao Wang, Tong Chen, and Wei Wang. Research in Veterinary Science 135 (2021): 523–531.
Wong, Alice ST, Chi-Ming Che, and Kar-Wah Leung. Natural Product Reports 32, no. 2 (2015): 256–272.
World Health Organization (WHO). World Malaria Report 2015. 20 Avenue Appia, 1211 Geneva 27. Switzerland. [Google Scholar] (2015).
Wu, Zhao, Xinyu Deng, Qichao Hu, Xiaolin Xiao, Jing Jiang, Xiao Ma, and Mingquan Wu. Frontiers in Pharmacology 12 (2021).
Xie, Jun, Bo Liu, Qunlan Zhou, Yongteng Su, Yijin He, Liangkun Pan, Xianping Ge, and Pao Xu. Aquaculture 281, no. 1–4 (2008): 5–11.
Xie, Songbo, and Jun Zhou. Frontiers in Plant Science 8 (2017): 720.
Xu, Haitao, and Songyan Zhang. Phytotherapy Research 27, no. 10 (2013): 1524–1528.
Xu, Hao, Liang Zhong, Jiaxin Deng, Jiakuan Peng, Hongxia Dan, Xin Zeng, Taiwen Li, and Qianming Chen. International Journal of Oral Science 12, no. 1 (2020): 1–5.
Xu, Xueqin, Hongzhi Ye, Wei Wang, Lishuang Yu, and Guonan Chen. Talanta 68, no. 3 (2006): 759–764.
Xu, Zhe, Lei Shi, Yijin Wang, Jiyuan Zhang, Lei Huang, Chao Zhang, Shuhong Liu et al. The Lancet Respiratory Medicine 8, no. 4 (2020b): 420–422.
Yimam, Mesfin, Jifu Zhao, Brandon Corneliusen, Mandee Pantier, Lidia Brownell, and Qi Jia. Diabetology & Metabolic Syndrome 6, no. 1 (2014): 1–8.
Yip, E. C. H., A. S. L. Chan, Haihong Pang, Y. K. Tam, and Yung Hou Wong. Cell Biology and Toxicology 22, no. 4 (2006): 293–302.
Yu, Jing, Masato Eto, Masahiro Akishita, Akiyo Kaneko, Yasuyoshi Ouchi, and Tetsuro Okabe. Biochemical and Biophysical Research Communications 353, no. 3 (2007): 764–769.
Zhang, Feng-Xiang, Zi-Ting Li, Chang Li, Min Li, Zhi-Hong Yao, Xin-Sheng Yao, and Yi Dai. Journal of Pharmacy and Pharmacology 72, no. 12 (2020b): 1879–1892.
Zhang, Qian, Xinhua Xiao, Kai Feng, Tong Wang, Wenhui Li, Tao Yuan, Xiaofang Sun, Qi Sun, Hongding Xiang, and Heng Wang. Evidence-Based Complementary and Alternative Medicine 2011 (2010).
Zhang, Shu, Z.H. Shi, J.J. Hao, and R.Q. Wang. Journal of Chengdu University Traditional Chinese Medicine 30 (2007): 54–55.
Zhengli, Shi. Cell Research 30, no. 3: 269–271.
Zhongzhen, Zhao, ed. School of Chinese Medicine, Hong Kong Baptist University, 2004.
Zhou, Jiang-Tao, Chen-Yang Li, Chun-Hua Wang, Yue-Fei Wang, Xiao-Dong Wang, Hong-Tao Wang, Yan Zhu, Miao-Miao Jiang, and Xiu-Mei Gao. Molecules 20, no. 8 (2015): 13725–13739.
Zhou, Wei, and Xiao-Yan Zhang. The American Journal of Chinese Medicine 41, no. 04 (2013): 743–764.
Zhou, Wei, Hong Chai, Peter H. Lin, Alan B. Lumsden, Qizhi Yao, and Changyi Chen. Cardiovascular Drug Reviews 22, no. 4 (2004): 309–319.

Zhou, Xia, Baoan Song, Linhong Jin, Deyu Hu, Chunling Diao, Guangfang Xu, Zhihui Zou, and Song Yang. Bioorganic & Medicinal Chemistry Letters 16, no. 3 (2006): 563–568.

Zhu, Dan, Lei Wu, Chun-Rong Li, Xiao-Wei Wang, Yong-Jie Ma, Zhi-yin Zhong, Hai-Bao Zhao, et al. Journal of Cellular Biochemistry 108, no. 1 (2009): 117–124.

Zhu, Lin, Y.M. Liang, Liang Xu, Ming Xie, Fan Xia, and Xin Liu. Chinese Medicine Modern Distance Education of China 15, no. 12 (2017): 150–153.

Zhuang, Jieqin, Xingzhen Dai, Qihua Wu, Hairong Cai, Xue Fu, Weizhang Zhang, and Bojun Chen. Complementary Therapies in Medicine 60 (2021): 102754.

Zirihi, Guédé Noël, Lengo Mambu, Frédéric Guédé-Guina, Bernard Bodo, and Philippe Grellier. Journal of Ethnopharmacology 98, no. 3 (2005): 281–285.

3 Use of Nanoparticles in the Delivery of Plant-Based Therapeutics

*Indrani Manna, Debalina Das, Sejuty Mondal, and Maumita Bandyopadhyay**
Plant Molecular Cytogenetics Laboratory, Centre of Advanced Study, Department of Botany, Ballygunge Science College, University of Calcutta, Kolkata, India

3.1 INTRODUCTION

3.1.1 Phytonanotechnology: Mobilizing Nanotechnology to Potentiate Plant-Based Therapeutics

Nanotechnology is a multidisciplinary field that encompasses various aspects of physics, chemistry, biology, and engineering and pertains to the preparation and utilization of specifically designed nanosized materials. Such engineered *nanomaterials/nanoparticles* (ENPs) can be easily crafted for high-end medical uses (Donia and Carbone, 2019). "Nano" stands for nano-size, which is one-billionth part in the metric scale (Ghasemi et al., 2021); therefore, any particle with at least one of its dimensions within this range can be broadly termed as a nanoparticle (NP). NPs have unique properties. When compared with their bulk counterparts, their large surface area-to-volume ratio provides them with an array of advantages, harnessed for various applications, and they participate in various interactions with biological systems (Ahmed et al., 2018).

NPs are omnipresent and produced incessantly either by natural forces or through biogenic actions. Biogenic NPs can be produced under various circumstances, but *green synthesis* of NPs particularly pertains to their targeted production in bacterial cultures, fungal mycelia, or extracts of higher plants, collectively designated as *green factories*. *Phytonanotechnology* refers to the synthesis of highly functional engineered NPs, primarily metal or metal oxide NPs using plant-based biomolecules, and their utilization in novel applications. Such ENPs have various shapes and sizes, large surface area-to-volume ratio, and great catalytic, optical, and antimicrobial activities, and are mostly easily prepared, inexpensive, and eco-friendly. These factors have encouraged their use in various important sectors, namely, agriculture, crop development, phytoremediation, pest control, medical purposes, nutraceuticals, and healthcare (Meena et al., 2021). Phytonanotechnology can potentially revolutionize the nutraceutical and agriculture sectors and fulfil the present need for safe and better pharmaceuticals harnessing natural ingredients. Therefore, it is imperative to elaborate the immense potential of phytonanotechnology in the field of medicine and therapeutics, which could revolutionize this field in the coming decades.

Recently, the uses of phytonanoparticles in the treatment of multifarious human ailments are advocated because of the following advantages:

- High surface area-to-volume ratio
- Improved solubility, stability, and bioavailability

DOI: 10.1201/9781003231721-3

- Better cellular uptake
- Improved site-specific penetration, absorption, and retention
- Effective at lower doses
- Safer, with lesser side effects

3.1.2 Types of Phytonanoparticles and Their Use as Therapeutics

Three major types of ENPs have been used in therapeutics, namely, *organic* (liposomes and polymers), *inorganic* (ceramic NPs, quantum dots, metal NPs, and metal oxide NPs), and *carbon-based.* Most ENPs are phytosynthesized. In pharmaceuticals, ENPs such as liposomes, polymers, ceramics, quantum dots, metal NPs, metal oxide NPs, and carbon-based NPs are extensively used (Din et al. 2017).

Liposomes—These are spherical, lipid bilayer-bound vesicles; for hydrophilic compounds, the aqueous phase contains active compounds, and for hydrophobic compounds, the active compounds could reside in the lipid bilayer. Size distribution and shelf-life are generally well controlled in liposomes, which make them the vector of choice for delivering important drugs or their cocktails. Good biodegradability and biocompatibility are added advantages. Multilamellar vesicles and small unilamellar vesicles are examples of important liposomes.

Polymers—They are quite useful for their high biodegradability, biocompatibility, and surface modifications. They are mainly made of chitosan or polylactides, poly(methyl methacrylate), or polyethylene glycol (PEG). Alginate and cellulose are other examples of polymers.

Ceramic—Ceramic NPs are made of silica, alumina, glass, or clay, containing a variety of cores. Their porous nature makes them a good vector for delivering drugs. They are least affected by changes in pH or temperature. Metal oxide-capped silica, such as mesoporous silica NPs, are quite popular in cancer studies. Bioceramics with high conductivity, biodegradability, compatibility, hydrophilicity, and re-absorbability are quite useful.

Quantum dots—These are semiconductors with fluorescing ability. Semiconductor cores with coated shells are quite useful in bioimaging and biosensing.

Metal nanoparticles—These are mostly gold or silver NPs with unique electronic and optical properties. These are mostly biocompatible and nontoxic and are useful as ligands for biomolecules.

Metal oxide nanoparticles—They have excellent catalytic and antioxidant properties. They are mostly stable and biocompatible and have good optical properties. Iron oxide, titanium oxide, zirconia oxide, and cerium oxide are mostly used for such purposes. Additionally, they can switch between the oxidation states; therefore, they are useful in oxidation studies (e.g., studies on CeO_2-NPs, SiO_2-NPs, and TiO_2-NPs).

Carbon-based nanoparticles—They are known for their polygonal structures (fullerenes and nanotubes) and are either single walled or multi walled, possessing antimicrobial activity. They have high mechanical strength, optical properties, and electrical conductivity and are thus well suited for biomedical imaging purposes. However, they are easily aggregated and are not biodegradable and require intricate surface modifications for drug delivery.

Figure 3.1 shows the various types of NPs. Metal NPs, silica NPs, and magnetic NPs are inorganic nanocarriers, whereas solid lipid NPs (SLNs) and liposomes are organic nanocarriers. Compared

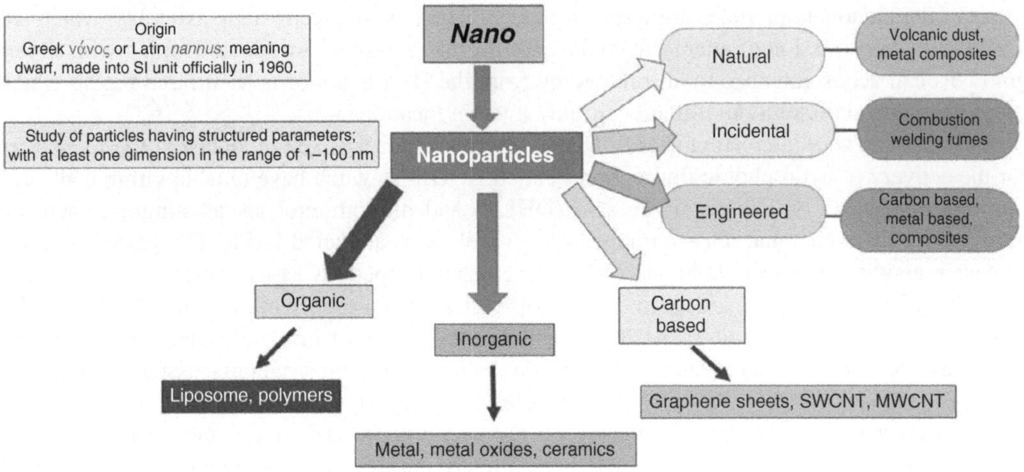

FIGURE 3.1 Schematic diagram showing various types of nanoparticles. (From Manna and Bandyopadhyay, 2019. With Permission.)

with inorganic nanocarriers, organic ones are better suited for *in vivo* applications owing to their nontoxic nature (Conte et al., 2017). Apart from inorganic and organic nanocarriers, the other nano-formulations commonly used include cell-derived nanovesicles such as exosomes, plant exosome–like nanovesicles, nanoemulsions, nanostructured lipid carriers, and dendrimers (León-Buitimea et al., 2021).

3.1.3 Important Uses of Phytonanoparticles

Phytonanoparticles have revolutionized the fields of medicine and therapeutics in recent years. Their major areas of application are as follows:

Bioimaging: Various high-end imaging techniques such as optical imaging, magnetic resonance imaging, ultrasound imaging, and positron emission tomography use NPs with luminescent and magnetic properties. Examples include quantum dots (QDs) and metal NPs such as AuNPs and AgNPs.

***In Vitro* Diagnostics:** Many nanowires and nanotubes are useful for *in vitro* diagnostics, for example, these nanocarriers are used as sensors in diagnostic devices such as in atomic force microscopy for improving their sensitivity and reducing production cost. Carbon nanotube-based sensors are useful for performing respiratory function tests. Atomic force microscopy-aided detection chips are useful for easy and early detection of viral infections.

In Cancer Therapy: Many engineered NPs are encapsulated in QDs for the immobilization of tumor-targeting agents and chemokines on the surfaces of the QDs, thus enabling ultraviolet light-aided detection in tumor tissues. Antibody-conjugated magnetic poly(D,L-lactide-co-glycolide) with doxorubicin is an example of polymeric NPs useful in cancer therapy. Regeneration of injured nerves could be visualized by nanotechnology-aided surface microscopy, silicon fabrication, and computational engineering.

As Drug Delivery Systems: Phytonanoparticles are better than conventional drug delivery systems (DDSs) because of their increased solubility, stability, efficacy, and unique physical properties.

Many of the phytonanoparticles are water-soluble structures with electrostatic assembly, which are used for the transmission of energy from the core for the release of drug load. Doping with silver/gold NPs can act as antennae to absorb energy from the laser beam of near-infrared region which cause heating and dissolution of microcapsules at target locations.

Among some very interesting uses of NPs as DDSs include a mixed monolayer of gold clusters for the delivery of hydrophobic fluorophores called BODIPY, which have tunable infrared absorption and high stability. Dihydrolipoic acid (DHLA) and dithiothreitol act as stimuli to activate mesoporous silica NPs that release trapped (drug) molecules adducted inside. Fluorescein isothiocyanate is another such formulation wherein magnetic NP–capped MCM-41 (a mesoporous material with a hierarchical structure) is seen in cancer cells with a large presence of intracellular DHLA. pH-responsive NPs provide an alternate mechanism for the release of inner adducted drugs, wherein the release depends on the lowered acidity inside the tumor and inflammatory tissue outside. This is seen in magnetic NPs covalently conjoined to DOX, an important anticancer drug, using an acid-labile hydrazone linker. The carrier, if encapsulated with thermosensitive polymers, can release the drug according to the surrounding temperature. Similarly, nitric oxide could be released from gold NPs in an acidic pH environment.

In Surgery: NPs are useful for repairing blood leaks in arteries during kidney or heart transplantation procedures. Gold-coated nanoshells are used to weld biological samples.

Figure 3.2 summarizes the multifarious uses of phytonanoparticles.

The use of plant-based NPs could easily delve into and provide relief against a variety of diseases and pathological conditions. The present chapter discusses and focuses on the scientific breakthroughs wherein phytonanoparticles were strategized into disease-fighting/preventing technologies. A comprehensive account of major human diseases has revealed the superior utility and application at a fraction of cost and minimal side effects.

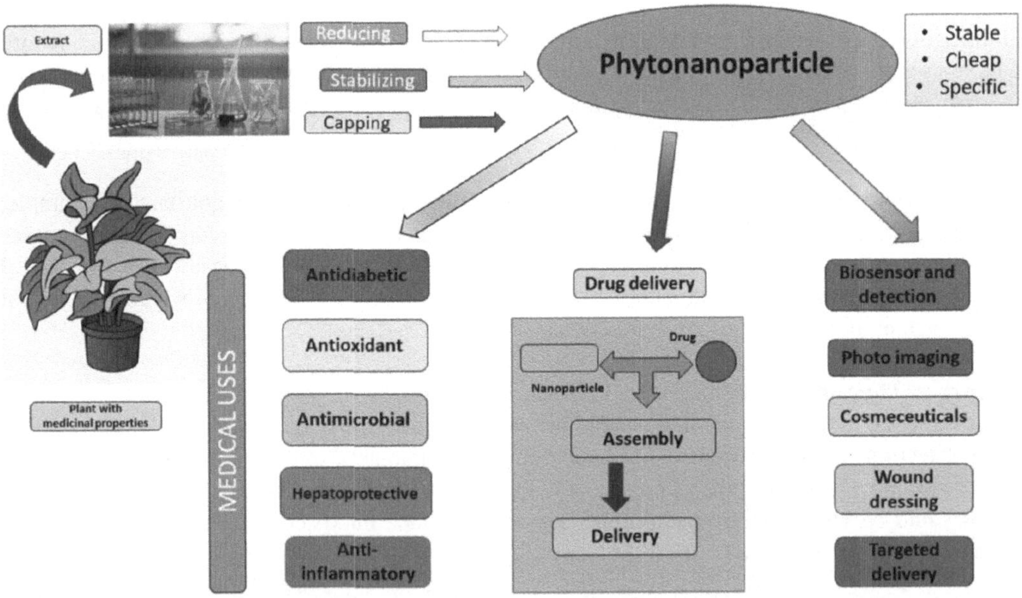

FIGURE 3.2 Schematic diagram showing the applications of plant-based nanoparticles and the advantages of plant-based nanoparticle synthesis over synthesis through conventional methods.

3.2 PHYTONANOPARTICLE TREATMENT FOR HUMAN DISEASES

Phytochemicals are low-molecular-weight compounds present abundantly in different parts of the plant. More than 10,000 phytochemicals belonging to various chemical categories have been identified in plants, such as polyphenols, alkaloids, terpenoids, saponins, steroids, and essential oils (Jeevanandam et al., 2018). By the virtue of their unique chemical properties, these phytochemicals exhibit a wide range of medicinal properties including antidiabetic, anticancer, antioxidant, antiulcer, antimicrobial, antilarval, antiaging, and neuro-protecting activities. Since the beginning of human history, plants have been the most consistent and exploited sources of medicinal compounds, and even in the modern era, their easy accessibility and superior biological activities have held their importance as reservoirs of novel medicinally important phytochemical constituents. Despite these advantages, the use of plant-based drugs has several drawbacks, most prominently low water solubility of most phytochemicals, as well as poor bioavailability, low retentivity, and undesired stability, all of which limit their clinical applications. Two conceivable solutions are available to resolve these problems: (1) chemical modification of the desired phytoconstituent and (2) formulation and application of nano-DDSs for targeted delivery of plant-based therapeutic compounds. While chemical modification might alter the intrinsic physiological activities of some phytochemicals, harnessing nanotechnology to encapsulate plant-based therapeutic agents and deliver them to specific organs and cells without degradation or adverse side effects has garnered scientific attention in recent times. Dosage regulation of plant-based secondary metabolites when used as drugs, as well as their side effects, especially at higher doses, has been a persistent problem plaguing medicinal uses of phytochemicals. Phytosynthesis of NPs and/or phytonano-encapsulation augments water solubility, bioavailability, and biostability of target phytochemicals and may lead to their better retentivity to enhance pharmacological activity (Malik et al., 2014; Conte et al., 2017; Khandel et al., 2018). Furthermore, for better target recognition, these designed NPs were functionalized with targeting molecules and conjugated particles, mostly on their surface, which increase their precision and effectiveness and aid in better uptake and internalization (Almeida et al., 2021).

In the last few decades, quite a few nanomaterials have been exploited as carriers for phytochemicals and drugs, which circulated more freely and specifically targeted disease sites. The *Vinca* alkaloid vincristine is one of the most potent anticancer agents available. Vincristine has been incorporated into liposome-bound gold NPs to increase its effectivity and ensure rapid release (Liu et al., 2015). Similarly, another *Vinca* alkaloid, vinblastine, was loaded onto polymeric NPs for sustained release (Zhu et al., 2019). Important taxanes such as paclitaxel has been developed as a pure NP form, which are smaller and more effective in terms of having improved targeted delivery and dose selectivity (Wu et al., 2018). Various other polymer-conjugated paclitaxel-bound NPs that have a better effect, with lesser side effects, have been developed and are already being tested in clinical trials (Ma and Mumper, 2013). Other anticancer drugs such as podophyllotoxin and its derivatives were reportedly encapsulated in lipid or aniline NPs and showed similar results and quite promising outcomes (Zhao et al., 2016; Fumagalli et al., 2017).

Fabrication of plant-based NPs for the treatment of various human diseases is a safe, environmentally friendly approach, as phytomolecules are cheaper and safer for daily use, together with higher stability and rapid mechanism of action being added advantages. The production of plant-based NPs is economical when compared with traditional laboratory-based fabrications, as the former does not require high temperature, pressure, or expensive chemicals and the plant extract can act as capping, reducing, and stabilizing agents (Puri et al., 2018). Plant-derived capping layers are sites of structural modifications, where genes, drugs, peptides, or any small molecules are attached and provide the active substrate for biological interactions (Anand et al., 2022). A recent comparative study has shown that biogenic phytonanoparticles act as antibacterial agents and are more effective than chemically synthesized ones against serious ailments caused by multidrug-resistant microorganisms (Singh et al., 2018). Thus, it can be asserted that nanotechnology has the potential to accelerate novel innovations in medicine and drug development in the near future.

The preparation of NPs using plant extracts and its application in human disease management or the encapsulation of phytochemicals as a plant therapeutic toward nanoformulation is much less hazardous and less toxic and has good bioavailability and better retentivity. Most studies on nanoformulations have assessed antibiotics, synthetic drugs, parts of dead pathogens, and utilization of plant-based drugs in nanoformulations is still at its early stages of development for treating different human microbial diseases. The main drawback resides in testing the efficacy of these phytochemicals against the relevant causal organisms. Therefore, many more phytochemicals are required to be evaluated for their activities against diseases, after which they can be used for nanoformulations, rendering better pharmacological effects. Targeted delivery and specific elicitation of immune responses is the basis for treating various infectious and noninfectious diseases. An extensive amount of research is being conducted on phytonanoparticle-disease interactions, and some of the notable examples are discussed in the following sections.

3.3 PHYTONANOPARTICLES FOR CANCER TREATMENT

Cancer is a severe life-threatening disease, and its treatment is hampered by a wide array of chemotherapy complexities such as cytotoxicity, nonspecificity, induction of multidrug resistance, and growth of stem-like cells (Wei et al., 2015). In recent years, studies on biomedical applications of NPs have garnered increasing attention because these nanomaterials target cancer cells, affect the tumor microenvironment, boost immune systems to overcome toxicity and lack of specificity, and enhance drug capacity and bioavailability; hence, these nanomaterials have been modified for application in a wide range of cancer therapies. Thus, to subvert the main challenges in cancer therapy such as drug resistance and toxicity, the utilization of phytonanotechnology for anticancer drug delivery has become an emerging platform for cancer treatment since the last decade or so (Chauhan et al., 2013; Davis et al., 2010).

As chemical synthesis of NPs poses a threat to the environment either directly or indirectly, researchers are focusing on eco-friendly, simple, and reproducible phytosynthesis of stable NPs that have the potential for cancer therapy. This shift from chemical to biogenic synthesis of NPs involves the use of phytoextracts from various plant parts, which make this approach cost-effective, eco-friendly, less toxic, and easier to synthesize (Manna et al., 2020). Information on the fabrication process and unique attributes of the synthesized phytonanoparticles responsible for anticancer activity is listed in Figure 3.3. Some popular NPs that are synthesized as anticancer phytofabricated nanocomplexes have been discussed in the following sections. Such NPs are favorable for application because of their easy availability, low toxicity, cost-effectiveness, and rapid formation of stable phytonanocomplexes, followed by effective performance of target functions.

3.3.1 Silver Nanoparticles

Like their chemically synthesized counterparts, phytosynthesized silver NPs are potent anticancer agents, both *in vitro* and *in vivo*, and are effective against cervical cancer, breast cancer, lung cancer, hepatocellular carcinoma, nasopharyngeal carcinoma, glioblastoma, colorectal adenocarcinoma, and prostate carcinoma (Pei et al., 2019; Wang et al., 2019b; Xu et al., 2020). The anticancer activity of silver NPs is attributed to the inherent properties of this NP, such as size, shape, and surface charge (Jo et al., 2015; Xu et al., 2020).

The *modus operandi* of phytosynthesized silver NPs' anticancer effect started with the disruption of cellular membrane integrity, decreased cell growth, induction of cytoplasmic condensation, cell shrinkage, and nuclear condensation and fragmentation, as seen in the human breast cancer cell line MCF-7. The phytonanocomplex induced free radical generation, resulting in oxidative damage and caspase-mediated apoptosis through DNA fragmentation and the subsequent cell cycle arrest at the G2/M phase (Das et al., 2013a). The anticancer activity of silver NPs synthesized from extracts of unripe fruits of *Solanum trilobatum* against MCF-7 demonstrated downregulated Bcl-2 expression and increased Bax expression, indicating the involvement of mitochondria in cell death induced by silver NPs (Ramar et al., 2015). Chung et al. (2016) carried out a detailed review on the physico-chemical conditions necessary for the green synthesis of plant-based silver NPs, followed by characterization of the synthesized nano-silvers and their therapeutic application in cancer control. Nath et al. (2020) described the advantages of green synthesis of phyto-silver NPs in terms of being simple, cost-effective, and nontoxic when compared with chemical synthesis.

The strong antiproliferative effects of phyto-nanosilver against A375 skin melanoma cell line was attributed to their irregular structures and shape-dependent properties (Das et al., 2013b). Regulation of apoptosis through the intrinsic pathway by silver NPs based on the aqueous extract of *Coptis chinensis* was reported as the reason for the inhibition of the migration and invasion of A549 lung alveolar carcinoma cells (Pei et al., 2019). Along the same lines, Ag-NPs synthesized from aqueous leaf extracts of *Adenium obesum* and *Sesbania grandiflora* induced autophagy and apoptosis through DNA damage and reactive oxygen species (ROS) generation in human breast cancer cells (Farah et al., 2016; Jeyaraj et al., 2013).

Gul et al. (2021) characterized a rapid, eco-friendly method of biogenic synthesis of silver NPs using *Poa annua*. These NPs demonstrated high antioxidant potential and radical scavenging activity that led to a decrease in the number of inflammatory monocytes and lymphocytes during cancer treatment. The authors touted this alternative as a highly potent biogenic drug, as it demonstrated an increase in drug encapsulation and enhanced drug-loading efficiency in cancer therapy.

Xu et al. (2020) reviewed the anticancer mechanisms of silver NPs synthesized from some plant extracts and concluded that destruction of the ultrastructure of cancer cells, induction of ROS generation and DNA damage, promotion of apoptosis and autophagy are some of the pathways of cancer control executed by the synthesized phyto-silver NPs. The biological performance of phyto-fabricated silver NPs as an anticancer agent is summarized in Table 3.1.

3.3.2 Gold Nanoparticles

The major reasons for using gold NPs for biomedical application, cancer research, and imaging procedure are its ease of synthesis, biocompatibility, and low toxicity (Rad et al., 2011; Iqbal et al., 2016). Biosynthesis of gold NPs from naturally occurring plant extracts has become very popular owing to the presence of flavonoids and polyphenols that act as reducing, stabilizing, and capping agents, accompanied by other advantages such as biocompatibility, cost-effectiveness, and less labor requirement, providing a green alternative for cancer treatment (Keshavamurthy et al., 2018; Li et al., 2016; Quelemes et al., 2013). Gold NPs generally demonstrate aggregation and size-dependent cytotoxic activity against cancer cells, but these inorganic phytosynthesized NPs have shown minimal risks and side effects in noncancer cells (Cui et al., 2012). Phytosynthesized gold NPs have promising anticancer potential against different types of human cancers such as human breast cancer, skin carcinoma, lung cancer, cervical cancer, ovarian cancer, leukemia, liver cancer, and colon cancer (Khan et al., 2019). Green-synthesized gold NPs reportedly generated ROS through oxidative stress and inflammatory pathways (Giljohann et al., 2020), which induced DNA damage and toxicity in cancer cell lines, resulting in apoptosis and cell death (Ashokkumar et al., 2014). Nath et al. (2020) summarized gold NPs prepared from several plants and discussed their anticancer properties on some cell lines. Some examples of gold NPs prepared from plant extracts with anticancer properties are summarized in Table 3.1.

Kajani et al. (2016) described a cost-effective method for preparing gold NPs from *Taxus baccata* extracts and studied their effectiveness on several cancer cell lines including breast (MCF-7), cervical (HeLa), and ovarian (Caov-4) cancer cell lines. Their investigation indicated that caspase-independent death program was the most probable anticancer mechanism of phytosynthesized gold NPs. Vaid et al. (2020) discussed anticancer pathways initiated by green-synthesized gold

NPs affecting breast and lung cancers. The antiproliferative effects of green-synthesized gold NPs against breast cancer cell lines (MDA-MB-231 and MCF-7) were observed with cell shrinkage, blebbing, cell cycle arrest at the G1/G0 and S phases, chromatin fragmentation, and nuclear condensation, eventually leading to apoptosis (Suganya et al., 2016).

Khan et al. (2019) presented an elaborate review discussing plant-based synthesis strategies, mechanistic actions, and applications of phyto-gold-NPs in cancer therapy. A very interesting mode of antiproliferative activity of gold NPs synthesized from *Moringa oleifera* was demonstrated to be through the modulation of oncogenes, tumor suppressor genes, and caspase-9 splice variants in A549 lung cancer cells (Tiloke et al., 2016). Gold NPs conjugated with curcumin, the main bioactive component of *Curcuma longa*, induced significant cytotoxicity and morphological alterations in cervical cancer cell line (HeLa cell) and promoted apoptosis and antimigratory effect on these cells. Because of the luminescent property of curcumin in this formulation, tracking and bioimaging during chemotherapy were facilitated by this approach (Govindaraju et al., 2018).

3.3.3 Copper Nanoparticles

Copper NPs synthesized from various plant extracts are cytotoxic against many kinds of cancers. Deshmukh and Beom (2020) reviewed synthetic procedures, characterization, and anticancer and cytotoxic activities of several phytosynthesized copper NPs against hepatic, breast, and kidney cancer cells. Copper oxide NPs synthesized from *Phaseolus vulgaris* were cytotoxic to cervical cancer cells through the activation of apoptotic cell death pathways initiated by intracellular ROS (Nagajyothi et al., 2017). According to Rehana et al. (2017a), copper oxide NPs synthesized using different leaf extracts were cytotoxic to MCF-7 breast cancer cells and lung cancer, with apoptotic cell death showing the characteristic features of cell shrinkage, membrane blebbing, cytoplasmic vacuolation activation of caspases, and fragmentation of the nucleus. Copper NPs synthesized from *Ficus religiosa* leaf extract showed cytotoxic activity on human lung cancer cells, possibly by the induction of apoptotic cell death pathway through ROS generation and mitochondrial depolarization of the cells (Sankar et al., 2014). Some plant-based CuNPs and their anticancer activity are listed in Table 3.1.

3.3.4 Zinc Nanoparticles

Zinc NPs are mainly used in their oxide form, that is, zinc oxide (ZnO). ZnO NPs when synthesized from *Eclipta prostrata* extracts demonstrated significant cytotoxicity on human hepatocellular carcinoma (HepG2) cells by caspase-mediated apoptosis (Chung et al., 2015) as was the case of those synthesized from *Pandanus odorifer* leaf extract (Hussain et al., 2019). Cytotoxicity of MCF-7 breast cancer cells by ZnO NPs synthesized from *Gymnema sylvestre* extracts was reportedly induced by ROS, matrix metalloproteinase injury, and apoptotic morphological changes (Bangroo et al., 2021). Similarly, apoptosis induction through enhanced oxidative stress and mitochondrial damage leading to significant cytotoxicity in MCF-7 cells was observed by ZnO NPs prepared from *Cynara scolymus* (Rajapriya et al., 2020) and from *Leea asiatica* (Ali et al., 2021). Evaluation of green-synthesized ZnO NPs using *Alhagi maurorum* demonstrated dose-dependent antitumor activity against several osteosarcoma cell lines such as G-292, HOS, MG-63, Saos-2, Hs 707 (A), and clone A141B1, through augmentation of ROS, eventually leading to the apoptotic signaling pathway (Chinnathambi and Alahmadi, 2021). Anjum et al. (2021), in their study, summarized some of the green-synthesized plant-based zinc oxide NPs, in terms of their physical properties as well as anti-cancer and cytotoxic potential on the liver, breast, lung, bone, and colon cancer cell lines. The authors revealed that the mechanism underlying the cytotoxic activity of these zinc NPs involved the induction of ROS, downregulated activity of matrix metalloproteinases, and intense morphological alteration triggering apoptosis.

3.3.5 Iron Nanoparticles

Ever since the first report on tomato-based synthesis of iron oxide NPs (Fe_2O_3-NP) that exhibited anticancer activity against human lung carcinoma cell line (A549) in a dose-dependent manner along with apoptosis (Bharathi et al., 2020), some very intriguing reports have been published. Shejawal et al. (2021) synthesized iron NPs using lycopene extracted from tomato and evaluated their cytotoxic potential against colorectal and cervical cancer cell lines (COLO320DM, HT29, and HeLa). Studies on the phytofabrication of α-Fe_2O_3 NPs and their promising cancer-treating property have been discussed by Al-Hakkani et al. (2021). According to Naz et al. (2019), green synthesis of Fe_2O_3 NPs from the *Rhus punjabensis* extract exhibited inhibitory activity against various cancer cell lines such as HL-60 and DU-145, depicting their potential as a chemo-preventive agent against human promyelocytic leukemia and prostate cancer. The mechanism of action of these NPs includes inhibition of the transcription factor nuclear factor-κB, which further inhibits the cascade of cellular reactions involved in cancer.

3.3.6 Cerium Nanoparticles

Cerium oxide NPs (CeO-NP) synthesized from the *Origanum majorana* leaf extract were evaluated for their cytotoxic effects in thin human breast carcinoma cells (MDA-MB-231 cell line), and they appeared to be a promising therapeutic agent for breast cancer (Aseyd Nezhad et al., 2020). These phytofabricated cerium NPs enhanced ROS, thereby inducing significant morphological changes such as plasma membrane blebbing, cell shrinkage, and destruction. CeO_2 NPs doped with bimetal Ag-Au were synthesized from *Justicia adhatoda* by Nithya and Sundrarajan (2020). The authors tested the biosynthesized CeO_2 NPs against human adenocarcinoma epithelial (HeLa) cancerous cell line and reported dose-dependent significant anticancer activity. As for the mechanisms underlying the anticancer activity, the bimetallic CeO-NPs bind to cancer cells and release Ag^+, Au^+, and Ce^{4+}, which disrupt the cancer cells. In addition, these cerium NPs enhanced the formation of ROS, resulting in cancer cell death.

3.3.7 Selenium Nanoparticles

Green-synthesized selenium NPs (SeNPs) exhibited anticancer properties. Recent reports on the conditions during NP synthesis, such as selenium precursor, temperature, time, and extract concentration, combined with characteristics of the synthesized phytonanocomplex such as size, morphology, stability, and composition, have been reviewed by Pyrzynska and Aleksandra (2021). The authors also summarized several methods that were proposed for synthesizing plant-based selenium NPs along with reports of anticancer activity. Ikram et al. (2021) reviewed the fabrication methodologies for plant-based selenium NPs with a focus on possible mechanisms of anticancer and cytotoxic activity of the synthesized NPs on human cancer cell lines such as human cervical carcinoma, lung cancer, and liver cancer cells. Anu et al. (2020) elaborated the biogenesis of selenium NPs using *Cassia auriculata* and their anti-leukemic activity. Cui et al. (2018) reported that selenium NPs prepared from hawthorn fruit extract induced apoptosis in the human hepatocellular carcinoma cell line (HepG2) through the mitochondrial pathway by promoting ROS.

3.3.8 Platinum Nanoparticles

A detailed review on the green synthesis approaches for the preparation of phytofabricated platinum NPs, reaction parameters, physical characteristics of the synthesized NPs, and anticancer activity on various cell lines such as HeLa, SiHa, A549, MCF-7, HCT116, MDA-MB-231, SW480, and SW620 was carried out by Fahmy et al. (2020a). Jan et al. (2021) summarized the green synthesis of platinum NPs from various plant sources and their potential medical applications. Biogenic platinum NPs prepared from black cumin seed extract (*Nigella sativa*) displayed significant cytotoxicity toward MDA-MB-231 and HeLa cancer lines as the concentrations of the NPs increased; hence, they can serve as an effective anticancer agent (Aygun et al., 2020). The mode of action of these phytofabricated platinum NPs involves the formation and accumulation of ROS and downregulation of DNA protein kinase, leading to loss of cell integrity and finally apoptosis. Ghosh et al. (2015) reported the biosynthesis of platinum-palladium bimetallic NPs from *Dioscorea bulbifera.* They demonstrated an evident enhancement of anticancer activity of bimetallic Pt-PdNPs as compared with that of individual PtNPs or PdNPs through the production of ROS and induction of apoptosis in HeLa cells. The authors contended that as these phytosynthesized NPs exhibited both anticancer and antioxidant properties, this design could be of dual purpose of preventing cancer progression owing to its antioxidant properties and inducing apoptosis and cell death owing to their cytotoxic potential against cancer cells.

3.3.9 Palladium Nanoparticles

Among the most recent studies on the phytosynthesis of palladium NPs, factors influencing their synthesis and their promising chemotherapeutic application have been discussed by Fahmy et al. (2020b). The group also elaborated the synthesis mechanism and characterized phyto-fabricated palladium NPs. A green synthesis approach for palladium NPs from the *Evolvulus alsinoides* leaf extract was reported by Gurunathan et al. (2015). Their study demonstrated caspase-dependent apoptotic potential of the green-synthesized palladium NPs in ovarian cancer (A2780) cells. As the NPs suppressed cell viability, elevated ROS generation, and finally induced autophagy in ovarian cancer cells, the study suggested that these green-synthesized phytonanoparticles could serve as effective therapeutic agents for cancer. Green synthesis of palladium-platinum bimetallic NPs prepared from the tuber extract of *Gloriosa superba* was reported by Rokade et al. (2018). Their anticancer mechanism against human breast adenocarcinoma MCF-7 cell line revealed induction of apoptosis characterized by phosphatidyl serine externalization, membrane breakdown, and blebbing with chromosome condensation. A detailed review by Phan et al. (2019) included the biogenic synthesis strategies of plant-based palladium NPs and their potential anticancer applications as a DDS and a radiotherapeutic agent.

3.3.10 Other Metal Nanoparticles

Phytomediated synthesis of green titanium oxide (TiO_2) NPs from the root extract of *Withania somnifera* demonstrated significant reduction in the viability of HepG2 *in vitro* by enhancing the production of ROS triggering apoptosis and thus could prove effective in controlling liver cancer (Al-Shabib et al., 2020). Lead oxide (PbO) NPs synthesized from *Papaver somniferum* were reported to generate oxidative stress, leading to cell cycle arrest and redistributing the cell cycle generation, thereby exerting anticancer activity on human hepatocellular cell line (HepG2) (Muhammad et al., 2019). Detailed studies on the anticancer pathways adopted by many other phyto-fabricated metal NPs and their clinical trials are still ongoing.

3.3.11 Nanoparticles as Carriers of Anticancer Drugs

Nanocarriers such as nanomicelles, nanotubes, and liposomes are some popular approaches of nanotechnology-based DDSs and have gained immense interest in cancer therapy owing to their small size for effective tumor penetration and high drug-loading potency (Schroeder et al., 2012; Wei et al., 2015; Cui et al., 2016). Li et al. (2019) developed a phyto-nano-complex using green tea polyphenols, (–)-epigallocatechin-3-gallate (EGCG), and lanthanide metal ions (Sm^{3+}) to form

SmIII-EGCG nanocomplexes and characterized these particles; they showed enhanced tumor inhibitory properties by regulating metabolic pathways. These nanocomplexes were used to study the therapeutic effects on both subcutaneous melanoma and metastatic lung melanoma in a mouse model. The combination of green tea-based EGCG building blocks and lanthanide ions showed greater anticancer effects on melanoma cells than the individual components and promoted targeted apoptosis of tumor cells through mitochondrial dysfunction.

Elbialy et al. (2019) developed an eco-friendly nanocarrier to deliver curcumin by reducing gold NPs using curcumin, which acts as a stabilizing agent and an anticancer agent. The *in vitro* study concluded that these gold NP-curcumin complexes have higher antiproliferative and apoptotic effects against colon cancer cells (HCT-116) and breast cancer cells (MCF-7) than free curcumin. Such a nanocarrier for curcumin greatly enhanced its antitumor activity, indicating the promising utilization of the nutraceutical nanoformulation in cancer therapy. Zhang et al. (2016a) synthesized NPs from ginger and modified their lipids into ginger-derived nanovectors. These nanovectors can be effectively taken up by colon cancer cells, leading to a decrease in cell proliferation and an increase in apoptosis. The ginger derived nanovectors, when conjugated with the targeting ligand folic acid, could deliver doxorubicin to colon tumors, thus inhibiting colon tumor growth.

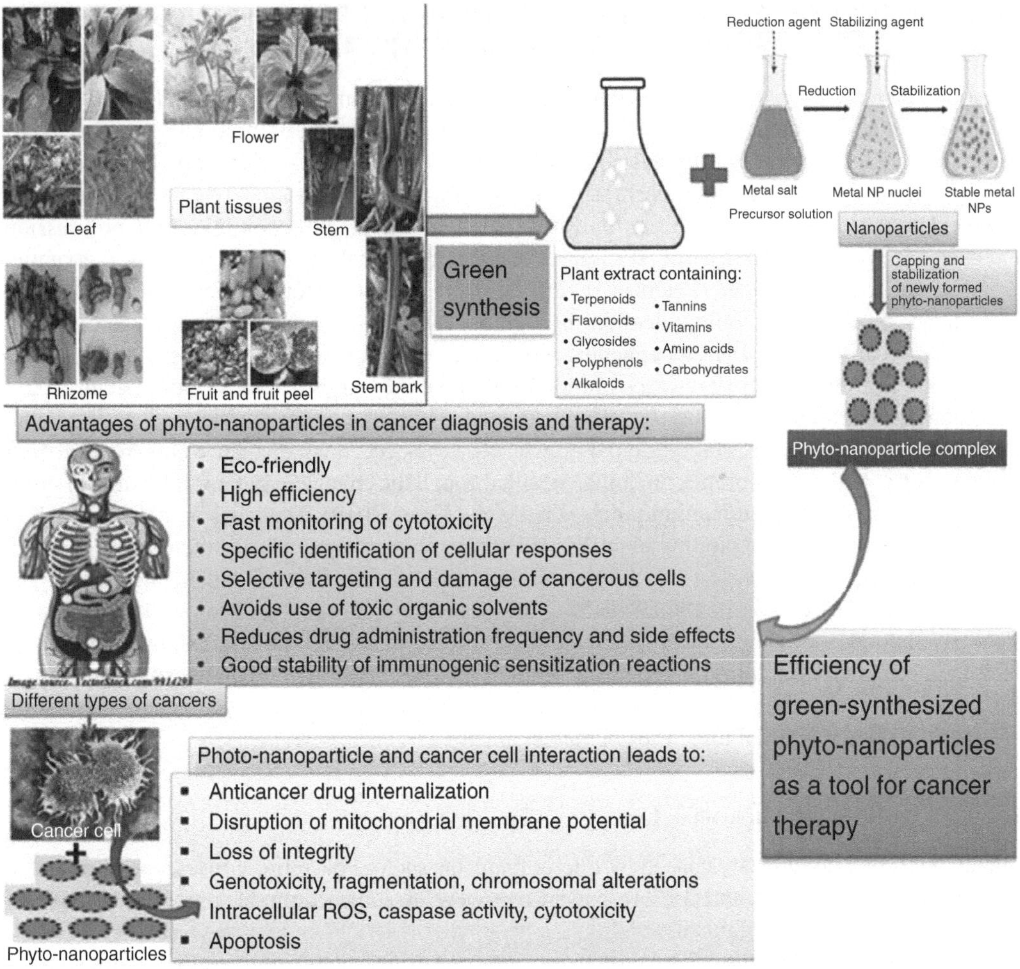

FIGURE 3.3 Schematic illustration of green synthesis and efficacy of the synthesized phytonanoparticles as a tool in cancer therapy.

Green synthesis of NPs has gained immense popularity, as this approach has led to a revolutionary alternative from chemical and physical synthetic methods by exercising environment friendly, cost, and labor-effective ways. Studies have already revealed the significant anticancer and cytotoxic potential of phyto-fabricated metal NPs, and further investigation on their underlying mechanism of action in cancer therapy is still ongoing. On the contrary, some limitations of this approach involve the formation of undesirable resultant nanomaterials initially until the preparatory technique is standardized. There might also be a problem if the concentration of the bioactive constituent in the selected plant extract and that of the NPs are disproportionate. Hence, extensive study is required on the preparation, identification, and characterization of phytonanoparticles. Increasing implementation of this concept can enable the discovery of new properties of the plant extracts used as well as the NP conjugate that will lead to rapid development in cancer research. In future prospects, more clinical trials using plant-based NPs to treat different cancers should be conducted so that wider avenues can open up for pharmaceutical development and clinical cancer therapy.

Figure 3.3 and Table 3.1 summarize the important points discussed in the previous sections.

Table 3.1 and Figure 3.3 outline the plant sources and metal NP conjugates with anticancer application.

3.4 NANOPARTICLES IN THE DELIVERY OF PLANT THERAPEUTICS FOR THE TREATMENT OF MICROBIAL DISEASES IN HUMANS

Infectious diseases have commonly occurred as epidemics or pandemics worldwide. Early microscopic studies of the human epithelial tissues revealed the colonization of multiple species in microbial communities. Some microorganisms have evolved to utilize human metabolic byproducts and thereby establish synergistic relationships, while some have shown competitive antagonistic strategies and co-colonization. How this colonization actually governs infection initiation, progression, and even severity of human microbial diseases remains unclear thus far, and targeted intervention strategies cannot be developed (Peters et al., 2012). However, very recent studies highlight the intriguing facts of cell-to-cell communication through quorum sensing, enhancement of colonization, and formation of augmented virulent phenotype(s) of microbial species as being important causes for the etiology as well as severity in cases of mono-microbial infectious diseases (Sartini et al., 2021).

Systemic microbial infections are usually treated with chemically synthesized antimicrobial compounds that exert either static or cidal effects. However, recent studies have shown negative or adverse effects of these antibiotics on human health through the change in gut microbiota and, simultaneously, compromising human immunity (Zhang and Chen, 2019). Thus, researchers have started gradually shifting toward exploiting plant-based antimicrobial compounds as potential drugs for the treatment of microbial diseases, but despite minimal side effects, these phytochemicals have certain limitations. There is limited bioavailability of these drugs; moreover, owing to the bulk size of plant-based drugs, they are unable to reach the site of infection; even if the drugs somehow reach the sites, their retentivity is minimum, rendering them incapable of treating human microbial infections. For this reason, green nanotechnology and nano-DDS are employed to deliver therapeutics for the treatment of these microbial infections (Conte et al., 2017).

3.4.1 Overview of Microbial Diseases in Humans

Human microbial diseases are caused by microorganisms such as bacteria, viruses, fungi, protozoa, and helminths. Tables 3.2 and 3.3 present an overview of some common infectious diseases in humans caused by microorganisms.

This section discusses the efficacy of the green-synthesized metal NPs against various mono-microbial diseases, namely, viral, bacterial, fungal, protozoan, and helminthic diseases, as well as polymicrobial diseases. A special focus has been paid to the nano-DDS to deliver plant drugs or

TABLE 3.1
Plant-Derived Metal Nanoparticles and Their Activity Against Various Cancers

Nanoparticle	Plant/Parts Used	Cancer Type and Cell Line	Refs
Silver	*Abutilon indicum*; leaf	Colon: COLO 205	Mata et al., 2015
	Acorus calamus; rhizome	Lung: A431	Nayak et al., 2015
	Adenium obesum; leaf	Breast: MCF-7	Farah et al., 2016
	Alternanthera sessilis; leaf	Prostate: PC3	Firdhouse and Lalitha, 2013
	Alternanthera tenella; leaf	Breast: MCF-7	Sathishkumar et al., 2016
	Azadirachta indica; leaf	Skin: HDFa; Lung: NCI-H460	Kummara et al., 2016
	Balanites aegyptiaca; fruit	Colon: Caco-2; Liver: HepG2	Yassin et al., 2017
	Bambusa arundinacea; leaf	Prostate: PC3	Kalaiarasi et al., 2015
	Calotropis gigantea; latex	Breast: MCF-7; Lymphoblastic leukemia: Jurkat	Maity et al., 2018
	Cassia fistula: flower	Breast: MCF-7	Remya et al., 2015
	Dimocarpus longan; fruit peel	Prostate: PC3	He et al., 2016
	Ficus krishnae; stem and bark	Ovary: SKOV-3	Kanjikar et al., 2018
	Gelsemium sempervirens; whole plant	Skin: A375	Das et al., 2013b
	Hydrastis canadensis; whole plant	Skin: A375	
	Justicia adhatoda; leaf	Cervical: HeLa	Kudle et al., 2014
	Moringa oleifera; stem and bark	Cervical: HeLa	Vasanth et al., 2014
	Phoenix dactylifera; pollen, root hair	Breast: MCF-7	Banu et al., 2018; Oves et al., 2018
	Phyllanthus emblica; fruit	Laryngeal carcinoma: Hep2	Rosarin et al.,2013
	Phytolacca decandra; whole plant	Skin: A375	Das et al., 2013b
	Punica granatum; fruit peel	Liver: HepG2; Lung: A549; Colon: HCT116	Padinjarathil et al., 2018
	Rheum rhabarbarum; Stem	Cervical: HeLa	Reddy et al., 2016
	Rosa indica; petal	Colon: HCT-15	Manikandan et al., 2015
	Rosmarinus officinalis; leaf	Leukemia: HL-60	Sulaiman et al., 2013
	Sesbania grandiflora; leaf	Breast: MCF-7	Das et al., 2013a
	Styrax benzoin; leaf	Lung: A549	Du et al., 2016
	Thuja occidentalis; whole plant	Skin: A375	Das et al., 2013b
	Zingiber zerumbet; rhizome	Kidney: Vero	Azizi et al., 2016
Gold	*Abies spectabilis*; aerial parts	Bladder: T24	Wu et al., 2019
	Acalypha indica; leaf	Breast: MDA-MB-231	
	Alternanthera bettzickiana; leaf	Lung: A549	Nagalingam et al., 2018
	Anacardium occidentale; leaf	Breast: MCF-7	Sunderam et al., 2019
	Antigonon letopus; aerial part	Breast: MCF-7	Balasubramani et al., 2015
	Areca catechu; nuts	Cervical: HeLa	Rajan et al., 2015
	Azadirachta indica; leaf	Cervical: HeLa; Kidney: MDCK	Dharmatti et al., 2014
	Backhousia citriodora; leaf	Breast: MCF-7; Liver: HepG2	Khandanlou et al., 2018

(Continued)

TABLE 3.1 (Continued)
Plant-Derived Metal Nanoparticles and Their Activity Against Various Cancers

Nanoparticle	Plant/Parts Used	Cancer Type and Cell Line	Refs
	Cassia tora; leaf	Colon: Col320	Abel et al., 2016
	Cajanus cajan; seed coat	Liver: HepG2	Ashokkumar et al., 2014
	Centaurea behen; leaf	Leukemia: THP-1	Abdoli et al., 2021
	Coleus aromaticus; leaf	Liver: HepG2	Boomi et al., 2019
	Couroupita guianensis; flower	Blood: HL-60	Geetha et al., 2013
	Curcumae kwangsiensis; leaf	Ovary: Sw-626, SK-0V, PA1	Chen et al., 2021
	Ferula persica; gum	Colon: CT26	Hosseinzadeh et al., 2020
	Ficus religiosa; leaf	Kidney: HEK239	Wani et al., 2013
	Genipa americana; leaf	Lung: A549; Cervical: HeLa	Kumar et al., 2016
	Hibiscus sabdariffa; stem and leaf	Glioblastoma: U87, HEK 293	Mishra et al., 2016
	Jasminum auriculatum; leaf	Cervical: HeLa	Balasubramanian et al., 2020
	Linum usitatissimum; seed	Breast: MCF-7	Al-Radadi, 2021
	Magnolia officinalis; leaf	Lung: A549	Zheng et al., 2019
	Mentha longifolia; leaf	Breast: HS319T, MCF-7, UACC-3133	Li et al., 2020
	Mimosa pudica; leaf	Breast: MDA-MB-231, MCF-7	Suganya et al., 2016
	Persicaria salicifolia; leaf	Breast: MCF7	Hosny and Fawzy, 2021
	Sesuvium portulacastrum; leaf	Lung: A549	Ramalingam et al., 2016
	Vites vinifera; fruit	Breast: HBL-100	Amarnath et al., 2011
Zinc	*Abutilon indicum*; aerial parts	Cervical: HeLa	Kulkarni et al., 2016
	Albizia lebbeck; stem bark	Breast: MCF-7	Umar et al., 2019; Loganathan et al., 2021
	Bergenia ciliata; leaf	Blood: HCT 116, K562	Ahlam et al., 2021
	Bergenia ciliate; rhizome	Colon: HT-29; Cervical: HeLa	Dulta et al., 2021
	Cucumis melo-inodorus; shell	Breast: MCF-7	Mahdizadeh et al., 2019
	Deverra tortuosa; aerial parts	Colon: Caco-2; Lung: A549	Selim et al., 2020
	Enicostema axillare; leaf	Lung: A549	Perumal et al., 2021
	Geranium wallichianum; leaf	Liver: HepG2	Abbasi et al., 2019
	Holigarna grahamii; leaf	Renal: A498	Lokapur et al., 2022
	Mangifera indica; leaf	Lung: A549	Rajeshkumar et al., 2018
	Marsdenia tenacissima; leaf	Laryngeal: Hep-2	Wang et al., 2019a
	Pongamia pinnata; seed	Breast: MCF-7	Malaikozhundan et al., 2017
	Raphanus sativus; leaf	Lung: A549	Umamaheswari et al., 2021
	Santalum album; leaf	Breast: MCF-7	Kavithaa et al., 2016
	Spondias pinnata; rhizome	Colon: HT-29, HCT-116	Dulta et al., 2021
	Withania somnifera; root	Breast: MCF-7	Prasad et al., 2021
Platinum	*Barleria prionitis*; leaf	Breast: MCF-7	Rokade et al., 2017
	Camellia sinensis; leaf	Cervical: Si-Ha	Alshatwi et al., 2015
	Gloriosa superb; tuber	Breast: MCF-7	Rokade et al., 2018
	Green tea; powder	Colon: SW480, SW620	Depciuch et al., 2020
	Ononis spinose; radix	Lung: A549	Dobrucka et al., 2019
	Maytenus royleanus; leaf	Lung: A549	Ullah et al., 2017
	Mentha piperita; leaf	Colon: HCT 116	Yang et al., 2017 (a)

TABLE 3.1 (Continued)
Plant-Derived Metal Nanoparticles and Their Activity Against Various Cancers

Nanoparticle	Plant/Parts Used	Cancer Type and Cell Line	Refs
	Phoenix dactylifera; fruit	Liver: HepG2; Breast: MCF-7; Colon: HCT	Al-Radadi., 2019
	Punica granatum; fruit peel	Breast: MCF-7	Sahin et al., 2018
Palladium	*Bauhinia variegate*; bark	Breast: MCF-7	Vaghela et al., 2018
	Couroupita guianensis; fruit	Lung: A549	Gnanasekar et al., 2018
	Dioscorea bulbifera; tuber	Cervical: HeLa	Ghosh et al., 2015
	Moringa oleifera; flower	Lung: A549	Anand et al., 2016
	Urtica sp.; leaf	Pancreatic: MIA PaCa-2; Colon: HT-29; Breast: MDA-MB-231	Gulbagca et al., 2021
Selenium	*Asteriscus graveolens*; leaf	Liver: HepG2	Zeebaree et al., 2020
	Carica papaya; latex	Breast: MCF-7	Rajasekar and Kuppusamy, 2020
	Crataegus hupehensis; fruit	Liver: HepG2	Cui et al., 2018
	Enicostema axillare; leaf	Lung: A549	Perumal et al., 2021
	Moringa oleifera; leaf	Rectal: Caco-2; Liver: HepG2; Breast: MCF-7	Hassanien et al., 2019
	Urtica dioica; leaf	Liver: HepG2	Hashem and Salem, 2022
	Withania somnifera; leaf	Lung: A549	Alagesan and Venugopal, 2019
Iron	*Carica papaya*; leaf	Kidney: BHK-21; Cervical: HeLa	Bhuiyan et al., 2020
	Cocos nucifera; pulp	Liver: HepG2	Rajendran et al., 2021
	Papaver somniferum; pod	Liver: HepG2	Muhammad et al., 2019
	Punica granatum; fruit peel	Nasopharyngeal : HONE1	Yusefi et al., 2020
	Rhamnus virgata; leaf	Liver: HepG2	Iqbal et al., 2020
Cerium	*Brassica napus*; pollen	Ovary: A2780	Ashna et al., 2020
	Bryophyllum daigremontianum; leaf	Breast: MCF-7	Sridharan et al., 2021
	Ceratonia siliqua; leaf	Breast: MCF-7	Javadi et al., 2019
	Salvadora persica; bark	Colon: HT-29	Miri et al., 2019
Copper	*Acalypha indica*; leaf	Breast: MCF-7	Sivaraj et al., 2014
	Azadirachta indica; leaf	Breast: MCF-7; Cervical: HeLa; Liver: Hep-2; Lung: A549	Rehana et al., 2017a
	Brassica oleracea; whole plant	Prostate: PC-3	Prasad et al., 2016
	Calotropis procera; latex	Cervical: HeLa; Lung: A549; Kidney: BHK21	Harne et al., 2015
	Duchesnea indica; leaf	Renal: A-498	Xue et al., 2018
	Eclipta prostrata; leaf	Liver: HepG2	Chung et al., 2017
	Hibiscus rosa-sinensis; leaf	Breast: MCF-7; Cervical: HeLa; Liver: Hep-2; Lung: A549	Rehana et al., 2017a

(Continued)

TABLE 3.1 (Continued)
Plant-Derived Metal Nanoparticles and Their Activity Against Various Cancers

Nanoparticle	Plant/Parts Used	Cancer Type and Cell Line	Refs
	Moringa oleifera; leaf	Breast: MCF-7; Cervical: HeLa; Liver: Hep-2; Lung: A549	
	Murraya koenigii; leaf	Breast: MCF-7; Cervical: HeLa; Liver: Hep-2; Lung: A549	
	Olea europaea; leaf	Ovarian: SKOV-3; Breast: AMJ-13	Sulaiman et al., 2018
	Prosopis cineraria; leaf	Breast: MCF-7	Jinu et al., 2018
	Syzygium aromaticum; bud	Prostate: PC-3	Rajesh et al., 2018
	Tamarindus indica; leaf	Breast: MCF-7; Cervical: HeLa; Liver: Hep-2; Lung: A549	Rehana et al., 2017a
	Tilia; leaf	Colon: Caco-2; Hepatic: HepG2; Breast: MCF-7	Hassanien et al., 2018
Titanium	*Coleus aromaticus*; leaf	Cervical: HeLa	Narayanan et al., 2021
	Glycyrrhiza glabra; root	Pharyngeal: HEP2; Kidney: Vero	Bavanilatha et al., 2019
	Nephelium lappaceum; fruit peel	Breast: MDA-MB-231	Isacfranklin et al., 2020

therapeutics to treat microbial diseases. These promising pharmaceutical technologies might provide current insights to the treatment of human microbial diseases. Figure 3.4 schematically outlines the antimicrobial activity of phytonanoparticles.

3.4.2 Green Nanotechnology in Combating Human Microbial Diseases

In the last 5 years, bioinspired fabrication of NPs using microorganisms and plants has been extensively documented. Plant-based NP synthesis has garnered extensive interest owing to easy availability, eco-friendly synthesis, low energy consumption, and, most importantly, its nontoxic nature (Rajkumari et al., 2019). Aqueous extracts or ethanolic extracts of different plant parts, including leaves, roots, stems, fruits, and seeds have been used for metal NP synthesis. The basic principle of the technology is that the phytoconstituents in the extract act as reducing and stabilizing agents for nontoxic NP production (Al-Shabib et al., 2020). These *plant-based metallic NPs*, or simply known as *phytonanoparticles*, were tested against and found to be highly efficient in controlling multiple microorganisms, exerting either cidal or static activities.

Various materials and structures have been used as nanocarriers for either passive or active targeting. Iron oxide or gold nanosized assemblies, for example, are subjected to surface modifications to act as drug carriers. However, organic materials are more versatile and are much more suited for application in the *in vivo* systems. Lipid-based nanoparticles such as liposomes and amphiphilic molecules can encapsulate hydrophobic drugs and deliver them to target sites. By contrast, polymers provide more versatility in nano-DDS technology. The success in the nano-drug delivery of plant therapeutics to treat various human microbial infectious diseases is opening new vistas in medicine (León-Buitimea et al., 2021).

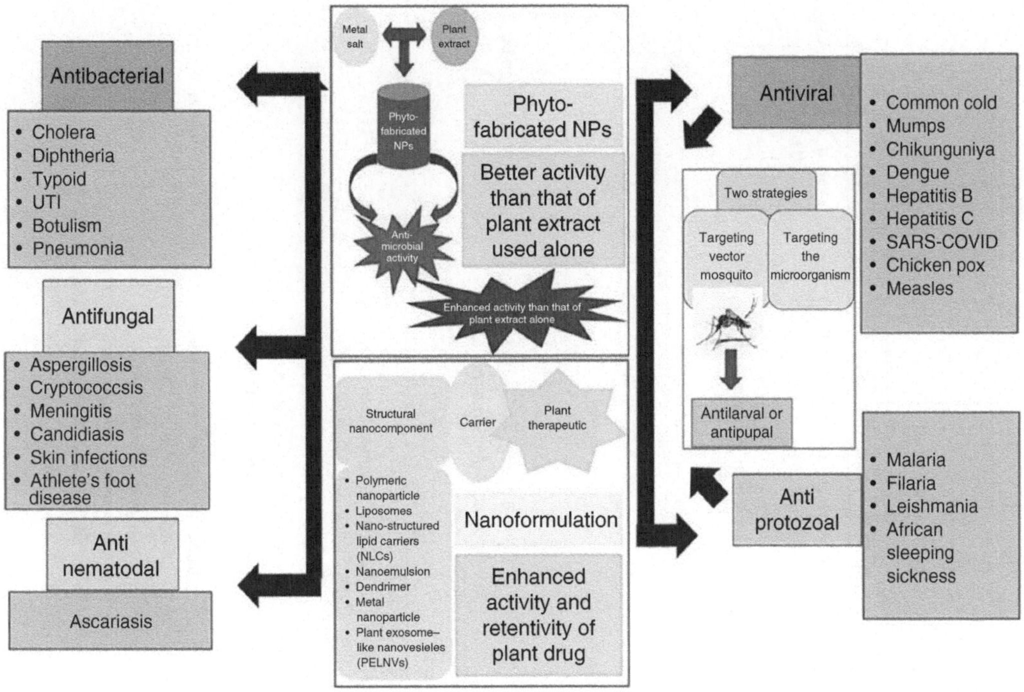

FIGURE 3.4 Efficacy of plant-based metal nanoparticles and nanoformulations in mitigating human microbial diseases.

3.4.3 Bacterial Diseases

Conventional drug targets against bacterial diseases act by inhibiting either cell wall synthesis by hydrolyzing the peptidoglycan layer (Sartini et al., 2021), suppressing nucleic acid synthesis by targeting proteins involved in DNA and RNA synthesis, or targeting 30S and 50S bacterial ribosomal subunits and different steps of protein synthesis. They can also act by modification or disruption of cell membrane function or by inhibiting different biosynthesis or metabolic pathways that are essential for bacterial survival. However, the development of multidrug resistance in bacteria critically undermines the efficiency of these conventional antimicrobial agents.

In a bid to develop drugs with novel antibacterial activity, phytonanoparticles have attracted intense scientific attention in the last few years for their unique ability to remodel bulk compounds into smaller nanosized particles. Because of their nano size (range: 1–100 nm), they can easily reach the site of microbial infection and can exert their effect against the pathogen, thereby exerting better activity than the plant extract or phytochemical compound alone. Details of the published reports on nanoparticles synthesized from plants with antibacterial activity are included in Table 3.2.

3.4.4 Fungal Diseases

Fungal diseases kill more than 1.5 million people and affect more than a billion people yearly worldwide. According to the degree of interactions between the pathogen and the host tissue, these diseases are broadly classified as superficial, subcutaneous, and systemic infections (Tiew et al., 2020). Superficial mycoses, estimated to occur in 25% of the world's population, are the most common form of fungal infection. Systemic mycosis, however, is most deadly, as it is usually associated with a high mortality rate, significant morbidity with limited chemotherapeutic options, and the diagnosis is frequently difficult and complex (Brunet et al., 2018).

TABLE 3.2
Efficacy of Phytonanoparticles in Combating Bacterial and Fungal Diseases

Nanoparticle	Plant Source or Phytochemical	Organisms Tested	Antimicrobial Activity	Refs
		Antibacterial activity		
Silver	Aqueous extracts of *Acacia arabica* (Arabic gum) and *Opophytum forsskalii* (Samh.) seed	*Staphylococcus aureus* (gram-positive bacterium) and *Pseudomonas aeruginosa* and *Escherichia coli* (gram-negative bacteria)	Effective against minor skin infections caused by *Staphylococcus* sp. and urinary tract infections caused by *E. coli*	Aabed and Mohammed, 2021
	Stem extract of *Caesalpinia pulcherrima*	Gram-positive bacteria (*Bacillus cereus*, *Staphylococcus aureus*, *Corynebacterium rubrum*, *Bacillus subtilis*), as well as gram-negative bacteria (*Escherichia coli*, *Klebsiella pneumoniae*, *Pseudomonas aeruginosa*, *Salmonella* typhimurium)	Potent antibacterial activity	Moteriya and Chanda, 2018
	Seed extract of mango	*Staphylococcus aureus*, *Corynebacterium rubrum*, *Bacillus subtilis*, *Bacillus cereus*, *Pseudomonas aeruginosa*, *Klebsiella pneumoniae*, *Escherichia coli*, *Salmonella* typhimurium	Effective against diseases such as typhoid and pneumonia	Donga and Chanda, 2021
	Bark extract of *Picea abies* and *Pinus nigra*	Methicillin-susceptible *Staphylococcus aureus*, methicillin-resistant *Staphylococcus aureus*, *Streptococcus pyogenes*, and gram-negative (*Escherichia coli*, *Pseudomonas aeruginosa*)	Effective against diseases such as fever, pneumonia, septicemia, urinary tract infection, minor skin and soft tissue infections such as abscesses (boils), furuncles, and cellulitis	Macovei et al., 2021
	Leaf extracts of *Martynia annua*	*Escherichia coli*, *Salmonella* typhimurium, *Vibrio cholerae* *Pseudomonas aeruginosa* and *Pseudomonas mirabilis*	Effective against urinary tract infection, typhoid, cholera, malignant external otitis, endocarditis, pneumonia endophthalmitis, meningitis, and septicemia	Sivakumar et al., 2018

		Salmonella typhimurium and *Brucella abortus*	Effective against brucellosis (fever, sweat, headache, back pain, and physical weakness)	Aabed and Mohammed, 2021
	Azadirachta indica and *Ocimum sanctum* leaf extract	*Escherichia coli*, *Klebsiella pneumoniae*, and *Staphylococcus aureus*	Effective against diseases such as urinary tract infection, pneumonia, and skin infections including cellulitis	Priyadarshini et al., 2019
	Carissa carandas (Karonda) berry water extract, *Origanum vulgare* (Oregano), *Coptidis* rhizome, and *Salicornia bigelovii* extract	Effective against various gram-positive and gram-negative bacteria	Promising antibacterial activity	Augustine and Hasan, 2020
	Nigella sativa seed extract as well as synthesis of nanoparticles from essential oils of *Nigella sativa*	Effective against various species of bacteria forming biofilm in gut microbiota as well as in food	Antibiofilm activity	Balyan et al., 2021
	Seed extract of *Durio zibethinus*	*Salmonella* typhi, *Salmonella* typhimurium, *Escherichia coli*, *Staphylococcus aureus*, *Staphylococcus haemolyticus*, *Bacillus subtilis*	Antimicrobial activity	Sumitha et al., 2018
Iron oxide	Leaf extract of *Cynometra ramiflora*	Effective against gram-positive and gram-negative bacteria	Antibacterial property	Groiss et al., 2017
	Fruit extract of *Terminalia bellirica* and leaf extract of *Moringa oleifera*	Effective against gram-positive and gram-negative bacteria	Antibacterial property	Jegadeesan et al., 2019
Zinc oxide		*Salmonella* typhimurium and *Brucella abortus*	Effective against brucellosis (fever, sweat, headache, back pain, and physical weakness)	Mohammad et al., 2021
	Nigella sativa	*Chromobacterium. violaceum*, *Pseudomonas aeruginosa*, *Listeria monocytogenes*, and *Escherichia coli*	Zinc nanoparticles reported to inhibit biofilm formation of four foodborne pathogens: *Chromobacterium. violaceum*, *Pseudomonas aeruginosa*, *Listeria monocytogenes*, and *E. coli*; effective against diseases such as botulism	Soltane et al., 2021

(*Continued*)

TABLE 3.2 (Continued)
Efficacy of Phytonanoparticles in Combating Bacterial and Fungal Diseases

Nanoparticle	Plant Source or Phytochemical	Organisms Tested	Antimicrobial Activity	Refs
		Antibacterial activity		
Titanium dioxide		*Salmonella* typhimurium and *Brucella abortus*	Effective against brucellosis (fever, sweat, headache, back pain, and physical weakness)	Mohammad et al., 2021
	Rhizome extract of *Gloriosa superba*	*Staphylococcus epidermidis* and *Pseudomonas aeruginosa*	Antibiofilm activity	Mahendran et al., 2021
Platinum (Ns-PtNPs)	*Nigella sativa*	Many gram-positive and gram-negative bacteria	Antibacterial activity	Soltane et al., 2021
Gold (NsEO-AuNPs)	*Nigella sativa*	Effective against biofilm-forming bacteria, namely, *Staphylococcus aureus* and *Vibrio harveyi*	Antibiofilm activity	Soltane et al., 2021
Palladium-silver bimetallic nanoparticles (Pd-Ag)	Extract of *Nigella sativa* seeds	Methicillin-resistant *Staphylococcus aureus*, *Bacillus subtilis*, and *Escherichia coli*	Antibacterial activity of Pd-Ag NPs against *Bacillus subtilis*, *Staphylococcus aureus*, and methicillin-resistant *Staphylococcus aureus* bacteria through 50%–76% lethality of bacterial strains	Gulbagca et al., 2022
		Antifungal activity		
Silver	Aqueous extract of *Ziziphus mauritiana* leaf	*Candida* species (*Candida krusei*, *Candida zeylanoides*, *Candida albicans*, and *Candida guilliermondii*)	Potent anticandidal activity shown	Al-Bahrani et al., 2018
	Trans-cinnamic acid from *Cinnamomum zeylanicum* bark extract	*Candida albicans*	Intermediate antifungal activity toward *Candida albicans*	Edis et al., 2020
	TTO oil (Terpinen-4-ol, 20.1% of γ-Terpinene, 9.8% of α-terpinene, and 6.0% of 1,8-cineole) from *Melaleuca alternifolia* extract	*Candida albicans*, *Candida glabrata*, *Candida tropicalis*, *Candida parapsilosis*, *Candida krusei*	Inhibition of biofilm formation	Souza et al., 2017
	peel of *Punica granatum*	*Candida albicans* and *Candida glabrata* oral clinical strains	Inhibition of biofilm formation	Souza et al., 2020

synthesized using curcumin	*Candida albicans*, *Candida glabrata*, *Candida tropicalis*, *Candida parapsilosis*, *Candida krusei*, *Candida kefyr*	Anticandidal-prominent antifungal activities against *Candida glabrata*, *Candida albicans*, and *Candida tropicalis*	Paul et al., 2018
Green synthesis using quercetin	*Candida* sp.	Effective antifungal activity	Tasca and Antiochia, 2020
Aqueous extracts of *Acacia arabica* (Arabic gum) (A-AgNPs) and *Opophytum forsskalii* seed (O-AgNPs)	*Candida albicans*	Significant anticandidal effect	Aabed and Mohammed, 2021
Seed extracts of *Syzygium cumini* (Sc-AgNPs)	*Candida* spp.	Effective anticandidal activity	Jalal et al., 2019
Extract of parsley, corn silk, and gum Arabic (*Petroselinum crispum, Stenocarpella maydis*), gum exudate of *Acacia senegal*, extract of *Mangifera indica*, root extract of *Arnebia hispidissima*	*Candida albicans*, *Candida tropicalis*, *Candida glabrata*	Significant antifungal activity	Mba and Nweze, 2020
Tea leaves	*Cryptococcus neoformans*	Good antifungal activity	Jasim et al., 2018
Stem extract of *Caesalpinia pulcherrima*	*Cryptococcus neoformans*	Inhibition of biofilm formation	Moteriya and Chanda, 2018
leaf extracts of *Shoreatum buggaia*, *Boswellia ovalifoliolata*, and *Svensonia hyderabadensis*	*Aspergillus niger*, *Aspergillus flavus*	Significant antifungal properties by destruction of fungal membrane integrity	Augustine and Hasan, 2020
from leaf extracts of *Ocimum sanctum*	Opportunistic human fungal pathogens including *Aspergillus niger* and *Candida albicans*	Strong antifungal activities, especially anticandidal and antiaspergillosis effects	Asha Monica and Senthikkumar, 2020

(*Continued*)

TABLE 3.2 (Continued)
Efficacy of Phytonanoparticles in Combating Bacterial and Fungal Diseases

Nanoparticle	Plant Source or Phytochemical	Organisms Tested	Antimicrobial Activity	Refs
		Antifungal activity		
Gold	Seed extract of mango	*Cryptococcus neoformans*	Anti-cryptococcoidal activity; prominent antifungal activity	Donga and Chanda, 2021
Zinc oxide	leaf extracts of *Azadirachta indica*, *Lycopersicon esculentum*, *Vitex trifolia*, *Cinnamomum* sp.	*Candida tropicalis*, *Candida glabrata*, *Candida albicans*	Good antifungal activities, especially anticandidal effects, against the pathogenic species tested	Murali et al., 2021
	Leaf extract of *Ziziphus nummularia*	*Cryptococcus neoformans*	Good antifungal property, especially anticryptococcoidal effect	Murali et al., 2021
Cr_2O_3	Fruit extract of *Hyphaene thebaica* (desert palm)	*Aspergillus fumigates*, *Aspergillus niger*, and *Aspergillus flavus*	Strong antifungal activity, especially antiaspergillosis effect	Ahmed Mohamed et al., 2020

TABLE 3.3
Efficacy of Phytonanoparticles in Combating Viral and Protozoal Diseases

Nanoparticle	Plant Source or Phytochemical	Organisms Tested	Antimicrobial Activity	Refs
		Antiviral activity		
Silver	Curcumin	HIV	Reduced replication of HIV by inhibiting the nuclear translocation factor NF-κB and reduced expression of downstream pro-inflammatory cytokines, namely, interleukin-6, interleukin-1β, tumor necrosis factor-α	Sharma et al., 2017
	Aqueous leaf extract of mangrove (*Rhizophora lamarckii*)	HIV	Inhibited HIV type 1 reverse transcriptase activity (IC_{50} of 0.4 μg/mL on HIV-1)	Kumar et al., 2017
	Curcumin	Respiratory syncytial virus	Reduced RSV replication	Jampílek and Kráľová, 2019
	Tannic acid	HSV-2	Reduced HSV-2 infection and inflammatory reaction in the mouse model	
	Glycyrrhizic acid	HSV-1	Reduced replication of HSV-1 virus	
	Aqueous extract of *Bruguiera cylindrica* leaves	Dengue virus	Inhibited the production of dengue viral envelope E protein in dengue virus and downregulated the expression of "E" gene of the virus	
	Leaf extract of *Carica papaya*	Dengue type 2 virus	Exhibited good binding affinity with the nonstructural protein of dengue virus	Renganathan et al. 2019
	Leaf extract of *Cleistanthus collinus* Leaf extract of *Strychnos nux-vomica*	Zika virus	Effective against the vector and controls growth at the larval and pupal stages of the vector mosquito species of Zika virus	Jinu et al., 2018
Gold	Mannose-type oligosaccharides	HIV	Increase HIV-specific CD4+ and CD8+ T-cell proliferation	Climent et al., 2018
	Sialic acid	Human influenza virus (H3N2)	Reduced influenza infection	Jampílek and Kráľová, 2019

(Continued)

TABLE 3.3 (Continued)
Efficacy of Phytonanoparticles in Combating Viral and Protozoal Diseases

Nanoparticle	Plant Source or Phytochemical	Organisms Tested	Antimicrobial Activity	Refs
		Antiprotozoal activity		
Zinc oxide	leaf extract of *Verbena officinalis* and *Verbena tenuisecta*	*Leishmania donovani*	Anti-leishmaniasis effect; Cytotoxic effect on the amastigote cells of the parasite	Saleem et al., 2019
	Verbena sp.	*Leishmania* spp.	Antileishmaniasis effect; Inhibited the proliferation of promastigote and amastigote cells	
Polymeric nanoparticle	Brazilian red propolis extract	*Leishmania* spp.	Antileishmaniasis Nanoparticles increased drug efficiency compared with the extract only	
Silver	*Cuminum cyminum* plant extract	*Leishmania* spp.	Antileishmaniasis effect; Effective against promastigote and amastigote cells	Bagirova et al., 2020
	Anethum graveolens leaf extract	*Leishmania* spp.	Antileishmaniasis effect; Nanoparticles increased drug efficiency compared with the extract only	Saleem et al., 2019
		Antilarval activity		
Silver	Aqueous leaf extract of *Leonotis nepetifolia*	*Aedes aegypti*, and *Culex quinquefasciatus*	Larvicidal activity against both *Aedes aegypti* and *Culex quinquefasciatus*	Manimegalai et al., 2020
	Moringa oleifera seed extract	*Aedes aegypti*	Effective against *Aedes aegypti* larvae	Sujitha et al., 2015
	Leaf extracts of *Aegle marmelos* and *Colocasia esculenta*	*Culex quinquefasciatus*	Larvae; instar stages II, III, and IV; and pupal stages affected	Dass, 2020
	Jatropha gossypiifolia, *Euphorbia tirucalli*, and *Pedilanthus tithymaloides* leaf extracts	*Aedes aegypti* and *Anopheles stephensi*	Instar stage IV affected	Borase and Patil, 2013
	Leaf extract of *Nyctanthes arbortristis*	Instar stage III of *Culex quinquefasciatus*, *Anopheles stephensi*, *Aedes aegypti*	High mortality rate of all mosquito species	Satsangi and Preet, 2021

	Leaf extract of *Moringa oleifera* and *Ficus exasperata*	Late-stage III instars *Anopheles gambiae*	Mortality of instar stages	Idowu et al., 2021
	Leaf extract of *Eupatorium odoratum*	*Culex quinquefasciatus*	Potent against III and IV larval stages	Elemike et al., 2017
	Extract of *Lippia citriodora*	*Culex quinquefasciatus*	Potent against instar stage IV of *Culex quinquefasciatus*	Elemike et al., 2017
	Extracts of various plant parts	Mosquito species responsible for causing malaria		Mohammadi et al., 2021
Zinc oxide	Aqueous leaf extract of *Lawsonia inermis*	Larval and pupal stages of *Anopheles stephensi*	Lethal at the larval and pupal stages	Amuthavalli et al., 2021
	Solanum lycopersicum aqueous leaf extract	*Aedes aegypti* and *Aedes albopictus*	Mortality of mosquito larvae	Nityasree et al., 2021
	Plant extracts of various plant parts	Mosquito species responsible for malaria		Mohammadi et al., 2021
Gold	Extracts of various plant parts	Mosquito species responsible for malaria	-	Mohammadi et al., 2021
Titanium	Extracts of various plant parts	Mosquito species responsible for malaria	-	Mohammadi et al., 2021

HIV: human immunodeficiency virus; RSV: respiratory syncytial virus; NF-κB: nuclear factor kappa B

During the last few years, the number of invasive filamentous fungal infections has increased considerably worldwide. Susceptibility to these diseases is higher in immunosuppressed individuals, persons with uncontrolled diabetes, hematological diseases, transplantation, and those undergoing chemotherapy. Nanomaterials (both metal and polymeric NPs) are being applied for treating several most prevalent invasive filamentous fungal infections (León-Buitimea et al., 2021).

Conventional antifungal drugs target the crucial steps of protein synthesis, more specifically inhibition of the activity of amino acyl tRNA synthases (Pfaller et al., 2007) or suppression of the inositol phosphorylceramide synthase activity, resulting in the lodgment of ceramides and degradation of sphingolipids from fungal cell wall (Prasad et al., 2016). These drugs also target the ergosterol biosynthetic pathway and inhibit the formation of important fungal metabolites such as lanosterol, fecosterol, zymosterol, and ergosterol (Prasad et al., 2016). Other drugs disrupt cell wall integrity through the inhibition of calcineurin synthesis or lead to oxidative damage by ROS generation and necrosis in the fungal pathogen (Dhamgaye et al., 2012, 2014). However, none of these drugs are very effective, and the search for a potent antifungal agent has led to the formulation and testing of several phytosynthesized NPs. Such NPs reportedly inhibited the production of various hydrolytic enzymes such as lipases, phospholipases, proteinases, and hemolysin, all of which are produced by the invading fungal pathogen and had strongly suppressed proliferation, germ tube formation, and biofilm development (Table 3.2).

3.4.5 Viral and Protozoal Diseases

Viruses are one of the most frequent causes of human diseases that cause mild illness to fatality. Although the pharmaceutical industry pays great attention to human viral infections, most of the chemical drugs are toxic to human cells, while plant-based antiviral drugs have limited bioavailability and very low retentivity. Thus, the development of new antiviral agents is of great importance. Most antiviral drugs target viral attachment to the host cell to prevent their entry or exert steric effects, disrupt membrane glycoproteins, and interfere with fusion of cell membrane bilayers (Jampílek and Kráľová, 2019). Only very few effective antivirals are clinically applied for human diseases; hence, the search for novel alternatives continues.

Malaria, filaria, and leishmania are the three major protozoal diseases occurring in humans, among which malaria and filaria are transmitted by mosquitoes. Leishmaniasis is an infectious vector-borne disease caused by a protozoan parasite, called *Leishmania* spp., which is transmitted through an infected sand fly; this epidemic has become an epidemic in humans, carnivores, and rodents. As for the drug target for leishmaniasis control, the drug interferes with the biochemical processes and, consequently, disturbs the parasite's metabolic pathways. Glycolysis is considered the major source of ATP in parasites; hence, blockade of any step of glycolysis hampers ATP production and subsequently causes death of the parasites. Some drugs or drug-like compounds act as uncouplers and interfere with the oxidative phosphorylation, thereby reducing the energy (ATP) of the parasites (Frayha et al., 1997). In addition to the inhibition of nucleic acid and protein synthesis, other mechanisms such as blockade of the synthesis of folic acid in some parasitic protozoan members were also reported to reduce the infection (Kiang et al., 2018).

A few viral diseases are transmitted by mosquitoes, such as dengue and chikungunya, and the strategy to combat these viral diseases is to target their vectors. Targeting the developmental stages of the vector mosquitoes, such as larval, instar stages (I, II, III, and IV), and pupal stages eventually lead to their lethality. The larvicidal and pupicidal effects of phytofabricated metallic NPs are also documented in this section. Table 3.3 shows the efficacy of the phytonanoparticles in combating the viral and protozoal diseases as well as those effective against vector-borne diseases.

3.4.6 Helminthic Diseases

Of all the infectious diseases that occur in humans and livestock, helminthic infections are less dangerous. Around 20% of the world's human population are infected with intestinal helminths and experience various health complications such as malabsorption, indigestion, diarrhea, blood loss, and reduced growth rates.

Drugs developed against helminths cause disturbance in their reproductive physiology and affect the survival of their larval stages, thereby inhibiting their multiplication. Other drugs interfere with the egg-laying process of the nematode worms or impede the reproductive physiology. Some drugs target the neuromuscular junctions and evoke muscular contraction or relaxation, finally leading to paralysis and death of the helminthic worms (Frahya et al., 1997). Other drugs interfere with either glycolysis, the major source of ATP production in helminthic worms, or other important metabolic activities (Kiang et al., 2018). Very few plant-based drugs have been evaluated for their antihelminthic activity, and only recently, green synthesis of sulfur and silver NPs has been performed using plant extracts to exert nematicidal activities (Al-Banna et al., 2020; Oluwatoyin et al., 2020) (Table 3.4).

3.4.7 Nanoparticles as Carriers of Antimicrobial Plant Therapeutics

In the last 5 years, almost 95% of reports available on Google Scholar and PubMed Central describe encapsulation of well-known or rarely used antibiotics as a nanoformulation in the form of nanocapsules or polymeric NPs or SLNs, and only very few studies have assessed the encapsulation of plant drugs for the treatment of bacterial diseases (Table 3.4). Various plant-based therapeutic compounds such as tea tree oil, clove oil, eugenol, quercetin, naringenin, and farnesol are successfully delivered to fungal infection sites, compared with drugs for bacterial diseases, through nanotechnology, especially structural nanocomponents such as liposomes, SLNs, and polymeric NPs (Table 3.4). Although chitosan or cell wall proteins have been encapsulated to prepare nanoformulations along with amphotericin B-nano-drugs effective against systemic fungal diseases, the use of phytofabrication or nanocomposites for the delivery of plant drugs for the treatment of these diseases has been reported thus far (Kischkel et al., 2020). Extensive research has been conducted on various nanoformulations for treating viral and protozoal diseases. NP formulations using popular antihelminthic drugs such as albendazole and mebendazoles, using oleic acid, palmitic acid, and bee wax, for the formulation of SLNs are well reported, but very few reports are available on the utilization of phytofabrication of NPs against nematodes. Similarly, encapsulation of plant-based drugs into nanoformulations for the treatment of these nematicidal infections has not been used extensively to date.

3.4.8 Beyond the "Single Microbe–Single Disease" Concept in Polymicrobial Diseases

Microorganisms generally thrive in complex polymicrobial biofilm communities, made of varied collections of fungi, bacteria, and viruses, attached to biotic and abiotic surfaces. In humans, for example, gastrointestinal (GI) tract and oral cavity harbor microbial communities, comprising thousands of unique bacterial species residing permanently or transiently on these mucosal membranes (Peters et al., 2012). With the advent of modern technologies and methodologies of culture-independent community analysis, most diseases that were previously characterized as monomicrobial are currently becoming increasingly recognized as polymicrobial infections. Diseases of the oral cavity, otitis media, diabetic foot wound infections, and cystic fibrosis are accordingly being categorized as polymicrobial infections. With the knowledge of true etiology of the diseases, innovative therapeutic measures are required to combat these polymicrobial diseases. Needless to say, NPs, either phytofabricated or nano-formulated with plant-based therapeutics, have shown revolutionary success in the treatment of polymicrobial diseases.

TABLE 3.4
Nanoformulations of Plant Therapeutics Designed for the Treatment of Various Microbial Diseases

Structural Nanocomponent	Carrier	Plant Therapeutic/Phytochemical	Outcome	Refs
		Bacterial diseases		
Nanocomposite	Hydroxyapatite	Ethanolic extract of triphala, secondary metabolites from triphala	Wound healing effect of ointment prepared from the synthesized nanocomposite in a Wistar rat model; inhibition of bacterial strains and enhanced inhibition of the biofilm	Namasivayam et al., 2021
Metal nanoparticle (silver nanoparticles)	Polyvinyl pyrrolidone polymer	*Ziziphus nummularia* leaf extract	Enhanced antibacterial activity against bacterial strains of *Escherichia coli* and *Staphylococcus aureus*	Arodiya et al., 2021
		Fungal diseases		
Nanohybrid hydrogel with ZnO nanoparticles	Dialdehyde cellulose developed from sugarcane bagasse cellulose	Quercetin	Treatment of Athlete's foot disease along with other skin fungal infections Commercial quercetin and quercetin in OPD (QE/OPD) loaded nanohybrid hydrogels inhibit the growth of *Staphylococcus aureus* and *Trichophyton rubrum* strains	George et al., 2019
	Cysteine conjugated dialdehyde cellulose	Naringenin	Treatment of skin fungal infections; Naringenin-loaded nanohybrid hydrogels inhibit the mycelial growth of *Staphylococcus aureus* and *Trichophyton rubrum* strains	George et al., 2020
Solid lipid nanoparticle	Glyceryl oleate	Graphene made from tea tree plant (*Melaleuca alternifolia*) extract	Treatment of coccoidal meningitis	Carbone et al., 2020
		Bacterial diseases		
Polymeric nanoparticle	Polylactic acid, oleic acid, cetyl alcohol	Curcumin	Treatment of oral candidiasis, vulvovaginal candidiasis	Araujo et al., 2020
	Chitosan with tripolyphosphate	Farnesol	Treatment of vulvovaginal candidiasis	
Liposomes	L-α-phosphatidylcholine, cholesterol	*Origanum dictamnus* extracts	Enhanced anticandidal activity against *Candida albicans*, *Candida tropicalis*, and *Candida glabrata*	

	Phosphatidylcholine	Cinnamaldehyde	Anticandidal effect; enhanced drug effectiveness and increased invasiveness	
	Soyabean lecithin	Quercetin and gallic acid	Anticandidal effect; effective against vulvovaginal candidiasis	
	Cholesterol phase	Phytochemical extracts of *Astronium urundeuva*, *Astronium fraxinifolium*, *Astronium graveolens*	Anticandidal effect; the nanocomposition reduced the fungal load from 6000 to less than 2000 colony-forming units in only 2 days, whereas free extract took 4 days for the same effect	
	Medium-chain triglycerides	*Pelargonium Graveolens* essential oil	Anticandidal effect; enhanced drug effectiveness	
	Sorbitan monooleate	*Ocimum basilicum* essential oil	Anticandidal effect; enhanced drug effectiveness	
	Propylene glycol	*Mentha spicata* essential oil	Anticandidal effect; enhanced drug effectiveness	
	Pluronic F127	Clove oil	Effective against cutaneous candidiasis	
	Chemophor	Tea tree oil	Effective against pulmonary candidiasis	
	Stearic acid	Eugenol	Effective against cutaneous candidiasis	
	Labrasol	*Lippia sidoides* essential oil	Anticandidal effect	
		Viral diseases		
Mesoporous silica nanoparticles (MSNs)	APTES molecules	Shikimic acid and quercetin	MSNs-NH2-QR nanoformulation and MSNs-NH2-SH nanoformulation reduce plaque formation caused by H5N1 virus	AbouAitah et al., 2021
Polyvinylpyrrolidone (PVP)-coated AgNPs			Promising microbicidal activity on vaginal/cervical agents to prevent HIV-1 transmission	Jampílek and Kráľová, 2019
Nanoformulation	Nonglycosylated neomucin	Oleanolic acid and betulinic acid	Inhibits the entry of influenza virus (H1N1, H3N2, and H9N2), hepatitis B viruses	Kaplan, 2022
	PLGA	Glycosyl flavonoid from the plant *Cecropia glaziovii*	Effective drug delivery in the treatment of herpes infections caused by herpes simplex virus	dos Santos et al., 2017
	Oleic acid and propylene glycol	Ethanolic leaf extract of *Cuphea ignea* containing the plant therapeutic compounds p-coumaric acid, gallic acid, syringic acid, rosmarinic acid, myricetin 3-O-rhamnoside, and rutin, quercetin, naringenin, and catechin	Effective against SARS-CoV-2 virus	Mahmoud et al., 2021

(Continued)

TABLE 3.4 (Continued)
Nanoformulations of Plant Therapeutics Designed for the Treatment of Various Microbial Diseases

Structural Nanocomponent	Carrier	Plant Therapeutic/Phytochemical	Outcome	Refs
	Herbo-mineral preparation Linga chenduram with nanoparticles of 18–25 nm diameter, spherical	Extract combination of *Vitis lanata* and Indian Sarsaparilla root (*Hemidesmus indicus*) to obtain purified cinnabar	Treatment against hepatitis C virus infection	Al-Ansari et al., 2021
Nanocomposite	Graphene oxide	Curcumin-loaded β-cyclodextrin	Reduced replication of RSV	Jampílek and Kráľová, 2019
Nanoparticles incorporated in carbomer hydrogels	PLGA	Volatile oil from the lemon grass *Cymbopogon citratus*	HSV was inhibited at 42.2-fold lower concentration than that of free oil, and the nanoformulation proved to be more efficient than the free oil alone	Almeida et al. 2018
Cationic nanoemulsion	Hydroxyethyl cellulose	Genistein	Notable increase in genistein retention in the mucosa vs. genistein used alone and showed antiviral activity against HSV-1 (strain 29R)	Jampílek and Kráľová, 2019
Nanoformulation, metal nanoparticles: AgNPs	Hydrogel	Tannic acid	Effective vaginal treatment of HSV-2 genital infection	Szymanska et al. 2018
Zinc oxide nanoparticles	TRP (3D organic molecules)	Ellagic acid	Nanoformulation is a potentially safe and low-cost natural agent against coronavirus and H1N1, as well as other RNA-based viruses, and it has the potential as an alternative therapeutic strategy for COVID-19	AbouAitah et al., 2020
		Protozoal diseases		
Lipid, metal-based, and polymeric nanoparticles	PLGA, carbon nanotubes, iron oxide, glucocorticoid prodrugs, fullerene, PEG	Artemisinin and derivatives such as artesunate and artemether	Antimalarial and antileishmanial activities with enhanced intracellular drug uptake	Alven and Aderibigbe, 2020
Polymeric nanoparticles	Zein with sodium caseinate	Artemether	Antimalarial with extended mean residence time of artemether and prolonged therapeutic duration	Boateng-Marfo et al., 2021
		Primaquine	Anti-leishmaniasis effect; 21-fold enhanced efficacy	Prasanna et al., 2021
	Antimony support	Meglumine	Anti-leishmaniasis effect; effective in the treatment of integumentary leishmaniasis	Carvalho et al., 2019

Polymeric nanoparticles, polymeric nanocapsule	PLGA, PCL	Curcumin, quinine	Antimalarial effect; enhanced activity with minimum toxicity	Patra et al., 2021
Dendrimer-based nanocarrier	PEG, PCL	Artemether	Antimalarial effect; increased drug efficiency	
Nanogel-based nanomedicine	Gum from *Acacia*	4-Aminoquinoline and curcumin	Antimalarial effect; faster dug release	
Nanosized micelles	PEG, PCL	Paclitaxel and docetaxel	Antimalarial effect; better drug release, low hemolysis rate	
Nanocapsule	Lipid nanocapsule	Quercetin	Antileishmaniasis effect against *Leishmania amazonensis*	
	PLA	Meglumine	Antileishmaniasis effect; effective in the treatment of integumentary leishmaniasis	Cosco et al., 2021
Solid lipid nanoparticle formulation	PLGA–D-α–tocopheryl PEG 1000 succinate nanoparticles	Oleanolic acid	Antileishmaniasis effect; effective in the treatment of visceral leishmaniasis caused by *Leishmania donovani*	Ghosh et al., 2016
	Polymeric hydrogel		Antifilariasis effect	Permana et al., 2019
Carbon nanotube	-	Betulin	Antileishmaniasis effect	Saleem et al., 2019
Nanoemulsion	-	Chalcone	Antileishmaniasis effect; effective against the treatment of cutaneous leishmaniasis	
Lipid nanoparticle	Mannose-conjugated chitosan supported	Curcumin	Antileishmaniasis against *Leishmania donovani*	Chaubey et al., 2018
	Poly-ε caprolactone	*Nigella sativa* oil	Antileishmaniasis effect; *Leishmania infantum* promastigotes and amastigotes were inhibited with prolonged drug delivery	Abamor et al., 2018
	PLGA	artemisinin	Antileishmaniasis effect; effective against *Leishmania donovani*	Nafari et al., 2020
	-	Andrographolide	Antileishmaniasis effect against *Leishmania donovani*	
	Chitosan oligosaccharides	Ursolic acid	Antileishmaniasis effect	Saleem et al., 2019
	-	Artemether	Antileishmaniasis effect; effective against *Leishmania major*	Rahnama et al., 2021

(*Continued*)

TABLE 3.4 (Continued)
Nanoformulations of Plant Therapeutics Designed for the Treatment of Various Microbial Diseases

Structural Nanocomponent	Carrier	Plant Therapeutic/Phytochemical	Outcome	Refs
Lipid nanostructure		Annatto oil from seeds of *Bixa orellana*	Antileishmaniasis effect; effective against cutaneous leishmaniasis	Ferreira et al., 2021
	-	Cedrol	Anti-leishmaniasis effect; effective in the treatment of their amastigotes	Saleem et al., 2019
Liposome		Ergosterol	Antileishmaniasis effects; reduced macrophage burden in the liver and spleen, increased drug penetrance, and enhanced drug tolerance	Prasanna et al., 2021
		Primaquine Cholesterol Artemisinin		
	Mannosylated chitosan	Curcumin	Antileishmaniasis effect against *Leishmania donovani*	Nafari et al., 2020
PEGylated liposomes	PEG	Artemisinin, curcumin	Antimalarial effect with increased loading efficiency and decreased antimalarial effect vs. free artemisinin/curcumin	Patra et al., 2021
Metal NPs	TiO_2 with zinc	Hypericin	Antileishmaniasis effect	Sepúlveda et al., 2020
		Helminthic diseases		
Phytofabrication Sulfur nanoparticles	*Rosmarinus officinalis* leaf extract	*Meloidogyne javanica*	Effective in killing the parasite	Al-Banna et al., 2020
Phytofabrication Silver nanoparticles	Bark and leaf extract of *Eucalyptus officinalis*	*Heterodera sacchari*	Nematicidal activity	Oluwatoyin et al., 2020
Phytofabrication Silver nanoparticles	*Lansium parasiticum* extract	*Haemonchus contortus*, a parasitic worm	Nematicidal activity	Goel et al., 2020
Nanoformulation Solid-lipid nanoparticles	Fruit peelings of pomegranate, orange, banana, and apple	Tested against earthworms; they are effective against ringworms	Nematicidal activity	Kiruthika and Somanathan, 2018
Nanoformulation Nanoencapsulation	Thiophene from *Tagetes patula*	*Haemonchus contortus*, a parasitic worm	Nematicidal activity	Politi et al., 2021

PLGA: poly(lactic-co-glycolic acid); RSV: respiratory syncytial virus; HSV: herpes simplex virus; PEG: polyethylene glycol; PCL: polycaprolactone

3.4.9 Diseases Due to Alteration in Oral and Gut Microbiota (Microbial Dysbiosis)

Microorganisms consider the intestine and the oral cavity of humans as their domain. Disruption of homeostasis in the gut or oral cavity can lead to microbial dysbiosis, which manifests as clinical symptoms such as inflammation, tooth decay, and corneal infections (Carding et al., 2015)

The complex interplay among the host, microorganisms, and environmental factors results in inflammatory bowel disease (IBD), including Crohn's disease, characterized by inflammation of the colon and GI tract and ulcerative colitis. Under normal physiological conditions, neutrophils undergo apoptosis at the inflamed site and macrophages ingest the apoptotic neutrophils, thereby leading to an anti-inflammatory state through the clearance of apoptotic neutrophils and enables macrophages to undergo tissue repair and regaining of their physiological functions. Lifestyle amendments and conventional drugs can assure a lasting treatment for IBD in most patients, and as a re-course, many medical professionals and scientists are recommending the use of phytosynthesized and/or loaded NPs as possible alternatives.

Under normal physiological conditions, saliva supplies the resident oral microbiota with nutrients, which maintains microbial homeostasis. Any alteration in the salivary bacterial profile leads to clinical conditions of oral diseases in humans, such as periodontal diseases and dental caries. Nonoral polymicrobial diseases include cystic fibrosis, rhinosinusitis, corneal infections, keratitis, and diabetic foot wound. Different types of structural nanocomponents (SLNs, polymeric NPs, and liposomes) are used to load the cargo plant drugs such as curcumin and apigenin toward the treatment of diseases due to oral dysbiosis.

Table 3.5 presents the details of published reports on the possible treatments of IBD and oral and nonoral diseases by utilizing nanoscience technologies.

3.5 PHYTONANOPARTICLES FOR THE TREATMENT OF LUNG AND RESPIRATORY AILMENTS

Lungs are frequently exposed to various chemicals and allergens that enter the respiratory system during inhalation by coming into direct contact with them, often leading to irritation, inflammation, and oxidative stress with far reaching effects. Many diseases such as asthma, chronic obstructive pulmonary disease, emphysema, and mesothelioma are quite common nowadays among people dwelling in large cities or working in factories, tanneries, or extraction sites and in the general population, as they are frequently exposed to polluting allergens such as smog, volatile organic compounds, paint fumes, and dust. Even far-reaching medical fall outs such as heart ailments and lung cancers are known to be caused in animal models under such circumstances. The usefulness of plant-based remedies in combating respiratory troubles is well known and recorded in various regional scriptures and medicinal practices (Popova et al., 2021).

Various plants and their isolated constituents are used to cure and protect the respiratory tract and lungs from infections. *Paeonia emodi* is a perennial Himalayan plant used for its effectiveness against asthma, uterine diseases, palpitations, colic, and hypertension. The aqueous extract of *Paeonia emodi* leaves was used for the green synthesis of Fe_2O_3 NPs as well as a capping and reducing agent. Fe_2O_3 NPs are known for their targeted drug delivery uses and mild cytotoxicity, harnessed by the scientists to target bacteria such as *Staphylococcus aureus*, which is the causal agent of pneumonia, and *Escherichia coli*. Through *in vitro* and *in vivo* assays on a mouse model, *Staphylococcus aureus* was more susceptible to *Paeonia emodi* phytoformulated Fe_2O_3 NP injection, thus showing the effectiveness of this nanoformulation to combat respiratory problems (Shah et al., 2021).

Nigella sativa, a common spice, is known for its anti-inflammatory, antioxidant, and antineoplastic properties because of the presence of thymoquinone (TQ). Although the therapeutic effect of this molecule on pulmonary problems is proven through both *in vivo* and *in vitro* studies, their low solubility

TABLE 3.5
Phytonanoparticles and/or Plant Drug Nanoformulations in the Treatment of IBD (Exhibiting Anti-inflammatory Activity) and Oral and Nonoral Diseases

Structural Nanocomponent/ Phytofabricated Nanoparticles	Carrier	Plant therapeutic/Phytochemical	Outcome	Refs
Treatment of IBD (caused by gut microbiota dysbiosis)				
Nanocapsules, solid lipid nanoparticles, nanoemulsion	PLGA, PEG, liposome	Bioactive components such as polyphenols, terpenoids, essential oils, and phytosterols	Enhanced anti-inflammatory activity	Conte et al., 2017
Nanoemulsion		resveratrol	RENE (**re**sveratrol **n**ano **e**mulsion) exhibited promising intestinal anti-inflammatory properties in the DSS model of colitis in rats	Mahmoud et al., 2021
Nanoformulations: cell-derived nanovesicles		6-Shogaol from ginger	Effective against ulcerative colitis	Chen et al., 2022
Phytofabricated NPs and metal NPs (AgNP, AuNP, ZnO NPs, titanium dioxide NPs, selenium NPs)		Extracts from various parts of plants	Potent anti-inflammatory properties	Agarwal et al., 2019
Phytofabricated NP $ZnFe_2O_4$ and $CrFe_2O_4$ nanoparticles		Resin of *Boswellia carteri*	Anti-inflammatory activity	Imraish et al., 2021
Phytofabricated AgNPs	-	Black currant extract	Potent anti-inflammatory properties in the DSS model of colitis in mice	Krajewska et al., 2020
Treatment of oral microbial diseases (caused by oral microbiota dysbiosis)				
Nanoformulation as solid lipid NPs	PEG	Essential oil from *Mentha* sp. (*Mentha spicata, Mentha piperita*)	Essential oils had increased efficacy as antibacterial agents against dental caries	Mostafa et al., 2020
Phytofabricated NPs AgNPs and AuNPs		Extract of *Cornus mas*	Effective against oral dysplasia	Baldea et al., 2020
AgNPs		Rice bran, rice husk, rice germ	Potent antimicrobial activity of phytofabricated AgNPs against *Streptococcus mutans*, one of the causal organisms of dental caries	Suwan et al., 2018
Phytofabricated NP–Sn, Cu, Hg, Ag, composite NPs in the form of nanopowder		Root extract of *Salvia miltiorrhiza*	Effective against dental bacteria *Streptococcus mutans* and *Lactobacillus acidophilus*	Lee et al., 2016

		Treatment of nonoral diseases (caused by oral microbiota dysbiosis)		
Biodegradable polymer carrier (NP)		Apigenin (4,5,7-trihydroxyflavone)	Enhanced bioavailability and biosolubility of apigenin and therapeutic efficacy against lung fibrosis	Pápay et al., 2017
Liposome		Phytoconstituents	Enhanced drug delivery for the treatment of cystic fibrosis	Malhotra et al., 2021
Liposomes	PEG	Curcumin	Chronic obstructive pulmonary disease, characterized by chronic bronchitis and emphysema; provided good protection against oxidant-induced damage to the lungs	Xu et al., 2020
Solid lipid NP	Lipid and surfactant	Proanthocyanidins	Reduced ROS production and protected the lungs from fibrosis	
Phytofabricated NPs: Ag and ZnO NPs			Antibacterial property against *Haemophilus influenzae* strains, responsible for cystic fibrosis, otitis media, rhinosinusitis	Salem et al., 2015
Phytofabricated NPs AgNPs		Quercetin from methanolic petal extract of *Clitoria ternatea* L.	Strong antibacterial activity against *Staphylococcus aureus* and *Shigella* sp., responsible for diabetic foot wound and keratitis, respectively	Vanaraj et al., 2017
Phytofabricated AuNPs		Leaf extract of parsley	Effective against gram-positive bacteria (*Bacillus subtilis* and *Enterococci faecalis*) and gram-negative bacteria (*Escherichia coli* and *Enterobacter ludwigii*), causing polymicrobial diseases	El-Borady et al., 2020

IBD: inflammatory bowel disease; PLGA: poly(lactic-co-glycolic acid); PEG: polyethylene glycol; NP: nanoparticle: ROS: reactive oxygen species; DSS: drug delivery system; ROS: reactive oxygen species and bioavailability posed problems in its medicinal use. Scientists have prepared nanoformulations of TQ, which could overcome these limitations. Recently, the methoxy poly(ethylene glycol)-b-poly (ε-caprolactone) copolymer was nano-precipitated with TQ to form nanodroplets of TQ-encapsulated NPs, which showed favorable TQ release under controlled conditions (Sunoqrot et al., 2020).

Withania somnifera is a wonder plant with widespread uses against an array of serious diseases such as asthma, arthritic diseases, diabetes, hypertension, stress, and various types of cancer. Porous TiO_2 NPs synthesized using aqueous extracts of *Withania somnifera* roots were tested against the formation of bacterial biofilms containing *Staphylococcus aureus*, *Streptococcus pyogenes*, and *Klebsiella pneumoniae*, all known for causing deadly respiratory problems (Al-Shabib et al., 2020). Selenium NPs photosynthesized from leaf extracts of *Withania somnifera* and selenious acid showed remarkable activity against bacteria causing respiratory distress (Alagesan and Venugopal, 2018). The same plant extract used for the fabrication of selenium NPs reduced the viability of the AJ549 cell line (adeno-carcinomic human alveolar basal epithelial cancer cells) (Ikram et al., 2021). Similarly, AgNPs synthesized from the leaf extract of *Gossypium hirsutum*, wherein it acts as a phytoreducer and a capping agent, have shown noticeable cytotoxicity against AJ549 cells (Kanipandian and Thirumurugan, 2014). In another report, silver NPs synthesized from *Beta vulgaris* extract showed similar positive effects against A549 cells (Venugopal et al., 2017).

Silver NPs synthesized from the leaf extract of *Gossypium hirsutum* acted as a phytoreducing and capping agent and showed remarkable cytotoxicity against the human lung cancer line A549, signifying the biological and medical significance of cotton leaf extract in the synthesis of medicinally active silver NPs (Kanipandian et al., 2014).

Gold phytonanoparticles extracted using *Moringa oleifera* hailed as a superfood, and when they were tested against the A549 cell line and SNO esophageal carcinoma cell line, they demonstrated potent antiproliferative effects. Apoptosis was induced in the A549 cell line through the activation of alternate splicing of caspase-9 (Tiloke et al., 2016).

Luteolin is a flavonoid widely found in cruciferous green vegetables. It activates apoptosis and suppression of cell survival pathways in various cancer cell lines, namely, lung, head and neck, prostrate, breast, colon, liver, skin, and cervical cancer cell lines (Majumdar et al., 2014). However, nonspecific delivery, poor efficacy, and less bioavailability limit the utility of this molecule; such problems were negated when polymeric NPs of luteolin were formulated and their efficacy proved in the treatment of the lung cancer cell line A549/T12 (Majumdar et al., 2014).

3.6 PHYTONANOPARTICLES FOR THE TREATMENT OF GASTROINTESTINAL AND INFLAMMATORY PROBLEMS

Stomach and gastric problems, as well as inflammatory responses, due to various infections and cytokine responses are common yet significant diseases that affect a large percentage of the population, leading to reduced quality of life and daily discomfort. Gastritis, stomach ulcer, inflammatory bowel syndrome (IBS), diarrhea, cramps, colitis, and stomach cancer are common stomach-related ailments. The mechanism of action involved in the treatment procedures involves

- Inhibition of inflammatory cytokines (e.g., anti–tumor necrosis factor–α antibodies), inducing T-lymphocyte apoptosis
- Downregulation of T-lymphocyte proliferation by targeting anti-inflammatory cytokines
- Development of selective adhesion molecule molecules, which suppress T-lymphocytes in the epithelium

Often, these drugs have significant side effects and require systemic administration and high maintenance.

NPs designed with known kinetics and specific detection provide the perfect DDS in various gastric ailments. Many biogenic NPs provide the optimized carrier system for suitable molecules

(Lautenschläger et al., 2014). Zhang et al., (2016b) reported that synthesized NPs can deliver drugs and siRNA at designated dosages at specific locations in IBS treatment. They formulated ginger-derived NPs (GDNPs) using super-high-speed centrifugation and ultrasonic dispersion. Being larger than the engineered NPs, these NPs have membrane lipids, few membrane proteins, miRNAs, and active components such as 6-gingerol and 6-shogaol. These orally administered GDNPs are drug-containing NPs with better targeted delivery. Similar GDNPs from grapefruit are known to target intestinal macrophages and intestinal stem cells (Ju et al., 2013).

Brown macroalgae are widely found in the Mediterranean, Baltic, and Atlantic Oceans. Extract of one such member, *Cystoseira baccata*, efficiently synthesized gold NPs. These phytonanoparticles were effective against colorectal cancer lines (HT-29 and Caco-2) as well as the fibroblast cell line PCS-201–010, with selective toxicity and good compatibility with healthy cells. The results showed apoptosis induction in cancer lines through the activation of the extrinsic and mitochondrial pathways (González-Ballesteros et al., 2017). In another study, gold NPs synthesized from the *Cystoseira tamariscifolia* extract showed cytotoxicity against fibroblast cell line (BJ5-ta cell line); at low concentration, they aided in normal cell regeneration (Machado et al., 2019).

Desert date, known as *Balanites aegyptiaca,* is predominantly found in African and Middle Eastern countries. The fruit extract of this plant aided the synthesis of silver NPs that play an effective role in apoptosis induction in colon and liver cancers, with gene expression of both extrinsic and intrinsic apoptotic pathways being affected (Mittal et al., 2018).

Abies spectabilis has wide uses against various types of respiratory disorders, cancers, neuronal problems, infections, and stomach ulcer (Majeed et al., 2013). Copper NPs synthesized from *Abies spectabilis* plant extract showed great effectiveness against inflammatory responses in mice, providing cues for drug development in the near future (Liu et al., 2020).

Terminalia species are good sources of flavonoids, phenols, tannins, alkaloids, and other such compounds and have major uses in the medical field. Silver NPs synthesized from the leaf extract of *Terminalia catappa*, *Terminalia muelleri*, *Terminalia bentzoe*, and *Terminalia bellirica* showed remarkable anti-inflammatory responses in mice (El-Rafie et al., 2014).

Indomethacin is a strong anti-inflammatory drug effective in soothing gastric ailments. Poly(ε-caprolactone) NPs synthesized using essential oil from *Nigella sativa* seeds and indomethacin through nanoprecipitation greatly enhanced the anti-inflammatory and analgesic property of indomethacin as well as improved topical application (Badri et al., 2018).

Use of argan oil as a nanocarrier for indomethacin in the treatment of acute rheumatoid arthritis has been reported (Badri et al., 2015). Silver NPs synthesized from *Piper nigrum* and *Eichhornia crassipes* leaf extract showed enhanced anti-inflammatory activity (Aparna et al., 2015; Prabakaran and Mani, 2019). Similarly, ZnO NPs synthesized using *Trianthema portulacastrum* Linn. extract showed enhanced anti-inflammatory properties *in vitro* and had rapid wound healing properties in case of skin cuts and eruptions (Yadav et al., 2018).

There is a myriad of plants with bioactive anti-inflammatory compounds, and the preparation of nanoformulations from these plants or compounds is an urgent need.

3.7 PHYTOFORMULATIONS WITH ANTIDIABETIC ACTIVITY

Diabetes is a chronic metabolic disorder closely related to lifestyle alterations (such as food habits, wherein an exponential increase occurs in the number of meals); it affects billions of people worldwide, and in 2014, approximately 8.5% of the world's population were living with diabetes. It is associated with several complications, named diabetic complications, such as retinopathy, cardiac and arterial diseases, nephropathy, and neurological problems. Many plant-derived secondary metabolites are potent antidiabetic molecules (Dewanjee et al., 2020).

Recently, many studies have described the synthesis of easy-to-use and stable phytonanoparticles (Table 3.6 and Figure 3.5).

TABLE 3.6
Phytonanoparticles and Their Mode of Action Against Type 2 Diabetes Mellitus

NP	Plant/Bioactive Compound	Mode of Action	Refs
Gold	Escin	Improvement in glucose uptake by skeletal muscle	Shamprasad et al., 2019
	Guavanoic acid from *Psidium guajava*	Phosphatase inhibitor activity	Dewanjee et al., 2020
Zinc and silver	Docosahexaenoic acid	Reduction of hyperglycemia and insulin resistance	Hussein et al., 2019 a, b
Silver	*Allium cepa*	α-Glucosidase inhibition	Jini et al., 2020
	Ananas comosus outer peel	As above	Das et al., 2019
	Argyreia nervosa	As above	Saratale et al., 2017
	Avicennia officinalis	As above	Das et al., 2019
	Bauhinia variegata	α-Amylase activity reduction	Johnson et al., 2018
	Botryodiplodia theobromae	As above	Janakiraman et al., 2018
	Breynia retusa	As above	Nithiya et al., 2014
	Caesalpinia coriaria	Better glucose uptake	Anandhi et al., 2019
	Calophyllum tomentosum	α-Amylase activity	Govindappa et al., 2018
	Centella asiatica	Better glucose uptake and α-amylase activity	Wilson et al., 2015
	Chromolaena odorata	Increase in glucose uptake	Anandhi et al., 2019
	Cinnamomum cassia	Better insulin sensitivity	Koffi et al., 2019
	Cinnamomum tsoi	α-Amylase and α-glucosidase activity affected	Maddinedi et al., 2017
	Cladosporium sp.	As above	Popli et al., 2018
	Clausena anisata	Glucose uptake affected	Yakoob et al., 2016
	Colpomenia sinuosa	α-Amylase and α-glucosidase activity affected	Manam et al., 2014
	Corchorus depressus	α-Glucosidase activity affected	Tahir et al., 2019
	Andrographis echioides	Plasma glucose and plasma insulin reduction	Kumar, 2018
	Costus pictus	Glucose uptake lesser and reduced nonenzymatic hemoglobin glycosylation, α-glucosidase, α-amylase	Vinotha et al., 2019
	Curcuma longa	Free radical quenching, α-amylase reduction	Dewanjee et al., 2020
	Cymbopogon citratus	α-Amylase reduction	Agarwal et al., 2021
	Enhalus acoroides	α-Glucosidase reduction	Senthilkumar et al., 2016
	Halymenia poryphyroides	Reduced α-amylase and α-glucosidase activity	Manam et al., 2014

	Eysenhardtia polystachya Ort. Sarg	Aids in pancreatic β-cell survival and insulin secretion; enhancement of hyperglycemia and hyperlipidemia in a glucose-induced hyperglycemic zebrafish model	Campoy et al., 2018
	Persea americana, Arachis hypogea, Beta vulgaris	α-Glucosidase inhibition	Das et al., 2020
Silver and zinc	*Heritiera fomes*	Reduced α-amylase activity	Thatoi et al., 2016
Silver	*Holoptelea integrifolia*	Free radical scavenging, α-amylase activity reduction	Kumar et al., 2019
	Ipomoea batatas	α-Glucosidase activity reduction	Das et al., 2019
Silver and cerium	*Lawsonia inermis*	Reduction in blood glucose	Kalakotla et al., 2019
Silver	*Lonicera japonica*	Reduction in α-amylase and α-glucosidase activity	Balan et al., 2016
	Manilkara zapota	Reduction in α-amylase	Mathur et al., 2017
Silver and zinc	*Momordica charantia*	Better insulin activity, reduced blood glucose	Elekofehinti et al., 2022; Shanker et al., 2017b
Silver	*Ocimum basilicum*	Reduction in α-amylase and α-glucosidase activity	Malapermal et al., 2017
	Ocimum sanctum	As above	Do
	Parmelia perlata	Reduction in α-amylase activity	Leela et al., 2017
	Pisum sativum	As above	Patra et al., 2019
	Psoralea corylifolia	antidiabetic activity	Shanker et al., 2017a
	Pterocarpus marsupium	α-Glucosidase reduction	Bagyalakshmi et al., 2017
	Rumex hymenosepalus	Better glucose tolerance	Virgen-Ortiz et al., 2015
	Solanum nigrum	Better glucose tolerance, reduced cholesterol and triglyceride levels	Sengottaiyan et al., 2015
	Punica granatum	Reduced α-amylase and α-glucosidase activity	Saratale at al., 2018
	Sonneratia apetala	Reduced α-amylase activity, free radical scavenging, inflammation	Thatoi et al., 2016
	Sphaeranthus amaranthoides	Reduced α-amylase activity	Swarnalatha et al., 2012
	Stoechospermum marginatum	Reduced α-amylase activity and cytotoxicity	Sabatini et al., 2018
	Syzygium cumini	Reduced free radical scavenging in glucose-stressed H9C2 cardiac cells	Bitencourt et al., 2015
	Tephrosia tinctoria	Reduced α-amylase and free radical scavenging	Rajaram et al., 2015
	Tinospora graecum	Reduced levels of blood glucose, serum urea, and serum creatinine	Ragavee et al., 2019
	Withania somnifera	Reduced α-amylase and α-glucosidase	Uddin et al., 2018
	Xylocarpus granatum	As above	Das et al., 2019
Gold	Resveratrol	Enhanced activation of retinal pigment epithelium–derived factor mRNA	Dong et al., 2019

(Continued)

TABLE 3.6 (Continued)
Phytonanoparticles and Their Mode of Action Against Type 2 Diabetes Mellitus

NP	Plant/Bioactive Compound	Mode of Action	Refs
	Cassia fistula	Better control of serum blood glucose concentrations, promotes favorable body weight with a better lipid profile, better transaminase activity, and changes in renal dysfunction	Burlacu et al., 2019
	Cassia auriculata	Effectively reduces hyperglycemia, suppresses hyperlipidemia, and improves insulin level	Dewanjee et al., 2020
	Cassia fistula	Serum glucose reduction	Daisy and Saipriya, 2012
	Chamaecostus cuspidatus	Blood glucose reduction	Ponnanikajamideen et al., 2019
	Couroupita guianensis	Decreased blood sugar level and improved vital stats	Sengani et al., 2017
	Gymnema sylvestre	Increased plasma insulin, reduced hyperglycemia	Nirmala et al., 2018
	Helianthus tuberosus	Increased glucose tolerance	Sengottaiyan et al., 2016
	Marsilea quadrifolia	Reduced glucose uptake	Chowdhury et al., 2017
	Moricandia nitens	α-Glucosidase activity reduction	Soliman et al., 2017
	Padina boergesenii	α-Glucosidase activity reduction	Senthilkumar et al., 2015
	Sambucus nigra	Blood glucose and cholesterol reduced; serum insulin increased	Opris et al., 2017
	Saraca asoca	α-Amylase reduction	Patra et al., 2018
	Sargassum swartzii	Reduced blood glucose and glycated hemoglobin; elevated hemoglobin and plasma insulin	Dhas et al., 2016
	Vinecin	As above	Chockalingam et al., 2015
Zinc	*Azadirachta indica*	α-Amylase and α-glucosidase activity affected	Rehana et al., 2017b
	Andrographis paniculata	α-Amylase and α-glucosidase activity affected	Rajakumar et al., 2018
	Hibiscus rosa-sinensis	α-Amylase activity reduced and free radical quenching	Rehana et al., 2017b
	Hibiscus sabdariffa	Reduced α-amylase activity and protection of the pancreas	Bala et al., 2015
	Moringa oleifera	α-Amylase and α-glucosidase activity reduction	Rehana et al., 2017b
	Murraya koenigii	As above	Do
	Nasturtium officinale	Increased insulin activity	Bayrami et al., 2019
	Pterocarpus santalinus	α-amylase and α-glucosidase activity reduction	Kitture et al., 2015
	Silybum marianum	Blood glucose and serum triglyceride reduction, serum insulin increase	Arvanag et al., 2019
	Tamarindus indica	α-Amylase and α-glucosidase activity reduction	Rehana et al., 2017b
	Vaccinium arctostaphylos	Reduced blood glucose, increased serum insulin	Bayrami et al., 2019
	Curcumin	Blood glucose reduction, improvement in serum insulin, activation of GLUT2 and glucokinase genes in the pancreas and liver in a rat model	Raslan et al., 2018

Copper	*Barleria prionitis*	α-Amylase and α-glucosidase activity affected	Bhagwat et al., 2018
	Dioscorea bulbifera	As above	Ghosh et al., 2015
	Gnidia glauca	As above	Jamdade et al., 2019
	Litchi chinensis	As above	Bhagwat et al., 2018
	Platanus orientalis	α-Amylase activity reduction	Do
	Plumbago zeylanica	α-Amylase and α-glucosidase activity reduction	Jamdade et al., 2019
Selenium	Berberine	Better glucose uptake	Yin et al., 2017
	Catathelasma ventricosum	Blood glucose reduction	Liu et al., 2018
	Morus alba and *Pueraria lobata*	Antioxidant properties; better glucose uptake and bioavailability	Deng et al., 2019
Cerium	*Morus nigra*	Better glucose uptake	Rajan et al., 2019
Iron	*Sesamum indicum*	α-Amylase reduction	Bano et al., 2017
Titanium	*Stevia rebaudiana*	Reduced blood glucose, glycated hemoglobin, serum triglycerides, and serum cholesterol; increased serum insulin	Langle et al., 2015
Platinum	*Withania somnifera*	Plasma glucose reduction	Li et al., 2017

NP: nanoparticle

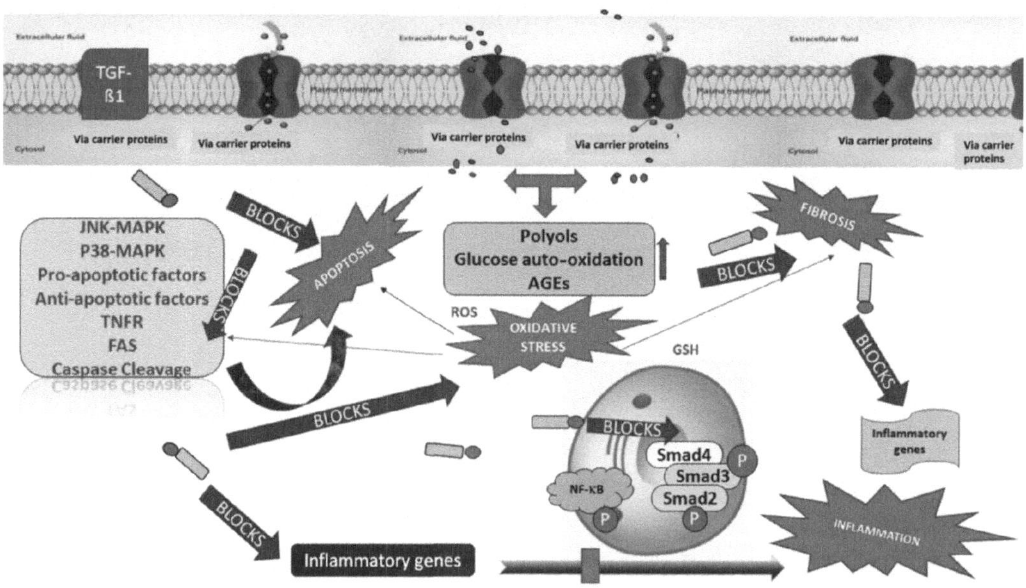

FIGURE 3.5 Schematic diagram showing the accepted mode of action of numerous plant-based nanoparticles against diabetes.

3.8 PHYTONANOPARTICLES WITH NEUROPROTECTIVE EFFECTS

Many plants containing useful metabolites have been harnessed for the synthesis of phytonanoparticles that exert a neuroprotective effect. The effect of these NPs on neurodegenerative diseases such as Alzheimer's disease and Parkinson's disease are most widely studied. NPs are used for drug delivery to the brain owing to their specific delivery because of their multifunctional, adaptable, and stable architectures (Bhattacharya et al., 2022).

Table 3.7 and Figure 3.6 summarize the types of metal phytonanoparticles with neuroprotective effects.

3.9 NANOFORMULATION FOR SKIN REJUVENATION

Skin, being the largest organ in the body, is often exposed to a host of external chemicals and pollutants such as radiation, smoke, and heat, all of which harm the organ by decreasing its softness, glaze, and suppleness. Skin is the first barrier that takes effect when a xenobiotic agent tries to enter the human body. Moreover, the symptoms of aging are prominent on the skin, showing bruises due to aging, pigmentation, and sagging. Natural, plant-based therapeutic products do not contain any harsh chemicals; hence, they are often supplemented with rich antioxidants. These days scientists have begun preparing nanoemulsions of plant-based topical therapeutics, as these nanocarriers are more stable and have easier access to target sites, with better permeation through the skin layers. Many modern treatments use nanoemulsion in skin regeneration. Plants such as *Tagetes patula*, *Phyllanthus emblica*, *Tamarindus indica*, *Punica granatum*, *Eugenia dysenterica*, and *Equisetum arvense* and important bioactive compounds such as quercetin, 3-O-methylquercetin, curcumin, and lycopene have been applied in the cosmeceutical industry. Lipid-based, PLGA-based, or chitosan-based NPs prepared from these plants and compounds are being explored through extensive research and are being developed by many commercial companies such as Revlon, Lancome, and Revitalift, among others (Romes et al., 2021).

TABLE 3.7
Phytonanoformulations and Their Mode of Action Against Neuronal Ailments

NP	Plant/Bioactive Compound	Disease	Mode of Action	Refs
Gold	*Terminalia arjuna*	Neuroprotective effect	Inhibition of AChE, reduction in Aβ fibrillation process, and destabilization of mature fibrils	Suganthy et al., 2018
Platinum	*Bacopa monnieri*	Neuroprotective effect	Antiamyloidogenic molecules	Bhattacharya et al., 2022
Selenium	*Thea sinensis*	Neuroprotective effect in Alzheimer's disease	Inhibitory effect against a host of amyloid-forming proteins	Bhattacharya et al., 2022
	Curcuma sp.	Alzheimer's disease	Antiamyloid binding action	Reddy et al., 2018
T80-coated RIN-loaded PLGA NP	*Uncaria* sp.	Do	Suppression of soluble A-induced hyperexcitability of hippocampus neurons by RIN	Xu et al., 2020
Silver	*Pulicaria undulata*	Antiepileptic activity, insect repellent, and antioxidant effect	Prevention of fibril formation by decreasing protein conformation and self-association	Dehvari et al., 2018
Iron-sapogenin	*Camellia oleifera var. monosperma*–defatted seeds	Neuroprotective effect	Alleviation of rotenone-induced neurotoxicity by hindering neurodegeneration in a rat model	Yang et al., 2017 (b)
Zinc oxide	*Aquilegia pubiflora*	Protection against Alzheimer's disease, anti-inflammatory, antidiabetic, antiaging	Inhibition of key enzymes responsible for most health issues, namely, AChE and BChE enzymes, AGEs of pentosidine and vesperlysine, sPLA2 and 15-LOX	Jan et al., 2021
Gold	Root extract of *Paeonia moutan*	Protection against Parkinson's disease	Alleviation of neuroinflammation, improvement of motor coordination in Parkinson's disease-induced mouse model	Xue et al., 2019

NP: nanoparticle; AChE: acetylcholinesterase; RIN: rhynchophylline; PLGA: poly(lactic-co-glycolic acid); BChE: butyl cholinesterase; AGE: advanced glycation end product

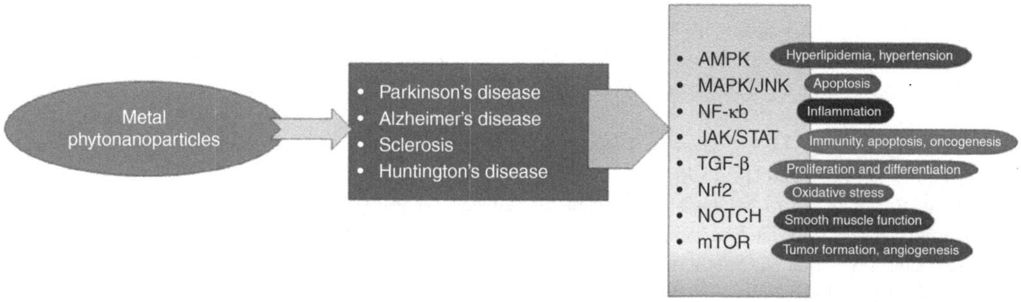

FIGURE 3.6 Schematic diagram illustrating the signaling cascades targeted by various plant-based nanoformulations in the treatment of neuronal diseases.

3.10 CONCLUSION AND FUTURE PROSPECTS

Phytonanoparticles are a boon to mankind, but their long-term safety is yet to be confirmed. The changes and alterations in their structure and chemical nature over time could lead to deleterious effects. Currently, we are in the early stages of such evaluations and yet to reach future directives. There are numerous spectacular applications that require further studies. Nanorobots/microrobots (NMRs) are recent discoveries that involve the use of miniature size and motility of NPs for adept biomedical applications. Incorporating plant-based NMRs could be an attractive proposition in the future.

As plant-based NPs have been accepted as an attractive alternative to conventional drugs in the future, it is necessary to focus on the synthesis of advanced phytonanoparticles that will have least side effects and hypersensitivity. It is important to identify active groups of important drugs that would attach to the NPs efficiently and stay stable for a convenient period of time because NPs that have an extended shelf-life would not be effective in eliciting a patient's innate immunity against a wide range of diseases and dose compromise.

REFERENCES

Aabed, Kawther, and Afrah E. Mohammed. *Nanomaterials* 11, no. 10 (2021): 2573.

Abamor, Emrah Sefik, Ozlem Ayse Tosyali, Melahat Bagirova, and Adil Allahverdiyev. *IET Nanobiotechnology* 12, no. 8 (2018): 1018–1026.

Abbasi, Banzeer Ahsan, Javed Iqbal, Riaz Ahmad, Layiq Zia, Sobia Kanwal, Tariq Mahmood, Canran Wang, and Jen-Tsung Chen. *Biomolecules* 10, no. 1 (2019): 38. https://doi.org/10.3390/biom10010038

Abdoli, Mohadese, Elham Arkan, Zahra Shekarbeygi, Salar Khaledian. *Inorganic Chemistry Communication* 129 (2021): 108649. https://doi.org/10.1016/j.inoche.2021.108649

Abel, Ezra Elumalai, Preetam Raj John Poonga, and Shirly George Panicker. *Applied Nanoscience* 6, no. 1 (2016): 121–129.

AbouAitah, Khaled, Abdou K. Allayh, Jacek Wojnarowicz, Yasser M. Shaker, Anna Swiderska-Sroda, and Witold Lojkowski. *Pharmaceutics* 13, no. 12 (2021): 2174.

AbouAitah, Khaled, Anna Swiderska-Sroda, Ahmed Kandeil, Asmaa MM Salman, Jacek Wojnarowicz, Mohamed A. Ali, Agnieszka Opalinska, Stanislaw Gierlotka, Tomasz Ciach, and Witold Lojkowski. *International Journal of Nanomedicine* 15 (2020): 5181.

Agarwal, Happy, Amatullah Nakara, and Venkat Kumar Shanmugam. *Biomedicine & Pharmacotherapy* 109 (2019): 2561–2572.

Agarwal, Happy, S. Venkat Kumar, and S. Rajeshkumar. *Journal of Microbiology, Biotechnology and Food Sciences* 2021 (2021): 371–376.

Ahlam, Abdul Aziz, V. S. Shaniba, P. R. Jayasree, and P. R. Manish Kumar. *Biological Trace Element Research* 199, no. 5 (2021): 1778–1801.

Ahmed Mohamed, Hamza Elsayed, Shakeeb Afridi, Ali Talha Khalil, Tanzeel Zohra, Muhammad Ali, Muhammad Masroor Alam, Aamer Ikram, Zabta Khan Shinwari, and Malik Maaza. *Nanomedicine* 15, no. 17 (2020): 1653–1669.

Ahmed, Iqbal, Kamisah Kormin, R. Rajput, M. H. Albeirutty, Z. A. Rehan, and J. Zeb. *Nanomaterials for Environmental Applications and their Fascinating Attributes* 2 (2018): 218.

Alagesan, Venkatesan, and Sujatha Venugopal. *Bionanoscience* 9, no. 1 (2019): 105–116. https://doi.org/10.1007/s12668-018-0566-8

Al-Ansari, Mysoon M., AJA Ranjit Singh, Fatimah S. Al-Khattaf, and J. S. Michael. *Saudi Journal of Biological Sciences* 28, no. 3 (2021): 1596–1606.

Al-Bahrani, Rusol M., Sura Muayad Abdel Majeed, Mustafa Nadhim Owaid, Abdullah B. Mohammed, and Duha A. Rheem. *ACTA Pharmaceutica Sciencia* 56, no. 3 (2018).

Al-Banna, Luma S., Nidá M. Salem, Gharam A. Jaleel, and Akl M. Awwad. *Chemistry International* 6, no. 3 (2020): 137–143.

Al-Hakkani, Mostafa F., Gamal A. Gouda, and Sedky HA Hassan. *Heliyon* 7, no. 1 (2021): e05806. https://doi.org/10.1016/j.heliyon.2020.e05806

Ali, Saheb, Kattakgoundar Govindaraj Sudha, Gopalu Karunakaran, Mariyappan Kowsalya, Evgeny Kolesnikov, and Mohan Prasanna Rajeshkumar. *Journal of Biotechnology* 329 (2021): 65–79. https://doi.org/10.1016/j.jbiotec.2021.01.022

Almeida, Kessiane B., Juliana L. Araujo, Jéssica F. Cavalcanti, Maria Teresa V. Romanos, Samanta C. Mourão, Ana Claudia F. Amaral, and Deborah Q. Falcão. *Revista Brasileira de Farmacognosia* 28 (2018): 495–502.

Al-Radadi, Najlaa S. *Journal of Saudi Chemical Society* 25, no. 6 (2021): 101243. https://doi.org/10.1016/j.jscs.2021.101243

Al-Radadi, Najlaa S. *Arabian Journal of Chemistry* 12, no. 3 (2019): 330–349. https://doi.org/10.1016/j.arabjc.2018.05.008

Al-Shabib, Nasser A., Fohad Mabood Husain, Faizan Abul Qais, Naushad Ahmad, Altaf Khan, Abdullah A. Alyousef, Mohammed Arshad, et al. *Frontiers in Microbiology* (2020): 1680.https://doi.org/10.3389/fmicb.2020.01680

Alshatwi, Ali A., Jegan Athinarayanan, and Periasamy Vaiyapuri Subbarayan. *Journal of Materials Science: Materials in Medicine* 26, no. 1 (2015): 1–9.

Alven, Sibusiso, and Blessing Atim Aderibigbe. *Pharmaceutics* 12, no. 8 (2020): 748.

Amarnath, Kanchana, Nina Liza Mathew, Jayshree Nellore, Chagam Reddy Venkat Siddarth, and Jayanthi Kumar. *Cancer Nanotechnology* 2, no. 1 (2011): 121–132. https://doi.org/10.1007/s12645-011-0022-8

Amuthavalli, Pandiyan, Jiang-Shiou Hwang, Hans-Uwe Dahms, Lan Wang, Jagannathan Anitha, Murugan Vasanthakumaran, Arumugam Dhanesh Gandhi et al. *Scientific Reports* 11, no. 1 (2021): 1–13.

Anand, Krishnan, Charlette Tiloke, Alisa Phulukdaree, B. Ranjan, A. Chuturgoon, S. Singh, and Robert Moonsamy Gengan. *Journal of Photochemistry and Photobiology B: Biology* 165 (2016): 87–95. https://doi.org/10.1016/j.jphotobiol.2016.09.039

Anand, Uttpal, M. Carpena, Monika Kowalska-Góralska, P. Garcia-Perez, Kumari Sunita, Elza Bontempi, Abhijit Dey, Miguel A. Prieto, Jarosław Proćków, and Jesus Simal-Gandara. *Science of The Total Environment* (2022): 153472. https://doi.org/10.1016/j.scitotenv.2022.153472

Anandhi, D., G. Kanaga, S. Jagatheesh, and K. Revathi. *Drug Invention Today* 12, no. 6 (2019).

Anjum, Sumaira, Mariam Hashim, Sara Asad Malik, Maha Khan, José M. Lorenzo, Bilal Haider Abbasi, and Christophe Hano. *Cancers* 13, no. 18 (2021): 4570. https://doi.org/10.3390/cancers13184570

Anu, Kasi, Sandhanasamy Devanesan, Ramesh Prasanth, Mohamad S. AlSalhi, Singaravelu Ajithkumar, and Ganesan Singaravelu. *Journal of King Saud University-Science* 32, no. 4 (2020): 2520–2526. https://doi.org/10.1016/j.jksus.2020.04.018

Aparna Mani KM, Seethalakshmi S, Gopal V. *Journal of Nanomedicine and Nanotechnology* 6, no. 2 (2015): 1–5.

Araujo, Victor Hugo Sousa, Jonatas Lobato Duarte, Gabriela Corrêa Carvalho, Amanda Letícia Polli Silvestre, Bruno Fonseca-Santos, Gabriel Davi Marena, Tais de Cassia Ribeiro, Matheus Aparecido dos Santos Ramos, Taís Maria Bauab, and Marlus Chorilli. *Critical Reviews in Microbiology* 46, no. 5 (2020): 508–547.

Arodiya, Faruk, Chirag Makvana, and Kokila Parmar. *Biosciences Biotechnology Research Asia* 18, no. 4 (2021): 691.

Arvanag, Farid Mohammadi, Abolfazl Bayrami, Aziz Habibi-Yangjeh, and Shima Rahim Pouran. *Materials Science and Engineering: C* 97 (2019): 397–405. https://doi.org/10.1016/j.msec.2018.12.058

Aseyd Nezhad, Saynaz, Ali Es-Haghi, and Masoud Homayouni Tabrizi. *Applied Organometallic Chemistry* 34, no. 2 (2020): e5314. https://doi.org/10.1002/aoc.5314

Asha Monica, A., and Senthilkumar SR. *Research J. Pharm. and Tech* 13, no. 12: December (2020).

Ashna, Mohammad, Ali Es-Haghi, Maryam Karimi Noghondar, Dheyaa Al Amara, and Mohammad Ehsan Taghavizadeh Yazdi. *Materials Technology* (2020): 1–8. https://doi.org/10.1080/10667857.2020.1863558

Ashokkumar, Thirunavukkarasu, Durai Prabhu, Ravi Geetha, Kasivelu Govindaraju, Ramar Manikandan, Chinnasamy Arulvasu, and Ganesan Singaravelu. *Colloids and Surfaces B: Biointerfaces* 123 (2014): 549–556. https://doi.org/10.1016/j.colsurfb.2014.09.051

Augustine, Robin, and Anwarul Hasan. *Journal of Drug Delivery Science and Technology* 56 (2020): 101516.

Aygun, Aysenur, Fulya Gülbagca, Lutfiye Yildiz Ozer, Buket Ustaoglu, Yasemin Celik Altunoglu, Mehmet Cengiz Baloglu, Mehmet Nuri Atalar, Mehmet Hakkı Alma, and Fatih Sen. *Journal of Pharmaceutical and Biomedical Analysis* 179 (2020): 112961. https://doi.org/10.1016/j.jpba.2019.112961

Azizi, Susan, Rosfarizan Mohamad, Raha Abdul Rahim, Amin Boroumand Moghaddam, Mona Moniri, Arbakariya Ariff, Wan Zuhainis Saad, and Farideh Namvab. *Applied Surface Science* 384 (2016): 517–524. https://doi.org/10.1016/j.apsusc.2016.05.052

Badri, Waisudin, A. El Asbahani, Karim Miladi, Abdellatif Baraket, Géraldine Agusti, Qand Agha Nazari, Abdelhamid Errachid, Hatem Fessi, and Abdelhamid Elaissari. *Journal of Drug Delivery Science and Technology* 46 (2018): 234–242. https://doi.org/10.1016/j.jddst.2018.05.022

Badri, Waisudin, Karim Miladi, Rkia Eddabra, Hatem Fessi, and Abdelhamid Elaissari. *Journal of Nanomaterials* 2015 (2015). https://doi.org/10.1155/2015/935439

Bagirova, Melahat, Sahar Dinparvar, Adil M. Allahverdiyev, Kubra Unal, Emrah Sefik Abamor, and Metanet Novruzova. *Acta Tropica* 208 (2020): 105498.

Bagyalakshmi, J., and H. Haritha. *American Journal of Advanced Drug Delivery ISSN* (2017).

Bala, Niranjan, S. Saha, M. Chakraborty, M. Maiti, S. Das, R. Basu, and P. Nandy. *RSC Advances* 5, no. 7 (2015): 4993–5003. https://doi.org/10.1039/C4RA12784F

Balan, Kannan, Weixia Qing, Youyou Wang, Xiuhua Liu, Thayumanavan Palvannan, Yong Wang, Fanyi Ma, and Yun Zhang. *RSC Advances* 6, no. 46 (2016): 40162–40168. https://doi.org/10.1039/C5RA24391B

Balasubramani, Govindasamy, Rajendiran Ramkumar, Narayanaswamy Krishnaveni, Annamalai Pazhanimuthu, Thillainathan Natarajan, Rajamani Sowmiya, and Pachiappan Perumal. *Journal of Trace Elements in Medicine and Biology* 30 (2015): 83–89. https://doi.org/10.1016/j.jtemb.2014.11.001

Balasubramanian, S., Kala, S.M.J., Pushparaj, T.L. *Journal of Drug Delivery Science and Technology* 57 (2020): 101620. https://doi.org/10.1016/j.jddst.2020.101620

Baldea, Ioana, Adrian Florea, Diana Olteanu, Simona Clichici, Luminita David, Bianca Moldovan, Mihai Cenariu, et al. *Nanomedicine* 15, no. 1 (2020): 55–75.

Balyan, Prairna, Shruti Shinde, and Ahmad Ali. *Journal of Phytonanotechnology and Pharmaceutical Sciences* 1, no. 2 (2021): 1–9.

Bangroo, A, Malhotra, A.; Sharma, U.; Jain, A. and Kaur, A. *Journal of Inorganic and Organometallic Polymers and Materials.* 2021, in press. https://doi.org/10.3390/cancers13184570

Bano, Farah, Muhammad Baber, Amjad Ali, Ziaullah Shah, and Syed Aun Muhammad. *Pharmacognosy Magazine* 13, no. Suppl 1 (2017): S33.

Banu, Hussaina, N. Renuka, S. M. Faheem, Raees Ismail, Vinita Singh, Zahra Saadatmand, Saad Sultan Khan et al. *Biological Trace Element Research* 186, no. 1 (2018): 122–134.

Bavanilatha, M., L. Yoshitha, S. Nivedhitha, S. Sahithya. *Biocatal. Agric. Biotechnol.* 19 (2019): 101131. https://doi.org/10.1016/j.bcab.2019.101131

Bayrami, Abolfazl, Sanaz Alioghli, Shima Rahim Pouran, Aziz Habibi-Yangjeh, Alireza Khataee, and S. Ramesh. *Ultrasonics Sonochemistry* 55 (2019): 57–66. https://doi.org/10.1016/j.ultsonch.2019.03.010

Bhagwat, Tanay R., Komal A. Joshi, Vijay Singh Parihar, Adersh Asok, Jayesh Bellare, and Sougata Ghosh. *World Journal of Pharmaceutical Research* 7, no. 4 (2018): 183–196.

Bharathi, Devaraj, Selvaraj Preethi, Kalimuthu Abarna, Manjunathan Nithyasri, Palaniswamy Kishore, and Kathirvel Deepika. *Biocatalysis and Agricultural Biotechnology* 27 (2020): 101698. https://doi.org/10.1016/j.bcab.2020.101698

Bhattacharya, T., Chopra, H., Rahman, M., Hasan, Z., Swain, S. S., and Cavalu, S. 15, no. 3 (2022): 804. https://doi.org/10.3390/ma15030804

Bhuiyan, Md Shakhawat Hossen, Muhammed Yusuf Miah, Shujit Chandra Paul, Tutun Das Aka, Otun Saha, Md Mizanur Rahaman, Md Jahidul Islam Sharif, Ommay Habiba, and Md Ashaduzzaman. *Heliyon* 6, no. 8 (2020): e04603. https://doi.org/10.1016/j.heliyon.2020.e04603

Bitencourt, Paula Eliete Rodrigues, Carolina dos Santos Stein, Raphaela Maleski Borges, and Maria Beatriz Moretto. In Diabetology & Metabolic Syndrome 7, no. 1 (2015): 1–2. *BioMed Central.* https://doi.org/10.1186/1758-5996-7-S1-A2

Boateng-Marfo, Yaa, Yuancai Dong, Wai Kiong Ng, and Hai-Shu Lin. *International Journal of Molecular Sciences* 22, no. 3 (2021): 1141.

Boomi, P., R. M. Ganesan, G. Poorani, H. Gurumallesh Prabu, S. Ravikumar, and J. Jeyakanthan. *Materials Science and Engineering: C* 99 (2019): 202–210. https://doi.org/10.1016/j.msec.2019.01.105

Borase, H. P., C. D. Patil, R. B. Salunkhe, C. P. Narkhede, B. K. Salunke, and S. V. Patil. *Journal of Nanomedicine and Biotherapeutic Discovery* 3, no. 1 (2013): 1–7.

Brunet, Kevin, Alexandre Alanio, Olivier Lortholary, and Blandine Rammaert. *Journal of Infection* 77, no. 6 (2018): 463–468.

Burlacu, Ema, Corneliu Tanase, Năstaca-Alina Coman, and Lavinia Berta. *Molecules* 24, no. 23 (2019): 4354. https://doi.org/10.3390/molecules24234354

Campoy, Abraham Heriberto Garcia, Rosa Martha Perez Gutierrez, Gabriela Manriquez-Alvirde, and Alethia Muñiz Ramirez. *International Journal of Nanomedicine* 13 (2018): 2601. https://dx.doi.org/10.2147%2FIJN.S163714

Carbone, C., V. Fuochi, A. Zielińska, T. Musumeci, Eliana B. Souto, A. Bonaccorso, C. Puglia, G. Petronio, and P. M. Furneri. *Colloids and Surfaces B: Biointerfaces* 186 (2020): 110705.

Carding, Simon, Kristin Verbeke, Daniel T. Vipond, Bernard M. Corfe, and Lauren J. Owen. *Microbial Ecology in Health and Disease* 26, no. 1 (2015): 26191.

Carvalho, Sílvia H, Frédéric Frézard, Neila P. Pereira, Alexandre S. Moura, Lucinéia MQC Ramos, Gabriel B. Carvalho, and Manoel OC Rocha. *Tropical Medicine & International Health* 24, no. 4 (2019): 380–391.

Chaubey, Pramila, Brahmeshwar Mishra, Shyam Lal Mudavath, Ravi R. Patel, Sundeep Chaurasia, Shyam Sundar, Vasanti Suvarna, and Marvis Monteiro. *International Journal of Biological Macromolecules* 111 (2018): 109–120.

Chauhan, Vikash P. and Rakesh K. Jain. *Nature Materials* 12, no. 11 (2013): 958–962.

Chen, Chaoxiang, Jialin Wang, Mengdi Sun, Jian Li, and Hui-Min David Wang. *Biomedicine & Pharmacotherapy* 145 (2022): 112416.

Chen, Jingqin, Yaoyan Li, Gang Fang, Zhiyong Cao, Yuzhi Shang, Saleh Alfarraj, Sulaiman Ali Alharbi, Jingjing Li, Siyuan Yang, Xuelin Duan. *Arabian Journal of Chemistry* 14 (2021): 103000. https://doi.org/10.1016/j.arabjc.2021.103000

Chinnathambi, Arunachalam, and Tahani Awad Alahmadi. *Arabian Journal of Chemistry* 14, no. 4 (2021): 103083. https://doi.org/10.1016/j.arabjc.2021.103083

Chockalingam, Shivashri, Rajarajeshwari Thada, Ramesh Kumar Dhandapani, and Rajasekar Panchamoorthy. *Biotechnology Progress* 31, no. 4 (2015): 1096–1106. https://doi.org/10.1002/btpr.2112

Chowdhury, Anindita, Selvaraj Kunjiappan, Chiranjib Bhattacharjee, Balasubramanian Somasundaram, and Theivendren Panneerselvam. *In Vitro Cellular & Developmental Biology-Animal* 53, no. 6 (2017): 483–493. https://doi.org/10.1007/s11626-017-0136-3

Chung, Ill-Min, Abdul Rahuman, Sampath Marimuthu, Arivarasan Vishnu Kirthi, Karunanithi Anbarasan, Parthasarathy Padmini, and Govindasamy Rajakumar. *Experimental and Therapeutic Medicine* 14, no. 1 (2017): 18–24. https://doi.org/10.3892/etm.2017.4466

Chung, Ill-Min, Abdul Rahuman, Sampath Marimuthu, Arivarasan Vishnu Kirthi, Karunanithi Anbarasan, and Govindasamy Rajakumar. *Nanomaterials* 5, no. 3 (2015): 1317–1330. https://doi.org/10.3390/nano5031317

Chung, Ill-Min, Inmyoung Park, Kim Seung-Hyun, Muthu Thiruvengadam, and Govindasamy Rajakumar. *Nanoscale Research Letters* 11, no. 1 (2016): 1–14.

Climent, Núria, Isabel García, Marco Marradi , Fabrizio Chiodo, Laia Miralles, María José Maleno, José María Gatell, Felipe García, Soledad Penadés, and Montserrat Plana. *Nanomedicine: Nanotechnology, Biology and Medicine* 14, no. 2 (2018): 339–351.

Conte, Raffaele, Valentina Marturano, Gianfranco Peluso, Anna Calarco, and Pierfrancesco Cerruti. *International Journal of Molecular Sciences* 18, no. 4 (2017): 709.

Cosco, Donato, Federica Bruno, Germano Castelli, Roberto Puleio, Sonia Bonacci, Antonio Procopio, Domenico Britti, Massimo Fresta, Fabrizio Vitale, and Donatella Paolino. *Macromolecular Bioscience* 21, no. 7 (2021): 2100046.

Cui, Dongxiao, Tingting Liang, Liqian Sun, Liqiang Meng, Congcong Yang, Liwei Wang, Taigang Liang, and Qingshan Li. *Pharmaceutical Biology* 56, no. 1 (2018): 528–534. https://doi.org/10.1080/13880209.2018.1510974

Cui, Wei, Junbai Li, and Gero Decher. *Advanced Materials* 28, no. 6 (2016): 1302–1311. https://doi.org/10.1002/adma.201502479

Cui, Wenjuan, Jinru Li, Yakun Zhang, Huilin Rong, Wensheng Lu, and Long Jiang. *Nanomedicine: Nanotechnology, Biology and Medicine* 8, no. 1 (2012): 46–53. https://doi.org/10.1016/j.nano.2011.05.005

Daisy, P., and K. Saipriya. *International Journal of Nanomedicine* 7 (2012): 1189.

Das, Gitishree, Han-Seung Shin, and Jayanta Kumar Patra. *International Journal of Nanomedicine* 15 (2020): 9075. https://doi.org/10.2147/IJN.S277625

Das, Gitishree, Jayanta Kumar Patra, Chethala N. Vishnuprasad Nagaraj Basavegowda, and Han-Seung Shin. *International Journal of Nanomedicine* 14 (2019): 4741. https://dx.doi.org/10.2147%2FIJN.S210517

Das, Gitishree, Jayanta Kumar Patra, Trishna Debnath, Abuzar Ansari, and Han-Seung Shin. *PloS One* 14, no. 8 (2019): e0220950.

Das, J., M. Paul Das, and P. Velusamy. *Spectrochimica Acta Part A: Molecular and Biomolecular Spectroscopy* 104 (2013a): 265–270. https://doi.org/10.1016/j.saa.2012.11.075

Das, Sreemanti, Jayeeta Das, Asmita Samadder, Soumya Sundar Bhattacharyya, Durba Das, and Anisur Rahman Khuda-Bukhsh. *Colloids and Surfaces B: Biointerfaces* 101 (2013b): 325–336. https://doi.org/10.1016/j.colsurfb.2012.07.008

Das, Swagat Kumar, Supriti Behera, Jayanta Kumar Patra, and Hrudayanath Thatoi. *Journal of Cluster Science* 30, no. 4 (2019): 1103–1113.

Dass, K. Larvicidal and pupicidal activity of green synthesized silver nanoparticles using selected plants extract against Culex quinquefasciatus (2020).

Davis, Mark E., Zhuo Chen, and Dong M. Shin. *Nanoscience and Technology: A Collection of Reviews from Nature Journals* (2010): 239–250. https://doi.org/10.1142/9789814287005_0025

de Almeida, Mauro Sousa, Eva Susnik, Barbara Drasler, Patricia Taladriz-Blanco, Alke Petri-Fink, and Barbara Rothen-Rutishauser. *Chemical Society Reviews* 50, no. 9 (2021): 5397–5434. https://doi.org/10.1039/D0CS01127D

Dehvari, M., and Ghahghaei, A. *Int. J. Biol. Macromol.* 2018: 108, 1128–1139.

Deng, Wenji, Huan Wang, Baojian Wu, and Xingwang Zhang. *Acta Pharmaceutica Sinica B* 9, no. 1 (2019): 74–86.

Depciuch, Joanna, Małgorzata Stec, Alexey Maximenko, E. Drzymała, M. Pawlyta, Jarosław Baran, and Magdalena Parlinska-Wojtan. *Applied Organometallic Chemistry* 34, no. 3 (2020): e5401. https://doi.org/10.1002/aoc.5401

Deshmukh, Aarti R., and Beom Soo Kim. In: *Phytonanotechnology*, pp. 97–121. Elsevier, 2020. https://doi.org/10.1016/B978-0-12-822348-2.00006-1

Dewanjee, Saikat, Pratik Chakraborty, Biswajit Mukherjee, and Vincenzo De Feo. *International Journal of Molecular Sciences* 21, no. 6 (2020): 2217. https://dx.doi.org/10.3390%2Fijms21062217

Dhamgaye, Sanjiveeni, Frédéric Devaux, Patrick Vandeputte, Nitesh Kumar Khandelwal, Dominique Sanglard, Gauranga Mukhopadhyay, and Rajendra Prasad. *PLoS One* 9, no. 8 (2014): e104554.

Dhamgaye, Sanjiveeni, Frederic Devaux, Raman Manoharlal, Patrick Vandeputte, Abdul Haseeb Shah, Ashutosh Singh, Corinne Blugeon, Dominique Sanglard, and Rajendra Prasad. *Antimicrobial Agents and Chemotherapy* 56, no. 1 (2012): 495–506.

Dharmatti, Roopa, Chinmay Phadke, Ashmi Mewada, Mukeshchand Thakur, Sunil Pandey, and Madhuri Sharon. *Materials Science and Engineering: C* 44 (2014): 92–98. https://doi.org/10.1016/j.msec.2014.08.006

Dhas, T. Stalin, V. Ganesh Kumar, V. Karthick, K. Vasanth, G. Singaravelu, and K. Govindaraju. *Enzyme and Microbial Technology* 95 (2016): 100–106. https://doi.org/10.1016/j.enzmictec.2016.09.003

Din, Fakhar, Waqar Aman, Izhar Ullah, Omer Salman Qureshi, Omer Mustapha, Shumaila Shafique, and Alam Zeb. *International Journal of Nanomedicine* 12 (2017): 7291. https://doi.org/10.2147%2FIJN.S146315

Dobrucka, Renata, Aleksandra Romaniuk-Drapała, and Mariusz Kaczmarek. *Biomedical Microdevices* 21, no. 3 (2019): 1–10.

Dong, Yi, Guangming Wan, Panshi Yan, Cheng Qian, Fuzhen Li, and Guanghua Peng. *Journal of Photochemistry and Photobiology B: Biology* 195 (2019): 51–57. https://doi.org/10.1016/j.jphotobiol.2019.04.012.

Donga, Savan, and Sumitra Chanda. *Artificial Cells, Nanomedicine, and Biotechnology* 49, no. 1 (2021): 292–302.

Donia, D. T., and M. Carbone. *International Journal of Environmental Science and Technology* 16, no. 1 (2019): 583–600. https://doi.org/10.1007/s13762-018-1960-z

Dos Santos, Talitha Caldas, Nicoletta Rescignano, Laurita Boff, Flávio Henrique Reginatto, Cláudia Maria Oliveira Simões, Angela Machado de Campos, and Carmen Mijangos. *Materials Science and Engineering: C* 75 (2017): 1214–1220.

Du, Juan, Hina Singh, and Tae-Hoo Yi. *Bioprocess and Biosystems Engineering* 39, no. 12 (2016): 1923–1931.

Dulta, Kanika, Gözde Koşarsoy Ağçeli, Parveen Chauhan, Rohit Jasrotia, and P. K. Chauhan. *Journal of Inorganic and Organometallic Polymers and Materials* 31, no. 1 (2021): 180–190.

Edis, Zehra, Samir Haj Bloukh, May Reda Ibrahim, and Hamed Abu Sara. *Pharmaceutics* 12, no. 4 (2020): 361.

Elbialy, Nihal Saad, Eman A. Abdelfatah, and Wafaa A. Khalil. *BioNanoScience* 9, no. 4 (2019): 813–820.

El-Borady, Ola M., Mahmoud S. Ayat, Mostafa A. Shabrawy, and Pierre Millet. *Advanced Powder Technology* 31, no. 10 (2020): 4390–4400.

Elekofehinti, Olusola Olalekan. *Egyptian Journal of Medical Human Genetics* 23, no. 1 (2022): 1–10. https://doi.org/10.1186/s43042-022-00282-0

Elemike, Elias E., Damian C. Onwudiwe, Anthony C. Ekennia, Richard C. Ehiri, and Nnaemeka J. Nnaji. *Materials Science and Engineering: C* 75 (2017): 980–989.

El-Rafie, Hanaa Mohamed, and Manal Abdel-Aziz Hamed. *Advances in Natural Sciences: Nanoscience and Nanotechnology* 5, no. 3 (2014): 035008.

Fahmy, Sherif Ashraf, Eduard Preis, Udo Bakowsky, and Hassan Mohamed El-Said Azzazy. *Molecules* 25, no. 21 (2020a): 4981. https://doi.org/10.3390/molecules25214981

Fahmy, Sherif Ashraf, Eduard Preis, Udo Bakowsky, and Hassan Mohamed El-Said Azzazy. *Materials* 13, no. 17 (2020b): 3661. https://doi.org/10.3390/ma13173661

Farah, Mohammad Abul, Mohammad Ajmal Ali, Shen-Ming Chen, Ying Li, Fahad Mohammad Al-Hemaid, Faisal Mohammad Abou-Tarboush, Khalid Mashay Al-Anazi, and Joongku Lee. *Colloids and Surfaces B: Biointerfaces* 141 (2016): 158–169. https://doi.org/10.1016/j.colsurfb.2016.01.027

Ferreira, Marianna Araújo, Renato Ferreira de Almeida Júnior, Thiago Souza Onofre, Bruna Renata Casadei, Kleber Juvenal Silva Farias, Patricia Severino, Camilo Flamarion de Oliveira Franco, Fernanda Nervo Raffin, Túlio Flávio Accioly de Lima e Moura, and Raquel de Melo Barbosa. *Pharmaceutics* 13, no. 11 (2021): 1912.

Firdhouse, M. Jannathul, and P. Lalitha. *Cancer Nanotechnology* 4, no. 6 (2013): 137–143.

Frayha, George J., J. D. Smyth, Jean Gerard Gobert, and Jean Savel. *General Pharmacology: The Vascular System* 28, no. 2 (1997): 273–299.

Fumagalli, Gaia, Michael S. Christodoulou, Benedetta Riva, Inigo Revuelta, Cristina Marucci, Veronica Collico, Davide Prosperi, et al. *Organic & Biomolecular Chemistry* 15, no. 5 (2017): 1106–1109.https://doi.org/10.1039/C6OB02591A

Geetha, Ravi, Thirunavukkarasu Ashokkumar, Selvaraj Tamilselvan, Kasivelu Govindaraju, Mohamed Sadiq, and Ganesan Singaravelu. *Cancer Nanotechnology* 4, no. 4 (2013): 91–98. https://doi.org/10.1007/s12645-013-0040-9

George, Dhanya, P. Uma Maheswari, and KM Meera Sheriffa Begum. *Carbohydrate Polymers* 236 (2020): 116101.

George, Dhanya, P. Uma Maheswari, and KM Meera Sheriffa Begum. *International Journal of Biological Macromolecules* 132 (2019): 784–794.

Ghasemi, Razieh, and Mahkameh Baghernia. *Personalized Medicine Journal* 6, no. 23 (2021): 30–34.

Ghosh, Santanu, Nabanita Kar, and Tanmoy Bera. *International Journal of Biological Macromolecules* 93 (2016): 961–970.

Ghosh, Sougata, Piyush More, Rahul Nitnavare, Soham Jagtap, Rohan Chippalkatti, Abhishek Derle, Rohini Kitture, et al. *Journal of Nanomedicine & Nanotechnology* S6 (2015): 1. http://doi.org/10.4172/2157-7439.S6-007

Ghosh, Sougata, Rahul Nitnavare, Ankush Dewle, Geetanjali B. Tomar, Rohan Chippalkatti, Piyush More, Rohini Kitture, Sangeeta Kale, Jayesh Bellare, and Balu A. Chopade. *International Journal of Nanomedicine* 10 (2015): 7477. https://doi.org/10.2147%2FIJN.S91579

Giljohann, David A., Dwight S. Seferos, Weston L. Daniel, Matthew D. Massich, Pinal C. Patel, and Chad A. Mirkin. *Spherical Nucleic Acids* 49, no. 19 (2020): 55–90. https://doi.org/10.1002/anie.200904359

Gnanasekar, Sathishkumar, Jeyaraj Murugaraj, Balakrishnan Dhivyabharathi, Varunkumar Krishnamoorthy, Pradeep K. Jha, Prabukumar Seetharaman, Ravikumar Vilwanathan, and Sivaramakrishnan Sivaperumal. *Journal of Applied Biomedicine* 16, no. 1 (2018): 59–65. https://doi.org/10.1016/j.jab.2017.10.001

Goel, Vanshita, Pawandeep Kaur, Lachhman Das Singla, and Diptiman Choudhury. *Frontiers in Molecular Biosciences* (2020): 396.

González-Ballesteros, N., S. Prado-López, J. B. Rodríguez-González, M. Lastra, and M.C. Rodríguez-Argüelles. *Colloids and Surfaces B: Biointerfaces* 153 (2017): 190–198. https://doi.org/10.1016/j.colsurfb.2017.02.020

Govindappa, M., B. Hemashekhar, Manoj-Kumar Arthikala, V. Ravishankar Rai, and Y. L. Ramachandra. *Results in Physics* 9 (2018): 400–408.

Govindaraju, Saravanan, Arunkumar Rengaraj, Roshini Arivazhagan, Yun-Suk Huh, and Kyusik Yun. *Bioconjugate Chemistry* 29, no. 2 (2018): 363–370. https://doi.org/10.1021/acs.bioconjchem.7b00683

Groiss, Silvia, Raja Selvaraj, Thivaharan Varadavenkatesan, and Ramesh Vinayagam. *Journal of Molecular Structure* 1128 (2017): 572–578.

Gul, Anam Rana, Farhat Shaheen, Rafia Rafique, Jyotiranjan Bal, Shahid Waseem, and Tae Jung Park. *Chemical Engineering Journal* 407 (2021): 127202. https://doi.org/10.1016/j.cej.2020.127202

Gulbagça, Fulya, Aysenur Aygun, Elif Esra Altuner, Muhammed Bekmezci, Tugba Gur, Fatih Sen, Hassan Karimi-Maleh, Najmeh Zare, Fatemeh Karimi, and Yasser Vasseghian. *Chemical Engineering Research and Design* 180 (2022): 254–264.

Gulbagca, Fulya, Ayşenur Aygün, Mehmet Gülcan, Sadin Ozdemir, Serpil Gonca, and Fatih Şen. *Applied Organometallic Chemistry* 35, no. 8 (2021): e6272. https://doi.org/10.1002/aoc.6272

Gupta, Ritu, and Huan Xie. *Journal of Environmental Pathology, Toxicology and Oncology* 37, no. 3 (2018). https://doi.org/10.1615/JEnvironPatholToxicolOncol.2018026009

Gurunathan, Sangiliyandi, EunSu Kim, Jae Woong Han, Jung Hyun Park, and Jin-Hoi Kim. *Molecules* 20, no. 12 (2015): 22476–22498. https://doi.org/10.3390/molecules201219860

Harne, Shrikant, Ashwinikumar Sharma, Mayur Dhaygude, Shreeram Joglekar, Kisan Kodam, and Manish Hudlikar. *Colloids and Surfaces B: Biointerfaces* 95 (2012): 284–288. https://doi.org/10.1016/j.colsurfb.2012.03.005

Hashem, Amr H., and Salem S. Salem. *Biotechnology Journal* 17, no. 2 (2022): 2100432. https://doi.org/10.1002/biot.202100432

Hassanien, Reda, Amro AI Abed-Elmageed, and Dalal Z. Husein. *Chemistry Select* 4, no. 31 (2019): 9018–9026. https://doi.org/10.1002/slct.201901267

Hassanien, Reda, Dalal Z. Husein, and Mostafa F. Al-Hakkani. *Heliyon* 4, no. 12 (2018): e01077. https://doi.org/10.1016/j.heliyon.2018.e01077

He, Yan, Zhiyun Du, Shijing Ma, Shupeng Cheng, Sen Jiang, Yue Liu, Dongli Li, Huarong Huang, Kun Zhang, and Xi Zheng. *Nanoscale Research Letters* 11, no. 1 (2016): 1–10.

Hochella Jr, Michael F., David W. Mogk, James Ranville, Irving C. Allen, George W. Luther, Linsey C. Marr, B. Peter McGrail, et al. *Science* 363, no. 6434 (2019): eaau8299. https://doi.org/10.1126/science.aau8299

Hosny, Mohamed, and Fawzy, Manal. *Advanced Powder Technology* 32, no. 8 (2021) 2891–2904. https://doi.org/10.1016/j.apt.2021.06.004

Hosseinzadeh, Neda, Tahoora Shomali, Saeid Hosseinzadeh, Fatemeh Raouf Fard, Maryam Pourmontaseri, and Mehdi Fazeli. *Journal of Pharmacy and Pharmacology* 72, no. 8 (2020): 1013–1025. https://doi.org/10.1111/jphp.13274

Hussain, Afzal, Mohammad Oves, Mohamed F. Alajmi, Iqbal Hussain, Samira Amir, Jahangeer Ahmed, Md Tabish Rehman, Hesham R. El-Seedi, and Imran Ali. *RSC Advances* 9, no. 27 (2019): 15357–15369. https://doi.org/10.1039/C9RA01659G

Hussein, Jihan, Mohamed F. Attia, Mona El Bana, Sherien M. El-Daly, Nadia Mohamed, Zakeria El-Khayat, and Mehrez E. El-Naggar. *International Journal of Biological Macromolecules* 140 (2019a): 1305–1314. https://doi.org/10.1016/j.ijbiomac.2019.08.201.

Hussein, J. S., W. Rasheed, T. Ramzy, M. Nabeeh, M. Harvy, S. El-Toukhy, O. Ali, J. Raafat, and M. El-Naggar. *Human & Experimental Toxicology* 38, no. 8 (2019b): 962–973. https://doi.org/10.1177/0960327119843586

Idowu, Emmanuel Taiwo, Adedapo O. Adeogun, Luqman A. Adams, Modupe A. Yusuf, Olawale W. Salami, Olakiigbe Abiodun Kanmi, Joseph A. Bello, Ifeoluwa Kayode Fagbohun, Olubunmi Adetoro Otubanjo, and Taiwo Samson Awolola. *The Journal of Basic and Applied Zoology* 82, no. 1 (2021): 1–9.

Ikram, Muhammad, Bilal Javed, Naveed Iqbal Raja, and Zia-ur-Rehman Mashwani. *International Journal of Nanomedicine* 2021, no. 16 (2021): 249. https://doi.org/10.2147%2FIJN.S295053

Imraish, Amer, Tuqa Abu Thiab, Wajdy Al-Awaida, Hamzeh J. Al-Ameer, Yasser Bustanji, Hana Hammad, Mays Alsharif, and Afnan Al-Hunaiti. *Journal of Food Biochemistry* 45, no. 6 (2021): e13730.

Iqbal, Javed, Banzeer Ahsan Abbasi, Riaz Ahmad, Amir Shahbaz, Syeda Anber Zahra, Sobia Kanwal, Akhtar Munir, Atiya Rabbani, and Tariq Mahmood. *Journal of Molecular Structure* 1199 (2020): 126979. https://doi.org/10.1016/j.molstruc.2019.126979

Iqbal, Muhammad, Gisele Usanase, Kafia Oulmi, Fairouz Aberkane, Tahar Bendaikha, Hatem Fessi, Nadia Zine, Géraldine Agusti, El-Salhi Errachid, and Abdelhamid Elaissari. *Materials Research Bulletin* 79 (2016): 97–104. https://doi.org/10.1016/j.materresbull.2015.12.026

Isacfranklin, M., R. Yuvakkumar, G. Ravi, P. Kumar, B. Saravanakumar, Dhayalan Velauthapillai, Tahani Awad Alahmadi, and Sulaiman Ali Alharbi. *Applied Nanoscience* 11, no. 2 (2021): 699–708.

Jalal, Mohammad, Mohammad Azam Ansari, Mohammad A. Alzohairy, Syed Ghazanfar Ali, Haris M. Khan, Ahmad Almatroudi, and Mohammad Imran Siddiqui. *International Journal of Nanomedicine* 14 (2019): 4667.

Jamdade, Dhiraj A., Dishantsingh Rajpali, Komal A. Joshi, Rohini Kitture, Anuja S. Kulkarni, Vaishali S. Shinde, Jayesh Bellare, Kaushik R. Babiya, and Sougata Ghosh. *Advances in Pharmacological Sciences* 2019 (2019). https://doi.org/10.1155/2019/9080279

Jampílek, Josef, and Katarína Kráľová. In: *Nanotheranostics*, pp. 137–178. Springer, Cham, 2019.

Jan, Hasnain, Muzamil Shah, Anisa Andleeb, Shah Faisal, Aishma Khattak, Muhammad Rizwan, Samantha Drouet, Christophe Hano, and Bilal Haider Abbasi. *Oxidative Medicine and Cellular Longevity* 2021 (2021). https://doi.org/10.1155/2021/4786227

Jan, Hasnain, Roby Gul, Anisa Andleeb, Sana Ullah, Muzamil Shah, Mehnaz Khanum, Imran Ullah, Christophe Hano, and Bilal Haider Abbasi. *Journal of Saudi Chemical Society* 25, no. 8 (2021): 101297. https://doi.org/10.1016/j.jscs.2021.101297

Janakiraman, Vardhana, Mercy Jenifer, and S. Ramarajan. *International Journal of Pharmaceutical Science Review and Research.* 53, no. 1 (2018): 72–76.

Jasim, Neeran Obied, Tabark Kalid Hasan, and Hassen Flieh. *Journal of Global Pharma Technology* 10, no. 10 (2018): 1423–427.

Javadi, Fatemeh, Mohammad Ehsan Taghavizadeh Yazdi, Mohsen Baghani, and Ali Es-Haghi. *Materials Research Express* 6, no. 6 (2019): 065408.

Jeevanandam, Jaison, Ahmed Barhoum, Yen S. Chan, Alain Dufresne, and Michael K. Danquah. *Beilstein Journal of Nanotechnology* 9, no. 1 (2018): 1050–1074. https://doi.org/10.3762/bjnano.9.98

Jegadeesan, Gautham B., K. Srimathi, N. Santosh Srinivas, S. Manishkanna, and D. Vignesh. *Biocatalysis and Agricultural Biotechnology* 21 (2019): 101354.

Jeyaraj, M., G. Sathishkumar, G. Sivanandhan, D. MubarakAli, M. Rajesh, R. Arun, G. Kapildev, et al. *Colloids and Surfaces B: Biointerfaces* 106 (2013): 86–92. https://doi.org/10.1016/j.colsurfb.2013.01.027

Jini, D., and S. Sharmila. *Materials Today: Proceedings* 22 (2020): 432–438.

Jinu, U., Ma Gomathi, Ib Saiqa, N. Geetha, G. Benelli, and P. Venkatachalam. *Microbial Pathogenesis* 105 (2017): 86–95. https://doi.org/10.1016/j.micpath.2017.02.019

Jinu, Udayabhanu, Sankar Rajakumaran, Sengottayan Senthil-Nathan, Natesan Geetha, and Perumal Venkatachalam. *Physiological and Molecular Plant Pathology* 101 (2018): 163–171.

Jo, Dong Hyun, Jin Hyoung Kim, Tae Geol Lee, and Jeong Hun Kim. *Nanomedicine: Nanotechnology, Biology and Medicine* 11, no. 7 (2015): 1603–1611. https://doi.org/10.1016/j.nano.2015.04.015

Johnson, Preethi, Vennila Krishnan, Chitra Loganathan, Kavitha Govindhan, Vijayan Raji, Penislusshiyan Sakayanathan, Sudha Vijayan, Palanivel Sathishkumar, and Thayumanavan Palvannan. *Artificial Cells, Nanomedicine, and Biotechnology* 46, no. 7 (2018): 1488–1494.

Ju, Songwen, Jingyao Mu, Terje Dokland, Xiaoying Zhuang, Qilong Wang, Hong Jiang, Xiaoyu Xiang, et al. *Molecular Therapy* 21, no. 7 (2013): 1345–1357. https://doi.org/10.1038/mt.2013.64

KS, Uma Suganya, K. Govindaraju, D. Prabhu, C. Arulvasu, V. Karthick, and Niranjan Changmai. *Applied Surface Science* 371 (2016): 415–424. https://doi.org/10.1016/j.apsusc.2016.03.004

Kajani, Abolghasem Abbasi, Abdol-Khalegh Bordbar, Sayyed Hamid Zarkesh Esfahani, and Amir Razmjou. *RSC Advances* 6, no. 68 (2016): 63973–63983. https://doi.org/10.1039/C6RA09050H

Kalaiarasi, Kalamegam, Govindaraj Prasannaraj, Shivendra Vikram Sahi, and Perumal Venkatachalam. *Turkish Journal of Biology* 39, no. 2 (2015): 223–232.

Kalakotla, Shanker, Naradala Jayarambabu, G. Krishna Mohan, Rabiatul Basria SMN Mydin, and V. Ramamohan Gupta. *Colloids and Surfaces B: Biointerfaces* 174 (2019): 199–206. https://doi.org/10.1016/j.colsurfb.2018.11.014

Kanipandian, Nagarajan, and Ramasamy Thirumurugan. *Industrial Crops and Products* 55 (2014): 1–10. https://doi.org/10.1016/j.indcrop.2014.01.042

Kanjikar, Amarvani P., Aruna L. Hugar, and Ramesh L. Londonkar. *Drug Development and Industrial Pharmacy* 44, no. 3 (2018): 377–384. https://doi.org/10.1080/03639045.2017.1386205

Kaplan, Ayse. *Beni-Suef University Journal of Basic and Applied Sciences* 11, no. 1 (2022): 1–15.

Kavithaa, K., Paulpandi, M., Ponraj, T., Murugan, K., Sumathi, S. *Karbala International Journal of Modern Science* 2 (2016): 46–55.

Keshavamurthy, M., B.S. Srinath, and V. Ravishankar Rai. *Particulate Science and Technology* 36, no. 7 (2018): 785–790. https://doi.org/10.1080/02726351.2017.1302533

Khan, Tariq, Nazif Ullah, Mubarak Ali Khan, and Akhtar Nadhman. *Advances in Colloid and Interface Science* 272 (2019): 102017. https://doi.org/10.1016/j.cis.2019.102017

Khandanlou, Roshanak, Vinuthaa Murthy, Dhananjaya Saranath, and Hetal Damani. *Journal of Materials Science* 53, no. 5 (2018): 3106–3118.

Khandel, Pramila, Ravi Kumar Yadaw, Deepak Kumar Soni, Leeladhar Kanwar, and Sushil Kumar Shahi. *Journal of Nanostructure in Chemistry* 8, no. 3 (2018): 217–254. https://doi.org/10.1007/s40097-018-0267-4

Kiang, Tony KL, Kyle John Wilby, and Mary HH Ensom. *Drug Interactions in Infectious Diseases: Antimicrobial Drug Interactions* (2018): 515–549.

Kiruthika, N., and T. Somanathan. *Research Journal of Pharmacy and Technology* 11, no. 10 (2018): 4308–4312.

Kischkel, Brenda, Suélen A. Rossi, Samuel R. Santos, Joshua D. Nosanchuk, Luiz R. Travassos, and Carlos P. Taborda. *Frontiers in Cellular and Infection Microbiology* (2020): 463.

Kitture, Rohini, Kalyani Chordiya, Shankar Gaware, Sougata Ghosh, Piyush A. More, Parag Kulkarni, Balu A. Chopade, and S. N. Kale. *Journal of Nanoscience and Nanotechnology* 15, no. 6 (2015): 4046–4051. https://doi.org/10.1166/jnn.2015.10323

Koffi Kouame, Aniekan Imo Peter, Edidiong Nnamso Akang, Roshila Moodley, Edwin Coleridge Naidu, and Onyemaechi Okpara Azu. *Bosnian Journal of Basic Medical Sciences* 19, no. 2 (2019): 138.

Krajewska, Julia B., Olga Długosz, Maciej Sałaga, Marcin Banach, and Jakub Fichna. *International Journal of Pharmaceutics* 585 (2020): 119549.

Krishnaraj, C., P. Muthukumaran, R. Ramachandran, M. D. Balakumaran, and P. T. Kalaichelvan. *Biotechnology Reports* 4 (2014): 42–49. https://doi.org/10.1016/j.btre.2014.08.002

Kudle, Karunakar Rao, Manisha R. Donda, Ramchander Merugu, Y. Prashanthi, and M. P. Rudra. *International Journal of Drug Development and Research* 6, no. 1 (2014): 113–119.

Kulkarni, Badarinath Druvarao, Samim Sultana, Mayuri Bora, Ishita Dutta, Padmaa Milaap Paarakh, and Vedamurthy Ankala Basappa. *Pharmacognosy Journal* 8, no. 2 (2016): 127–131. https://doi.org/10.5530/pj.2016.2.5

Kumar, Brajesh, Kumari Smita, Luis Cumbal, Javier Camacho, Elisabeth Hernández-Gallegos, María de Guadalupe Chávez-López, Marcelo Grijalva, and Kleber Andrade. *Materials Science and Engineering: C* 62 (2016): 725–731. https://doi.org/10.1016/j.msec.2016.02.029

Kumar, Nirubama. *International Journal of ChemTech Research* 11, no. 03 (2018): 265–283.

Kumar, Sekar Dinesh, Ganesan Singaravelu, Singaravelu Ajithkumar, Kadarkarai Murugan, Marcello Nicoletti, and Giovanni Benelli. *Journal of Cluster Science* 28, no. 1 (2017): 359–367.

Kumar, Vijay, Simranjeet Singh, Bhavana Srivastava, Ragini Bhadouria, and Ravindra Singh. *Journal of Environmental Chemical Engineering* 7, no. 3 (2019): 103094. https://doi.org/10.1016/j.jece.2019.103094

Kummara, Sivaiah, Mrityunjaya B. Patil, and Tiewlasubon Uriah. *Biomedicine & Pharmacotherapy* 84 (2016): 10–21. https://doi.org/10.1016/j.biopha.2016.09.003

Langle, Ariadna, Marco Antonio González-Coronel, Genaro Carmona-Gutiérrez, José Albino Moreno-Rodríguez, Berenice Venegas, Guadalupe Muñoz, Samuel Treviño, and Alfonso Díaz. *Revista Brasileira de Farmacognosia* 25 (2015): 145–151.

Lautenschläger, Christian, Carsten Schmidt, Dagmar Fischer, and Andreas Stallmach. *Advanced Drug Delivery Reviews* 71 (2014): 58–76. https://doi.org/10.1016/j.addr.2013.10.001

Lee, Jeong-Ho, Palanivel Velmurugan, Jung-Hee Park, Kui-Jae Lee, Jong-Sik Jin, Yool-Jin Park, Keuk-Soo Bang, and Byung-Taek Oh. *Journal of Photochemistry and Photobiology B: Biology* 159 (2016): 229–236.

Leela, K., and D. C. Anchana. *Int. J. Pharm. Sci. Invent* 6, no. 10 (2017): 42–59.

León-Buitimea, Angel, Javier A. Garza-Cervantes, Diana Y. Gallegos-Alvarado, Macario Osorio-Concepción, and José Ruben Morones-Ramírez. *Pathogens* 10, no. 10 (2021): 1303.

Li, Jiulong, Qinghao Li, Xiaoqiong Ma, Bing Tian, Tao Li, Jiangliu Yu, Shang Dai, Yulan Weng, and Yuejin Hua. *International Journal of Nanomedicine* 11 (2016): 5931. https://doi.org/10.2147%2FIJN.S119618

Li, Ke, Gao Xiao, Joseph J. Richardson, Blaise L. Tardy, Hirotaka Ejima, Wen Huang, Junling Guo, Xuepin Liao, and Bi Shi. *Advanced Science* 6, no. 5 (2019): 1801688. https://doi.org/10.1002/advs.201801688

Li, Shuiqin, Fahad A. Al-Misned, Hamed A. El-Serehy, and Linlin Yang. *Arabian Journal of Chemistry* 14, no. 2 (2021): 102931. https://doi.org/10.1016/j.arabjc.2020.102931

Li, Yuanyuan, Jiaqiang Zhang, Jiangning Gu, and Shi Chen. *Journal of Photochemistry and Photobiology B: Biology* 169 (2017): 96–100. https://doi.org/10.1016/j.jphotobiol.2017.02.017

Liu, Hongtao, SiMin Zheng, HongFei Xiong, Mona S. Alwahibi, and Xiaoli Niu. *Arabian Journal of Chemistry* 13, no. 9 (2020): 6995–7006. https://doi.org/10.1016/j.arabjc.2020.07.006

Liu, Ying, Man He, Mengmeng Niu, Yiqing Zhao, Yuanzhang Zhu, Zhenhua Li, and Nianping Feng. *International Journal of Nanomedicine* 10 (2015): 3081.https://doi.org/10.2147%2FIJN.S79550

Liu, Yuntao, Siqi Zeng, Yixi Liu, Wenjuan Wu, Yingbin Shen, Lan Zhang, Cheng Li et al. *International Journal of Biological Macromolecules* 114 (2018): 632–639.

Loganathan, Settu, Muthugounder Subaramanian Shivakumar, Sengodan Karthi, Sengottayan Senthil Nathan, and Kuppusamy Selvam. *Toxicology Reports* 8 (2021): 64–72. https://doi.org/10.1016/j.toxrep.2020.12.018

Lokapur, Vinayak, Varsha Jayakar, M. S. Divakar, Raju Krishna Chalannavar, Lyned Lasrado, and Manjula Shantaram. *Materials Today Communications* 31 (2022): 103338. https://doi.org/10.1016/j.mtcomm.2022.103338

Ma, P., & Mumper, R. J. *Journal of nanomedicine & nanotechnology* 4, no. 2 (2013): 1000164. https://doi.org/10.4172%2F2157-7439.1000164

Machado, Sofia, Anabela Ferreira Gonçalves, Noelia González-Ballesteros, Luana Magalhães, Marisa Sárria Pereira Passos, Maria Carmen Rodríguez-Argüelles, and Andreia C. Gomes. (2019). https://doi.org/10.3389/conf.fmars.2019.08.00155

Macovei, Irina, Simon Vlad Luca, Krystyna Skalicka-Woźniak, Liviu Sacarescu, Petronela Pascariu, Alina Ghilan, Florica Doroftei, et al. *Molecules* 27, no. 1 (2021): 217.

Maddinedi, Sireesh, Badal Kumar Mandal, and Swetha Kameswari Maddili. *Journal of Photochemistry and Photobiology B: Biology* 167 (2017): 236–241.

Mahdizadeh, Roya, Masoud Homayouni-Tabrizi, Ali Neamati, Seyed Mohammad Reza Seyedi, and Haleh Sadat Tavakkol Afshari. *Journal of Cellular Biochemistry* 120, no. 10 (2019): 17984–17993. https://doi.org/10.1002/jcb.29065

Mahendran, D., P. B. Kavi Kishor, N. Geetha, T. Manish, S. V. Sahi, and P. Venkatachalam. *Journal of Experimental Nanoscience* 16, no. 1 (2021): 11–30.

Mahmoud, Dina B., Walaa M. Ismail, Yassmin Moatasim, Omnia Kutkat, Aliaa N. ElMeshad, Shahira M. Ezzat, Kadriya S. El Deeb, et al. *Journal of Drug Delivery Science and Technology* 66 (2021): 102845.

Mahmoud, Riham A., Mohamed S. Naser Eldin, Ali A. Ali, and Mohammed A. Hussein. *Acta Scientific Nutritional Health* 5, no. 2 (2021): 18–29.

Maity, Pralay, Madhubanti Bepari, Ananya Pradhan, Rathindranath Baral, Sumita Roy, and Sujata Maiti Choudhury. *Colloids and Surfaces B: Biointerfaces* 161 (2018): 111–120. https://doi.org/10.1016/j.colsurfb.2017.10.040

Majeed, Humaira, Tasveer Zahra Bokhari, Sikandar Khan Sherwani, Uzma Younis, Muhammad Hasnain Raza Shah, and Benish Khaliq. *Journal of Pharmacognosy and Phytochemistry* 2, no. 3 (2013).

Majumdar, Debatosh, Kyung-Ho Jung, Hongzheng Zhang, Sreenivas Nannapaneni, Xu Wang, ARM Ruhul Amin, Zhengjia Chen, and Dong M. Shin. *Cancer Prevention Research* 7, no. 1 (2014): 65–73. https://doi.org/10.1158/1940-6207.CAPR-13-0230

Malaikozhundan, Balasubramanian, Baskaralingam Vaseeharan, Sekar Vijayakumar, Karuppiah Pandiselvi, Mohamed Ali Rajamohamed Kalanjiam, Kadarkarai Murugan, and Giovanni Benelli. *Microbial Pathogenesis* 104 (2017): 268–277. https://doi.org/10.1016/j.micpath.2017.01.029

Malapermal, Veshara, Izel Botha, Suresh Babu Naidu Krishna, and Joyce Nonhlanhla Mbatha. *Saudi Journal of Biological Sciences* 24, no. 6 (2017): 1294–1305. https://doi.org/10.1016/j.sjbs.2015.06.026

Malhotra, Hitesh, Anjoo Kamboj, and Rupesh K. Gautam. In: *Medicinal Plants for Lung Diseases*, pp. 583–608. Springer, Singapore, 2021.

Malik, Parth, Ravi Shankar, Vibhuti Malik, Nitin Sharma, and Tapan Kumar Mukherjee. *Journal of Nanoparticles* 2014 (2014). https://doi.org/10.1155/2014/302429

Manam, Dr, Vishnu Kiran, and S. Murugesan. *Journal of Chemical and Pharmaceutical Research* 5, no. 12 (2013): 1001–1008.

Manam, Dr, Vishnu Kiran, and S. Murugesan. *American Journal of Bio-pharmacology Biochemistry and Life Sciences* 3, no. 01 (2014): 01–07.

Manikandan, Ramar, Beulaja Manikandan, Thiagarajan Raman, Koodalingam Arunagirinathan, Narayanan Marimuthu Prabhu, Muthuramalingam Jothi Basu, Muthulakshmi Perumal, Subramanian Palanisamy, and Arumugam Munusamy. *Spectrochimica Acta Part A: Molecular and Biomolecular Spectroscopy* 138 (2015): 120–129. https://doi.org/10.1016/j.saa.2014.10.043

Manimegalai, Thulasiraman, Krishnan Raguvaran, Manickam Kalpana, and Rajan Maheswaran. *Environmental Science and Pollution Research* 27, no. 34 (2020): 43103–43116.

Manna, Indrani, and Maumita Bandyopadhyay. In: *Nanomaterials in Plants, Algae and Microorganisms*, pp. 323–352. Academic Press, 2019. https://doi.org/10.1016/B978-0-12-811488-9.00016-0

Manna, Indrani, Debalina Das, Sejuty Mondal, and Maumita Bandyopadhyay. In *Pharmacotherapeutic Botanicals for Cancer Chemoprevention*, pp. 221–281. Springer, Singapore, 2020.

Manosroi, Aranya, Pensak Jantrawut, Eri Ogihara, Ayako Yamamoto, Makoto Fukatsu, Ken Yasukawa, Harukuni Tokuda, Nobutaka Suzuki, Jiradej Manosroi, and Toshihiro Akihisa. *Chemistry & Biodiversity* 10, no. 8 (2013): 1448–1463. https://doi.org/10.1002/cbdv.201300149

Mariappan, N. *Biomedical and Pharmacology Journal* 12, no. 3 (2019): 1095–1127.

Mata, Rani, Jayachandra Reddy Nakkala, and Sudha Rani Sadras. *Colloids and Surfaces B: Biointerfaces* 128 (2015): 276–286. https://doi.org/10.1016/j.colsurfb.2015.01.052

Mathur, Ruchi, and Rekha Vijayvergia. *Int J Biol Med Res* 8, no. 4 (2017): 6142–6145.

Mba, Ifeanyi Elibe, and Emeka Innocent Nweze. *World Journal of Microbiology and Biotechnology* 36, no. 11 (2020): 1–20.

Meena, Mukesh, Andleeb Zehra, Prashant Swapnil, Avinash Marwal, Garima Yadav, and Priyankaraj Sonigra. *Frontiers in Chemistry* 9 (2021): 47. https://doi.org/10.3389/fchem.2021.613343

Miri, Abdolhossein, Majid Darroudi, and Mina Sarani. *Applied Organometallic Chemistry* 34, no. 1 (2020): e5308. https://doi.org/10.1002/aoc.5308

Mishra, Pratik, Sambit Ray, Sayantan Sinha, Bhaskar Das, Md Imran Khan, Susant K. Behera, Soon-Il Yun, Suraj K. Tripathy, and Amrita Mishra. *Biochemical Engineering Journal* 105 (2016): 264–272. https://doi.org/10.1016/j.bej.2015.09.021

Mittal, Rahul, Amit P. Patel, Vasanti M. Jhaveri, Sae-In S. Kay, Luca H. Debs, James M. Parrish, Debbie R. Pan, Desiree Nguyen, Jeenu Mittal, and Rahul Dev Jayant. *Expert Opinion on Drug Delivery* 15, no. 3 (2018): 301–318. https://doi.org/10.1080/17425247.2018.1420055

Mohammadi, Leili, Kaushik Pal, Muhammad Bilal, Abbas Rahdar, Georgios Fytianos, and George Z. Kyzas. *Journal of Molecular Structure* 1229 (2021): 129857.

Mohammed, Asmmaa Khalis, Khonaw Kader Salh, and Fattma A. Ali. *Cellular and Molecular Biology* 67, no. 3 (2021): 24–34. https://doi.org/10.14715/cmb/2021.67.3.4

Mostafa, Dalia A., Faten S. Bayoumi, Heba M. Taher, Basma H. Abdelmonem, and Tarek F. Eissa. *Research Journal of Pharmacy and Technology* 13, no. 9 (2020): 4415–4422.

Moteriya, Pooja, and Sumitra Chanda. *Journal of Genetic Engineering and Biotechnology* 16, no. 1 (2018): 105–113.

Muhammad, Wali, Muhammad Aslam Khan, Munazza Nazir, Aisha Siddiquah, Sadaf Mushtaq, Syed Salman Hashmi, and Bilal Haider Abbasi. *Materials Science and Engineering: C* 103 (2019): 109740. https://doi.org/10.1016/j.msec.2019.109740

Murali, Mahadevamurthy, Nataraj Kalegowda, Hittanahallikoppal G. Gowtham, Mohammad Azam Ansari, Mohammad N. Alomary, Saad Alghamdi, Natarajamurthy Shilpa, et al. *Pharmaceutics* 13, no. 10 (2021): 1662.

Nafari, Amir, Koroush Cheraghipour, Maryam Sepahvand, Ghazal Shahrokhi, Esraa Gabal, and Hossein Mahmoudvand. *Parasite Epidemiology and Control* 10 (2020): e00156.

Nagajyothi, P. C., P. Muthuraman, T. V. M. Sreekanth, Doo Hwan Kim, and Jaesool Shim. *Arabian Journal of Chemistry* 10, no. 2 (2017): 215–225. https://doi.org/10.1016/j.arabjc.2016.01.011

Nagalingam, M., V. N. Kalpana, and A. Panneerselvam. *Biotechnology Reports* 19 (2018): e00268. https://doi.org/10.2147%2FIJN.S108208

Namasivayam, S., Gayathri Venkatachalam, and R. S. Bharani. *Applied Nanoscience* 11, no. 5 (2021): 1511–1530.

Narayanan, Mathiyazhagan, Paramasivam Vigneshwari, Devarajan Natarajan, Sabariswaran Kandasamy, Mishal Alsehli, Ashraf Elfasakhany, and Arivalagan Pugazhendhi. *Environmental Research* 200 (2021): 111335. https://doi.org/10.1016/j.envres.2021.111335

Nath, Sunayana, Ritis K. Shyanti, and Bhawana Pathak. In *Green Nanoparticles*, pp. 163–186. Springer, Cham, 2020.

Nayak, Debasis, Sonali Pradhan, Sarbani Ashe, Pradipta Ranjan Rauta, and Bismita Nayak. *Journal of Colloid and Interface Science* 457 (2015): 329–338. https://doi.org/10.1016/j.jcis.2015.07.012

Naz, Sania, Mohammad Islam, Saira Tabassum, Nelson Freitas Fernandes, Esperanza J. Carcache de Blanco, and Muhammad Zia. *Journal of Molecular Structure* 1185 (2019): 1–7. https://doi.org/10.1016/j.molstruc.2019.02.088

Nirmala, Sankaradoss, Velayutham Ravichandiran, A. Vijayalakshmi, and P. Nadanasabapathi. *Research Journal of Pharmacy and Technology* 11, no. 3 (2018): 1193–1206.

Nithiya, S., & Sangeetha, R. World J*ournal of Pharmaceutical Research* 3 (2014) 1055–1066.

Nithya, Pandiyan, and Mahalingam Sundrarajan. *Journal of Photochemistry and Photobiology B: Biology* 202 (2020): 111706. https://doi.org/10.1016/j.jphotobiol.2019.111706

Nityasree, B. R., Raju Krishna Chalannavar, Sabia Kouser, M. S. Divakar, Ramesh Sabu Gani, K. Sowmyashree, and Ravindra B. Malabadi. *Advances in Natural Sciences: Nanoscience and Nanotechnology* 12, no. 1 (2021): 015009.

Oluwatoyin, Fabiyi, Olatunji Gabriel, Atolani Olubunmi, and Olawuyi Olanike. *JAPS: Journal of Animal & Plant Sciences* 30, no. 5 (2020).

Opris, Razvan, Corina Tatomir, Diana Olteanu, Remus Moldovan, Bianca Moldovan, Luminita David, Andras Nagy, Nicoleta Decea, Mihai Ludovic Kiss, and Gabriela Adriana Filip. *Colloids and Surfaces B: Biointerfaces* 150 (2017): 192–200. https://doi.org/10.1016/j.colsurfb.2016.11.033

Oves, Mohammad, Mohammad Aslam, Mohd Ahmar Rauf, Shariq Qayyum, Huda A. Qari, Mohd Shahnawaz Khan, Mohammad Zubair Alam, Shams Tabrez, Arivalagan Pugazhendhi, and Iqbal MI Ismail. *Materials Science and Engineering: C* 89 (2018): 429–443. https://doi.org/10.1016/j.msec.2018.03.035

Padinjarathil, Himabindu, Manu M. Joseph, B. S. Unnikrishnan, G. U. Preethi, R. Shiji, M. G. Archana, S. Maya, H. P. Syama, and T. T. Sreelekha. *International Journal of Biological Macromolecules* 118 (2018): 1174–1182. https://doi.org/10.1016/j.msec.2018.03.035

Pápay, Zsófia Edit, Annamária Kósa, Béla Böddi, Zahra Merchant, Imran Y. Saleem, Mohammed Gulrez Zariwala, Imre Klebovich, Satyanarayana Somavarapu, and István Antal. *Journal of Aerosol Medicine and Pulmonary Drug Delivery* 30, no. 4 (2017): 274–288.

Parvathy, Subramanian, Ganesan Manjula, Ramalingam Balachandar, and Ramasamy Subbaiya. *Materials Letters* 314 (2022): 131811. https://doi.org/10.1016/j.matlet.2022.131811

Patra, Jayanta Kumar, Gitishree Das, and Han-Seung Shin. *International Journal of Nanomedicine* 14 (2019): 6679. https://doi.org/10.2147%2FIJN.S212614

Patra, Nabanita, Dattatreya Kar, Abhisek Pal, and Anindita Behera. *Advances in Natural Sciences: Nanoscience and Nanotechnology* 9, no. 3 (2018): 035001.

Patra, Sukanya, Monika Singh, Kirti Wasnik, Divya Pareek, Prem Shankar Gupta, Sudip Mukherjee, and Pradip Paik. *ACS Applied Bio Materials* 4, no. 10 (2021): 7342–7365.

Paul, Sony, Kalyani Mohanram, and Iyanar Kannan. *Ayu* 39, no. 3 (2018): 182. https://doi.org/10.4103%2Fayu.AYU_24_18

Pei, Junwen, Binfan Fu, Lifeng Jiang, and Taizhen Sun. *International Journal of Nanomedicine* 14 (2019): 1969. https://doi.org/10.2147%2FIJN.S188235

Permana, Andi Dian. PhD dissertation, Queen's University Belfast. Faculty of Medicine, Health and Life Sciences, December, 2019.

Perumal, Sasidharan, Madhana Vigneshwari Gopal Samy, and Dharaneedharan Subramanian. *Bioprocess and Biosystems Engineering* 44, no. 9 (2021): 1853–1863.

Peters, Brian M., Mary Ann Jabra-Rizk, Graeme A. O'May, J. William Costerton, and Mark E. Shirtliff. *Clinical Microbiology Reviews* 25, no. 1 (2012): 193–213.

Pfaller, M. A., D. J. Diekema, D. L. Gibbs, V. A. Newell, J. F. Meis, I. M. Gould, W. Fu, A. L. Colombo, and E. Rodriguez-Noriega. *Journal of Clinical Microbiology* 45, no. 6 (2007): 1735–1745.

Phan, Thi Tuong Vy, Thanh-Canh Huynh, Panchanathan Manivasagan, Sudip Mondal, and Junghwan Oh. *Nanomaterials* 10, no. 1 (2019): 66. https://doi.org/10.3390/nano10010066

Politi, Flávio Augusto Sanches, Renata Vieira Bueno, Lucas André Zeoly, Rafaela Regina Fantatto, Josimar de Oliveira Eloy, Marlus Chorilli, Fernando Coelho, Rafael Victório Carvalho Guido, Ana Carolina de Souza Chagas, and Maysa Furlan. *Acta Tropica* 219 (2021): 105920.

Ponnanikajamideen, MohamedIbrahim, Shanmugam Rajeshkumar, Mahendran Vanaja, and Gurusamy Annadurai. *Canadian Journal of Diabetes* 43, no. 2 (2019): 82–89. https://doi.org/10.1016/j.jcjd.2018.05.006

Popli, Dimple, Vishaka Anil, Akshatha B. Subramanyam, Namratha MN, Ranjitha VR, Saroja N. Rao, Ravishankar V. Rai, and Melappa Govindappa. *Artificial Cells, Nanomedicine, and Biotechnology* 46, no. sup 1 (2018): 676–683.

Popova, Aneta, Dasha Mihaylova, and Alexander Spasov. *The Open Biotechnology Journal* 15, no. 1 (2021). http://doi.org/10.2174/1874434602015010046

Prabakaran, A., and N. Mani. *Journal of Pharmacognosy and Phytochemistry* 8, no. 4 (2019): 2556–2558.
Prasad, Kollur Shiva, Shashanka K. Prasad, Ravindra Veerapur, Ghada Lamraoui, Ashwini Prasad, M. N. Prasad, Sandeep Kumar Singh, Najat Marraiki, Asad Syed, and Chandan Shivamallu. *Separations* 8, no. 1 (2021): 8. https://doi.org/10.3390/separations8010008
Prasad, Rajendra, Abdul Haseeb Shah, and Manpreet Kaur Rawal. *Yeast Membrane Transport* (2016): 327–349.
Prasanna, Pragya, Prakash Kumar, Saurabh Kumar, Vinod Kumar Rajana, Vishnu Kant, Surendra Rajit Prasad, Utpal Mohan, V. Ravichandiran, and Debabrata Mandal. *Biomedicine & Pharmacotherapy* 141 (2021): 111920.
Priyadarshini, Sonali, Sushree Sulava, Rasmita Bhol, and Somanatha Jena. *Current Science.* 117 (2019): 1300–1307.
Puri, Abhijeet V., Swati S. Patil, and Mandar B. Mulik. *International Journal of Botany* 6, no. 3 (2021): 534–540.
Pyrzynska, Krystyna, and Aleksandra Sentkowska. *Journal of Nanostructure in Chemistry* (2021): 1–14.
Quelemes, Patrick V., Felipe B. Araruna, Bruna EF De Faria, Selma AS Kuckelhaus, Durcilene A. Da Silva, Ronaldo Z. Mendonça, Carla Eiras, Dos S. Soares, Maria José, and José Roberto SA Leite. *International Journal of Molecular Sciences* 14, no. 3 (2013): 4969–4981. https://doi.org/10.3390/ijms14034969
Rad, Adeleh Granmayeh, Hamed Abbasi, and Mohammad Hossein Afzali. *Physics Procedia* 22 (2011): 203–208. https://doi.org/10.1016/j.phpro.2011.11.032
Ragavee, A., and Selvaraj Asha Devi. *Pharmacognosy Magazine* 15, no. 64 (2019): 218.
Rahnama, Vahid, Mohammad Hossein Motazedian, Soliman Mohammadi-Samani, Qasem Asgari, Parisa Ghasemiyeh, and Meisam Khazaei. *Research in Pharmaceutical Sciences* 16, no. 6 (2021): 623.
Rajakumar, Govindasamy, Muthu Thiruvengadam, Govindarasu Mydhili, Thandapani Gomathi, and Ill-Min Chung. *Bioprocess and Biosystems Engineering* 41, no. 1 (2018): 21–30.
Rajan, Angel Rose, Anish Rajan, Annamma John, and Daizy Philip. In: *AIP Conference Proceedings* 2105, no. 1 (2019): 020008. AIP Publishing LLC.
Rajan, Anish, Vidya Vilas, and Daizy Philip. *Journal of Molecular Liquids* 212 (2015): 331–339. https://doi.org/10.1016/j.molliq.2015.09.013
Rajapriya, Mariappan, Sundararaj Aruna Sharmili, Raju Baskar, Ravichandran Balaji, Naiyf S. Alharbi, Shine Kadaikunnan, Jamal M. Khaled, Khalid F. Alanzi, and Baskaralingam Vaseeharan. *Journal of Cluster Science* 31, no. 4 (2020): 791–801.
Rajaram, K., D. C. Aiswarya, and P. Sureshkumar. *Materials Letters* 138 (2015): 251–254. https://doi.org/10.1016/j.matlet.2014.10.017
Rajasekar, Srinivasan, and Santhi Kuppusamy. *Journal of Cluster Science* 32, no. 4 (2021): 907–915.
Rajendran, Abinaya, Murad Alsawalha, and Thamer Alomayri. *Journal of Saudi Chemical Society* 25, no. 9 (2021): 101307. https://doi.org/10.1016/j.jscs.2021.101307
Rajesh, K. M., B. Ajitha, Y. Ashok Kumar Reddy, Y. Suneetha, and P. Sreedhara Reddy. *Optik* 154 (2018): 593–600. https://doi.org/10.1016/j.ijleo.2017.10.074
Rajeshkumar, S., S. Venkat Kumar, Arunachalam Ramaiah, Happy Agarwal, T. Lakshmi, and Selvaraj Mohana Roopan. *Enzyme and Microbial Technology* 117 (2018): 91–95. https://doi.org/10.1016/j.enzmictec.2018.06.009
Rajkumari, J., C. Maria Magdalane, B. Siddhardha, J. Madhavan, G. Ramalingam, Naif Abdullah Al-Dhabi, Mariadhas Valan Arasu, A. K. M. Ghilan, V. Duraipandiayan, and K. Kaviyarasu. *Journal of Photochemistry and Photobiology B: Biology* 201 (2019): 111667.
Ramalingam, V., S. Revathidevi, T. Shanmuganayagam, L. Muthulakshmi, and R. Rajaram. *RSC Advances* 6, no. 25 (2016): 20598–20608. https://doi.org/10.1039/C5RA26781A
Ramar, Manikandan, Beulaja Manikandan, Prabhu Narayanan Marimuthu, Thiagarajan Raman, Anjugam Mahalingam, Palanisamy Subramanian, Saravanan Karthick, and Arumugam Munusamy. *Spectrochimica Acta Part A: Molecular and Biomolecular Spectroscopy* 140 (2015): 223–228. https://doi.org/10.1016/j.saa.2014.12.060
Raslan, M. M., S. Mohamed, M. D. E. Abd El Maksoud, and K. El Nesr. *Journals of Biotechnology and Biomaterials* 8 (2018): 55.
Reddy, P. H., M. Manczak, X. Yin, M. C. Grady, A. Mitchell, S. Tonk, C. S. Kuruva, J. S. Bhatti, R. Kandimalla, M. Vijayan, et al. *Journals of Alzheimers Disease* 61 (2018): 843–866.
Reddy, Palem Ramasubba, Shimoga D. Ganesh, Nabanita Saha, Oyunchimeg Zandraa, and Petr Sáha. *Journal of Nanotechnology* 2016 (2016). https://doi.org/10.1155/2016/4964752

Rehana, Dilaveez, D. Mahendiran, R. Senthil Kumar, and A. Kalilur Rahiman. *Biomedicine & Pharmacotherapy* 89 (2017a): 1067–1077. https://doi.org/10.1016/j.biopha.2017.02.101

Rehana, Dilaveez, D. Mahendiran, R. Senthil Kumar, and A. Kalilur Rahiman. *Bioprocess and Biosystems Engineering* 40, no. 6 (2017b): 943–957. https://doi.org/10.1007/s00449-017-1758-2

Remya, R. R., SR Radhika Rajasree, L. Aranganathan, and T. Y. Suman. *Biotechnology Reports* 8 (2015): 110–115. https://doi.org/10.1016/j.biopha.2017.02.101

Renganathan, Sathiyapriya, Vincent Aroulmoji, Gnanendra Shanmugam, Geetha Devarajan, Kalagadda Venkateswara Rao, Vanga Rajendar, and Si-Hyun Park. *Materials Research Express* 6, no. 3 (2018): 035028.

Rokade, Shalaka S., Komal A. Joshi, Ketakee Mahajan, Geetanjali Tomar, Dnyanesh S. Dubal, Vijay Singh, R. Kitture, Jayesh Bellare, and Sougata Ghosh. *Global Journal of Nanomedicine* 2, no. 5 (2017): 555600.

Rokade, Shalaka S., Komal A. Joshi, Ketakee Mahajan, Saniya Patil, Geetanjali Tomar, Dnyanesh S. Dubal, Vijay Singh Parihar, Rohini Kitture, Jayesh R. Bellare, and Sougata Ghosh. *Bioinorganic Chemistry and Applications* 2018 (2018). https://doi.org/10.1155/2018/4924186

Romes, Nissha Bharrathi, Roswanira Abdul Wahab, and Mariani Abdul Hamid. *Biotechnology & Biotechnological Equipment* 35, no. 1 (2021): 711–729. https://doi.org/10.1080/13102818.2021.1915869

Rosarin, Fathima Stanley, Vadivel Arulmozhi, Samuthira Nagarajan, and Sankaran Mirunalini. *Asian Pacific Journal of Tropical Medicine* 6, no. 1 (2013): 1–10. https://doi.org/10.1016/S1995-7645(12)60193-X

Sabatini, P., and C. Anchana Devi. *International Journal of Pharmaceutical Sciences and Research* 9, no. 4 (2018): 1555–1564.

Sahin, B., A. Aygun, H. Gunduz, K. Sahin, E. Demir, S. Akocak, and F. Sen. *Colloids and Surfaces B Biointerfaces* 163 (2018) 119–124. https://doi.org/10.1016/j.colsurfb.2017.12.042

Saleem, Kiran, Zainab Khursheed, Christophe Hano, Iram Anjum, and Sumaira Anjum. *Nanomaterials* 9, no. 12 (2019): 1749.

Salem, Wesam, Georg Schneditz, Elisabeth Dornisch, and Stefan Schild. *Journal of Experimental and Applied Animal Sciences* 1, no. 3 (2015): 369–383.

Sankar, Renu, Ramasamy Maheswari, Selvaraju Karthik, Kanchi Subramanian Shivashangari, and Vilwanathan Ravikumar. *Materials Science and Engineering: C* 44 (2014): 234–239. https://doi.org/10.1016/j.msec.2014.08.030

Saratale, Ganesh Dattatraya, Rijuta Ganesh Saratale, Giovanni Benelli, Gopalakrishnan Kumar, Arivalagan Pugazhendhi, Dong-Su Kim, and Han-Seung Shin. *Journal of Cluster Science* 28, no. 3 (2017): 1709–1727.

Saratale, Rijuta G., Han Seung Shin, Gopalakrishnan Kumar, Giovanni Benelli, Dong-Su Kim, and Ganesh D. Saratale. *Artificial Cells, Nanomedicine, and Biotechnology* 46, no. 1 (2018): 211–222. https://doi.org/10.1080/21691401.2017.1337031

Sartini, Sartini, Andi Dian Permana, Saikat Mitra, Abu Montakim Tareq, Emil Salim, Islamudin Ahmad, Harapan Harapan, Talha Bin Emran, and Firzan Nainu. *Pathogens* 10, no. 2 (2021): 245.

Sathishkumar, Palanivel, Krishnan Vennila, Rajarajeswaran Jayakumar, Abdull Rahim, Mohd Yusoff, Tony Hadibarata, and Thayumanavan Palvannan. *Bioprocess and Biosystems Engineering* 39, no. 4 (2016): 651–659.

Satsangi, Neh, and Shabad Preet. *Journal of Plant Biochemistry and Biotechnology* 30, no. 3 (2021): 608–612.

Schroeder, Avi, Daniel A. Heller, Monte M. Winslow, James E. Dahlman, George W. Pratt, Robert Langer, Tyler Jacks, and Daniel G. Anderson. *Nature Reviews Cancer* 12, no. 1 (2012): 39–50.

Selim, Yasser A., Maha A. Azb, Islam Ragab, and Mohamed HM Abd El-Azim. *Scientific Reports* 10, no. 1 (2020): 1–9.

Sengani, Manimegalai. *Environmental Toxicology and Pharmacology* 50 (2017): 11–19.

Sengottaiyan, Arumugam, Adithan Aravinthan, Chinnapan Sudhakar, Kandasamy Selvam, Palanisamy Srinivasan, Muthusamy Govarthanan, Koildhasan Manoharan, and Thangaswamy Selvankumar. *Journal of Nanostructure in Chemistry* 6, no. 1 (2016): 41–48. https://doi.org/10.1007/s40097-015-0178-6

Senthilkumar, P., D. S. Santhosh Kumar, B. Sudhagar, M. Vanthana, M. Hajistha Parveen, S. Sarathkumar, Jeslin Cheriyan Thomas, A. Sandhiya Mary, and Chandramouleeswaran Kannan. *Journal of Nanostructure in Chemistry* 6, no. 3 (2016): 275–280.

Senthilkumar, P., Lakshmi Priya, Ranjith Santhosh Kumar, and D. S. Bhuvaneshwari. *International Journal of Advanced Multidisciplinary Research* 2, no. 11 (2015): 0917–0923.

Sepúlveda, Alex Arbey Lopera, Angela Maria Arenas Velásquez, Irwin Alexander Patiño Linares, Leticia de Almeida, Carla Raquel Fontana, Claudia Garcia, and Marcia Aparecida Silva Graminha. *Photodiagnosis and Photodynamic Therapy* 30 (2020): 101676.

Shah, Amreen, Isfahan Tauseef, Manel Ben Ali, Muhammad Arfat Yameen, Amine Mezni, Amor Hedfi, Syed Kashif Haleem, and Sirajul Haq. *Toxics* 9, no. 5 (2021): 105. https://doi.org/10.3390/toxics9050105

Shamprasad, Bhanuvalli R., Sivakumar Keerthana, Sengan Megarajan, Robert Lotha, Sivasubramanian Aravind, and Anbazhagan Veerappan. *Materials Letters* 241 (2019): 198–201. https://doi.org/10.1016/j.matlet.2019.01.086

Shanker, Kalakotla, Gottumukkala Krishna Mohan, Md Ashwaq Hussain, Naradala Jayarambabu, and Poka Lakshmi Pravallika. *Pharmacognosy Magazine* 13, no. 49 (2017a): 188. https://doi.org/10.4103%2F0973-1296.197642

Shanker, Kalakotla, Jayarambabu Naradala, G. Krishna Mohan, G. S. Kumar, and P. L. Pravallika. *RSC Advances* 7, no. 59 (2017b): 37158–37167. https://doi.org/10.1039/C7RA05693A

Sharma, Rakesh Kumar, Katherine Cwiklinski, Ravikumar Aalinkeel, Jessica L. Reynolds, Donald E. Sykes, Elizabeth Quaye, James Oh, Supriya D. Mahajan, and Stanley A. Schwartz. *Immunological Investigations* 46, no. 8 (2017): 833–846.

Shejawal, Kiran P., Dheeraj S. Randive, Somnath D. Bhinge, Mangesh A. Bhutkar, Sachin S. Todkar, Anjum S. Mulla, and Namdeo R. Jadhav. *Journal of Materials Science: Materials in Medicine* 32, no. 2 (2021): 1–12.

Singh, Priyanka, Abhroop Garg, Santosh Pandit, V. R. S. S. Mokkapati, and Ivan Mijakovic. *Nanomaterials* 8, no. 12 (2018): 1009. https://doi.org/10.3390/nano8121009

Sivakumar, Padmanaban, Minjong Lee, Yoon-Seok Kim, and Min Suk Shim. *Journal of Materials Chemistry B* 6, no. 30 (2018): 4852–4871.

Sivaraj, Rajeshwari, Pattanathu K. S. M. Rahman, P. Rajiv, S. Narendhran, and R. Venckatesh. *Spectrochimica Acta Part A: Molecular and Biomolecular Spectroscopy* 129 (2014): 255–258. https://doi.org/10.1016/j.saa.2014.03.027

Soliman, Nadia, Eman H. Ismail, Heba I. Abd El-Moaty, D. Y. Sabry, and Mostafa M. H. Khalil. *Egyptian Journal of Chemistry* 61, no. 4 (2017): 691–703.

Soltane, Raya, Dalila Mtat, Amani Chrouda, Noof Alzahrani, Youssef O. Al-Ghamdi, Hussam El-Desouky, and Khaled Elbanna. In: *Black Cumin (Nigella sativa) Seeds: Chemistry, Technology, Functionality, and Applications*, pp. 179–190. Springer, Cham, 2021.

Souza, José António Santos, Marta M. Alves, Debora Barros Barbosa, Maria Manuel Lopes, Eugénia Pinto, Maria Helena Figueiral, Alberto Carlos B. Delbem, and Nuno Pereira Mira. *Medical Mycology* 58, no. 4 (2020): 564–567.

Souza, M. E., L. Q. S. Lopes, P. C. Bonez, A. Gündel, D. S. T. Martinez, M. R. Sagrillo, J. L. Giongo, et al. *Microbial Pathogenesis* 104 (2017): 125–132.

Sridharan, M., P. Kamaraj, J. Arockiaselvi, T. Pushpamalini, P. A. Vivekanand, and S. Hari Kumar. *Materials Today: Proceedings* 36 (2021): 914–919. https://doi.org/10.1016/j.matpr.2020.07.031

Suganthy, Natarajan, Vijayan Sri Ramkumar, Arivalagan Pugazhendhi, Giovanni Benelli, and Govindaraju Archunan. *Environmental Science and Pollution Research* 25, no. 11 (2018): 10418–10433.

Sujitha, Vasu, Kadarkarai Murugan, Manickam Paulpandi, Chellasamy Panneerselvam, Udaiyan Suresh, Mathath Roni, Marcello Nicoletti, et al. *Parasitology Research* 114, no. 9 (2015): 3315–3325.

Sulaiman, Ghassan M., Amer T. Tawfeeq, and Marwa D. Jaaffer. *Biotechnology Progress* 34, no. 1 (2018): 218–230. https://doi.org/10.1002/btpr.2568

Sulaiman, Ghassan M., Arieg AW Mohammad, Hamssa E. Abdul-Wahed, and Mukhlis M. Ismail. *Digest Journal of Nanomaterials & Biostructures* 8, no. 1 (2013).

Sumitha, Samuggam, Sethu Vasanthi, Sivadasan Shalini, Suresh V. Chinni, Subash C. B. Gopinath, Periasamy Anbu, Mohammed Baidi Bahari, Rajak Harish, Sathasivam Kathiresan, and Veerasamy Ravichandran. *Molecules* 23, no. 12 (2018): 3311.

Sunderam, Veena, Devasena Thiyagarajan, Ansel Vishal Lawrence, Sathak Sameer Shaik Mohammed, and Arokiyaraj Selvaraj. *Saudi Journal of Biological Sciences* 26, no. 3 (2019): 455–459. https://doi.org/10.1016/j.sjbs.2018.12.001

Sunoqrot, Suhair, Malek Alfaraj , Ala'A. M. Hammad, Violet Kasabri, Dana Shalabi, Ahmad A. Deeb, Lina Hasan Ibrahim, Khaldoun Shnewer, and Ismail Yousef. *Pharmaceutics* 12, no. 9 (2020): 811. https://doi.org/10.3390/pharmaceutics12090811

Suwan, Temsiri, Sakornrat Khongkhunthian, and Siriporn Okonogi. *Drug Discoveries & Therapeutics* 12, no. 4 (2018): 189–196.

Swarnalatha, L., S. Rachela, P. Ranjan, and P. Baradwaj. *International Journal of Nanomaterials and Biostructures* 2 (2012): 25–29.

Szymańska, Emilia, Piotr Orłowski, Katarzyna Winnicka, Emilia Tomaszewska, Piotr Bąska, Grzegorz Celichowski, Jarosław Grobelny, Anna Basa, and Małgorzata Krzyżowska. *International Journal of Molecular Sciences* 19, no. 2 (2018): 387.

Tahir, Humna, Jameel Rahman, Muhammad Ashraf, Shazia Anjum, Irshad Hussain, and Muhammad Farooq Warsi. *Bioinspired, Biomimetic and Nanobiomaterials* 9, no. 1 (2019): 1–6.

Tasca, Federico, and Riccarda Antiochia. *Nanomaterials* 10, no. 5 (2020): 909.

Thatoi, Priyabrata, Rout George Kerry, Sushanto Gouda, Gitishree Das, Krishna Pramanik, Hrudayanath Thatoi, and Jayanta Kumar Patra. *Journal of Photochemistry and Photobiology B: Biology* 163 (2016): 311–318.

Tiew, Pei Yee, Micheál Mac Aogain, Nur Ali, Kai Xian Thng, Karlyn Goh, Kenny J. X. Lau, and Sanjay H. Chotirmall. *Mycopathologia* 185, no. 2 (2020): 207–231.

Tiloke, Charlette, Alisa Phulukdaree, Krishnan Anand, Robert M. Gengan, and Anil A. Chuturgoon. *Journal of Cellular Biochemistry* 117, no. 10 (2016): 2302–2314. https://doi.org/10.1002/jcb.25528

Tiloke, Charlette, Krishnan Anand, Robert M. Gengan, and Anil A. Chuturgoon. *Biomedicine & Pharmacotherapy* 108 (2018): 457–466. https://doi.org/10.1016/j.biopha.2018.09.060

Uddin, Md Imad, G. Kavya, G. Soundarya, G. Ropa, and C. H. Sridhar. *Asian Journal of Pharmaceutical Education and Research* 7, no. 2 (2018): 45–59.

Ullah, Sadeeq, Aftab Ahmad, Aoke Wang, Muslim Raza, Amin Ullah Jan, Kamran Tahir, Aziz Ur Rahman, and Yuan Qipeng. *Journal of Photochemistry and Photobiology B: Biology* 173 (2017): 368–375. https://doi.org/10.1016/j.jphotobiol.2017.06.018

Umamaheswari, A., S. Lakshmana Prabu, S. Adharsh John, and A. Puratchikody. *Biotechnology Reports* 29 (2021): e00595. https://doi.org/10.1016/j.btre.2021.e00595

Umar, Huzaifa, Doga Kavaz, and Nahit Rizaner. *International Journal of Nanomedicine* 14 (2019): 87. https://doi.org/10.2147%2FIJN.S186888

Vaghela, Hiral, Rahul Shah, and Amanullakhan Pathan. *Current Nanomaterials* 3, no. 3 (2018): 168–177. https://doi.org/10.2174/2405461504666190131142303

Vaid, Prachi, Pankaj Raizada, Adesh K. Saini, and Reena V. Saini. *Sustainable Chemistry and Pharmacy* 16 (2020): 100247. https://doi.org/10.1016/j.scp.2020.100247

Vanaraj, Sekar, B. Bhargavi Keerthana, and Kathirvel Preethi. *Journal of Inorganic and Organometallic Polymers and Materials* 27, no. 5 (2017): 1412–1422.

Vasanth, Karunamoorthy, Kaliappan Ilango, Ramasamy MohanKumar, Aruna Agrawal, and Govind Prasad Dubey. *Colloids and Surfaces B: Biointerfaces* 117 (2014): 354–359. https://doi.org/10.1016/j.colsurfb.2014.02.052

Venugopal, K., H. Ahmad, E. Manikandan, K. Thanigai Arul, K. Kavitha, M. K. Moodley, K. Rajagopal, R. Balabhaskar, and M. Bhaskar. *Journal of Photochemistry and Photobiology B: Biology* 173 (2017): 99–107. https://doi.org/10.1016/j.jphotobiol.2017.05.031

Vinotha, Viswanathan, Arokiadhas Iswarya, Rajagopalan Thaya, Marimuthu Govindarajan, Naiyf S. Alharbi, Shine Kadaikunnan, Jamal M. Khaled, Mohammed N. Al-Anbr, and Baskaralingam Vaseeharan. *Journal of Photochemistry and Photobiology B: Biology* 197 (2019): 111541.

Virgen-Ortiz, A., S. Limón-Miranda, M. A. Soto-Covarrubias, A. Apolinar-Iribe, E. Rodríguez-León, and R. Iñiguez-Palomares. *Digest Journal of Nanomaterials and Biostructures* 10 (2015): 927–933.

Wang, Yu, Yayuan Zhang, Yan Guo, Jianguang Lu, Vishnu Priya Veeraraghavan, Surapaneni Krishna Mohan, Chao Wang, and Xiaoguang Yu. *Journal of Photochemistry and Photobiology B: Biology* 201 (2019a): 111624. https://doi.org/10.1016/j.jphotobiol.2019.111624

Wang, Lei, Jianwei Xu, Ye Yan, Han Liu, and Feng Li. *Artificial cells, nanomedicine, and biotechnology* 47, no. 1 (2019b): 1216–1223. https://doi.org/10.1080/21691401.2019.1593852

Wani, Kirti, A. Choudhari, Rajeev Chikate, and Ruchika Kaul-Ghanekar. *Carbon Sci. Technol* 5, no. 1 (2013): 203–210.

Wei, Tuo, Chao Chen, Juan Liu, Cheng Liu, Paola Posocco, Xiaoxuan Liu, Qiang Cheng et al. *Proceedings of the National Academy of Sciences* 112, no. 10 (2015): 2978–2983. https://doi.org/10.1073/pnas.1418494112

Wilson, Stephina, S. Cholan, U. Vishnu, Mohammed Sannan, R. Jananiya, S. Vinodhini, S. Manimegalai, and D. V. Rajeswari. *Der Pharmacia Lettre* 7, no. 12 (2015): 194–205.

Wu, Chao, Yu Gao, Ying Liu, and XiaoYan Xu. *International Journal of Nanomedicine* 13 (2018): 6189. https://doi.org/10.2147%2FIJN.S169209

Wu, Tao, Xi Duan, Chunyan Hu, Changqiang Wu, Xiaobin Chen, Jing Huang, Junbo Liu, and Shu Cui. *Artificial Cells, Nanomedicine, and Biotechnology* 47, no. 1 (2019): 512–523. https://doi.org/10.1080/21691401.2018.1560305

Xu, Li, Yi-Yi Wang, Jie Huang, Chun-Yuan Chen, Zhen-Xing Wang, and Hui Xie. *Theranostics* 10, no. 20 (2020): 8996. https://doi.org/10.7150%2Fthno.45413

Xu, R., J. Wang, J. Xu, X. Song, H. Huang, Y. Feng, C. Fu. *International Journal of Nanomedicine* 15 (2020): 1149–1160.

Xu, You, Hongmei Liu, and Lei Song. *Journal of Nanobiotechnology* 18, no. 1 (2020): 1–25.

Xue, Jinwei, Tongtong Liu, Yongdan Liu, Ye Jiang, Vidya Devanatha Desikan Seshadri, Surapaneni Krishna Mohan, and Li Ling. *Journal of Photochemistry and Photobiology B: Biology* 200 (2019): 111635. https://doi.org/10.1016/j.jphotobiol.2019.111635

Xue, Yifeng, Guoxin Yu, Zhengfei Shan, and Zhonglian Li. *Journal of Photochemistry and Photobiology B: Biology* 186 (2018): 131–136. https://doi.org/10.1016/j.jphotobiol.2018.07.004

Yadav, Ekta, Deepika Singh, Pankajkumar Yadav, and Amita Verma. *RSC Advances* 8, no. 38 (2018): 21621–21635. https://doi.org/10.1039/C8RA03500H

Yakoob, Arsia Tarnam, Nargis Begum Tajuddin, Muhammad Ilyas Mohammed Hussain, Shilu Mathew, Archunan Govindaraju, and Ishtiaq Qadri. *Pharmacognosy Journal* 8, no. 6 (2016).

Yang, Chuang, Minghua Wang, Junde Zhou, and Qiang Chi. *Materials Science and Engineering: C* 77 (2017a): 1012–1016. https://doi.org/10.1016/j.msec.2017.04.020

Yang, Qian, Chuang Zhao, Jun Zhao, and Yong Ye. *Pharmaceutical Biology* 55, no. 1 (2017b): 428–434. https://doi.org/10.1080/13880209.2016.1245346

Yassin, Abdelrahman M., Nehal M. El-Deeb, Ahmed M. Metwaly, Gomaa F. El Fawal, Mohamed M. Radwan, and Elsayed E. Hafez. *Applied Biochemistry and Biotechnology* 182, no. 4 (2017): 1675–1693.

Yin, Juntao, Yantao Hou, Yuyun Yin, and Xiaoyong Song. *International Journal of Nanomedicine* 12 (2017): 8671. https://doi.org/10.2147/IJN.S144615

Yusefi, Mostafa, Kamyar Shameli, Roshafima Rasit Ali, Siew-Wai Pang, and Sin-Yeang Teow. *Journal of Molecular Structure* 1204 (2020): 127539. https://doi.org/10.1016/j.molstruc.2019.127539

Zeebaree, Samie Yaseen Sharaf, Aymn Yaseen Sharaf Zeebaree, and Osama Ismail Haji Zebari. *Sustainable Chemistry and Pharmacy* 15 (2020): 100210. https://doi.org/10.1016/j.scp.2019.100210

Zhang, Mingzhen, Bo Xiao, Huan Wang, Moon Kwon Han, Zhan Zhang, Emilie Viennois, Changlong Xu, and Didier Merlin. *Molecular Therapy* 24, no. 10 (2016a): 1783–1796. https://doi.org/10.1038/mt.2016.159

Zhang, Mingzhen, James F. Collins, and Didier Merlin. *Nanomedicine* 12, no. 23 (2016b): 3035–3037. https://doi.org/10.2217/nnm-2016-0353

Zhang, Sheng, and De-Chang Chen. *Chinese Medical Journal* 132, no. 10 (2019): 1135–1138.

Zhao, Jihui, Xianghua Piao, Xiaoqin Shi, Aiyong Si, Yongtan Zhang, and Nianping Feng. *Molecules* 21, no. 11 (2016) 1549. https://doi.org/10.3390/molecules21111549

Zheng, Yuanyuan, Jianwu Zhang, Rui Zhang, Zhuang Luo, Chu Wang, and Shaoqing Shi. *Artificial Cells, Nanomedicine, and Biotechnology* 47, no. 1 (2019): 3101–3109. https://doi.org/10.1080/21691401.2019.1645152

Zhu, Yongyan, Ruixuan Liu, Haoji Huang, and Quanhong Zhu. *Molecular Pharmaceutics* 16, no. 6 (2019): 2675–2689.https://doi.org/10.1021/acs.molpharmaceut.9b00243

4 Nanocarrier Systems for Delivering Plant-Based Anticancer Agents

Antony V Samrot[1*], *Shobana N*[2], *Krithika Shree S*[2], *Rajalakshmi D*[2], *Sathiyasree M*[2], *and Sanjay Preeth R*[2]
[1]School of Bioscience, Faculty of Medicine, Bioscience and Nursing, MAHSA University, Jenjarom 42610, Selangor, Malaysia
[2]Department of Biotechnology, School of Bio and Chemical Engineering, Sathyabama Institute of Science and Technology, Chennai 600119, Tamil Nadu, India
*Correspondence should be addressed to Dr. Antony V Samrot; antonysamrot@gmail.com

4.1 INTRODUCTION

Since ancient times, plant-derived natural chemicals have been utilized to cure various diseases owing to their broad spectrum of biological functions. Natural chemicals with antibacterial and antioxidant activities are now being widely used in medicines, cosmetics, food products, and pharmaceuticals (Oguz et al., 2017). Submicrometric or microparticulate systems might exhibit a better potential for medication delivery (Acharya & Sahoo, 2011). The difference between microparticles and nanoparticles in particulate drug delivery is commonly made (Acharya & Sahoo, 2011). The main obstacles encountered in targeted therapy are physiological barriers, biochemical hinderances in discovering and confirming molecular targets, and pharmacological challenges in designing suitable approaches for conjugating targeted ligands to nanosystems (Jaspreet et al., 2005; Elanthendral et al., 2021). The primary goal of nanotechnology is to deliver nanomedicines by utilizing non-invasive therapies for diagnosing diseases and maintaining good health, with minimal or no adverse effects. Nanomedicines provide a plethora of new tools and skills to help achieve these objectives (Chávez et al., 2012). They are mainly delivered using the most extensively researched nanocarriers in diverse nanocarrier systems including liposomes, polymeric nanoparticles, lipid nanoparticles, polymeric micelles, submicron lipid emulsions, microemulsions, and nanogels (Samrot et al., 2018a; Rhee & Mansour, 2011). Plant-based chemicals have been used to develop several potent anticancer drugs (Samrot et al., 2018; Shah et al., 2013). With the discovery of vinca alkaloids, namely, vinblastine and vincristine, and with the isolation of the cytotoxic agent podophyllotoxins, the hunt for anticancer medicines from plant sources began in earnest in the 1950s. Phytochemicals refer to vitamins, minerals, and bioactive substances derived from plants or plant parts such as fruits, vegetables, or grains (Raji et al., 2019; Davidson et al., 2016). Phytochemicals exert their bioactivity by interfering with signal transduction pathways of malignant cells, thereby reducing tumor cell proliferation, differentiation, angiogenesis, and apoptosis as observed in *in vitro* and *in vivo* investigations in animals (Samrot et al., 2021; Singh et al., 2016). Nanoparticles such as liposomes, polymeric micelles, lipoplexes, and polyplexes have been intensively investigated as targeted drug carrier systems. Bioactive substances, ranging from low-molecular-weight drugs to macromolecules such as proteins and nucleic acids, can be incorporated into or complexed with

DOI: 10.1201/9781003231721-4

these nanoparticles (Solaro et al., 2010). Plant metabolites with antitumor activity exert cytotoxicity in the first phase, most likely because of their origin as natural pesticides for plant self-defense (Kintzios, 2006; Shobana et al., 2022). Nanobiopolymers have helped refine the range of pharmacological and biological applications. Dendrimers have emerged as one of the most promising novel polymeric nanocarriers of several therapeutic bio actives in the field of nanotechnology (Kesharwani et al., 2014).

4.2 PLANT-BASED ANTICANCER AGENTS

Since ancient times, plant extracts and their bioactive components have been used for their medicinal effects (Iqbal et al., 2017). Natural products play a major role in the development of pharmaceuticals and medications (Pezzuto, 1997). Several clinically significant anticancer drugs have been developed using plant-based chemicals (Shah et al., 2013; Samrot et al., 2018c). Plant-based natural products with antitumor property can be classified into following chemical groups: alkaloids, phenylpropanoids, terpenoids, aldehydes, annonaceous acetogenins, glycosides, lignans, lipids, unsaponified lipids, nucleic acids, and polysaccharides (Kintzios, 2006). Natural products are becoming increasingly prominent in cancer treatment because they are considered more biocompatible, which means that they are more specific and adaptable to their targeted areas and are less harmful to normal cells (Seca & Pinto, 2018; Shobana et al., 2019). Furthermore, anticancer medicines prepared from natural products induce cancer cell death through diverse mechanisms (Gali-Muhtasib et al., 2015; Khalid, 2016).

4.2.1 Anticancer Phytochemicals

4.2.1.1 Polyphenols

Polyphenols are plant-based secondary metabolites that play a key role in growth regulation, metabolism, and protection from ultraviolet radiation and pathogens. Based on the number of phenol rings and structural elements that bind these rings to one another, polyphenols are divided as phenolic acids, stilbenes, lignans, and flavonoids. Fruits, legumes, spices, chocolate, vegetables, coffee, nuts, beer, wine, fruits, vegetables, tea, oils, cereals, and olive oil are all dietary sources of polyphenols (Abdel Dayem et al., 2016). Polyphenols affect many cancer-preventive mechanisms including prevention of oxidation, detoxification of xenobiotics, induction of apoptosis, estrogenic/anti-estrogenic activity, stimulation of immune system function, anti-inflammatory activity, and cellular signaling system (Niedzwiecki, 2016). Formation of genotoxic molecules and mutagens-transforming enzymes blocking, heme-containing phase I enzymes, such as cytochrome P450s regulation, carcinogen-detoxifying phase II enzymes regulation, such as NADPH-quinone oxidoreductase-1, glutathione *S*-transferase, and uridine diphospho (UDP) glucuronosyl transferase, formation of DNA adducts prevention, and quinone reductase (Abdel Dayem et al., 2016).

4.2.1.2 Taxanes

Taxanes are most commonly found in yew trees (*Taxus* sp. [Taxaceae]) (Škubník et al., 2021). They exert anticancer activity by blocking the mitotic spindle—they bind to microtubules and prevent them from depolymerizing, thus suppressing mitosis and inducing cellular apoptosis (Stratulat et al., 2017). Taxanes bind to tubulin and stimulate microtubule assembly while concurrently inhibiting disassembly, which stabilizes microtubule dynamics, resulting in apoptosis (Ojima et al., 2015). Both paclitaxel, a diterpenoid natural product, and docetaxel, a structural analog of paclitaxel, are first-generation taxanes used as anticancer agents, and they have an important role in today's cancer treatment (Ojima et al., 2015). Paclitaxel is the best known and widely used taxane (Škubník et al.,

2021). Taxanes are used to treat malignancies such as non-small cell lung, breast, prostate, and ovarian cancers by inhibiting microtubule function (Monje et al., 2008; Schlom et al., 2013; Samrot et al., 2021; Lee et al., 2015).

4.2.1.3 Flavonoids

Flavonoids, a plant-based secondary metabolite, are polyphenolic components used for coloring and flavoring and for their pharmacological activity (Kopustinskiene et al., 2020; Veeramuthu et al., 2017). Flavonoids are broadly classified into major groups: flavanols, flavones, anthocyanidins, and isoflavonoids (Raji et al., 2019, Veeramuthu et al., 2017). They have a broad-spectrum anticancer activity including modulating reactive oxygen species-scavenging enzyme activity, participating in cell cycle arrest, inducing apoptosis and autophagy, suppressing cancer cell proliferation and invasiveness, triggering apoptotic pathways, and downregulating pro-inflammatory signaling pathways (Kopustinskiene et al., 2020; Abirami et al., 2021a, b). Strawberry, apple, chocolate, cocoa, beans, cherry, green, black tea, Siberian larch tree, onion, milk thistle, acai palm, lemon juice, orange juice, grape juice, kale, cherries, leek, Brussels sprouts, pepper, broccoli, capsicum, parsley, soybeans, soy flour, soy milk, beer, tempeh, and celery are dietary sources of flavonoids (Veeramuthu et al., 2017). Thus, flavonoids are used in the treatment of oral, rectal, breast, lung, leukemia, thyroid, stomach, laryngeal, colon, and prostate cancers in humans (Batra et al., 2013).

4.2.1.4 Resveratrol

Resveratrol is one of the main phytoalexins, which are plant-based secondary metabolites to counter pathogenic infections (Singh et al., 2015). Resveratrol influences signal transduction pathways related to cell division and development, apoptosis, inflammation, angiogenesis, and metastasis; in this way, resveratrol influences all three stages of carcinogenesis (initiation, promotion, and progression). Resveratrol exerts anticancer effects that are mediated by a variety of mechanisms such as apoptosis in cancer cells through several pathways related to cell death and survival (Udenigwe et al., 2008). This compound is primarily found in peanuts and grapes, followed by blueberries, cranberries, turmeric, mulberries, bilberries, and aromatic hops (Udenigwe et al., 2008). Thus, resveratrol is used in the treatment of skin, breast, prostate, gastrointestinal tract, lung, esophageal, thyroid, uterine, bladder, kidney, eye, brain, bone, and blood cancers (Rauf et al., 2017; Bishayee, 2009).

4.2.1.5 Vinca Alkaloids

The vinca alkaloids vinblastine and vincristine present in Madagascar periwinkle (*Catharanthus roseus* L.; Apocynaceae) were the first anticancer agents to enter clinical trials, and they have been used to treat leukemias, lymphomas, advanced testicular tumors, breast cancer, Kaposi's sarcoma, and non-small cell lung cancer (Shah et al., 2013; Martino et al., 2018). Vinca alkaloids are categorized under microtubule-targeting agents. These molecules attach to tubulin proteins in the mitotic spindle and prevent them from polymerizing into microtubules, which consequently suppresses cell division and leads to cell death through the activation of p53-dependent and/or p53-independent apoptotic pathways (Martino et al., 2018).

4.2.1.6 Camptothecin

Camptothecin is a quinoline alkaloid derived from the plant *Camptotheca acuminata* and is used to treat leukemia as a chemotherapeutic drug (Pund et al., 2017; Willey et al., 2016). Camptothecin binds to the topoisomerase I enzyme–DNA cleavable complex and stabilizes the bound complex, after which it interacts with the DNA replication fork, causing double-strand breaks in the DNA. Subsequently, endonucleases are activated, resulting in further DNA fragmentation and cell death

(Dancey & Eisenhauer, 1996). Camptothecin and its derivatives are currently used to treat individuals with endocrine-resistant breast cancer (Tesauro et al., 2019).

4.2.1.7 Podophyllotoxin

Podophyllotoxin is an aryltetralin-type lignan found in the *Podophyllum* species *Podophyllum peltatum* and *Sinopodophyllum hexandrum* Royle (Shah et al., 2021). It binds to tubulin and prevents microtubulin formation by suppressing tubulin polymerization. This event results in cell cycle arrest and obstruction of microtubule formation in the mitotic-spindles (Hamidreza et al., 2017). The podophyllotoxin derivatives etoposide and teniposide are widely used as antineoplastic agents (Zhang et al., 2018). They are also used as chemotherapeutic agents against cervical carcinoma, osteosarcoma, nasopharyngeal carcinoma, colon cancer, breast cancer, prostate tumors, small cell lung cancer, and testicular carcinoma (Zhang et al., 2018).

4.2.1.8 Anthracyclins

Anthracyclin exerts cytostatic activity, wherein it blocks the cell cycle during the S phase and mitosis by intercalating with the DNA and inhibiting topoisomerase II enzyme activity. It is used for the treatment of acute lymphoblastic leukemia, acute myeloblastic leukemia, Wilms tumor, neuroblastoma, soft tissue and bone sarcomas, breast carcinoma, ovarian carcinoma, transitional cell bladder carcinoma, thyroid carcinoma, gastric carcinoma, Hodgkin's disease, malignant lymphoma, and bronchogenic carcinoma.

4.2.1.9 Curcumin

Curcumin, also known as diferuloylmethane, is a major natural polyphenol present in *Curcuma longa* (turmeric) rhizomes and other *Curcuma* sp. (Elanthendralet al., 2021; Hewlings et al., 2017). Curcumin's anticancer efficacy is primarily mediated by inducing apoptosis and decreasing tumor cell proliferation and invasion through the suppression of a wide range of cellular signaling pathways (Tomeh et al., 2019). Curcumin has demonstrated anticancer activity against breast cancer, lung cancer, head and neck squamous cell carcinoma, prostate cancer, and brain tumors (Anand et al., 2008).

4.2.1.10 Salvicine

Salvicine is a diterpenoid quinone chemical derived from *Salvia prionitis* Hance (Labiatae). It exerts a significant cytotoxic effect on multidrug-resistant cells (Meng & Ding, 2007). It inhibits tumor development by causing DNA strand breaks facilitated by topoisomerase II enzyme activity (Lichota & Gwozdzinski, 2018). It is mainly used for the treatment of breast cancer (Lang, 2005).

4.2.1.11 Ellipticine

Ellipticine, a plant alkaloid found in several Apocynaceae plant species, and its derivatives are DNA-damaging prodrugs whose pharmacological efficacy and genotoxic side effects are determined by cytochrome P450 activation (Stiborová et al., 2014). Ellipticines form covalent bonds with DNA and topoisomerase II enzyme, resulting in cell cycle disruption by regulating the expression of certain proteins, including cyclin B1 and CDC2 (Lichota et al., 2018). Ellipticine is used in the treatment of human breast cancer cells, leukemia, neuroblastoma cells, and glioblastoma cells (Poljaková et al., 2007). Ellipticine induces tumor growth remission (Lichota & Gwozdzinski, 2018).

4.2.1.12 Roscovitine

Roscovitine is a purine-based semi-synthesized derivative of R-roscovitine that can be isolated from the cotyledons of *Raphanus sativus* L. (Brassicaceae) and is an olomoucine derivative (Lichota

& Gwozdzinski, 2018). It is a potent cyclin–dependent kinase/cyclin E inhibitor and exerts a wide range of biological activities, namely, anticancer, antineurodegenerative, antiviral, and antiglomerulonephritis effects (Meijer et al., 1997).

4.2.2 Availability of Plant-Based Anticancer Drugs

Despite the increasing interest in drug development using molecular modeling, combinatorial chemistry, and other synthetic chemistry approaches, compounds derived from natural products continue to be a useful source of medications for human diseases. Recent reports have emphasized the relevant use of plants in modern medicine. Plant-derived secondary metabolites have been utilized as drug precursors, templates for synthetic alteration, and pharmacological probes, among other applications (Salim et al., 2008). Phytochemical-based cancer chemopreventive agents have demonstrated promising outcomes in the treatment of various cancers. Secondary metabolites derived from plants might provide an endless supply of molecules for drug development. Although the chemicals extracted from plants may not immediately function as a medication, it does pave the way for the development of new agents (Muhtasib et al., 2015). Hartwell compiled a list of more than 3000 plants that have been used to treat cancer. According to the report, more than 60% of currently used anticancer medicines are derived from natural resources such as plants and microorganisms (Varsha et al., 2017). The vinca alkaloids vincristine and vinblastine have been used for more than 30 years to treat hematological malignancies such multiple myeloma, acute lymphoblastic leukemia, Hodgkin's lymphoma and nonlymphoma, Hodgkin's tumors, and Wilms tumor. Vinca alkaloids have semisynthetic counterparts that are less toxic but more active. Vincristine was used to make vinglycinate, which was first reported in 1967, whereas vinblastine was used to make vindesine in 1980 (Nadine et al., 2007). Vindesine (eldisine, wherein the C(23) acetyl group in vinblastine was converted to an amido group) was the first semisynthetic vinca alkaloid to enter human clinical trials. Vindesine is primarily used to treat acute lymphocytic leukemia. It is used less commonly in breast cancer, chronic myelogenous leukemia blast crisis, colorectal cancer, non-small cell lung cancer, and renal cell cancer (kidney cancer) (Khazir et al., 2014).

4.3 TARGETED DELIVERY OF ORGANIC DRUGS

Cancer diagnostics and treatment rely on the ability to develop effective drugs and gene delivery systems that specifically target only the diseased cells without being toxic to healthy cells and tissues. To accomplish this feat, anticancer agents should be efficiently delivered to the tumor microenvironment, which would then enter the tumor cells by passing through several physiological and biological barriers without being degraded. The occurrence of these events depends on the size, biocompatibility, and surface chemistry of the anticancer agents for preventing unspecific interactions and promoting specific binding to their targets (Attia et al., 2019). It is important that the nanocarriers remain steady and stable in the bloodstream until they reach the target site and do not pass through the reticuloendothelial system (Samrot et al., 2018d). Moreover, the nanocarrier should not affect other tissues and nontarget organs *en route* and selectively kill the tumor cells. In addition, the agent should penetrate the tumor interstitial fluid and not be eliminated from the body by mononuclear phagocytes (Samrot et al., 2020; Barar & Omid, 2013). Furthermore, the nanocarriers are required to release the organic drug at an appropriate dose and at the required time. It is also essential for the remnant drug particles to be expelled from the body to ensure normal functioning of other normal tissues. It is imperative for the anticancer drug to reach the targeted tumor tissues, with only minimal losses in volume or activity in the blood circulation. Nanocarriers for cancer treatment can be used in two different ways, namely, passive and active targeting, both of which occur simultaneously, and one process does not exclude the other (Figure 4.1A).

4.3.1 Passive Targeting

In passive targeting, nanocarriers accumulate at the tumor site because of the altered permeability of blood vessels, which is called the enhanced permeability retention effect (Maeda, 2015). This phenomenon causes inflammation or hypoxia, and the vasculature does not provide sufficient oxygen supply to the cells for active proliferation, leading to cell death. Such a reaction promotes the secretion of growth factors that induce blood vessels to grow and become more permeable than that under normal conditions, leading to the formation of leaky blood vessels, which allows selectively enhanced vascular permeability of macromolecules larger than 40 kDa (Figure 4.1B) and drug-loaded nanocarriers into the tumor cells (Torchilin, 2011). The absence of proper lymphatic drainage in tumor cells makes way for the nanocarriers to percolate through the holes in the blood vessels and contribute to the accumulation of the nanocarriers (Bates et al., 2002). The high concentration of drug buildup excites the healing process. A drug-encapsulated nanocarrier is engineered to increase pharmacokinetics and tumor selectivity and to reduce treatment side effects.

4.3.2 Active Targeting

Nanocarriers with a modified surface are used to deliver drugs and theranostic agents to the tumor site for active targeting (Figure 4.1C). Extensive utilization of nanocarriers limits the side effects of the drug by avoiding targeting normal tissues and enhances the therapeutic effect (Bartlett et al., 2007). Nanocarriers can be modified by adding antibodies, antibody fragments, ligands, and peptides, which attach to the surface receptors present on the tumor cells; in this way, the nanocarriers gain

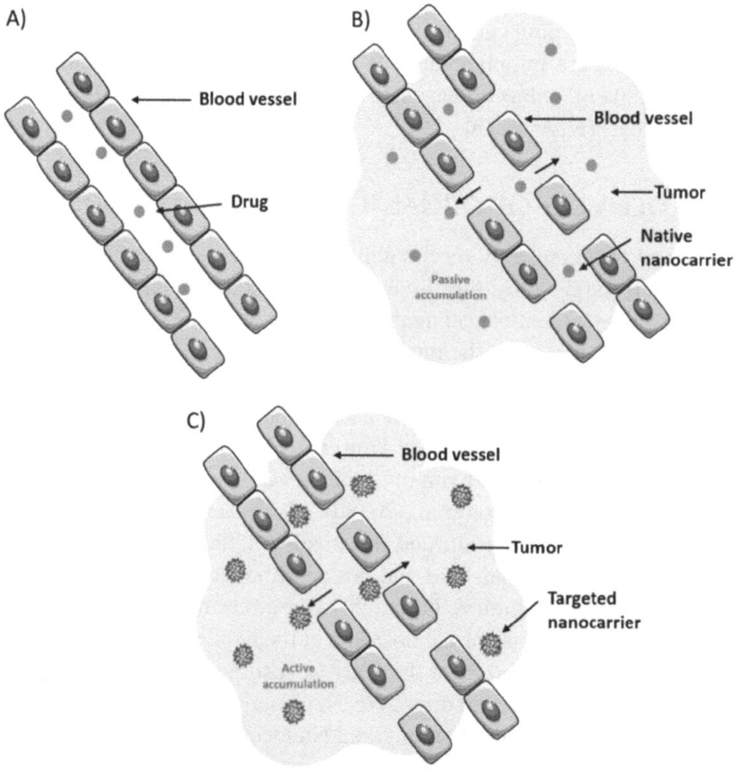

FIGURE 4.1 Targeted delivery with nanocarriers. A) Normal blood vessel, B) passive targeting in the diseased tissue, C) active targeting in the diseased tissue.

increasing access to binding to the cell surfaces, which enhances drug penetration (Peer et al., 2007). The molecules used in active targeting usually include a drug-loaded/encapsulated lipid or biopolymer carrier with a ligand- or antibody-targeting material.

4.4 NANOCARRIERS TO DELIVER PLANT METABOLITES FOR CANCER TREATMENT

4.4.1 Drug Delivery Using Various Nanocarriers

In nanotechnology, nanoparticles play an important role as drug delivery carriers, and they have gained substantial research interest in the field of cancer (Zielinska et al., 2009). As a drug delivery system, nanoparticles can be used to alter and improve the pharmacokinetics and pharmacodynamics of various types of drugs (Glassman et al., 2009). The primary goal of designing nanoparticles as drug delivery systems is to control particle characteristics, surface properties, and release of pharmaceutically active substances; therefore, nanoparticles help in accurately delivering a drug to the tumor site at an appropriate dose and at the required time (Li et al., 2020). With advances in nanotechnology, a variety of nanocarriers are currently available for application, such as carbon nanotubes, viral-based nanoparticles, polymeric nanoparticles, polymeric micelles, dendrimers, and liposomes.

4.4.1.1 Carbon Nanotubes

A carbon nanotube (Figure 4.2) is a tube-like structure made of graphite and is an allotrope of pure carbon. This tube usually contains a minimum of two layers (multiwalled), and the outer diameter varies in the range of 3–30 nm (Saifuddin et al., 2013). Almost all of the tubes have pentagonal carbon rings attached to their ends. Although the tubes can exist in various morphologies, in extremely rare cases, the caps appear hemispherical (Dresselhaus et al., 2008). These nanotubes are of two types: single-walled carbon nanotubes (SWCNTs) and multi-walled carbon nanotubes (MWCNTs). In SWCNTs, the hexagonal carbon rings are arranged as a rolled-up tubular shell, similar to a benzene ring (Samrot et al., 2023). MWCNTs appear as a bundle of graphene sheets rolled into concentric cylinders. One nanotube consists of millions of atoms, and the tube diameter can be as small as 0.7 nm, with a length of tens of micrometers (Saifuddin et al., 2013). MWCNTs can be manufactured from carbon soot using the arc-discharge method (Iijima et al., 1991). Additionally, according to their basic structures, carbon nanotubes can be categorized as armchairs, zigzags, and chiral (Saifuddin et al., 2013; Dadwal et al., 2018).

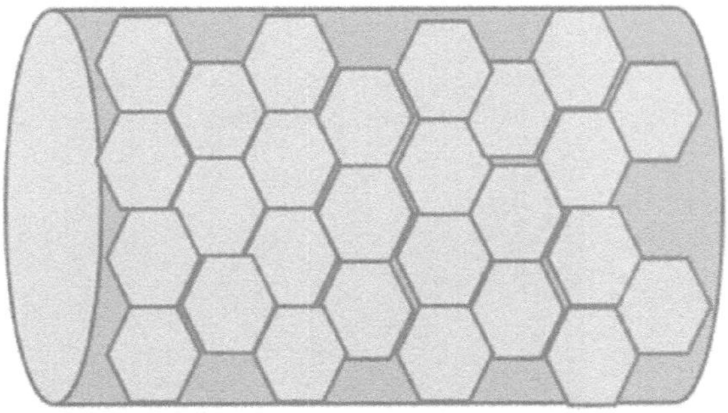

FIGURE 4.2 Structure of a carbon nanotube. (From Dadwal et al., 2018. With Permission.)

4.4.1.2 Virus-Based Nanoparticles

In nanotechnology, different viruses have been utilized for various purposes including biomaterials, vaccines, chemical tools, imaging, and molecular electronic materials (Singh et al., 2006). Nanometer-sized virus particles (less than 100 nm) are made of rare protein cages. Nanoscale fabrication of these virus particles is made easier by their well-aligned geometry and striking uniformity. Virus nanoparticles can be utilized to target cancer cells and deliver imaging agents or chemotherapeutics directly to tumors, which results in greater sensitivity and reduces systemic toxicity (Manchester et al., 2006). Polymeric nanomaterials have the potential to improve the delivery of chemotherapeutics.

4.4.1.3 Polymeric Nanoparticles

The emergence of challenges in drug targeting and drug delivery using polymeric nanoparticles will enable the discovery of promising approaches in drug delivery in the future (Figure 4.3). Nanoparticles prepared using this polymeric method can be used to encapsulate various kinds of drug molecules. The type of emulsion imposes a remarkable effect on the nature of drug molecules present in the nanoparticles (Vauthier et al., 2009). The benefits include reducing side effects, increasing its residence time in the body, offering a sustained release, and delivering its beneficial contents to the target site in the body (Brewer et al., 2011). If the drug can remain entrapped in the nanoparticle until the release time, then the pharmacokinetic properties of the drug will be determined by the pharmacokinetic properties of the nanoparticles (Allen et al., 2004). These drug carriers have the ability to circulate in the body and selectively target tumors and endosomes based on the low pH at these sites (Brewer et al., 2011).

4.4.1.4 Polymeric Micelles

Micelles are microscopic or nanoscopic structures made of polymeric macromolecules (Croy et al., 2006) such as amphiphilic copolymers, primarily consisting of two or three synthetic blocks of copolymers (poloxamers and poloxamines) or graft copolymers with different degrees of hydrophobicity (Almeida et al., 2018). Oral chemotherapy is advantageous in terms of sustained and mild plasma drug concentration, convenient and flexible administration, and low cost and is one of the most important advances in 21st century medicine (Feng et al., 2011). Furthermore, polymeric micelle drug delivery systems can solubilize, selectively target tumors, reduce systemic side effects, and show insensitivity to multidrug-resistant cells, all of which have been achieved in the light of advances in pharmaceuticals, molecular biology, bioimaging, and oncology (Almeida et al., 2018; Gagliardi et al., 2021).

FIGURE 4.3 Polymeric nanoparticle drug delivery system. (From Gagliardi et al., 2021. With Permission.)

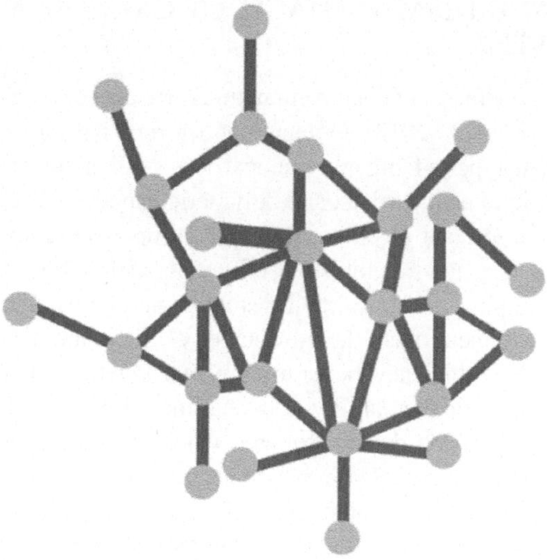

FIGURE 4.4 Dendrimers for drug delivery. (From Zeb et al., 2017. With Permission.)

4.4.1.5 Dendrimers

Dendrimers (Figure 4.4) are highly branched polymeric macromolecules with uniform size and shape. The basic structure of dendrimers includes three components, namely, a central core, repeated branching units, and terminal groups, which provide surface-modifiable characteristics (Palmerston et al., 2017). Dendrimer properties including size, structure, water-solubility, and mono-dispersity are easily controllable, and they have shown high potency as drug delivery systems, as their various terminal functional groups can be modified (Noriega et al., 2014; Zeb et al., 2017).

4.4.1.6 Liposomes

Liposomal drug carriers are a possible method for improving the delivery of therapeutic agents through liposomes, which are microparticulate lipoidal vesicles. With recent developments in liposome technology in the field of nanotechnology, several liposome-based drug formulations are currently undergoing clinical trials (Sharma et al., 1997). Because of their unique characteristics, liposomes have been used in clinical treatments as one of the leading nanocarriers (Li et al., 2020). When phospholipids are dissolved in water, the membrane of the phospholipid bilayers forms spheres containing internal water-filled compartments, which is the actual structure of a liposome (Alavi et al., 2017).

A liposome nanocarrier loads the drugs through two main methods: encapsulation of hydrophilic drugs and encapsulation of hydrophobic drugs. When hydrophilic drugs are encapsulated in the phospholipid bilayer of liposomes, a lipid–hydrophilic drug mixture is formed. During this process, the drug enters the liposome core, whereas the other components remain in the liposome (Chatterjee et al., 2012). When encapsulating hydrophobic drugs in the phospholipid bilayer of liposomes, drug diffusion toward the outer aqueous and inner lipid layers of the liposomes decreases. Phospholipids and organic solvents are used to solubilize the drugs required to be encapsulated (Akbarzadeh et al., 2013).

4.5 ADVANTAGES AND DISADVANTAGES OF ORGANIC NANOCARRIERS IN DRUG DELIVERY

Organic nanoparticles are generally obtained from natural products or are synthesized chemically for biomedical applications (Liu et al., 2013). Organic nanocarriers have a variety of shapes and sizes; however, the most frequent types of organic nanocarriers studied for drug delivery include solid lipid nanocarriers (SLNs), lipid nanocarriers, and polymeric nanocarriers (Wicki et al., 2015). SLNs, which involves the use of a tolerant physiological lipid component, combine the benefits of polymeric nanoparticles, liposomes, and fat emulsions (Hu et al., 2015). Because of their nontoxic composition, SLNs have unique physiochemical abilities and could be used for drug delivery (Almeida and Souto, 2007). As already mentioned, the advantages of SLNs are their strong biocompatibility, adaptability, high drug-loading efficiency, better drug release kinetics, and long-term stability (Mulik et al., 2010). Because of their excellent ability to crystalline, SLNs acquire a low drug-loading efficacy and the risk of drug expulsion during storage, which is a disadvantage (Yoon et al., 2013). Another disadvantage is the first burst release, which is common with these formulations (Makwana et al., 2015). Liposome nanocarriers, another type of nanocarrier, could be used as drug delivery systems (Caracciolo, 2015). Liposomes have a high biocompatibility and are less prone to toxicity, which make them ideal systems for drug delivery. Liposomes are vesicular structures comprising a hydrophobic lipid bilayer with an internal water-containing compartment that allows the incorporation of both soluble and insoluble drugs without any resultant chemical alterations to the core structure (Yingchoncharoen et al., 2016). Furthermore, liposomes' lipid bilayer structure can also be fused with other cell bilayers, which enable the drug content to be released to target sites. *In vitro* investigations on liposomes have revealed considerable ceramide formation and release into cells, resulting in tumor cell death; this feature suggests that nanocarriers have enhanced drug release and delivery mechanisms (Ma & Mumper, 2013). However, because of their colloidal stability and biodegradability, SLNs are still considered a better carrier over liposomes. Currently, a wide variety of drug delivery applications with increased drug load but reduced drug toxicity are available (Almeida & Souto, 2007). The main demerit is the toxicity caused by cationic lipids. The mononuclear phagocyte system rapidly degrades the nanocarriers (Ekambaram et al., 2012). Polymeric nanocarriers have a distinct structure comprising a polymer core with a drug-loading site (Prakash et al., 2011). By adding a polymer shell with a modifiable surface that can accommodate ligands and targeting moieties, polymeric carriers can be further altered to enable multifunctional drug delivery (Yang et al., 2010). Each polymeric nanocarrier can be designed through complete modification so as to reveal the desired chemical groups and control the release of bioactive compounds, giving it a flexible advantage. Degradation of the carrier is the main limitation of polymeric nanocarriers (Wong et al., 2006).

4.6 CONCLUSION

Considering the major challenges in cancer treatment, researchers in the field of nanotechnology have reported that active cancer therapy approaches require a crucial strategy. Traditional cancer treatment approaches such as chemotherapy, radiation, and surgery are still regarded effective, but because of the unique negative effects on healthy body parts, more research into cancer treatment modalities is urgently needed. Nanocarrier-based techniques play an important role in biological applications, particularly in chemotherapeutic drug delivery. Various modifications are combined to overcome the limitations of traditional chemotherapy and reduce drug toxicity in various nanocarriers. In conclusion, this review described the anticancer drug delivery systems that use polymeric-based nanoparticles, magnetic nanoparticles, lipid-based carriers, dendrimers, and quantum dots. It also summarized the mechanisms by which few types of nanocarriers deliver anticancer drugs, drug targeting processes, and cancer diagnosis. Based on the study findings, it is possible to conclude that nanocarriers prepared from biomaterials that are nontoxic to normal cells, when used in combination with tumor cell markers for targeted and enhanced drug delivery at the target site, have shown superior effects over traditional nanocarrier systems.

REFERENCES

Abdal Dayem A, Choi H, Yang G.-M, Kim K, Saha S, Cho S.-G. 2016. Nutrients 8(9):581.

Abirami, S., Jabesta, J., Renitta, R.E., Anand, D.A. and Samrot, A.V. 2021a. Antimicrobial Activity of Flower Extracts against Wound Pathogens and Fungi. Current Research in Green and Sustainable Chemistry 4:100076.

Abirami, S., Priyalakshmi, M., Soundariya, A., Samrot, A.V., Saigeetha, S., Emilin, R.R., Dhiva, S. and Inbathamizh, L. 2021b. Antimicrobial Activity, Antiproliferative Activity, Amylase Inhibitory Activity and Phytochemical Analysis of Ethanol Extract of Corn (Zea mays L.) Silk. Current Research in Green and Sustainable Chemistry 4:100089.

Acharya S, Sahoo SK. 2011. Advanced Drug Delivery Reviews 63(3):170–183.

Akbarzadeh A, Rezaei-Sadabady R, Davaran S, Joo SW, Zarghami N, Hanifehpour Y, Samiei M, Kouhi M, Nejati-Koshki K. 2013. Nanoscale Research Letters 8(1):102.

Alavi M, Karimi N, Safaei M. 2017. Advanced Pharmaceutical Bulletin 7(1):3–9.

Alexa-Stratulat T, Luca A, Bădescu M, Bohotin C-R, Alexa ID. 2017. Nutritional Modulators in Chemotherapy-Induced Neuropathic Pain pp. 9–33.

Allen TM, Cullis PR. 2004. Science 303(5665):1818–1822.

Almeida AJ, Souto E. 2007. Advanced Drug Delivery Reviews 59(6):478–490.

Almeida M, et al. 2018. Journal of Polymer Research 25(1):1–14.

Anand SC, Jhurani S, Kunnumakkara AB, Aggarwal BB. 2008. Cancer Letters 267:133–164.

Attia MF, Anton N, Wallyn J, Omran Z, Vandamme TF. 2019. Journal of Pharmacy and Pharmacology 71(8):1185–1198.

Barar J, Omidi Y. 2013. Bioimpacts 3(4):149–162.

Bartlett DW, Su H, Hildebrandt IJ, Weber WA, Davis ME. 2007. Proceedings of the National Academy of Sciences of the United States of America 104:15549–15554.

Bates DO, Hillman NJ, Williams B, Neal CR, Pocock TM. 2002. Journal of Anatomy 200:581–597.

Batra P, Sharma AK. 2013. 3 Biotech 3:439–459.

Bishayee A. 2009. Cancer Prevention Research 2(5):409–418.

Brewer E, Coleman J, Lowman A. 2011. Journal of Nanomaterials 2011.

Caracciolo G. 2015. Nanomedicine 11(3):543–557.

Chatterjee SN, Devhare PB, Lole KS. 2012. Journal of Clinical Microbiology 50(4):1467–1470.

Cragg GM, Newman DJ. 2005. Journal of Ethnopharmacology 100(1–2):72–79.

Croy S, Kwon G. 2006. Current Pharmaceutical Design 12(36):4669–4684.

Dadwal A, Baldi Ashish, Kumar Narang Raj. 2018. Artificial Cells, Nanomedicine, and Biotechnology 46(2):1–11.

Dancey J, Eisenhauer E. 1996. British Journal of Cancer 74(3):327–338.

Davidson KT, Zhu Z, Fang Y. 2016. Pathology Oncology Research 22(4):655–660.

Ekambaram P, Sathali AA, Priyanka K. 2012. Scientific Reviews and Chemical Communications 2(1):80–102.

Elanthendral G, Shobana N, Meena R, Prakash P, Samrot AV. 2021. Biocatalysis and Agricultural Biotechnology 38:102212.

Escobar-Chávez JJ, Rodríguez-Cruz IM, Domínguez-Delgado CL, Díaz-Torres R, Revilla-Vázquez AL, Casas Aléncaster N. 2012. Recent Advances in Novel Drug Carrier Systems 201.

Feng S-S, Zhao L, Tang J. 2011. Nanomedicine 6(3):407–410.

Gagliardi A, Giuliano E, Eeda V, Massimo F, Stefania B, Vibhudutta A, Donato C. 2021. Frontiers in Pharmacology 12:17.

Gali-Muhtasib H, Hmadi R, Kareh M, Tohme R, Darwiche N. 2015. Apoptosis, 20:1531–1562.

Glassman PM, Muzykantov VR. 2019. The Journal of Pharmacology and Experimental Therapeutics 370(3):570–580.

Hamidreza Ardalani, Amir Avan, Majid Ghayour-Mobarhan. 2017. Avicenna Journal of Phytomedicine 7(4): 285–294.

Hewlings S, Kalman D. 2017. Foods. 6(10), 92.

Hu L, Tang X, Cui F. 2004. Journal of Pharmacy and Pharmacology 56(12):1527–1535.
Iijima S. 1991. Nature 354(6348):56–58.
Iqbal J, Abbasi BA, Mahmood T, Kanwal S, Ali B, Shah SA, Khalil AT. 2017. Asian Pacific Journal of Tropical Biomedicine 7(12):1129–1150.
Jaspreet KV, Maram KR, Vinod DL. 2005. Current Nanoscience 1(1):47–64.
Jorio A, Dresselhaus G, Dresselhaus MS, eds. 2008. Carbon Nanotubes Vol. 111. Berlin: Springer.
Kesharwani P, Jain K, Kumar Jain N. 2014. Progress in Polymer Science 39(2):268–307.
Khalid EB, Ayman EE, Rahman H, Abdelkarim G, Najda A. 2016. Tumor Biology, 37:14513–14536.
Khazir J, Mir BA, Pilcher LA, Riley DL. 2014. Phytochemistry Letters 7(1):173–181.
Kintzios SE. 2006. Critical Reviews in Plant Science 25(2):79–113.
Kopustinskiene DM, Jakstas V, Savickas A, Bernatoniene J. 2020. Nutrients 12(2):457.
Lang J-Y. 2005. Clinical Cancer Research 11(9):3455–3464.
Lee, Y.F.; Sridewi, N.; Ramanathan, S.; Sudesh, K. 2015. The Influence of Electrospinning Parameters and Drug Loading on Polyhydroxyalkanoate (PHA) Nanofibers for Drug Delivery. Int. J. Biotechnol. Wellness Ind. 4:103–113.
Li Y, Cong H, Wang S, Yu B, Shen Y. 2020. Biomaterials Science 8:6442–6468.
Lichota A, Gwozdzinski K. 2018. International Journal of Molecular Sciences 19(11):3533.
Liu BR, Winiarz JG, Moon JS, et al. 2013. Colloids and Surfaces B: Biointerfaces. 111:162–170.
Ma P, Mumper RJ. 2013. Journal of Nanomedicine and Nanotechnology 4(2):1000164.
Maeda H. 2015. Advanced Drug Delivery Reviews 91:3–6.
Makwana V, Jain R, Patel K, Nivsarkar M, Joshi A. 2015. International Journal of Pharmaceutics 495(1):439–446.
Manchester M, Singh P. 2006. Advanced Drug Delivery Reviews 58(14):1505–1522.
Martino E, Casamassima G, Castiglione S, Cellupica E, Pantalone S, Papagni F, … Collina S. 2018. Bioorganic & Medicinal Chemistry Letters 28(17):2816–2826.
Meijer L, Borgne A, Mulner O, Chong JPJ, Blow Inagaki N, Inagaki M, Delcros JG, Moulinoux JP. 1997. Biochemistry 243(1–2): 527–536.
Meng LH, Ding J. 2007. Acta Pharmacologica Sinica 28(9):1460–1465.
Monje M, Wen PY. 2008. Handbook of Neuro-Oncology Neuroimaging. pp. 102–115.
Mulik RS, Mönkkönen J, Juvonen RO, Mahadik KR, Paradkar AR. 2010. International Journal of Pharmaceutics 398(1–2):190–203.
Nadine D, El-Banna S, Gali-Muhtasib H. 2007. Expert Opinion on Drug Discovery 2(3):361–379.
Niedzwiecki A, Roomi M, Kalinovsky T, Rath, M. 2016. Nutrients. 8(9):552.
Noriega-Luna B, Godínez LA, Rodríuez FJ, Rodríguez A, Zaldívar-Lelo de Larrea G, Sosa-Ferreyra CF, Mercado-Curiel RF, Manríquez J, Bustos E. 2014. Journal of Nanomaterials 2014:507273
Oguz B, Ipek E, Kose MD, and Kalmaz G. 2017. 395–410, chapter 17.
Ojima I, Lichtenthal B, Lee S, Wang C, Wang X. 2015. Expert Opinion on Therapeutic Patents 26(1):1–20.
Palmerston Mendes L, Pan J, and Torchilin VP. 2017. Molecules 22(9): 1401.
Peer D, Karp JM, Hong S, Farokhzad OC, Margalit R, Langer R. 2007. Nature Nanotechnology 2 :751–760.
Pezzuto JM. 1997. Biochemical Pharmacology 53(2):121–133.
Poljaková J, Frei E, Gomez JE, Aimova D, Eckschlager T, Hrabeta J, Stiborová M. 2007. Cancer Letters 252:270–279.
Prakash S, Malhotra M, Shao W, Tomaro-Duchesneau C, Abbasi S. 2011. Advanced Drug Delivery Reviews 63(14–15):1340–1351.
Pund S, Joshi A. 2017. Nano- and Microscale Drug Delivery Systems 439–480.
Raji P, Samrot AV, Keerthana D. et al. 2019. Journal of Cluster Science 30:881–895.
Raji P, Samrot AV, Rohan B, Kumar P, Geetika R, Sharma V, Dhanabal K. 2019. Rasayan Journal of Chemistry 1:123–137.
Rauf A, Imran M, Butt MS, Nadeem M, Peters DG, Mubarak MS. 2017. Critical Reviews in Food Science and Nutrition 58(9):1428–1447.
Rhee Y-S and Mansour HM. 2011. International Journal of Nanotechnology. 8(1–2):84–114.
Saifuddin N, Raziah AZ, Junizah AR. 2013. Journal of Chemistry 2013:1–18.
Salim AA, Chin Y-W, Kinghorn AD. 2008. Bioactive Molecules and Medicinal Plants. Springer, Berlin, Heidelberg. 1–24.
Samrot AV, Kudaiyappan T, Bisyarah U, Mirarmandi A, Faradjeva E, Abubakar A, Selvarani JA, Kumar Subbiah S. 2020. International Journal of Nanomedicine 15:7097–7115.

Samrot AV, Saigeetha S, Mun CY, Abirami S, Purohit K, Cypriyana PJJ, Dhas TS, Inbathamizh L, Kumar SS. 2021. Scientific Reports 11(1):24511.

Samrot, A.V., Burman, U., Philip, S.A., Shobana, N. and Chandrasekaran, K. 2018a. Synthesis of Curcumin Loaded Polymeric Nanoparticles from Crab Shell Derived Chitosan for Drug Delivery. Informatics in Medicine Unlocked, 10:159–182.

Samrot AV, Shobana N, Jenna R. 2018, BioNanoScience 8(2):632–646.

Samrot, A.V., Suvedhaa, B., Sahithya, C.S. and Madankumar, A., 2018c. Purification and Utilization of Gum from Terminalia Catappa L. for Synthesis of Curcumin Loaded Nanoparticle and its In Vitro Bioactivity Studies. Journal of Cluster Science 29:989–1002.

Samrot, A.V., Shobana, N., Philip, S.A., Burman, U. and Chandrasekaran, K. 2018d. Utilization of Crab Shell Derived Chitosan for Production of Gallic Acid Loaded Nanocomposites for Drug Delivery. Journal of Pharmaceutical Sciences and Research 10(9):2169–2174.

Samrot, A.V., Sathiyasree, M., Rahim, S.B.A., Renitta, R.E., Kasipandian, K., Krithika Shree, S., Rajalakshmi, D., Shobana, N., Dhiva, S., Abirami, S. and Visvanathan, S. 2023. Scaffold Using Chitosan, Agarose, Cellulose, Dextran and Protein for Tissue Engineering—A Review. Polymers 15(6):1525.

Schlom J, Hodge JW, Palena C, Greiner JW, Tsang K-Y, Farsaci B, … Gulley JL. 2013: 309–331.

Seca A, Pinto D. 2018b. International Journal of Molecular Sciences 19(1):263.

Shah U, Shah R, Acharya S, Acharya N. 2013. Chinese Journal of Natural Medicines 11(1):16–23.

Shah Z, Gohar U. F, Jamshed I, Mushtaq A, Mukhtar H, Zia-UI-Haq M, … Popovici B. 2021. Biomolecules 11(4):603.

Sharma A, Sharma US. 1997. International Journal of Pharmaceutics 154(2):123–140.

Shobana N, Kumar PS, Raji P, Samrot AV. 2019. Utilization of Crab Shell-derived Chitosan in Nanoparticle Synthesis for Curcumin Delivery 48(8):1183–1188.

Shobana N, Prakash P, Samrot AV, Jane Cypriyana PJ, Kajal P, Sathiyasree M, Saigeetha S, Stalin Dhas, T, Alex Anand D, Sabesan GS, Muthuvenkatachalam BS. 2022. Purification and Characterization of Gum-derived Polysaccharides of Moringa Oleifera and Azadirachta Indica and their Applications as Plant Stimulants and Bio-pesticidal Agents. Molecules 27(12):3720.

Singh AK, Sharma N, Ghosh M, Park YH, Jeong DK. 2016. Critical Reviews in Food Science and Nutrition 57(16):3449–3463.

Singh CK, Ndiaye MA, Ahmad N. 2015. Biochima et Biophysica Acta 1852(6):1178–1185.

Singh P, Gonzalez MJ, Manchester M. 2006. Drug Development Research 67:23–41.

Škubník J, Pavlíčková V,RumlT, Rimpelová S. 2021. Plants 10:569.

Solaro R, Chiellini F , Battisti A. 2010. Materials 3(3):1928–1980.

Stiborová M, Černá V, Moserová M, Mrízová I, Arlt V, Frei E. 2014. International Journal of Molecular Sciences 16(1):284–306.

Tesauro C, Simonsen AK, Andersen MB, et al. 2019. BMC Cancer 19:1158.

Tomeh M, Hadianamrei R, Zhao X. 2019. International Journal of Molecular Sciences 20(5):1033.

Torchilin V. 2011. Advanced Drug Delivery Reviews 63:131–135.

Udenigwe CC, Ramprasath VR, Aluko RE, Jones PJ. 2008. Nutrition Reviews 66(8):445–454.

Varsha K, Sharma A, Kaur A, Madan J, Pandey RS, Jain UK, Chandra R. 2004. Natural Plant-Derived Anticancer Drugs Nanotherapeutics: A Review on Preclinical to Clinical Success pp. 775–809.

Vauthier C, Bouchemal K. 2009. Pharmaceutical Research 26(5):1025–1058.

Veeramuthu D, Raja WR, Al-Dhabi WA, Savarimuthu I. 2017. Flavonoids – From Biosynthesis to Human Health.

Wicki A, Witzigmann D, Balasubramanian V, Huwyler J. 2015. Journal of Controlled Release 200:138–157.

Willey CD, Yang ES-H, Bonner JA. in Clinical radiation oncology, ed. J. E. Tepper, J. A. Bogart and L. L. Gunderson, Elsevier, Philadelphia, PA, 2016, pp. 63–79.e4.

Wong HL, Rauth AM, Bendayan R, et al. 2006. Pharmaceutical Research 23(7):1574–1585.

Yang D, Yu L, Van S. 2010. Cancers (Basel) 3(1):17–42.

Yingchoncharoen P, Kalinowski DS, Richardson DR. 2016. Pharmacological Reviews 68(3):701–787.

Yoon G, Park JW, Yoon IS. 2013. Journal of Pharmaceutical Investigation 43(5):353–362.

Zeb A, Din Ud F, Aman W, Qureshi O. 2017. International Journal of Nanomedicine 12:7291–7309.

Zhang X, Rakesh KP, Shantharam CS, Manukumar HM, Asiri AM, Marwani HM, Qin H-L. 2018. Bioorg Med Chem 26(2), 340–355.

Zielinska A, Skwarek E, Zaleska A, Gazda M, Hupka J. 2009. Procedia Chemistry 1(2):1560–1566.

5 Antidiabetic Plant Products and Their Nanoparticle-Based Delivery

TD Alabi[1] *and JA Badmus*[2*]
[1]Department of Life Sciences, Baze University, Abuja, Nigeria
[2]Department of Biochemistry, College of Health Sciences, Ladoke Akintola University of Technology, Ogbomosho, Nigeria
[*]Corresponding Author: jabadmus@lautech.edu.ng

LIST OF ABBREVIATIONS

FBS: fasting blood sugar, FBG: fasting blood glucose, TG: triglycerides, TC: total cholesterol, HDL: high-density lipoprotein, LDL: low-density lipoprotein, VLDL: very low-density lipoprotein, HbA1c: glycosylated hemoglobin, SOD: superoxide dismutase, CAT: catalase, TBARS: thiobarbituric acid reactive substances, GSH: reduced glutathione, DCM: dichloromethane, MDA: malondialdehyde, TAC: total antioxidant capacity, NO: nitric oxide, BH4: tetrahydrobiopterin, PAB: pro-oxidant-antioxidant balance, BUN: blood urea nitrogen, AGE: advanced glycation end products, TP: total protein, NEFA: nonesterified fatty acids, HOMA-IR: homeostatic model assessment for insulin resistance, HMG-CoA: 3-hydroxyl, 3-methyl glutaryl_coA reductase, PPAR-α/γ: peroxisome proliferator–activated receptor alpha/gamma, KIM-1: kidney injury molecule-1, La: lanthanum

5.1 INTRODUCTION

The increasing rate and debilitating effect of diabetes in today's world are worrisome or distressing issues, with more than 463 million people being diagnosed with diabetes, and this number is estimated to increase to 642 million within two decades[1,2]. Diabetes is a chronic endocrine disorder that induces austere dysregulation in the metabolism of proteins, carbohydrates, and lipids[3]. This disorder is characterized by elevated blood glucose levels owing to impaired function of pancreatic β-cells, leading to reduced insulin secretion or lack of insulin sensitivity in the cells of the liver, muscles, and adipose tissue[4]. The ailment is categorized as type I, II, and gestational diabetes (occurs in pregnant women). This disorder is associated with several complications such as hypertension, neurological disorders, cardiac failure, impaired vision, polydipsia, and kidney diseases, among others[5]. Diabetes is managed through various approaches such as lifestyle modifications, which includes exercise, diet, and medication. Conventional drugs such as biguanides (metformin), sulfonylureas (glibenclamide), dipeptidyl peptidase 4 inhibitors (linagliptin), and thiazolidinediones are commonly used for diabetes management. These drugs stimulate insulin secretion and release, enhance glucose uptake, minimize insulin resistance, and hepatic glucose production but exert a wide range of adverse effects or contraindications[6]. Gastric discomfort, water retention, chest pain, nausea, vomiting, headaches, diarrhea, high risk of genital and urinary infection, and heart failure are common side effects of the use of conventional drugs, as these side effects dampen their therapeutic efficiency. Therefore, extensive attention has been

DOI: 10.1201/9781003231721-5

devoted to research on phytomedicine because of its affordability, broad-range therapeutic effect, and low toxicity[6]. Medicinal plants contain bioactive compounds that are effective in mitigating the pathological effect of diabetes and other diseases; however, there are some limitations in the delivery and optimal efficacy of these bioactive compounds owing to low specificity and stability, steric hinderance, and so on[2,7]. Nanoparticles have therefore been synthesized and used to enhance drug delivery efficiency and dosage efficacy because of their precise specificity in targeting diseased cells and their ability to overcome transportation barriers[7]. Non-green-synthesized (physically and chemically synthesized) nanoparticles have been widely used for different applications in science, including the diagnosis of diseases and the development of nanosurgical and medical devices, but they have certain constraints such as being expensive and toxic[2,8,9]. The lethal potency of non-green-synthesized nanoparticles in biological systems among other factors has narrowed its advancement, which projects the importance of green-synthesized nanoparticles, as they are eco-friendly, cost-effective, readily available, and biocompatible[10-13]. Green synthesis of nanoparticles can be achieved using microorganisms (bacteria, fungi, and yeast), templates (membranes, viral DNAs, or diatoms), and plant materials, which are the most commonly explored material[10,13,14]. Bioactive compounds present in the extracts of different plant parts used for nanoparticle synthesis also serve as capping agents and confer additional biological properties to the nanoparticles. Collaboratively, the nanoparticles have enhanced delivery and target-specificity of these compounds, achieving this feature at a lesser drug concentration[7,11]. This chapter focuses on nanotechnological applications that use plant-based materials and their mechanisms in diabetes management.

5.2 GREEN SYNTHESIS OF NANOPARTICLES

Nanoparticle synthesis involves two major approaches: top-down and bottom-up approaches. The top-down method is a deconstructive approach, wherein the nanoparticles are synthesized from bulk materials by physical methods and techniques. These techniques include milling (mechanical/ball), laser ablation, evaporation-condensation, lithography, plasma arcing, spluttering, and thermal decomposition[10,12,13]. The bottom-up method is a constructive approach wherein nanoparticles are synthesized by assembling smaller units such as atoms. This mainly involves two approaches: the chemical method, such as chemical vapor deposition, sol-gel process, and spinning, among others; and biological method, such as the use of plants or microorganisms[2,13,15,16]. The bottom-up method is mostly used and preferred for the green synthesis of nanoparticles because it involves less energy and involves less robust processes that do not require an extensive or long time period such as in the case of mechanical milling for top-down synthesis[17,18].

5.3 TYPE OF MATERIALS FOR THE SYNTHESIS OF PHYTONANOPARTICLES

Currently, numerous nanomaterials have been used in green nanotechnology; however, metal nanoparticles (MNPs) are preferred and broadly used (especially for plants) owing to their unique properties relating to their optics, electrical and magnetic features, and atomic ordering at nanoscale. Additionally, metals have innate pharmacological activities including antioxidant, anti-inflammatory, and hypoglycemic effects[14,19-21]. Shaheen et al. reported the ability of gold and silver nanoparticles to significantly reduce blood glucose levels, elevate insulin levels while eliminating insulin resistance, and mitigate inflammation and oxidative stress[22]. In addition to their biological activities, green synthesis of MNPs is cost-efficient, less complicated, and environment-friendly[20,23,24]. Gold, copper, zinc, and silver are the metals commonly used for synthesizing phytonanoparticles; however, several other metals such as titanium (Ti), iron (Fe), nickel (Ni), magnesium (Mg), platinum (Pt), palladium (Pd), lead (Pb), lanthanum (La), selenium (Se), and silicon (Si) are sometimes used for green synthesis[23].

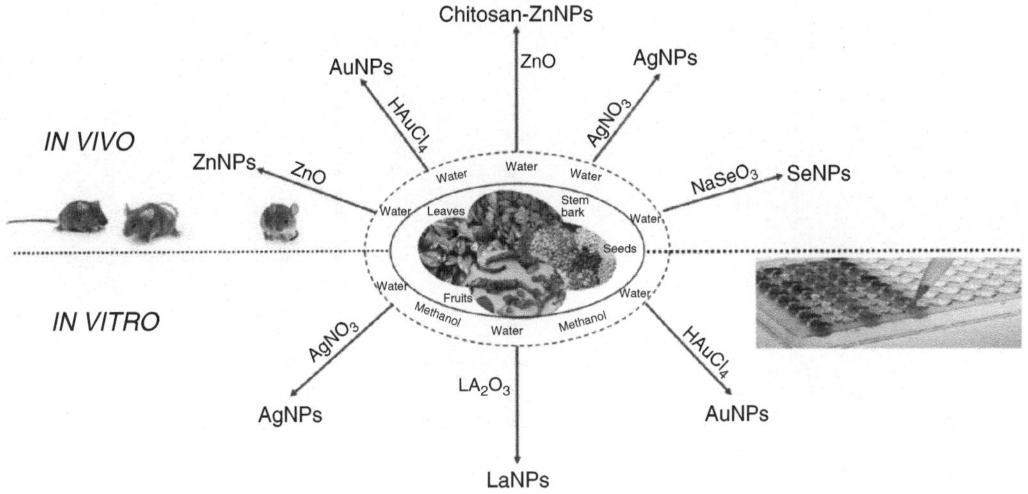

FIGURE 5.1 Green-synthesized nanoparticles and their metal precursors in *in vivo* and *in vitro* studies. The dotted circular lines indicate the extraction of solvents from the plant material before synthesis in each study model. The dotted straight lines represent a partition between the *in vivo* and *in vitro* models.

5.4 PHYTONANOTHERAPY IN DIABETES MANAGEMENT

MNPs have been investigated for their antidiabetic potential using different models: *in vitro*, *in vivo*, and *ex vivo*. Parameters used for measuring the antidiabetic potential of MNPs in *in vitro* assessments usually include α-glucosidase, α-amylase, glucose uptake, hemoglobin glycosylation, and bovine serum albumin denaturation (Table 5.1). Metrics used for measuring the antidiabetic ability of MNPs in *in vivo* models were fasting blood glucose, glycogen content, HbA1c estimation, and peroxisome proliferator-activated receptor (PPAR)-γ/α expression for glycemic indices; lipid profile (total cholesterol [TC], triglyceride [TG], high-density lipoprotein [HDL], low-density lipoprotein [LDL], and very low-density lipoprotein [VLDL]); and β-hydroxy β-methylglutaryl-CoA (HMG-CoA; Table 5.2). Additionally, the parameters used for measuring insulin resistance are HOMA-IR, insulin levels, and oral glucose tolerance test, while the indices for measuring nephropathy include urea, creatinine, and kidney injury molecule-1 (KIM-1). Furthermore, alanine transaminase and aspartate transaminase are estimated for determining liver injury, and superoxide dismutase, catalase, reduced glutathione (GSH), total antioxidant capacity, thiobarbituric acid reactive substances, advanced glycation end products (AGEs), and thiols are assessed to determine antioxidant and oxidative stress levels. Measures of endothelial dysfunction include tetrahydrobiopterin, albumin, globulin, and total protein as indices of protein catabolism as well as tumor necrosis factor alpha (TNF-α) as a marker of inflammation (Table 5.2). Several plants have been used in the bio-reduction of metal ions to form nanoparticles[23]. However, nanoparticles made from selected plants with a history and empirical evidence of antidiabetic potential are discussed in the sections below.

5.4.1 *Ananas comosus*

Ananas comosus (AC) fruit, popularly known as pineapple, are grown in tropical and sub-tropical regions and are mostly consumed as food; its mesocarp (peel), which is an agricultural waste/byproduct, has been used as a reducing agent for the synthesis of silver nanoparticles (AgNPs). AC-AgNPs have been investigated for their antidiabetic potential through the inhibition of α-glucosidase[20]. AC-AgNPs has shown 100% inhibition against α-glucosidase at concentrations ranging from 0.0625 mg/mL to 1 mg/mL, indicating that they are a promising tool to decrease glucose release

TABLE 5.1
***In Vitro* Studies on Metal Nanoparticles Synthesized From Antidiabetic Plant Products as Reducing Agents**

Plant Source and Parts	Region of Harvest/ Common Name	Extract, Fraction, and Compounds Used	Precursor	Size and Shape of the Nanoparticles Used	Dose	Parameters/Assays	Findings	Refs
Ananas comosus (fruits' exocarp)	Republic of Korea Pineapple	Aqueous extract	$AgNO_3$	- Clusters of spherical structure	0.0625–1 mg/mL	α-glucosidase	100% inhibition	20
Annona muricata (leaves)	Ondo State, Nigeria	Aqueous extract	$AgNO_3$	35 nm Spherical	-	α-amylase α-glucosidase	IC_{50} 0.90±0.01 µg/mL IC_{50} 3.32 ± 0.032 µg/mL	28
Chrysophyllum albidum (seed)	Osun State, Nigeria	Methanolic extract	$AgNO_3$	-	0.0–0.3 mg/mL	α-amylase	123.07 µg/mL	37
Costus pictus (leaves)	-	Methanolic extract	$AgNO_3$	-	0.2–1 mg/mL 0.2–1 mg/mL 0.2–1 mg/mL 0.04–0.2 mg/mL	HbA1c α-glucosidase α-amylase glucose uptake	IC_{50} 609.61 µg/mL IC_{50} 501.34 µg/mL IC_{50} 534.39 µg/mL IC_{50} 78.39 µg/mL	47
Moringa oleifera (leaves)	India	Aqueous extract	La_2O_3	52.05–73.64 nm - Agglomerated, irregular cubic, and rectangular	10–500 µg/mL 50–500 µg/mL 10–500 µg/mL	α-amylase inhibition, total antioxidant activity bovine serum albumin denaturation	80% Inhibition of α-amylase at 500 µg/mL 75% inhibition of N_2 radicals at 500 µg/mL 94% inhibition at 500 µg/mL	76
Helichrysum foetidum (leaves and flowers)	Worcester, South Africa	Helichrysin	$AuCl_4$	6.7±1.68nm hexagonal, spherical, and triangular	0.0625–1 mg/mL 0.0625–1 mg/mL -	α-amylase α-glucosidase glucose uptake	IC_{50} No activity, corresponds with its precursor, no activity. IC_{50} 173.3±0.1 µg/mL exhibited lesser activity than that of its precursor (48.3±0.1). Decreased glucose uptake, compound only was more effective than its nanoparticles	51

		Helichrysetin	$AuCl_4$	2.02±0.53 Hexagonal, spherical, and triangular	0.0625–1 mg/mL 0.0625–1 mg/mL -	α-amylase α-glucosidase glucose uptake	IC_{50} 35.6±0.1 μg/mL less inhibitory activity than that of its precursor IC_{50} 40.0±0.2 μg/mL, which has lesser activity than that of its compound only (19.4±0.3) Decreased glucose uptake Its precursor showed more efficacy	
		Methanolic extract	$AuCl_4$	12.5±0.43 Hexagonal, spherical, triangular, and rod	0.0625–1 mg/mL 0.0625–1 mg/mL -	α-amylase α-glucosidase glucose uptake	IC_{50} 44.5±0.1 μg/mL while its precursor had no activity IC_{50} 391±0.1 μg/mL had far less activity than the extract only (27.3±0.3). Decreased glucose uptake when compared with that of its corresponding extract	
Leucosidea sericea (aerial parts)	-	Procyanidin fraction (F1)	$AgNO_3$	2.9±0.8 nm Spherical	12.5–200 μg/mL 12.5–200 μg/mL	α-amylase α-glucosidase	IC_{50} 13.24±0.60 had activity while its fraction only had no activity. IC_{50} 18.76±1.0, less inhibitory potential to its fraction F1 only (7.30±0.50)	12
		Procyanidin fractions (F2)	$AgNO_3$	3.3±1.2 nm Spherical	12.5–200 μg/mL 12.5–200 μg/mL	α-amylase α-glucosidase	IC_{50} 19.13±0.80, the same inhibitory activity as that of its precursor IC_{50} 8.75±0.70, comparable activity as that of its precursor (7.10±0.40)	
		Aqueous-ethanolic extract	$AgNO_3$	7.8±2.9 nm Dendritic shapes	12.5–200 μg/mL 12.5–200 μg/mL	α-amylase α-glucosidase	14.92±1.0, lesser inhibitory activity than that of the crude extract IC_{50} 21.48±0.90, not as active as its precursor (8.10±0.60)	

TABLE 5.2
***In Vivo* Studies on Metal Nanoparticles Using Antidiabetic Plant Products as Reducing Agents**

Plant Source and Parts	Region of Harvest	Precursor	Size and Shape of the Nanoparticles Used	Dose/Mode and Period of Administration	Experimental Animals	Parameters/Assays	Findings	Refs
Chamaecostus cuspidatus (leaves)	Tirunelveli, Courtallam, Tamil Nādu in India	$AuCl_4$	2 µm Cubic and square	0.75 and 1.5 mg/kg orally for 10 days	Wistar mouse Male 150–180 g	FBG, insulin, glycogen, body weight, TP, TC	Decreased glucose levels, normalized body weight loss due to hyperglycemia, and restored insulin and glycogen content, cholesterol, and protein levels to near-normal, similar to glibenclamide (50 mg/kg)	71
Cinnamomum cassia	Warren Chem, South Africa	$AgNO_3$	-	5 mg/kg, 10 mg/kg orally for 56 days	Sprague Dawley rats Male 250±20 g	FBG, body weight, MDA, GSH, urea, creatinine, ALT, AST, albumin, kidney histology	10 mg/kg NPs exhibited more efficacy in reducing blood glucose, urea, and creatinine and improving weight loss and antioxidant levels than 5 mg/kg NPs and extract (200 mg/kg) only	44
Cassia fistula (stem bark)	India	$AuCl_4$	55.2–98.4 nm Rectangular, triangular	60 mg/kg, gastric intubation for 30 days	Wistar rats Male 150–200 g	Blood glucose, HbA1c, body weight, TC, TG, HDL, LDL, VLDL, AST, ALT, albumin, globulin, total protein, urea, creatinine, uric acid	Weight loss repair, reduced hyperglycemia and hyperlipidemia, HbA1c, normalized protein catabolism, mitigated kidney, and liver injury in a comparable trend as that of insulin (3 IU/kg)	30
Cleome droserifolia	Haraz Herbal Store, Cairo, Egypt	$NaSeO_3$	50–190 nm Spherical	3.5 mg/kg, gastric intubation for 45 days	Wistar rats Male 150–180 g	Blood glucose, insulin, ALT, AST, TC, TG, HDL, VLDL, LDL, NEFA, urea, creatinine	Significant decrease in blood glucose, hyperlipidemia, liver and kidney injury; restored pancreatic architecture to near-normal. However, Galvus-Met; (5 mg:100 mg) showed more efficacy	39

Curcumin	-	ZnO-chitosan	43 nm -	100 mg/kg, oral administration for 28 days	Wistar rats Male 200±30g	FBS, OGTT, Insulin, TG, HDL, LDL, VLDL, GSH, SOD, TBARS, CAT, pharmacokinetics, DNA damage	Improved absorption and bioavailability of curcumin, reversed dyslipidemia, minimized DNA damage in a similar trend as that of glibenclamide (0.6 mg/kg)	65
Curcumin	-	Nanoprecipitation using DCM and Tween 60	40–50 nm spherical	30 mg/mL, 60 mg/mL orally for 30 days	Albino rats Male 140±20 g	FBS, MDA, TAC, insulin, NO, biopterin, BH4	Declined lipid peroxidation while boosting antioxidant defense and repaired endothelial dysfunction (BH4, biopterin) in a dose-dependent manner. 60 mg/mL exhibited more efficiency, as it restored parameters to normal levels	72
Momordica charantia	Ondo State, Nigeria	$AgNO_3$	-	50 mg/kg for 12 days	Wistar rats Male -	FBG, HMG-CoA, PPAR-α, PPAR-γ, KIM-1	Blood glucose level was reduced while HMG-CoA, PPAR-α, PPAR-γ, and KIM-1 expression was significantly upregulated	42
Nasturtium officinale (leaves)	-	ZnO	4.8 nm Spindle-shaped and spherical	8 mg/dL, intraperitoneal injection for 16 days	Wistar rats Male 200 g	FBG, TG, TC, HDL, Insulin, body weight	Lowered blood glucose, TG, and TC more effectively significantly reduced HDL than insulin injection (10 IU/kg); it raised insulin levels to normal in a similar trend as that of insulin (10 IU/kg)	41
Polydatin	-	Chitosan	56.33±1.55–144.25±3.37 nm Irregular spherical	50 mg/kg, gastric intubation for 28 days	Wistar rats Male 110–130 g	Body weight, FBS, HbA1c, insulin, HOMA-IR, liver glycogen	FBS, HbA1c and HOMA-IR declined while body weight, insulin, and hepatic glycogen levels increased when compared with metformin (100 mg/kg) and more efficiently than polydatin only (50 mg/kg)	73

(Continued)

TABLE 5.2 (Continued)
***In Vivo* Studies on Metal Nanoparticles Using Antidiabetic Plant Products as Reducing Agents**

Plant Source and Parts	Region of Harvest	Precursor	Size and Shape of the Nanoparticles Used	Dose/Mode and Period of Administration	Experimental Animals	Parameters/Assays	Findings	Refs
Smilax glabra (rhizome)	China	$AuCl_4$	50–90 nm Hollow shaped	50 mg/kg, orally using translucent fabric gauze for 15 days	Wistar rats Male -	FBG, insulin, HbA1c, AGEs, OGTT, AST, ALT, ALP, TC, TG, HDL, LDL, leptin, resistin, adiponectin, TNF-α, histology of heart and liver	NPs modulated hormones implicated in obesity and caused a reduction in escalated cholesterol, leading to decreased obesity, blood glucose, liver injury markers, inflammation, comparable to metformin (300 mg/kg)	74
Urtica dioica (leaves)	-	ZnO	19–20 nm Spherical	8 mg/dL, intraperitoneal injection for ≥16 days	Wistar rats Male 200 g	FBS, insulin, TC, TG, HDL	Significant lowering of FBS, improved insulin levels, and modulation of lipid profile when compared with the extract (150 mg/dL) or insulin (10 U/kg) only	69
Ziziphus jujube (fruits)	South Khorasan, Iran	$AuCl_4$	7–27 nm Spherical	0.5 mg/kg and 1.0 mg/kg for 21 days	Sprague Dawley rats Male 180–200 g	FBG, insulin, body weight, ALT, AST, BUN, creatinine, TC, TG, HDL, LDL, HOMA-IR, Vitamin D, PAB, MDA, thiol	Both dosages of NPs from *Z. jujuba* had a positive effect on body weight, significantly reduced hyperglycemia, liver injury markers, oxidative stress, and modulated lipid metabolism, with the 1 mg/kg dose being more effective	75

FBG: fasting blood glucose; TP: total protein; TC: total cholesterol; MDA: malondialdehyde; GSH: reduced glutathione; ALT: alanine transaminase; AST: aspartate transaminase; ALP: alkaline phosphatase; NP: nanoparticle; TG: triglyceride; HDL: high-density lipoprotein; LDL: low-density lipoprotein; VLDL: very low-density lipoprotein; NEFA: nonesterified fatty acid; DCM: dichloromethane; FBS: fasting blood sugar; TAC: total antioxidant capacity; NO: nitric oxide; HMG-CoA: β-hydroxy β-methylglutaryl-CoA; PPAR-γ: peroxisome proliferator-activated receptor gamma; KIM-1: kidney injury molecule 1; AGE: advanced glycation end product; TNF: tumor necrosis factor; BUN: blood urea nitrogen; PAB: pro-oxidant-antioxidant balance; OGTT: oral glucose tolerance test; BH4: tetrahydrobiopterin; CAT: catalase; TBARS: thiobarbituric acid reactive substance into the blood in non–insulin-independent diabetes[25]. Das et al. further explored the antioxidant abilities (ABTS [2,2′-azino-bis-3-ethylbenzthiazoline-6-sulfonic acid], DPPH [1,1-diphenyl-2-picrylhydrazyl], NADPH oxidase [NOX], and reducing power) of AC-AgNPs. AC-AgNPs have showed antioxidant potential in scavenging DPPH radicals but with less effectiveness for ABTS radicals and NOX[20].

5.4.2 *Annona muricata*

Annona muricata (AM), commonly referred to as "soursop" or "prickly custard apple," is predominantly grown in tropics and sub-tropics. The ethnomedicinal uses of different parts of AM range from treatment for hypertension, diabetes, and inflammation to arthritis, cancer, and respiratory complications[26,27]. The antidiabetic potential of AgNPs synthesized from the leaves of AM as reducing and capping agents has been reported[28]. AgNPs prepared from the leaves of AM (AM-AgNPs) have been investigated for their *in vitro* antioxidant (ABTS, DPPH) and antidiabetic (α-amylase and α-glucosidase) properties. AM-AgNPs demonstrated high inhibitory activity against α-amylase and α-glucosidase, with IC_{50} values of 0.90 μg/mL and 3.32 μg/mL, respectively, while acarbose (standard) had an IC_{50} value of 10.20 μg/mL for α-amylase and showed no effective activity against α-glucosidase. The lower IC_{50} value of AM-AgNPs highlights that AM-AgNPs are more efficient in inhibiting both enzymes than acarbose. Badmus et al. showed that AM-AgNPs inhibited α-glucosidase in an uncompetitive kinetic manner. This implies that the nanoparticles did not compete with the substrate at the active site of the enzyme, which may have been the reason for their high inhibitory efficacy[28].

5.4.3 *Cassia fistula*

Cassia fistula (CF), commonly known as golden shower, purging cassia, and Indian laburnum, belongs to the legume family Fabaceae. It is native to the Indian subcontinent, where it is being traditionally used for the treatment of liver conditions, skin diseases, and diabetes[29,30]. Ali[29] highlighted the numerous pharmacological properties of CF, including anti-inflammatory, antipyretic, antitumor, antimicrobial, hepatoprotective, gastroprotective, and antidiabetic effects. The efficacy of gold nanoparticles (AuNPs) synthesized from an aqueous extract of CF stem bark (CF-AuNPs) was tested in male Wistar rats. The study showed that bio-reduction of gold ions (chloroaurate) to AuNPs by CF enhanced their hypoglycemic and hypolipidemic potential (CF-AuNPs) when compared with the CF extract only. The CF-AuNPs (60 mg/kg) conferred protection to the animals against weight loss, ameliorated hepatopathy, and alleviated kidney dysfunction with the same efficacy as that of insulin (3 IU/kg)[30].

5.4.4 *Catharanthus roseus*

Catharanthus roseus (CR) is a plant of indigenous importance and has remarkable ornamental and medicinal values. CR has several essential bioactive components, some of which have been isolated and their hypoglycemic potential reported[31]. The bioactive compounds, most of which are identified as alkaloids, might be responsible for the synthesis and capping of zinc oxide nanoparticles (ZnONPs) as reported by Govindan et al.[32]. The antidiabetic potential of the nanoparticles was evaluated based on the *in vitro* α-amylase inhibition. The activities of the nanoparticles are comparable with those of acarbose (standard)[32].

5.4.5 *Chamaecostus cuspidatus*

Chamaecostus cuspidatus (CC), a herbaceous plant, is also known as *Costus cuspidatus*, *Costus igneus*, and *Globba cuspidatus*. CC belongs to the *Costaceae* family, and its origin traces back to

Eastern Brazil, with its prevalence recently reported in India. It is commonly called "insulin plant" in India owing to its antidiabetic potential[33]. AuNPs were synthesized from the aqueous extract of CC leaves (CC-AuNPs). Treatment with CC-AuNPs (0.75 mg/kg and 1.5 mg/kg) has shown the same efficacy as that of glibenclamide (50 mg/kg) but higher efficacy than the extract only (300 mg/kg and 600 mg/kg) in significantly lowering blood glucose concentrations, elevating insulin levels, increasing glycogen synthesis, and restoring body's total proteins to normal levels. Wound healing is a major, challenging issue in individuals with diabetes; however, CC-AuNPs (97%) have shown greater wound-healing ability than the plant extract and vehicle (78% and 66%, respectively). Velisetty and Chandrakala proposed that the wound-healing potential of CC-AuNPs is suggestive of the gold precursor enhancing the efficacy of CC-AuNPs because AuNPs have wound-healing properties and ameliorate wounds that affect the skin[33,34].

5.4.6 *Chrysophyllum albidum*

Chrysophyllum albidum (CA) is commonly known as African cherry or African star apple. Its fruit has a sweet-sour pulp that contains four to five seeds. The seeds and pulp of CA are used in traditional medicine to treat diabetes and hypertension[35,36]. As for the mechanism of α-amylase inhibition by AgNPs synthesized from a methanolic extract of CA seeds (CA-AgNPs), the synthesis of AgNPs in the presence of CA resulted in 33% enhanced inhibition of α-amylase when compared with

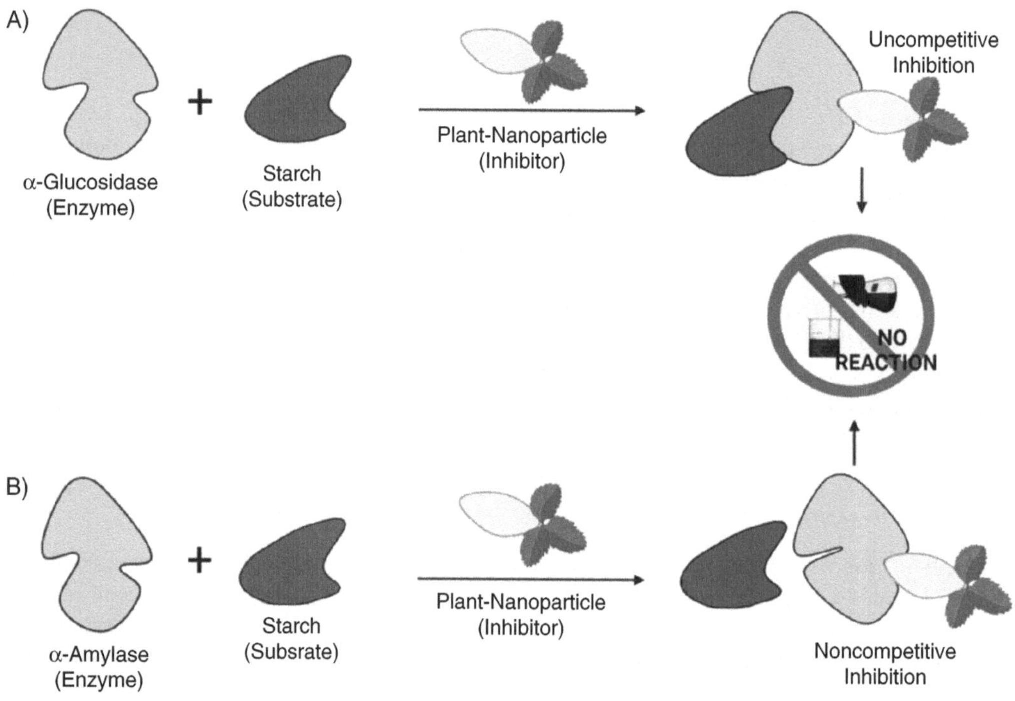

FIGURE 5.2 Mechanisms of inhibitory actions of plant-based nanoparticles on α-glucosidase and α-amylase. A) Plant-based nanoparticles bind to the secondary sites of α-glucosidase with stronger affinity and does not alter the active site; therefore, the substrate still binds, but no reaction occurs. B) Plant-based nanoparticles have a stronger binding affinity for α-amylase and, therefore, binds to the allosteric site of the enzyme. This alters the arrangement of tyrosine and tryptophan residues, thereby changing the active site conformation of α-amylase and resulting in the lack of binding to its substrate. (The figure was created using BioRender.)

that of the methanolic extract[37]. Kinetic and thermodynamic assays (KTA), fluorescence quenching mechanism (FQM), and synchronous fluorescence quenching (SFQ) analyses were applied to investigate the inhibitory mechanism of CA-AgNPs. KTA characterized the nature of binding force that existed between CA-AgNPs/CA-extract and α-amylase and revealed that spontaneous bonding and hydrogen bonding were the major interaction forces. FQM analysis strongly supported the interaction between the ligand (CA-AgNPs/CA-extract) and α-amylase at very low concentrations: 0.06–0.36 mg/mL for CA-AgNPs and 0.071–0.426 mg/mL for CA-extract. FQM analysis also indicated the higher stability of α-amylase–CA-AgNPs than that of α-amylase–CA-extract at increasing temperatures. The output from SFQ analysis revealed conformational changes to the tyrosine (Tyr) and tryptophan (Trp) residues in α-amylase in the presence CA-AgNPs and CA-extract, suggesting noncompetitive inhibition. However, Tyr residues demonstrated more significant changes than Trp residues[37]. The observed differences in binding concentrations, enzyme stability, and degree of conformational changes relate to the inhibitory activities of CA and CA-AgNPs against α-amylase. This unveiled the mechanism of action of CA-extract and CA-AgNPs, providing insights into the design and development of possible hypoglycemic agents from the extract.

5.4.7 *Cleome droserifolia*

Cleome droserifolia (CD), which is commonly referred to as "samwa" in Egypt, is widely used as a hypoglycemic agent and often grows in hot deserts, rocks, and wadis. It is found in some countries of North Africa (Egypt and Libya), Western Asia (Syria, Palestine, and Arabia), and the Middle East (Jordan)[38-40]. CD nanoparticles prepared with selenium as the precursor (CD-selenium nanoparticles [CD-SeNPs]) were assessed for their effect against streptozotocin (STZ)-induced diabetes in male Wistar rats, with Galvus-Met as the reference drug[39]. CD-SeNPs significantly reduced hyperglycemia, hyperlipidemia, liver and kidney injury markers and elevated insulin levels in diabetic rats when compared with diabetic controls, but not as much as Galvus-Met (5:100 mg/kg). Furthermore, the combination of Galvus-Met and CD-SeNPs demonstrated a better outcome than that with CD-SeNPs only. Perhaps, if there were treatment groups with only CD-extract and/or Se-NPs as treatment controls, then it would have shown a better impact of nanosizing the plant using selenium[41,42]. Another point of interest is the dosage used; if the dosage of CD-SeNPs had been upturned, then it could have shown comparable activities to those of Galvus-Met. This is, however, only a hypothesis, and a more certain option is to determine the effective concentration before extensive treatment[43].

5.4.8 *Cinnamomum cassia*

Cinnamomum cassia (CCa) is native to Southern China, and it is widely grown in South and Southeastern regions of Asia (India, Indonesia, Laos, Malaysia, Thailand, and Vietnam). CCa is frequently known as "Chinese cinnamon" or "Chinese cassia"[44]. Kouame et al. evaluated the potential of CCa and its AgNPs (CCa-AgNPs) to ameliorate diabetic nephropathy in Sprague Dawley rats. CCa-extract (300 mg/kg) and CCa-AgNPs (5 mg/kg and 10 mg/kg) prevented weight loss, lowered blood glucose levels to normal range, curtailed high concentrations of creatinine and uremia, restored GSH levels, and significantly minimized liver injury markers[44]. Periodic acid–Schiff staining and hematoxylin and eosin staining of the kidney tissue sections evidenced that both CCa-extract and CCa-AgNPs mitigated glomerular distortions and ruptured capsular walls, hypertrophy and vacuolization of the tubules, and infiltration of macrophages into the mesangial spaces and interstitium in the diabetic kidney. The 10-mg/kg dose of CCa-AgNPs demonstrated the highest potency, followed by the 300-mg/kg dose of CCa-extract and the 5-mg/kg dose of CCa-AgNPs. The results indicated the beneficial effects of synthesizing nanoparticles from CCa[44].

5.4.9 *Costus pictus*

Costus pictus (CP) D. Don belongs to the family Costaceae and is indigenous to South America, Mexico, and West Indies. It has been mainly used for medicinal purposes, especially against diabetes, by the traditional medical practitioners of the Upper Assam region[45]. The plant boosted insulin secretion in STZ- and alloxan-induced diabetic animals in addition to its ability to inhibit the activities of α-amylase and α-glucosidase[46]. Aruna et al.[47] compared the *in vitro* antidiabetic potential between the methanolic leaf extract of CP and its fabricated AgNPs. The study reported the improved performance of AgNPs against the methanolic extract of the plant. The AgNPs showed significant inhibitions of α-amylase and α-glucosidase, improved glucose uptake by tissues, and nonenzymatic glycosylation. The observed activities of the AgNPs could be related to the higher contents of phenolic, flavonoids, and vitamin C in the nanoparticles. The results imply that the nanoparticles increased the surface area of the leaf extract, which contributed to the higher concentrations of the identified bioactive compounds[47].

5.4.10 *Fritillaria cirrhosa*

Fritillaria cirrhosa (FC) is a traditional Chinese medicine reportedly used for several ailments owing to its anti-asthmatic, expectorant, and antitussive activities[48]. However, the plant alone has not been evaluated for its antidiabetic effect and is also not known in traditional medicine as an antidiabetic plant. Guo et al.[49] evaluated the potential of AuNPs prepared using FC against STZ-induced diabetes in rats. This study revealed the potential of AuNPs as an antidiabetic agent and concluded that the plant has an inherent bioactive compound responsible for the antidiabetic activity.

5.4.11 *Helichrysum foetidum*

Helichrysum foetidum (HF) is an indigenous plant in South Africa and is popularly known as "stinging star flower" owing to its pungent odor[50]. In traditional medicine, the plant has gained importance for wound treatment and herpes influenza. The plant has shown anti-proliferative property in cancer cell lines, antibacterial, antifungal, antiviral, and antioxidant activities[50]. The *in vitro* antidiabetic effect of the plant and its nanoparticles was first reported by Omolaja et al.[51]. The crude extract of HF showed significant α-amylase activity when compared with its fabricated AuNPs, while the latter had better α-glucosidase inhibitory activity. Furthermore, Helichysetin, an aglycone compound isolated from the plant, is responsible for the antidiabetic potential in the crude plant extract[51]. The compound and its fabricated AuNPs showed compelling inhibitory activities against α-amylase and α-glucosidase when compared with the crude and its nanoparticles. This shows that aglycone alone and its nanoparticles can be further explored for their antidiabetic activity in multicellular organisms and clinical trials.

5.4.12 *Leucosidea sericea*

Leucosidea sericea (LS) Eckl. and Zeyh is an indigenous plant in Southern Africa. It is a rich source of essential phytochemicals, which accounts for its diverse medicinal property. Its antimicrobial, antiparasitic, antioxidant, acetylcholinesterase inhibitory, and inflammatory properties are medicinal properties reported for the plant[52]. The plant is not known to be used in traditional parlance, and no scientific finding has reported its antidiabetic property before this study. AuNPs synthesized from fractions of LS displayed effective inhibitory activities against α-amylase amylase and α-glucosidase better than standard drug acarbose[12]. The procyanidin fractions of this plant were responsible for the synthesized nanoparticles and could be important agents with antidiabetic property. Badeggi et al.[12] is the only study that links the antidiabetic effect of LS. This study, however, showed that the plant is not used traditionally for treating ailments but could possibly have antidiabetic activity.

5.4.13 *Moringa oleifera*

Moringa oleifera belongs to the family Moringaceae. The plant is widely distributed in Africa, Asia, and America. The flower, pod, seed, and leaves of the plant are important medicinal entities[53]. They are used to treat inflammatory diseases, microbial infections, and hyperglycemia. It also contains known constituents of Ayurvedic medicine for the treatment of cardiac and circulatory diseases[54]. Kiran et al.[55] reported the synthesis of AuNPs from an aqueous extract of *Moringa oleifera.* In their study, the antidiabetic potential of the nanomaterial was evaluated using the α-amylase inhibitory protocol. The findings showed that the nanoparticles inhibited α-amylase activity in a dose-dependent manner, with an IC_{50} value of 130 μg/mL. The study, however, demonstrated that the AuNPs have a good prospect for diabetes treatment[55].

5.4.14 *Momordica charantia*

Momordica charantia (MC), also known as "bitter melon," "bitter gourd," "karela," and "balsam pear," is used in ethnomedicine for the treatment of diabetes and its complications in South America, Asia, India, and East Africa[56]. Elekofehinti and Akinjiyan[42] investigated the antidiabetic potential of the plant and its nanoparticles and measured the effect of MC-extract and its AgNPs (MC-AgNPs) on lipid metabolism (PPAR-α and HMG-CoA reductase), glycemic control, and insulin resistance (PPAR-α and PPAR-γ). PPAR-α plays an essential role in the metabolism of lipids and lipoproteins, thereby reducing diabetes-induced dyslipidemia, while PPAR-γ enhances insulin sensitivity by impeding the action of TNF-α on adipocytes[57]. The expression of PPAR-α and PPAR-γ was downregulated in diabetes, whereas treatment with MC-extract and MC-AgNPs upregulated their levels in the liver of diabetic rats. The upregulated expression of PPAR-α and PPAR-γ by the extract and its nanoparticles is indicative of the blood glucose–reducing potential of the extract. The expression of HMG-CoA reductase was significantly reduced in the liver of diabetic rats, which was reversed after treatment with MC-extract and MC-AgNPs. The reduction in the expression of HMG-CoA reductase may have been the result of dysregulation in lipid metabolism, and MC-AgNPs moderated this effect by regulating cholesterol synthesis through a feedback mechanism. The mechanism of MC-extract and MC-AgNPs in diabetic nephropathy was further explored by determining KIM-1 expression. MC-extract and MC-AgNPs lowered hyperglycemia-induced KIM-1 expression[42]. MC-extract and MC-AgNPs had superior efficacy over metformin (100 mg/kg). The study also showed that AgNPs prepared from the plant extract had enhanced bioavailability of the bioactive constituents, resulting in lesser dosage being effective than that of the extract[42].

5.4.15 *Morus alba* L.

Morus alba (MA) is an important traditional Chinese herbal medicinal plant used for diabetes treatment. The plant has a compendium of bioactive phytochemicals that play different significant roles in lowering postprandial blood sugar levels[58]. The aqueous extracts of different parts of the plant contain polyphenol and polysaccharide constituents. These bioactive constituents of the plant have shown antihyperglycemic and antihyperlipidemic effects[59]. The molecular mechanism underlying the antidiabetic effect of this plant is difficult to decipher, probably because of the presence of different phytochemicals in the plant[58]. Synthesis of nanoparticles from this plant might lower the cross-talk between the phytochemicals in exhibiting the hypoglycemic effect because of the increase in surface area. Xu et al.[60] reported the antidiabetic effect of AgNPs synthesized from MA leaves following the green synthesis approach. The nanoparticles were found to exhibit antidiabetic effect in rats during perinatal life[61].

5.4.16 *Nasturtium officinale*

Nasturtium officinale (NO), commonly known as "watercress," is a perennial plant with its nativity in Europe and Asia[62]. The antidiabetic properties of NO have been documented[63,64]. The antidiabetic effect of zinc nanoparticles synthesized from NO leaves (NO-ZnNPs) was examined in alloxan-induced diabetes and compared with that of the NO-extract and insulin[41]. The finding unveiled that a lesser dose of NO-ZnNPs (8 mg/dL) demonstrated more efficiency in attenuating hyperglycemia and hyperlipidemia, especially by enhancing HDL levels, than that of NO-extract (150 mg/kg) and insulin (10 U/kg). This finding underlines the benefits of NO-ZnNPs as an antidiabetic agent. However, zinc is known to activate insulin action and is an important metal for the activity of enzymes involved in lipid digestion and absorption[41,65]. The antidiabetic activity of the nanoparticles might have been due to a synergistic effect between the plant compounds and zinc.

5.4.17 *Nigella sativa*

Nigella sativa (NS) is an important medicinal plant in folklore medicine. It is commonly called "miracle plant" because of its extensive use in traditional medicine worldwide. Apart from NS being used in traditional parlance, scientific studies have shown that NS seed has a broad-spectrum pharmacological action. It has antidiabetic, anti-inflammatory, antimicrobial, analgesic, antioxidant, and hepatoprotective, effects[66]. The green-synthesized AgNPs prepared using the NS extract were tested against diabetic-induced neuropathy by Alkhalaf et al.[67]. Their results showed that the nanoparticles significantly ameliorated diabetes-related disrupted biomarkers of inflammation, antioxidants, and AGEs. The study concluded that the nanoparticles could be an important agent with neuroprotective activity in diabetes.

5.4.18 *Urtica dioica*

Urtica dioica (UD), commonly referred to as "stinging nettle" or "nettle leaf," has its origin from Europe and is grown in other places such as India and Malaysia. Its extract has demonstrated potency against diabetes[68,69]. The aqueous extract of the leaves was used to synthesize ZnONPs, which were administered to alloxan-induced diabetic Wistar rats for 16 days[69]. The extract-ZnNPs showed significant antidiabetic activity by lowering blood glucose levels and increasing insulin levels. Increased TC and TG levels due to hyperglycemia were reduced to near-normal levels by extract-ZnNPs, while HDL levels were significantly elevated. Interestingly, the HDL level was increased (62%) in diabetic rats that received extract-ZnNPs when compared with the levels in negative control and nondiabetic rats. The HDL-boosting property of extract-ZnNPs was associated with zinc-containing metalloenzymes involved in the lipid digestion and absorption processes[70]. These findings indicated that the ZnNPs prepared from the UD leaf extract exhibited more efficacy than the extract, zinc oxide (ZnO), or insulin. This implies that a synergistic potential exists between ZnO and the plant material against diabetes occurrence.

5.5 MECHANISMS OF PHYTONANOPARTICLES IN AMELIORATING DIABETES AND ITS COMPLICATIONS

Plant MNPs exhibit similar mechanism of action as that of their plant extract, fractions, or compounds. However, because of the presence of metals in MNPs and their innate pharmacological properties, their mechanisms are significantly enhanced. Some of the proposed mechanisms of action of MNPs in attenuating diabetes are discussed below.

5.5.1 Glycemic Control by Impeding Glucose Absorption Through the Inhibition of α-Amylase and α-Glucosidase

α-Amylases and α-glucosidases are involved in the breakdown of starch, glycogen, and other polysaccharides to glucose in the intestine (α-glucosidase) and other organs (α-amylases). As pointed out in *in vitro* studies, one of the mechanisms by which plant MNPs regulate glucose homeostasis is through the inhibition of α-amylase and α-glucosidase. These studies confirmed that plant MNPs possess stronger affinity for these enzymes than for their substrate, and the inhibition could be either uncompetitive or noncompetitive[28,37].

5.5.2 Regulation of Lipid and Glucose Metabolism to Moderate Hyperlipidemia and Hyperglycemia Through Increased Activities of Metalloproteins

As for the action of MNPs on hyperlipidemia in diabetes, MNPs increase HDL levels, wherein they transport excess cholesterol to peripheral tissues; this finding is consistent among studies[30,69]. An increase in HDL levels may have been due to the increased activation of lecithin cholesterol acyltransferase, which, in turn, induces the activation of reverse cholesterol transport by HDL[30]. Additionally, metal deficiency affects the activities of metalloenzymes and leads to hyperlipidemia[77]. Bayrami et al.[69] proposed that the activities of metalloproteins, particularly those involved in lipid digestion and absorption, are amplified by MNPs. Furthermore, metals such as zinc are actively involved in glucose utilization, insulin secretion, and improved glucose tolerance in diabetes[78-80].

5.5.3 Repair of Body Weight Loss and Prevention of Protein Glycosylation by Moderating Protein Catabolism and Increasing Antioxidant Levels

Plant-based MNPs impede loss of body weight in diabetes by their ability to control muscle wasting (through reversal of antagonism) and normalize the turnover rate of protein catabolism[27]. Glycosylation and alkylation of proteins are promoted in the presence of elevated production of reactive oxygen species. The protein glycosylation-lowering mechanism of plant-based MNPs is exhibited through their ability to increase antioxidant levels and deplete free radicals, thereby reducing alkylation and glycosylation of proteins[30]. Hemoglobin is one such protein that is frequently glycosylated in diabetes as demonstrated by the increased levels of HbA1c, and plant-based MNPs have demonstrated a significant reduction in HbA1c levels in diabetes[73,74].

5.5.4 Plant-Based MNPs Enhance Bioactivity Through Increased Bioavailability and Prolonged Release

Bio-reduction of metal ions using plant materials reduces the size of the bioactive compounds to nanometers, which increases their bioavailability and bioactivity. For example, in a study where curcumin nanoparticles were synthesized using ZnO, curcumin retention, bioavailability, and bioactivity was significantly enhanced in STZ-induced diabetes[65]. The antidiabetic property of green-synthesized MNPs prepared from polydatin was strengthened by its prolonged release and improved absorption. The nanocarrier particles gradually released the bioactive compound and enabled its circulation in the blood for a longer period, thus maintaining glucose levels at the minimal level[81].

5.5.5 Upregulated mRNA Expression of PPAR-γ and PPAR-α to Mediate Glucose Homeostasis and Hyperlipidemia

One of the suggested mechanisms by which MNPs prepared from plants regulate glucose and lipid homeostasis was through the upregulated mRNA expression of PPAR-γ and PPAR-α in diabetic animals[42]. PPAR-γ induces insulin-mediated glucose uptake in the liver and muscle by minimizing

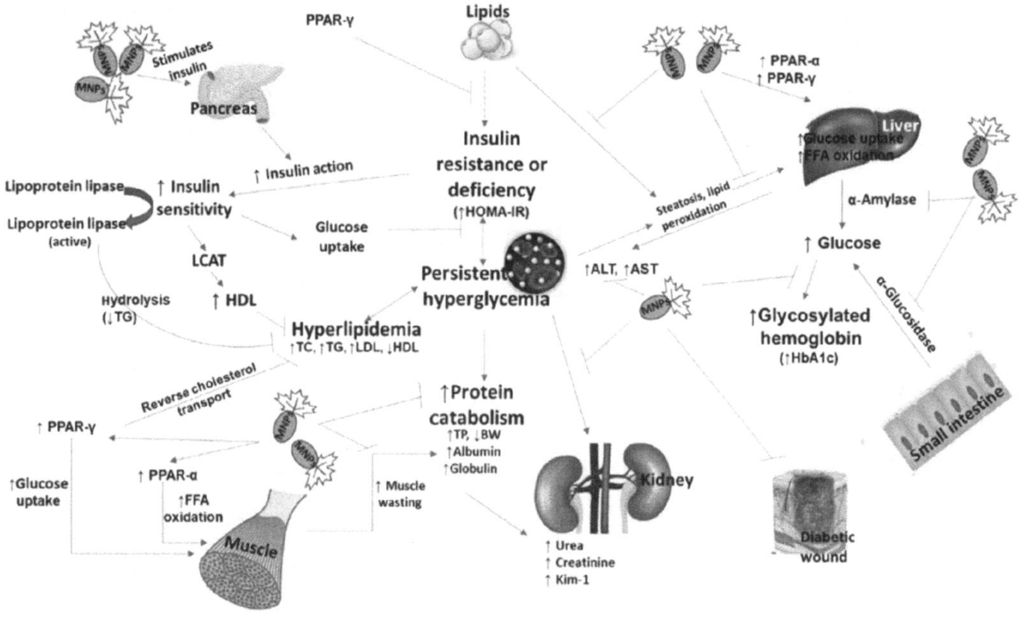

FIGURE 5.3 The role of plant-based nanoparticles and their mechanistic actions in glucose, lipid, and protein metabolism in diabetes and diabetic complications. MNP: metal nanoparticle, TP: total protein, LCAT: lecithin-cholesterol acyltransferase.

insulin resistance and enhancing fatty acid storage. Furthermore, PPAR-α triggers fatty acid oxidation in the liver and muscle, and both PPAR-α and PPAR-γ stimulate reverse cholesterol transport in the blood circulation[42,82].

5.6 CONCLUSION

This chapter describes the effect of synthesizing nanoparticles from plants as reducing and capping agents in diabetes management. Most MNPs prepared from plants have demonstrated higher efficacy than their precursors at comparable levels as those of reference drugs in terms of ameliorating diabetes and its complications. This underscores the synergistic effect between plant components and MNPs. More interestingly, plant nanoparticles have higher antidiabetic potential than insulin. However, findings from certain studies have shown that plant extracts at higher doses were as effective as their nanoparticles. Nanoparticles with equal efficacy and at lower doses are more preferable when considering drug design through dose refinement. Some studies have reported lesser activities of nanoparticles synthesized from a single compound when compared with those of its counterpart from the total crude extract[51]. This may be attributed to the synergistic effect of the bioactive compounds in the crude extract/fractions that resulted in higher activity[12]. In summary, green-synthesized nanoparticles prepared from antidiabetic plants can significantly increase the bioavailability of plant bioactive constituents, thereby enhancing the antidiabetic potential of these plant compounds through various mechanisms. The improved bioavailability of plant bioactive constituents through the utilization of their nanoparticles led to higher efficacy including prolonged insulin action; improved insulin sensitivity; downregulation of glucose release; increase in glucose uptake; normalized dyslipidemia; and attenuation of accompanying complications in the liver, kidney, and heart. These findings suggest the need for a paradigm shift from conventional drugs or plant materials only to the design and development of novel antihyperglycemic agents based on green nanotechnology.

5.6.1 Future Perspectives/Recommendation

The development of antidiabetic nanoformulations at the market scale remains in infancy owing to the paucity of empirical data that establish their optimal doses, validate long-term therapeutic efficacy, and elucidate possible toxicological challenges for prolonged use. This opens up areas of research interests/opportunities for future studies in green nanotechnology for the management of diabetes and diabetic complications. This gap also spotlights the importance of closer collaboration between industries and research hubs, especially in African countries, to avoid duplicating research data and channelizing resources toward transforming and translating these findings from the desk to shelf.

REFERENCES

1. Zheng Y, Ley SH, Hu FB. *Nat Rev Endocrinol.* 2018;14(2):88–98. doi:10.1038/nrendo.2017.151
2. Bhardwaj M, Yadav P, Dalal S, Kataria SK. *Biomed Pharmacother.* 2020;127. doi:10.1016/j.biopha.2020.110198
3. Kitture R, Chordiya K, Gaware S, Ghosh S, More PA, Kulkami P, Chopade BA, Kale SN. *J Nanosci and Nanotech* 2015;15:4046–4051. doi:10.1166/jnn.2015.10323
4. Anushka MM, Mahomooda F. *J Ethnopharmacol* 2014;151:413–444. doi:10.1016/j.jep.2013.10.069
5. Xianghong C, Yahui X, Qian J, Ning G, Zhenzhen W, Yu W. *Drug Deliv.* 2020;27(1):1263–1270. doi:10.1080/10717544.2020.1809555
6. Simos YV, Spyrou K, Patila M, et al. *Asian J Pharm Sci.* 2021;16(1):62–76. doi:10.1016/j.ajps.2020.05.001
7. Nasimi P, Haidari M. *Int J Green Nanotechnol.* 2013;5(1):1–5. doi:10.1177/1943089213506978
8. Bar H, Bhui DK, Sahoo GP, Sarkar P, De SP, Misra A. *Colloids Surfaces A Physicochem Eng Asp.* 2009;339(1–3):134–139. doi:10.1016/j.colsurfa.2009.02.008
9. Khan I, Saeed K, Khan I. *Arab J Chem.* 2019;12(7):908–931. doi:10.1016/j.arabjc.2017.05.011
10. Rafique M, Sadaf I, Rafique MS, Tahir MB. *Artif Cells, Nanomedicine Biotechnol.* 2017;45(7):1272–1291. doi:10.1080/21691401.2016.1241792
11. Badeggi UM, Badmus JA, Botha SS, et al. *J Funct Biomater.* 2020;11(3):1–20. doi:10.3390/JFB11030066
12. Badeggi UM, Ismail E, Adeloye AO, et al. *Biomolecules.* 2020;10(3):452. doi:10.3390/biom10030452
13. Ijaz I, Gilani E, Nazir A, Bukhari A. *Green Chem Lett Rev.* 2020;13(3):59–81. doi:10.1080/17518253.2020.1802517
14. Silva LP, Bonatto CC, Polez VLP. Green synthesis of metal nanoparticles by fungi: Current trends and challenges. *Advances and Applications through Fungal Nanobiotechnology.* 2016:71–89. doi:10.1007/978-3-319-42990-8_4
15. Panigrahi S, Kundu S, Ghosh S, Nath S, Pal T. *J Nanoparticle Res.* 2004;6(4):411–414. doi:10.1007/s11051-004-6575-2
16. Hoeppener S, Maoz R, Cohen SR, Chi LF, Fuchs H, Sagiv J. *Adv Mater.* 2002;14(15):1036. doi:10.1002/1521-4095(20020805)14:15<1036::AID-ADMA1036>3.0.CO;2-J
17. Jamkhande PG, Ghule NW, Bamer AH, Kalaskar MG. *J Drug Deliv Sci Technol.* 2019;53:101174. doi:10.1016/j.jddst.2019.101174
18. Tavakoli A, Sohrabi M, Kargari A. *Chem Pap.* 2007;61(3). doi:10.2478/s11696-007-0014-7
19. Nguyen NHA, Padil VVT, Slaveykova VI, Černík M, Ševců A. *Nanoscale Res Lett.* 2018;13(1):159. doi:10.1186/s11671-018-2575-5
20. Das G, Patra JK, Debnath T, Ansari A, Shin H-S. Mukherjee A, ed. *PLoS One.* 2019;14(8):e0220950. doi:10.1371/journal.pone.0220950
21. Singh D, Reeta KH, Sharma U, Jagannathan NR, Dinda AK, Gupta YK. *Neurochem Int.* 2019;126:96–108. doi:10.1016/j.neuint.2019.03.010
22. Shaheen TI, El-Naggar ME, Hussein JS, et al. *Biomed Pharmacother.* 2016;83:865–875. doi:10.1016/j.biopha.2016.07.052
23. Khandel P, Yadaw RK, Soni DK, Kanwar L, Shahi SK. *J Nanostructure Chem.* 2018;8(3):217–254. doi:10.1007/s40097-018-0267-4

24. Mikhailova EO. *J Funct Biomater*. 2020;11(4):84. doi:10.3390/jfb11040084
25. Balan K, Qing W, Wang Y, et al. *RSC Adv*. 2016;6(46):40162–40168. doi:10.1039/C5RA24391B
26. Gavamukulya Y, Wamunyokoli F, El-Shemy HA. *Asian Pac J Trop Med*. 2017;10(9):835–848. doi:10.1016/j.apjtm.2017.08.009
27. Abdul Wahab SM, Jantan I, Haque MA, Arshad L. *Front Pharmacol*. 2018;9: 661. doi:10.3389/fphar.2018.00661
28. Badmus JA, Oyemomi SA, Adedosu OT, et al. *Heliyon*. 2020;6(11):e05413. doi:10.1016/j.heliyon.2020.e05413
29. Ashraf Ali M. *Int J Pharm Sci Res*. 2014;5(6):2125–2130. doi:10.13040/IJPSR.0975-8232
30. Daisy P, Saipriya K. *Int J Nanomedicine*. 2012;7:1189–1202. doi:10.2147/IJN.S26650
31. Goboza M, Aboua YG, Chegou N, Oguntibeju OO. *Biomed Pharmacother*. 2019;112:108638. doi:10.1016/j.biopha.2019.108638
32. Govindan N, Vairaprakasam K, Chinnasamy C, Sivalingam T, Mohammed MKA. *Mater Adv*. 2020;1(9):3460–3465. doi:10.1039/D0MA00698J
33. Velisetty VP, Chandrakala K. *Sch J Appl Med Sci*. 2017;5(8A):2976–2979.
34. Pan A, Zhong M, Wu H, et al. *Nanomed Nanotechnol Biol Med*. 2018;14(5):1619–1628. doi:10.1016/j.nano.2018.04.007
35. Ibrahim MA, Mohammed A, Isah MB, Aliyu AB. *J Ethnopharmacol*. 2014;154(1):26–54. doi:10.1016/j.jep.2014.04.012
36. Asagba SO, Kadiri HE, Ezedom T. *J Basic Appl Zool*. 2019;80(1):42. doi:10.1186/s41936-019-0118-y
37. Avwioroko OJ, Anigboro AA, Atanu FO, et al. *Chem Data Collect*. 2020;29. doi:10.1016/j.cdc.2020.100517
38. Shtaiwi M, Rawi S, Alshibly N, Al-Hazimi M. *J Med Plants Res*. 2013;7(37):2769–2776. doi:10.5897/JMPR2013.5088
39. Abdel Maksoud HA, Abou Zaid OAR, Elharrif MG, Omnia MA, Alaa EA. *Clin Nutr ESPEN*. 2020;40:383–391. doi:10.1016/j.clnesp.2020.07.016
40. Moustafa, Abdel Raouf A, Mansour, Samira R, Alotaibi MO. *Asian J Plant Sci Res*. 2019;9(1):14–21.
41. Bayrami A, Ghorbani E, Rahim Pouran S, Habibi-Yangjeh A, Khataee A, Bayrami M. *Ultrason Sonochem*. 2019;58:104613. doi:10.1016/j.ultsonch.2019.104613
42. Elekofehinti O, Akinjiyan M. *Niger J Biotechnol*. 2021;37(2):126–133. doi:10.4314/njb.v37i2.12
43. Thaipitakwong T, Supasyndh O, Rasmi Y, Aramwit P. *Complement Ther Med*. 2020;49:102292. doi:10.1016/j.ctim.2019.102292
44. Kouame K, Peter AI, Akang EN, Moodley R, Naidu EC, Azu OO. *Bosn J Basic Med Sci*. 2019;19(2):138–145. doi:10.17305/bjbms.2019.3481
45. Biman B, Dipak C. *Ind J Clin Biochem* 2020; 35(2):225–231. doi: 10.1007/s12291-019-00819-y
46. Ngan T, Bao P, Ly L. Biol. 2020; 9:252. doi: 10.3390/biology9090252
47. Aruna A, Nandhini R, Karthikeyan V, Bose P, Vijayalakshmi K. *Indo Am J Pharm Res*. 2014;4(07):3217–3230.
48. Wang Y, Hou H, Ren Q, Hu H, Yang T, Li X. *Chin Med*. 2021;16(1):40. doi:10.1186/s13020-021-00450-1
49. Guo Y, Jiang N, Zhang L, Yin M. *Arab J Chem*. 2020;13(4):5096–5106. doi:10.1016/j.arabjc.2020.02.009
50. Akinyede KA, Cupido CN, Hughes GD, Oguntibeju OO, Ekpo OE. *Plants*. 2021;10(8):1566. doi:10.3390/plants10081566
51. Omolaja AA, Pearce B, Omoruyia SI, et al. *Surf Interfaces*. 2021. doi:10.1016/j.surfin.2021.101251
52. Mafole TC, Aremu AO, Mthethwa T, Moyo M. *J Ethnopharmacol*. 2017;203:288–303. doi:10.1016/j.jep.2017.03.044
53. Al-Malki AL, El Rabey HA. *Biomed Res Int*. 2015;2015:1–13. doi:10.1155/2015/381040
54. Gupta R, Mathur M, Bajaj VK, et al. *J Diabetes*. 2012;4(2):164–171. doi:10.1111/j.1753-0407.2011.00173.x
55. Kiran MS, Rajith Kumar CR, Shwetha UR, Onkarappa HS, Betageri VS, Latha MS. *Chem Data Collect*. 2021;33:100714. doi:10.1016/j.cdc.2021.100714
56. Joseph B, Jini D. *Asian Pacific J Trop Dis*. 2013;3(2):93–102. doi:10.1016/S2222-1808(13)60052-3
57. Tyagi S, Sharma S, Gupta P, Saini A, Kaushal C. *J Adv Pharm Technol Res*. 2011;2(4):236. doi:10.4103/2231-4040.90879

58. Tian S, Tang M, Zhao B. *J Tradit Chinese Med Sci*. 2016;3(1):3–8. doi:10.1016/j.jtcms.2016.04.001
59. Jiao Y, Wang X, Jiang X, Kong F, Wang S, Yan C. *J Ethnopharmacol*. 2017;199:119–127. doi:10.1016/j.jep.2017.02.003
60. Xu L, Li W, Shi Q, et al. *J Photochem Photobiol B Biol*. 2019;196:111502. doi:10.1016/j.jphotobiol.2019.04.011
61. Mohamed AA-R, Khater SI, Hamed Arisha A, et al. *Gene*. 2020;768:145288. doi:10.1016/j.gene.2020.145288
62. Hadjzadeh M-A-R, Rajaei Z, Moradi R, Ghorbani A. *Indian J Physiol Pharmacol*. 59(2):223–230. www.ncbi.nlm.nih.gov/pubmed/26685512.
63. Romina S, Norozi M. *J Islam Iran Tradit Med*. 2017;8(3):315–321. http://jiitm.ir/article-1-905-en.html. Accessed December 1, 2021.
64. Fenton-Navarro B, Urquiza Martinez MV, Fiscal Castro BB, et al. *African J Tradit Complement Altern Med*. 2018;15(2):68–79. doi:10.21010/ajtcam.v15i2.9
65. Chauhan P, Mahajan S, Prasad GBKS. *J Drug Deliv Sci Technol*. 2019;52:738–747. doi:10.1016/j.jddst.2019.05.020
66. Ahmad A, Husain A, Mujeeb M, et al. *Asian Pac J Trop Biomed*. 2013;3(5):337–352. doi:10.1016/S2221-1691(13)60075-1
67. Alkhalaf MI, Hussein RH, Hamza A. *Saudi J Biol Sci*. 2020;27(9):2410–2419. doi:10.1016/J.SJBS.2020.05.005
68. Jan KN, Zarafshan K, Singh S. *J Food Meas Charact*. 2017;11(2):423–433. doi:10.1007/s11694-016-9410-4
69. Bayrami A, Haghgooie S, Rahim Pouran S, Mohammadi Arvanag F, Habibi-Yangjeh A. *Adv Powder Technol*. 2020;31(5):2110–2118. doi:10.1016/j.apt.2020.03.004
70. Ranasinghe P, Wathurapatha W, Ishara M, et al. *Nutr Metab (Lond)*. 2015;12(1):26. doi:10.1186/s12986-015-0023-4
71. Ponnanikajamideen M, Rajeshkumar S, Vanaja M, Annadurai G. *Can J Diabetes*. 2019;43(2):82–89.e6. doi:10.1016/j.jcjd.2018.05.006
72. Abu-Taweel GM, Attia MF, Hussein J, et al. *Biomed Pharmacother*. 2020;131: 110688. doi:10.1016/j.biopha.2020.110688
73. Abdel-Moneim A, Mahmoud B, Nabil A, Negeem Z. *Diabetes Metab Syndr Clin Res Rev*. 2019;13(4):2365–2373. doi:10.1016/j.dsx.2019.06.014
74. Ansari S, Bari A, Ullah R, Mathanmohun M, Veeraraghavan VP, Sun Z. *J Photochem Photobiol B Biol*. 2019;201:111643. doi:10.1016/j.jphotobiol.2019.111643
75. Javanshir R, Honarmand M, Hosseini M, Hemmati M. *Clin Phytoscience*. 2020;6(1):74. doi:10.1186/s40816-020-00224-6
76. Maheshwaran G, Malai Selvi M, Selva Muneeswari R, Nivedhitha Bharathi A, Krishna Kumar M, Sudhahar S. *Adv Powder Technol*. 2021;32(6):1963–1971. doi:10.1016/j.apt.2021.04.004
77. Jones AA, DiSilvestro RA, Coleman M, Wagner TL. Metabolism. 1997;46(12):1380–1383. doi:10.1016/S0026-0495(97)90135-9
78. Sun W, Wang Y, Miao X, et al. *Free Radic Biol Med*. 2014;68:22–34. doi:10.1016/j.freeradbiomed.2013.11.015
79. Zhang C, Lu X, Tan Y, et al. *PLoS One*. 2012;7(12):e49257. doi:10.1371/journal.pone.0049257
80. Umrani RD, Paknikar KM. *Nanomedicine*. 2014;9(1):89–104. doi:10.2217/nnm.12.205
81. Abdel-Moneim A, El-Shahawy A, Yousef AI, Abd El-Twab SM, Elden ZE, Taha M. *Int J Biol Macromol*. 2020;154:1496–1504. doi:10.1016/j.ijbiomac.2019.11.031
82. Chigurupati S, Dhanaraj SA, Balakumar P. *Eur J Pharmacol*. 2015;755:50–57. doi:10.1016/j.ejphar.2015.02.043

6 Phytochemical Delivery Through Nanocarriers in Pulmonary Diseases

Çiğdem Kahraman[1] *and Nadire Özenver*[1]
[1]Department of Pharmacognosy, Faculty of Pharmacy, Hacettepe University, 06100 Sıhhiye, Ankara, Turkey

LIST OF ABBREVIATIONS

ASC, ascorbic acid; Bcl-2, B-cell lymphoma-2; BSA-Api-NP, apigenin-loaded bovine serum albumin nanoparticle; CF, cystic fibrosis; CS, celastrol; Curcumin-DPIs, curcumin spray-dried powder for inhalation; Curcumin-SLNs, curcumin-solid lipid nanoparticles; DTX, docetaxel; DPPC, dipalmitoylphosphatidylcholine; EGCG, epigallocatechin gallate; EGF, epidermal growth factor; GA, gallic acid; HA, hyaluronic acid; IPF, idiopathic pulmonary fibrosis; LPN, lipid-polymer hybrid nanoparticle; LPPC, liposomal polyethylene glycol–polyethyleneimine complex; MCN, mesoporous carbon nanoparticle; NAR-HA@CH-PCL-NP, naringenin-loaded hyaluronic acid-decorated polycaprolactone nanoparticle; NPDPI, naringenin-loaded dipalmitoylphosphatidylcholine phytosomes for dry powder inhalation; NSCLC, non-small cell lung carcinoma; QR, quercetin; RIF, rifampin; RSV, resveratrol; RSV-LNC, resveratrol-loaded lipid core nanocapsule; SA, stearic acid; PCL, polycaprolactone; PEG, polyethylene glycol; PE-PEG, phosphoethanolamine-polyethylene glycol; PEG-PE, polyethylene glycol–phosphoethanolamine; PEG-PEI, polyethylene glycol–polyethyleneimine; PF, pulmonary fibrosis; PEG-PLA, poly(ethylene glycol)-co-poly(D,L-lactic acid); PLA-PEG, polylactic acid–polyethylene glycol; PLGA[1], poly(lactic-co-glycolic acid); PLGA[2], poly-D,L-lactide-co-glycolide nanoparticle; RIF/ASC NP, rifampin and ascorbic acid co-loaded into alginate/chitosan nanoparticle; SCLC, small cell lung carcinoma; VEGF, vascular endothelial growth factor; ZnO, zinc oxide.

6.1 INTRODUCTION

Natural products are an undeniable source of chemical compounds that have a wide range of structural and biological diversity. Phytochemicals generally include secondary metabolites of plants and comprehend various phytocompounds from different classes such as flavonoids, saponins, phenolics, and alkaloids, among others. Phytochemicals have garnered greater attention recently, providing either conventional or complementary therapy options with mitigated side effect profiles and, thus, rendering therapeutic potential in the management of many critical diseases. Considering the common use of phytochemical-containing medicinal plants as traditional medicine in various civilizations since prehistoric times (Schmidt and Cheng 2017; Suleria et al. 2019), phytochemicals used either in their native form or along with their phytoconstituents play a prominent role in the discovery of potential drug leads. For instance, Newman and Cragg reported that only 24.6% of all drugs and 33.3% of all small-molecule drugs approved during 1981–2019 had an entirely synthetic origin (2020). On this note, medicinal plants have gained clinical value among all natural products in terms of providing therapeutic agents, which emphasizes their importance in the pharmaceutical

DOI: 10.1201/9781003231721-6

industry (Kinghorn et al. 2011; Newman and Cragg 2020). Historically, medicinal plants have been used in traditional medicine applications by various cultural groups for decades, suggesting clinical tests beyond reaching single-compound-based drugs (Schmidt et al. 2007). Nevertheless, medicinal plants and their phytochemicals may cause some intrinsic difficulties owing to the presence of multicomponent mixtures of substances with broad-ranging activity that can be active, partially active, or inactive. Furthermore, their behavior may differ based on the co-existing compound, exhibiting either synergism or antagonism. Their chemical complexity and structural diversity are other issues, leading to variation among the extracts; therefore, the relevant pharmacological activity cannot be always regenerated. Pharmacokinetic characteristics including absorption, distribution, metabolism, and excretion as well as safety (or toxicity) profiles may become challenging issues in natural drug discovery owing to their poor bioavailability (Atanasov et al. 2015; Di 2015; González-Medina et al. 2016; Heinrich 2010; Méndez-Lucio and Medina-Franco 2017; Yao et al. 2017). In such circumstances, efforts devoted to interdisciplinary research play an important role in the development of natural product-based therapeutics, which enhances their drug-like properties. Thus, it is possible to overcome the poor drug-likeness character of natural products.

Nanotechnology, a branch of science dealing with the study of substances with a size of 100 nm or smaller (Nishiyama 2007), has recently emerged as an innovative approach in the field of phytochemical drug delivery. Nanotechnological applications in the pharmaceutical and medical fields, termed "nanomedicine," have gained prominence, as they provide a rational drug design of phytochemicals. Phytochemicals are secondary metabolites that exert beneficial effects in various diseases and human health. Bioactive phytochemicals and herbal drugs usually have poor solubility in aqueous media, poor stability, and poor bioavailability. Moreover, the molecular size and polar character of these compounds obstruct their crossover of the gastrointestinal tract, mucosa, blood–brain barrier, or endothelial line of blood vessels. Considering these limitations of phytochemicals, nanocarriers have emerged as an alternative method to overcome these drawbacks, demonstrating improved efficacy and enabling the delivery of the phytochemicals of interest to target tissues (Gunasekaran et al. 2014; Manickam et al. 2021; Sarker and Nahar 2017). Additionally, the nanocarriers have attenuated systemic toxicity and improved bioavailability through controlled and sustained release of the drug to the target site; nanoformulations such as liposomes, polymeric nanoparticles, polymeric micelles, and nanodispersion are widely used nano-phytomedicines (Dhupal and Chowdhury 2020; Sajid et al. 2019).

This chapter focuses on phytochemical delivery through nanocarrier systems in pulmonary diseases. It exemplifies useful phytochemicals encapsulated or conjugated with nanoscale drug delivery systems that are under preclinical investigations. This chapter also provides a comprehensive coverage of the recent novel nano-phytomedicines in the field of lung disorders. Thus, we present a general concept of the applications, advantageous, and potential benefits of nanoscale-designed phytochemicals in pulmonary diseases.

6.2 IMPORTANCE OF NANOCARRIER-CONJUGATED PHYTOCHEMICALS IN PHARMACEUTICALS

Ongoing studies have addressed the potential of natural products as efficient drug candidates in the prevention and/or management of many diseases, which was verified through both *in vitro* and *in vivo* experimental systems. However, thus far, the available evidence could not be converted into clinical application owing to several factors including insufficient bioavailability and ineffective systemic delivery. Recent developments and advances have increased the possibility of delivering phyto-compounds through novel drug delivery systems, especially through organic, inorganic, and hybrid nanoparticle encapsulation of plant active metabolites. Thus, phytochemical formulations have been improved in terms of bioavailability, solubility, pharmacological activity, sustained delivery, toxicity, and so on (Ahmad et al. 2021; Aqil et al. 2013; Rahman et al. 2020).

Phytochemical nanocarriers are constructed and used to deliver phytochemicals to target tissues with improved bioefficacy. They can encapsulate both hydrophobic and hydrophilic molecules based on the carrier material used. Many nanocarrier systems including organic and inorganic nanoparticles are generally taken into account when designing nanophytomedicines. Inorganic nanoparticles encompass an inorganic core made of gold, silver, silica, iron oxide, or other metals (mesoporous silica nanoparticles, gold/silver nanoparticles, etc.), while organic nanocarriers may be either lipid-based (micelles, liposomes, niosomes, solid lipid nanoparticles, etc.) or polymer-based (synthetic or natural polymers) systems. Furthermore, other nanocarriers include carbon nanotubes, dendrimers, and quantum dots (Ahmad et al. 2021; Aqil et al. 2013). The applications of phytochemical nanocarriers and their health benefits are exemplified below.

Nanocarrier-based phytomedicines have been extensively applied for treating various disorders including cancer, infectious diseases, inflammatory conditions, and so on. For instance, curcumin nanoparticles, wherein curcumin is enclosed in polymer-based nanoparticles of size ranging from 30 nm to 100 nm, demonstrated enhanced cardioprotective effects when compared with curcumin only in an isoproterenol-induced myocardial ischemia model of rats, thus limiting cardiac injury after myocardial infarction (Boarescu et al. 2019).

In one study, resveratrol (RSV)-conjugated gold nanoparticles augmented the bioavailability of RSV, exhibiting greater anticancer effects toward human breast (MDA-MB-231), pancreatic (PANC-1), and prostate (PC-3) cancers than those of RSV (Thipe et al. 2019).

Pretreatment of phosphatidylcholine and polyamidoamine–conjugated gallic acid (GA) dendrimers markedly alleviated carbon tetrachloride-induced oxidative injury in rat liver when compared with free GA pretreatment, which evidenced the increase in bioavailability and enhanced the hepatoprotective effect of conjugated GA dendrimers (Abdou and Masoud 2018).

Additionally, genistein-loaded chitosan nanoparticles boosted genistein penetration through the nasal mucosa when compared with pure genistein. Furthermore, the nanoparticles protected the cell vitality of rat pheochromocytoma–derived cell lines, demonstrating their safety profile (Rassu et al. 2019).

As shown in the examples, nanocarrier-conjugated phytochemicals used in the pharmaceutical field have diverse aspects. Within the scope of our chapter, we draw attention to the nanomedicine applications of phytochemicals with a special emphasis on pulmonary diseases.

6.3 PRECLINICAL STUDIES OF NANO-FORMULATED PHYTOCHEMICALS IN PULMONARY DISEASES

Pulmonary lung diseases include asthma, cystic fibrosis (CF), chronic obstructive pulmonary disease, acute lung injury, pulmonary tuberculosis, idiopathic pulmonary fibrosis (IPF), and lung cancer (Lelli et al. 2017; Yetisgin et al. 2020). These diseases may cause devastating effects, and efficient therapeutic approaches for the complete recovery of lung function are not available.

Phytochemicals with effective anti-inflammatory and antioxidative properties may be promising agents, as they help maintain lung function and in the management of pulmonary diseases. Preclinical and clinical investigations have demonstrated that many phytochemicals such as vitamins, carotenoids, assorted flavonoids, RSV, and omega-3 fatty acids may positively influence lung health; further well-designed clinical trials are required to strengthen the transition of individualized nutrient treatment into the health policies (Hosseini et al. 2021; Majnooni et al. 2020; Zhai et al. 2018).

Conventional medicines are usually delivered to the lungs in a local (e.g., inhalers) or systemic manner. When bioactive molecules are applied as aerosols, burst release may occur, with a likelihood of excessive toxicity. Additionally, molecules with a particle size higher than 5 μm have difficulty in crossing the upper respiratory tract, which may hinder drug efficacy, whereas molecules of particle size in the range of 1–5 μm are generally deposited in the lower respiratory tract, and particles of size less than 1 μm remain suspended in air and are exhaled (Jabbal et al. 2017). For instance, the natural compound apigenin, a flavonoid well known for its antioxidant and anti-inflammatory properties in the lungs, has been examined for its delivery into the pulmonary organs when administered as nanoparticles, namely apigenin-loaded bovine serum albumin nanoparticles (BSA-Api-NPs). Spray-dried BSA-Api-NPs have shown great aerodynamic properties contrary to their dissolution feature. Thus, BSA-Api-NPs represent a novel drug delivery system against pulmonary damage and exert probable antioxidant activity (Pápay et al. 2017). Additionally, nanoformulations of the stilbene compound RSV are an alternative drug for acute lung injury. RSV-loaded lipid core nanocapsules (RSV-LNCs) increased pulmonary activity and obstructed total leukocyte and neutrophil accumulation equally in both bronchoalveolar lavage fluid and lung tissue when administered 4 hours before lipopolysaccharide challenge. Thus, RSV-LNCs presented as an effective procedure against acute lung injury and respiratory failure (de Oliveira et al. 2019). Polymer-glycerosomes containing curcumin, a polyphenol compound obtained from the rhizomes of *Curcuma* species (Itokawa et al. 2008), exerted improved physicochemical properties and accumulation in the lungs, which supports their potential use in the treatment of lung diseases (Manca et al. 2015). Similarly, curcumin spray-dried powders for inhalation (curcumin-DPIs) demonstrated remarkably increased drug dissolution and exhibited good aerosol performance. Plasma curcumin concentrations increased by inhalation, and most of the curcumin-DPIs were distributed in the lungs, indicating that curcumin-DPI is a promising strategy for lung cancer therapy (Hu et al. 2015).

Nanoparticle-based drug delivery of phytochemicals is rational, as it provides enhanced bioavailability, controlled release, attenuated dosage, and frequent application in lung disorders (Figure 6.1). A variety of nanophytomedicine formulations have been designed for the management of pulmonary diseases, which have been summarized and exemplified along with their benefits on lungs in the following sections below.

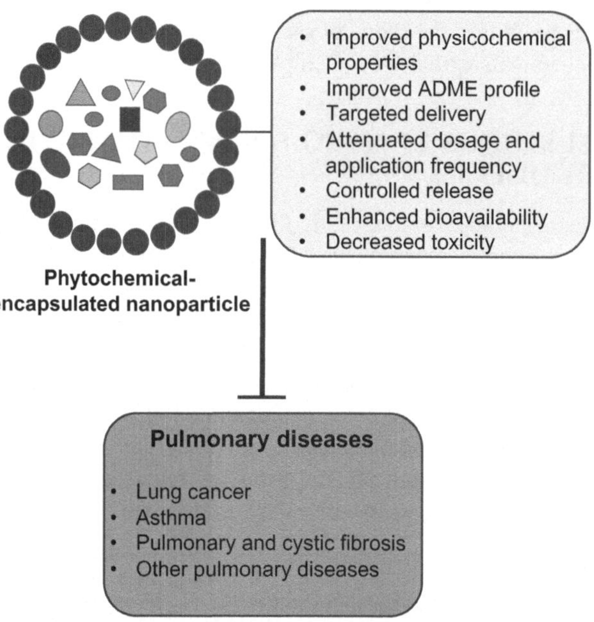

FIGURE 6.1 Potential benefits of phytochemical-encapsulated nanoparticles in the treatment of pulmonary diseases.

6.3.1 Lung Cancer

Lung cancer is the leading cause of cancer-related deaths worldwide, and its incidence in the year 2020 was 2.21 million cases (World Health Organization 2021b). Recent evidence on the development of cancer chemotherapeutics has drawn increasing attention owing to the anticancer potential of phytochemicals either individually or complementarily, and alternative medicine can be used for the prevention and/or treatment of lung cancer (Chen et al. 2007; Lian et al. 1998; Shukla et al. 2016; Wu et al. 2010).

Lung cancer is divided into two main groups, namely, small cell lung carcinoma (SCLC) and non-SCLC (NSCLC), and the main subtypes are adenocarcinoma, squamous cell carcinoma, and large cell carcinoma (American Cancer Society 2021). Approximately 85% of all lung cancer cases include NSCLCs (Molina et al. 2008). Although early diagnosis is crucial for selecting the best therapy option, unfortunately, lung cancer is frequently diagnosed only after progression to an advanced stage (Abdulbaqi et al. 2021; Hammerschmidt and Wirtz 2009). According to the SEER Program of National Cancer Institute, the 5-year survival rate of patients diagnosed with lung and bronchus cancers is only 21.7% in the U.S. based on 2011–2017 data (National Cancer Institute 2021). Moreover, this rate ranged from 10% to 20% in most countries during 2010–2014 (Sung et al. 2021). Surgical treatment, chemotherapy, and radiation therapy, as well as targeted therapy, are suitable treatment options for lung cancer; even so, the most appropriate one is chosen depending on the cancer subtype or stage (Molina et al. 2008; Zappa and Mousa 2016). Chemotherapeutic agents cause many undesirable side effects, as they exert toxicity on healthy proliferating cells and combat cancer cells (Pérez-Herrero and Fernández-Medarde 2015). Although conventional medications are currently available and prevailing, nanoformulations are an emerging therapeutic strategy to overcome chemotherapeutic barriers such as toxicity, drug resistance, and bioavailability and to improve bioaccumulation at tumor sites (Liang et al. 2010). Several studies have evaluated the treatment approaches for lung cancer and have indicated the preclinical efficacy of nanoformulations carrying phytoconstituents (Figure 6.2), some of which are summarized in the corresponding table in this section (Table 6.1).

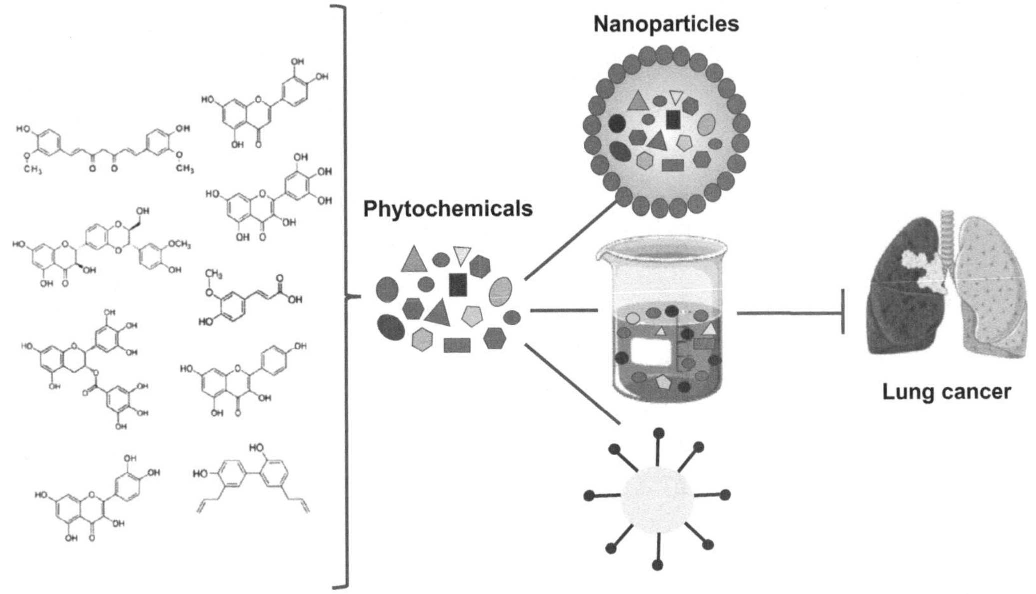

FIGURE 6.2 Phytoconstituent-carrying nanoformulations used in the management of lung cancer.

TABLE 6.1
Phytochemical-Encapsulated Nanocarriers for the Treatment of Lung Cancer

Phytochemical	Nanocarrier	Formulation	Disease Model	Finding	Refs
Carvacrol	Nanoemulsion	Nanoemulsion containing polysorbate 80 as emulsifier	A549	The viability of A549 cells was 52.7% at a dose of 100 μg/mL carvacrol nanoemulsion. LDH activity, accumulation of sub-G1 population, and intracellular ROS level were elevated by the nanoformulation in A549 cells	Khan et al. 2018
			A549 tumor model	Carvacrol nanoformulation reduced tumor volume by 34.2% and 62.1% at doses of 50 and 100 mg/kg	
Cineole	Liposome	DOTAB cationic liposome	A549	*Increased cytotoxicity of liposomal* AsODN *compared with* AsODN liposome alone	Moghimi et al. 2015
			A549 tumor model	Further reduction in tumor volume twice more ($p<0.05$) and further increase in its doubling time by more than twofold over AsODN ($p<0.01$)	
Curcumin	Liposome	Phosphoethanolamine-poly(ethylene glycol)-modified liposome	A549	Liposomal curcumin showed greater water solubility and more cytotoxicity with a lower EC_{50} value ($p<0.05$) than native curcumin	Rahman et al. 2012
Curcumin	Liposome	PEG-polyethyleneimine cationic liposome	A549 and LL-2	IC_{50} value of curcumin/LPPC for both A549 and LL-2 cells was 1.4 μM, which was lower than that of nonencapsulated curcumin	Lin et al. 2012
Curcumin	Liposome	-	LL-2, LL-2 tumor and zebrafish embryo models	Liposomal curcumin treatment reduced tumor volume and could block angiogenesis *in vivo*, possibly resulting from its cytotoxicity and apoptotic effects on endothelial cells (but not on LL-2 cells) *in vitro*	Wang et al. 2012
Curcumin	Micelle	-	Lung metastasis model	Nanomicellar curcumin reduced metastatic colonies in the lung and extended the survival rates of mice with B16 F10 lung metastatic tumors, implying that liposomal curcumin may have the ability to prevent lung metastasis	Mardani et al. 2020
Curcumin	Carbon nanotube	Oxidatively modified and polysaccharide-encapsulated (alginate and chitosan) single-walled carbon nanotube	A549	Curcumin-loaded nanotubes showed more cytotoxicity than free curcumin, inhibiting A549 cells in a dose-dependent manner at tested concentrations (0–20 μg/mL), whereas cytotoxicity was less for NIH 3T3 cells, and there was no significant difference with curcumin. Curcumin-loaded nanotubes dose-dependently induced apoptosis in 549 cells	Singh et al. 2018

Curcumin	Exosome	-	A549 and H1299	Growth inhibition of A549 and H1299 cells was enhanced from 34% to 61% and from 23% to 66%, respectively, by curcumin versus exo-curcumin	Aqil et al. 2017
Curcumin	SLNs	SA, lecithin, and Myrj53-based nanoparticles	A549 and NCL-H1299	SLN-curcumin (2:1) enhanced cytotoxicity as the IC_{50} value decreased from 185 to 32 µM and from 78 to 18 µM in NCL-H1299 and A549 cells, respectively. Cytotoxicity was further increased in case of SLN-curcumin (4:1; 9 and 4 µM)	Wang et al. 2013
			A549 tumor model	SLN-curcumin increased drug accumulation in lungs and tumors and significantly inhibited tumor growth	
EGCG	Polymeric nanoparticle	Poly(lactic-co-glycolic acid)	A549 and H1299	EGCG-loaded nanoparticles effectively induced inhibitory activity on the growth of A549 and H1299 cells when compared with native EGCG and had higher ability to induce apoptosis as well as suppress nuclear factor-κB	Zhang et al. 2020
Ferulic acid	Polymeric nanoparticles	PLGA	NCI-H460	Ferulic acid-PLGA nanoparticles were more effective than ferulic acid in decreasing both cell proliferation and mitochondrial membrane potential and in enhancing the levels of ROS and DNA damage as well as the percentage of apoptotic cell death	Merlin et al. 2012
Galbanic acid	SLN	Glyceryl palmitostearate–based SLNs	A549	Galbanic acid nanoparticles could inhibit the growth of A549 cells, especially after 48 h, having a higher effect than that of free galbanic acid, without any toxicity on the HUVEC cell line. It also induced a higher degree of apoptosis than galbanic acid in the early phase	Eskandani et al. 2015
Ginsenoside Rg3	Liposome	Phosphatidyl choline (egg yolk) liposome	A549	Liposomal ginsenoside showed antiproliferative effects on A549 cells with an IC50 value of 84.68 µg/mL, which had lower efficacy than that of free ginsenoside (126.43 µg/mL)	Yu et al. 2013
Ginsenoside Rh2	Nanocomposite	HA-functionalized ZnO nanoparticle	A549	Ginsenoside-HA-functionalized ZnO enhanced the cytotoxic effects of ginsenoside Rh2 against A549 cells	Kim et al. 2019
HK and cisplatin	Liposome	PEG-phosphoethanolamine–modified liposome	A549 tumor model	Combination of liposomal honokiol at a dose of 25 mg/kg with cisplatin resulted in a significant reduction in tumor volume as well as increased survival over single treatment groups. Liposomal HK enabled longer plasma circulation than that of the free form	Jiang et al. 2008

(Continued)

TABLE 6.1 (Continued)
Phytochemical-Encapsulated Nanocarriers for the Treatment of Lung Cancer

Phytochemical	Nanocarrier	Formulation	Disease Model	Finding	Refs
HK	Micelle	Poly(ε-caprolactone)-poly(ethylene glycol)-poly(ε-caprolactone) copolymer micelle	A549	While HK and HK-loaded micelle had dose-dependent cytotoxicity, which were comparable, HK-loaded micelle showed a sustainable drug release for several weeks	Wei et al. 2009
Kaempferol	Gold nanoclusters	-	A549	Kaempferol-conjugated gold nanoclusters showed a selective cytotoxicity at the tested concentrations of 12.5 and 25 μg/mL on A549 cells while HK-2 cells were not affected	Govindaraju et al. 2019
β-Lapachone	Micelle	Poly(ethylene glycol)-co-poly(D,L-lactic acid) polymer micelle	A549 and LLC tumor models	β-Lapachone micelles provided prolonged blood concentration, increased accumulation, and diminished toxicity as well as considerable tumor growth inhibition and enhanced survival rate	Blanco et al. 2010
β-Lapachone and paclitaxel	Micelle	poly(ethylene glycol)-*b*-poly(D,L-lactic acid) polymer micelle	A549	Cytotoxicity of the β-lapachone and paclitaxel micelle combination, wherein the combination index was <1, was shown to be further stronger over those of β-lapachone micelle and paclitaxel micelle alone	Zhang et al. 2015
Luteolin	Polymeric nanoparticles	Polylactic acid-PEG nanoparticles	H292	More effective inhibition of colony formation of H292 cells at a concentration of 10 μM. IC_{50} values of nano-lutein and free luteolin against H292 cells were 14.96 μM and 15.56 μM, respectively	Majumdar et al. 2014
Myricetin	SLN	Vitamin E and gelucire-based SLNs	A549	Compared with myricetin, myricetin-loaded nanoparticles considerably inhibited both viability and colony formation of A549 cells and enhanced necrosis by upregulating the gene expression of RIPK3 and MLKL while neither myricetin nor myricetin-loaded nanoparticles affected MRC5 cells	Khorsandi et al. 2020
Paclitaxel	Liposome	HSPC liposome	*In vivo* dog model	Higher accumulation in lungs than in other organs and tissues	Zhao et al. 2010
Quercetin	Micelle	PEG-phosphatidylethanolamine-PE copolymer micelle	A549	Antiproliferative activity was significantly improved by nanomicellar quercetin at a maximum concentration of 100 μg/mL compared with that of free quercetin, without meaningful toxicity of the empty micelle	Tan et al. 2012
			A549 tumor model	Nanomicellar formulation improved quercetin activity at a dose of 30 mg/kg while none of them caused notable changes in body weight	

Quercetin	Magnetic nanoparticle	Fe_3O_4 magnetic nanoparticles modified with PLGA	A549	Nanoparticles decreased the viability of A549 cells *in vitro* by nebulization after quercetin loading	Verma et al. 2013
Resveratrol and docetaxel	Lipid-polymer hybrid nanoparticle	EGF-conjugated PEG-SA nanoparticle	HCC827/NCI-H2135	EGF-conjugated resveratrol and paclitaxel dual encapsulated nanoparticles significantly inhibited HCC827 and NCI-H2135 cells when compared with dual loaded nanoparticle without EGF conjugation and free mixture of two drugs	Song et al. 2018
			HCC827 tumor model	Drug accumulation of dual drug-containing EGF nanoparticle was higher than that of the nonconjugated counterpart and free drug mixture	
Silibinin	Magnetic nanoparticles	Fe_3O_4 magnetic nanoparticles modified with PLGA-PEG copolymers	A549	Cytotoxicity of nanosilibinin was greater than that of free silibinin. Nanosilibinin intensively reduced hTERT expression	Amirsaadat et al. 2017

LDH: lactate dehydrogenase; ROS: reactive oxygen species; LPPC: liposomal polyethylene glycol–polyethyleneimine complex; PLGA: poly-D,L-lactide-co-glycolide nanoparticle; AsODN: antisense oligodeoxynucleotide; LPPC: liposomal polyethylene glycol–polyethyleneimine complex; SLN: solid lipid nanoparticle; EGCG: epigallocatechin gallate; LLC: Lewis lung carcinoma; HSPC: hydrogenated soybean phospholipid; HUVEC: human umbilical vein endothelial cell; hTERT: human telomerase reverse transcriptase; HA: hyaluronic acid; ZnO: zinc oxide; HK: honokiol; PEG: Polyethylene glycol; PE: phosphoethanolamine; EGF: epidermal growth factor; SA: stearic acid

Phytochemical-based nanomedicines further promote the anticancer property of phytochemicals. For instance, curcumin possesses antitumor activity against lung cancer (Anand et al. 2008; Mehta et al. 2014). Curcumin encapsulated in krill lipid-based liposomes (marinosomes) exhibited a greater cytotoxic effect on A549 lung cancer cells than free curcumin owing to enhanced solubility, stability, and controlled delivery of free curcumin, thus suggesting a promising drug delivery system of curcumin against lung cancer (Ibrahim et al. 2018).

Naringenin, a flavanone predominantly found in citrus fruits and grapefruit (Hernández-Aquino and Muriel 2017), exhibits antitumor effects against lung cancer (Chang et al. 2017; Shi et al. 2021). Naringenin-loaded hyaluronic acid (HA)-decorated polycaprolactone (PCL) nanoparticles (NAR-HA@CH-PCL-NP) conferred protection to the encapsulated drug against a broad-spectrum pH variability throughout the gastrointestinal tract and established the potential for NSCLC treatment. NAR-HA@CH-PCL-NP further markedly attenuated tumor development and biochemical species levels when compared with naringenin in a rat model of urethane-induced NSCLC (Parashar et al. 2018).

Combination formulations were prepared for further enhancement of cytotoxic activity, and successful results were achieved. Liposomal formulation of honokiol, a small-molecular-weight lignan isolated from the genus *Magnolia* (Wang et al. 2004; Bai et al. 2003), and cisplatin, a widely used cytotoxic chemotherapeutic agent (The International Adjuvant Lung Cancer Trial Collaborative Group 2004), were successfully used in combination for the treatment of lung cancer. In a A549 tumor model, honokiol-loaded PEG-modified liposomes (size range: 130±20 nm) at a concentration of 25 mg/kg, when used in combination with cisplatin, significantly suppressed tumor growth and increased patients' lifespan, when compared with free honokiol, cisplatin, and free liposome, individually. This combination inhibited angiogenesis and apoptosis induction, which was more than those observed in the control group (Jiang et al. 2008).

In another study, β-lapachone, a naphthoquinone obtained from *Tabebuia* sp. (Castellanos et al. 2009), showed cytotoxic properties (Li et al. 2000), and paclitaxel combination micelle exhibited superior cytotoxicity against A549 cells, with an IC value of 0.16 μM, which was much lower than those of free compounds and their micelle alone (Zhang et al. 2015).

Receptor targeting such as epidermal growth factor (EGF) is an important approach for designing a nanocarrier system. EGF-conjugated lipid-polymer hybrid nanoparticles (LPNs) were used for loading RSV and docetaxel (DTX) (Song et al. 2018). DTX, a hemisynthetic diterpene derived from *Taxus baccata*, is an antineoplastic agent used for treating advanced or metastatic NSCLCs (Fossella et al. 2003; He et al. 2015), and stilbene RSV, which is highly abundant in grapes, displays a wide range of biological activities including chemopreventive effects (Paredes-López et al. 2010). EGF-conjugated DTX/RSV LPNs, whose mean particle size was approximately 189.6 nm, markedly exhibited a dose-dependent inhibitory activity on HCC827 and NCI-H2135 cells when compared with the nonconjugated counterpart. Additionally, DTX and RSV demonstrated a synergistic effect in LPNs when compared with the free mixture of DTX and RSV. In a mouse model of HCC827, EGF-conjugated DTX/RSV LPNs showed the best growth inhibitory activity, consistent with *in vitro* results (Song et al. 2018).

Betulinic acid, ginsenoside Rh2, honokiol, and parthenolide, in PE-PEG liposome is another example of combination application (Jin et al. 2020). Betulinic acid is a pentacyclic triterpenoid isolated from a wide variety of plant species and exhibits various kinds of biological activities, the most important of which can be considered anticancer (Hordyjewska et al. 2019). Ginsenoside Rh2, a triterpene saponin, belongs to the protopanaxadiol group of ginsenosides of *Panax ginseng* and is known for its anticancer efficacy, especially against lung cancer (Kim et al. 2021). Parthenolide, derived from *Tanacetum parthenium*, is a sesquiterpene that contains an α-methylene-γ-lactone ring and an epoxide group. Parthenolide has demonstrated potential in complementary therapy against cancer, which makes it a good candidate for such a therapeutic approach (Sztiller-Sikorska and Czyz 2020). The cocktail application of these four bioactive compounds in a co-delivery system resulted in a better synergistic activity than the free cocktail, with combination indexes of 0.37 and 0.65, respectively. The liposomal cocktail formulation displayed cytotoxicity against A549 cells, which was greater than that of both free individual compounds and free cocktail, and was comparable to that of cisplatin. In a xenograft tumor model, the liposomal cocktail could significantly decrease tumor growth when compared with the free cocktail (Jin et al. 2020).

6.3.1.1 *IN VIVO* STUDIES ON INHALABLE LIPOSOMAL FORMULATIONS

Various inhalable liposomal formulations have been evaluated in preclinical studies for the management of lung cancer, some of which are exemplified below.

After the administration of liposomal curcumin powder, free curcumin powder, and gemcitabine into the lungs through the trachea of rats using an insufflator, the tumor nodes and bleeding in the lungs decreased, which were an indicator of the antitumoral effect; additionally, significant inhibition of vascular endothelial growth factor (VEGF), malondialdehyde, tumor necrosis factor-α, caspase-3, and Bcl-2 was observed after the application of liposomal curcumin powder when compared with the other experimental groups. Liposome was conventionally prepared by soybean lecithin:cholesterol (5:1 [mol/mol]) (Zhang et al. 2018). In another study on curcumin, an inhalable liposomal formulation was prepared with lecithin, cholesterol, stearyl amine, and poloxamer 188, with hydroxypropyl-beta-cyclodextrin being an effective carrier. Dry powder of proliposomal curcumin was administered intratracheally to rats, and according to the results of pharmacokinetic studies, proliposomal curcumin powder was more effectively absorbed and preserved in the lung than curcumin powder (Adel et al. 2021). Quercetin (QR), a flavonoid widely distributed in natural products and possessing a wide range of biological activities including anticancer properties (Rauf et al. 2018), was prepared in a liposomal formulation for lung cancer therapy, which targeted transferrin receptors. T7-targeted liposomes were prepared by conjugating T7 peptide with distearoylphosphoethanolamine-PEG-MAL micelles and loading QR. In a A549-Luc orthotopic lung cancer mice model, T7-QR-liposome, QR-liposome, or free-QR solutions were administered using the MicroSprayer® Aerosolizer Pulmonary Aerosol Kit for Mouse. The strongest inhibition of tumor growth was observed in the T7-QR-liposome-treated group as compared with that in the free-QR and nontargeted QR-liposome groups (Riaz et al. 2019). Tumor-targeted bacteria, used either alone or as a combination, is a novel approach in cancer treatment (Sedighi et al. 2019). In a primary lung cancer rat model, liposomal paclitaxel in *Escherichia coli*, intratracheally administered to rats, displayed the strongest therapeutic activity on tumors by downregulating VEGF, hypoxia-inducible factor-1α, and Bcl-2 (Zhang et al. 2020).

6.3.2 Asthma

Asthma is one of the most prevalent chronic inflammatory diseases globally, affecting 262 million people in the year 2019 according to the World Health Organization (World Health Organization 2021a). Airway hyperresponsiveness, airway obstruction, and multicellular inflammation are well-known clinical characteristics of asthma, and therapeutic intervention mainly relies on inhaled corticosteroids (Chung et al. 2014; Murdoch and Lloyd 2010).

As for treatment with phytochemicals, curcumin holds a therapeutic clinical value against various diseases including asthma (Abidi et al. 2014; Karaman et al. 2012; Suzuki et al. 2005). Assessing the effect of curcumin nanoformulations on asthma, curcumin-solid lipid nanoparticles (curcumin-SLNs) significantly blocked airway hyperresponsiveness and inflammatory cell infiltration when compared with the curcumin treatment in a rat model of asthma, rendering curcumin-SLNs as a favorable drug delivery system to improve the therapeutic effect of curcumin in asthma management (Wang et al. 2012). Casula et al. (2021) prepared a curcumin nanosuspension and a multicomponent water-based nanosuspension formulation of curcumin and beclomethasone dipropionate nanocrystals and observed that the multicomponent nanosuspension had optimum dimensional characteristics and aerodynamic parameters, which enabled proper and efficient delivery to the interior parts of the lungs.

Fisetin, a flavonoid abundant in assorted fruits and vegetables, is beneficial against anti-inflammatory diseases (Huang et al. 2018; Khan et al. 2012; Molagoda et al. 2021). As for the anti-inflammatory activity, fisetin-loaded mesoporous carbon nanoparticles (MCNs) mitigated oxidative and nitrosative stress and modified the occurrence of the composite asthma phenotype in a murine preclinical model, indicating the ability of fisetin as a lead compound in the management and modification of clinical asthma (Mitra et al. 2015).

Celastrol (CS), a natural substance present in *Tripterygium wilfordii*, exhibits anti-inflammatory properties and has the potential to relieve asthmatic symptoms (Kim et al. 2009; Zeng et al. 2018). Because of its poor bioavailability, CS-containing liquid crystalline nanoparticles were designed, and their effect was tested against asthma. They significantly reduced interleukin-1β generation *in vitro*, possibly contributing to the alleviation of asthmatic symptoms. Thus, the study rendered a basis for future investigations not only on asthma but also on other respiratory diseases (Chan et al. 2021).

6.3.3 Pulmonary and Cystic Fibrosis

Pulmonary fibrosis (PF) is a condition wherein the lungs are scarred and breathing becomes more difficult (American Lung Association 2021). PF is induced by a heterogeneous group of lung disorders causing progressive and irreversible devastation of the lung architecture, leading to abnormalities in lung function and disturbance in gas exchange, resulting in death from respiratory loss (Wynn 2011). IPF is a destructive form of PF with unknown etiology and has no remedial intervention. Therefore, substantial medical therapy should be immediately provided and is of critical importance. When the probable beneficial effect of fisetin (3,3′,4′,7-tetrahydroxyflavone), curcumin, and MCN-loaded fisetin was evaluated, all three compounds displayed a regenerative effect in a mouse model of IPA. Additionally, the administration of MCN-loaded fisetin further augmented the beneficial effects, proposing that it is a probable anti-inflammatory agent for IPA management (Kar et al. 2015; Kar et al. 2016).

HA, an extracellular matrix glycosaminoglycan, is a signaling molecule that modulates inflammatory responses and tissue recovery after harmful stimuli (Esposito et al. 2017; Singleton and Lennon 2011). HA nanoparticles administered to lungs before gamma irradiation led to shortening of destructive radiation-induced injury in the lung tissue *in vivo*, thereby potentially attenuating extracellular matrix degradation associated with lung fibrosis (Lierova et al. 2020).

CF is a genetic disease characterized by very thick and sticky mucus as well as digestion and breathing problems. People with CF are more susceptible to other health conditions (Centers for Disease Control and Prevention 2021). The intervention of CF by medicines depends on the maintenance and improvement of lung function, protection against infections, breathing assistance, mucus clearance, and amelioration of impaired CF transmembrane conductance regulator protein, all of which are involved in the disease etiology (National Heart, Lung and Blood Institute 2021). Although advanced hepatic fibrosis therapy through nanoparticle applications has been comparatively an undiscovered area, some promising nano-formulated phytochemicals are exemplified below.

An example of phyto-based nanomedicine for CF recovery is poly(lactic-co-glycolic acid) (PLGA) nanoparticles encapsulating curcumin. Cartiera et al. (2010) reported that these nanoparticles can improve the effects of curcumin intervention and partially correct CF symptoms in *in vivo* mouse models of CF when orally administered, thus confirming that curcumin nanoparticles are more promising agents than free curcumin.

6.3.4 Other Pulmonary Diseases

The effect of phytochemical-based nanoformulations has been further investigated and assessed in other pulmonary diseases; some of the examples are stated below.

Rifampin (RIF) and ascorbic acid (ASC) co-loaded into alginate/chitosan nanoparticles (RIF/ASC NPs) exhibited a synergistic effect against *Staphylococcus aureus* test strains. RIF/ASC co-loaded NPs were a convenient system for antibiotic lung administration, showing encouraging perspectives for effective interventions for lung infections when compared with the effects of known antibiotics that are problematic owing to gaining antimicrobial resistance (de Oliveira et al. 2019). Likewise, RIF in combination with ASC, namely made of sodium alginate coated with chitosan and Tween 80 (RIF/ASC NPs), showed a markedly enhanced activity against nine clinical strains of *Mycobacterium tuberculosis* when compared with the free drug, showing high stability at different physiological pH values. Thus, the research highlighted the likelihood of the RIF/ASC NP-based formulation as an effective antibiotic regimen administered through the pulmonary route, thus enhancing drug distribution in lung tissues and the primary infection site as well as reducing the risk of systemic toxicity (Scolari et al. 2019).

Naringenin-loaded dipalmitoylphosphatidylcholine (DPPC) phytosomes for dry powder inhalation (NPDPI) treatment showed efficacy in a rat model of acute lung injury. They remarkably suppressed p38 phosphorylation in the mitogen-activated protein kinase pathway and oxidative stress. The DPPC vehicle displayed potential effects against acute lung injury by preserving respiratory function. Thus, NPDPIs enabled prolonged drug release, promoted pulmonary bioavailability of the drug, and stimulating defense against acid-associated acute lung injury *in vivo*, confirming that NPDPIs have the potential to be used in the clinical setting as effective dry powder inhalers for treating acute lung injury (Yu et al. 2020).

6.4 DISCUSSION

A wide range of phytonanomedicines have been developed during the past decades. The integration of phytomedicine into nanotechnology practices has recently gained prominence. Nevertheless, their profitable application still has a few problematic issues. Emerging data have unraveled that inhaled nanoparticles may increase risk in individuals with pre-existing respiratory diseases (Bonner 2010; Lu et al. 2014). Selecting proper nanomaterials for the treatment of lung diseases is critical. Particularly, inorganic nanoparticles are not biodegradable and are not easily eliminated from the delivery site (Yang et al. 2019). Biodegradable polymers should be preferred, as they possess fewer adverse effects on the lung. Metal nanoparticles and carbon nanotubes exert side effects on pulmonary tissues (Duke and Bonner 2018; Ko et al. 2018). Thus, the evaluation of safety profiles of all nanoparticles is essential before their application. Selecting appropriate nanomaterials may minimize nanomedicine-related adverse effects. Assorted properties including their physicochemical characteristics, administration routes, doses, exposure periods, and particle size should be highlighted for the estimation of potential risk. Furthermore, their behavior in different media or environments should be examined in detail, which will provide a model for determining their original specialties.

The mucus barrier in the respiratory tract is a hinderance, given that chronic lung disease is usually characterized by hypersecretion of mucus in the airway, which constitutes a major physical impediment during the application of inhaled nanoparticles. Likewise, in CF, highly viscous sputum may adhere to the nanoparticles, which obstructs their penetration into the respiratory epithelium (Sanders et al. 2000; Schuster et al. 2013; Suk et al. 2009). Assorted approaches have been developed to improve the penetration and delivery of nanoparticles into the pulmonary tissues (Chen et al. 2021; Schuster et al. 2013).

The clinical application of phytochemicals requires designing of proper formulations for efficient drug delivery. In such cases, nanoparticles offer a rational aspect, wherein they enhance the solubility, stability, absorption, and bioavailability of phytochemicals, as well as improving their penetration and delivery into the target tissues (Wang et al. 2014). Lipid nanocarriers, nanoliposomes, nanoemulsions, micelles, and PLGA nanoparticles are among the most commonly used biodegradable and biocompatible nanoparticles (Sarker and Nahar 2017).

Nanotechnological applications of phytochemicals have recently been popular owing to compensation in the restrictions of phytochemicals (poor solubility, low bioavailability, toxicity, and so on) and enhancement of their delivery to the targeted tissue (Kim et al. 2021). Nanoparticles can be designed; prepared in different shapes, sizes, and compositions; and functionalized and modified physicochemically to achieve specific properties depending on the characteristics of both the bioactive molecule and the target organ. Many investigations have pointed out the potential of phytochemical-based nanotechnological applications, emphasizing their promising use in various diseases including lung diseases (Amirsaadat et al. 2017; Singh et al. 2018; Wang et al. 2012).

Within the scope of our chapter, we discussed the nanoparticle-based drug delivery of phytochemicals for pulmonary diseases to present a concept about current therapeutic strategies. The development of innovative phytonanomaterials is in progress, with the aim of obtaining complete cure from lung diseases.

6.5 Future Perspectives

Phytochemicals are of substantial importance as promising drug leads or currently used drugs for the management of various disorders. There are various routes of delivery to the particular tissue of interest. Because of their usual characteristics such as insufficient bioavailability, poor pharmacokinetic profile, poor solubility, poor stability, and so on, ongoing research has focused on nanotechnological applications for the past decades. Emerging data are promising and draw attention to the benefits of nano-based formulations of phytochemicals against many diseases. Some phytochemical-encapsulated nanocarriers (e.g., silibinin) for pulmonary diseases have been under clinical trials. As for the evaluation of the potential of many phytochemicals in preclinical trials, many of the phytochemicals hold remarkable potential to be designed for clinical trials and deserve further investigation.

Lung diseases are diverse and commonly affect individuals' quality of life adversely. The convention protocols along with phytonanomedicines are yet to be addressed. Outcome-based therapeutic attitude should be formulated in a way that comprises phytochemicals and their encapsulated material to overcome lung diseases.

The nanocarrier-based approaches of phytochemicals have demonstrated encouraging outcomes in lung diseases. Nevertheless, their use in the clinical setting exhibits toxicity in human subjects, requiring further validation. Smaller nanoparticles can usually cross cell barriers. However, their deposition and clearance may be restricted. In case of pulmonary diseases, particles of size less than 1 μM are exhaled from the respiratory system. Therefore, determining the target specificity of large- or medium-sized nanoparticles is of importance. Phytochemical-based nanoformulations are usually administered through the intravenous or subcutaneous routes in preclinical models. Yet, further investigations are needed regarding phytochemical-based nanoformulations for the management of pulmonary diseases.

Conflict of interest: The authors declare that they have no conflicts of interest to report regarding the present study.

REFERENCES

Abdou, E. M., and M. M. Masoud. (2018). *Pharm Dev Technol* 23 (1):55–66.

Abdulbaqi, I. M., R. A. Assi, A. Yaghmur, et al. 2021. *Pharmaceuticals* 14 (8):725.

Abidi, A., S. Gupta, M. Agarwal, et al. 2014. *Clin Diagnostic Res* 8 (8):19–24.

Adel, I. M., M. F. ElMeligy, M. E. A. Abdelrahim, et al. 2021. *Int J Nanomed* 16:2667–87.

Ahmad, R., S. Srivastava, S. Ghosh, et al. 2021. *Colloids Surf B Biointerfaces* 197:111389.

Amirsaadat, S., Y. Pilehvar-Soltanahmadi, F. Zarghami, et al. 2017. *Artif Cells Nanomed Biotechnol* 45 (8):1649–56.

Anand, P., C. Sundaram, S. Jhurani, et al. 2008. *Cancer Lett* 267 (1):133–64.

Aqil, F., R. Munagala, J. Jeyabalan, et al. 2017. *AAPS J* 19 (6):1691–1702.

Aqil, F., R. Munagala, J. Jeyabalan, et al. 2013. *Cancer Lett* 334 (1):133–41.

American Cancer Society. 2021. accessed 27.11.2021 www.cancer.org/cancer/lung-cancer/about/what-is.html

American Lung Association. 2021. accessed 29.11.2021 www.lung.org/lung-health-diseases/lung-disease-lookup/pulmonary-fibrosis

Atanasov, A. G., B. Waltenberger, E.-M. Pferschy-Wenzig, et al. 2015. *Biotechnol Adv* 33 (8):1582–614.

Bai, X., F. Cerimele, M. Ushio-Fukai, et al. 2003. *J Biol Chem* 278 (37):35501–7.

Blanco, E., E. A. Bey, C. Khemtong, et al. 2010. *Cancer Res* 70 (10):3896–904.

Boarescu, P.-M., I. Chirilă, A. E. Bulboacă, et al. 2019. *Oxid Med Cell Longev* 2019:Article ID 7847142.

Bonner, J. C. 2010. *Proc Am Thorac Soc* 7 (2):138–41.

Cartiera, M. S., E. C. Ferreira, C. Caputo, et al. 2010. *Mol Pharm* 7 (1):86–93.

Casula, L., F. Lai, E. Pini, et al. 2021. *Pharmaceutics* 13 (8).

Centers for Disease Control and Prevention. 2021. accessed 29.11.2021 www.cdc.gov/genomics/disease/cystic_fibrosis.htm

Chan, Y., S. W. Ng, D. K. Chellappan et al. 2021. *Int J Polym Mater* 70 (11):754–763.

Chang, H.-L., Y.-M. Chang, S.-C. Lai et al. 2017. *Exp Ther Med* 13 (2):739–44.

Chen, D., J. Liu, J. Wu et al. 2021. *Expert Opin Drug Deliv* 18 (5):595–606.

Chen, W., X. Wang, J. Zhuang et al. 2007. *Carcinogenesis* 28 (10):2114–21.

Chung, K. F., S. E. Wenzel, J. L. Brozek, et al. 2014. *Eur Respir J* 43 (2):343–73.

de Oliveira, M. T. P., D. de Sá Coutinho, É. T. de Souza, et al. 2019. *Int J Nanomedicine* 14:5215–28.

Dhupal, M. and D. Chowdhury. 2020. *Int J Nanomedicine* 15:9125–57.

Di, L. 2015. *AAPS J* 17 (1):134–43.

Duke, K. S., and J. C. Bonner. 2018. *Wiley Interdiscip Rev Nanomed Nanobiotechnol* 10 (3):e1498.

Eskandani, M., J. Barar, J. E. N. Dolatabadi et al. 2015. *Pharm Biol* 53 (10):1525–38.

Esposito, A. J., P. K. Bhatraju, R. D. Stapleton et al. 2017. *Crit Care* 21 (1):304.

Fossella, F., J. R. Pereira, J. von Pawel et al. 2003. *J Clin Oncol* 21 (16):3016–24.

Castellanos, J. R. G., J. M. Prieto, and M. Heinrich. 2009. *J Ethnopharmacol* 121 (1):1–13.

González-Medina, M., F. D. Prieto-Martínez, J. J. Naveja, et al. 2016. *Future Med Chem* 8 (12):1399–412.

Govindaraju, S., A. Roshini, M. H. Lee, et al. 2019. *Int J Nanomedicine* 14:5147–57.

The International Adjuvant Lung Cancer Trial Collaborative Group. 2004. *N Engl J Med* 350 (4):351–60.

Gunasekaran, T., T. Haile, T. Nigusse, et al. 2014. *Asian Pac J Trop Biomed* 4:S1–S7.

Hammerschmidt, S., and H. Wirtz. 2009. *Dtsch Arztebl Int* 106 (49):809–18.

He, X., J. Wang, and Y. Li. 2015. *Onco Targets Ther* 8:2023–31.

Heinrich, M. 2010. *Front Pharmacol* 1:8–8.

Hernández-Aquino, E., and P. Muriel. 2017. In: *Liver pathophysiology*, ed. P. Muriel, 633–51. Boston: Academic Press.

Hordyjewska, A., A. Ostapiuk, A. Horecka, et al. 2019. *Phytochem Rev* 18 (3):929–51.

Hosseini, S. A., F. Zahedipour, T. Sathyapalan, et al. 2021. *BioFactors* 47 (3):250–69.

Hu, L., D. Kong, Q. Hu, et al. 2015. *Nanoscale Res Lett* 10 (1):381.

Huang, W., M.-L. Li, M.-Y. Xia, et al. 2018. *Int J Mol Med* 42 (1):208–18.

Ibrahim, S., T. Tagami, T. Kishi, et al. 2018. *Int J Pharm* 540 (1):40–9.

Itokawa, H., Q. Shi, T. Akiyama, et al. 2008. *Chin Med* 3:11.

Jabbal, S., G. Poli, and B. Lipworth. 2017. *J Allergy Clin Immunol* 139.

Jiang, Q. Q., L. Y. Fan, G. L. Yang, et al. 2008. *BMC Cancer* 8:242.

Jin, X., Qing Y., Ning C., et al. 2020. *Nanomedicine* 15 (1):41–54.

Kar, S., S. Biswas, and E. R. Banerjee. 2016. *Biomed Res Ther* 3 (7):707–22.
Kar, S., S. Konsam, G. Hore, et al. 2015. *Bioml Res Ther* 2 (4):250–262.
Karaman, M., F. Firinci, S. Cilaker, et al. 2012. *Allergol Immunopathol* 40 (4):210–4.
Khan, I., A. Bahuguna, P. Kumar, et al. 2018. *Sci Rep* 8 (1):144.
Khan, N., D. N. Syed, N. Ahmad, et al. 2012. *Antioxid Redox Signal* 19 (2):151–62.
Khorsandi, L., E. Mansouri, M. Rashno, et al. 2020. *Int J Pept Res Ther* 26 (2):899–910.
Kim, B., J.-E. Park, E. Im, et al. 2021. *Int J Mol Sci* 22 (7):3571.
Kim, D. Y., J. W. Park, D. Jeoung, et al. 2009. *Eur J Pharmacol* 612 (1–3):98–105.
Kim, S., N. Kim, J.Y. Jeong, et al. 2021. *Processes* 9 (8):1344.
Kim, Y. J., H. Perumalsamy, V. Castro-Aceituno, et al. 2019. *Int J Nanomedicine* 14:8195–208.
Kinghorn, A. D., L. Pan, J. N. Fletcher, et al. 2011. *J Nat Prod* 74 (6):1539–55.
Ko, J.-W., H.-J. Lee, N.-R. Shin, et al. 2018. *Molecules* 23 (9):2247.
Lelli, D., A. Sahebkar, T. P. Johnston, et al. 2017. *Pharmacol Res* 115:133–48.
Li, Y., C. J. Li, D. Yu, et al. 2000. *Mol Med* 6 (12):1008–15.
Lian, F., M. Bhuiyan, Y. W. Li, et al. 1998. *Nutr Cancer* 31 (3):184–91.
Liang, X.-J., C. Chen, Y. Zhao, et al. 2010. In: *Multi-Drug Resistance in Cancer*, ed. J. Zhou, 467–88. Totowa, NJ: Humana Press.
Lierova, A., J. Kasparova, J. Pejchal, et al. 2020. *Front Pharmacol* 11:1199.
Lin, Y.-L., Y.-K. Liu, N.-M. Tsai, et al. 2012. *Nanomed: Nanotechnol Biol Med* 8 (3):318–27.
Lu, X., T. Zhu, C. Chen, et al. 2014. *Int J Mol Sci* 15 (10):17577–600.
Majnooni, M. B., S. Fakhri, Y. Shokoohinia, et al. 2020. *Front Pharmacol* 11 (1744).
Majumdar, D., K. H. Jung, H. Zhang, et al. 2014. *Cancer Prev Res (Phila)* 7 (1):65–73.
Manca, M. L., J. E. Peris, V. Melis, et al. 2015. *RSC Advances* 5 (127):105149–59.
Manickam, V., V. Sundar, R. L. Panchangam, et al. 2021. In: *Nanopharmaceuticals: Principles and Applications Vol. 2*, ed. V. K. Yata, S. Ranjan, N. Dasgupta and E. Lichtfouse, 211–64. Cham: Springer International Publishing.
Mardani, R., M. R. Hamblin, M. Taghizadeh, et al. 2020. *Pathol Res Pract* 216 (9):153082.
Mehta, H. J., V. Patel, and R. T. Sadikot. 2014. *Target Oncol* 9 (4):295–310.
Méndez-Lucio, O., and J. L. Medina-Franco. 2017. *Drug Discov Today* 22 (1):120–26.
Merlin, J. P. J., N. R. Prasad, S. M. A. Shibli, et al. 2012. *Biomed Prev Nutr* 2 (1):69–76.
Mitra, S., P. Paul, K. Mukherjee, et al. 2015. *J Nanomed Biother Discovery* 5:2.
Moghimi, H. R., F. H. Shirazi, M. S. Ardestani et al. 2015. *J Nanomater* 2015:Article ID 967473.
Molagoda, I. M. N., J. A. C. C. Jayasingha, Y. H. Choi, et al. 2021. *Sci Rep* 11 (1):8377.
Molina, J. R., P. Yang, S. D. Cassivi, et al. 2008. *Mayo Clin Proc* 83 (5):584–94.
Murdoch, J. R., and C. M. Lloyd. 2010. *Mut Res* 690 (1–2):24–39.
National Cancer Institute. Surveillance, Epidemiology, and End Results Program (SEER). 2021. accessed 27.11.2021. https://seer.cancer.gov/statfacts/html/lungb.html
National Heart, Lung and Blood Institute. accessed 21.11.2021. www.nhlbi.nih.gov/health-topics/cystic-fibrosis
Newman, D. J., and G. M. Cragg. 2020. *J Nat Prod* 83 (3):770–803.
Nishiyama, N. 2007. *Nat Nanotechnol* 2 (4):203–4.
Pápay, Z. E., A. Kósa, B. Böddi, et al. 2017. *J Aerosol Med Pulm Drug Deliv* 30 (4):274–88.
Parashar, P., M. Rathor, M. Dwivedi, et al. 2018. *Pharmaceutics* 10 (1):33.
Paredes-López, O., M. L. Cervantes-Ceja, M. Vigna-Pérez, et al. 2010. *Plant Foods Hum Nutr* 65 (3):299–308.
Pérez-Herrero, E., and A. Fernández-Medarde. 2015. *Eur J Pharm Biopharm* 93:52–79.
Rahman, H. S., H. H. Othman, N. I. Hammadi, et al. 2020. *Int J Nanomedicine* 15:2439–83.
Rahman, S., S. Cao, K. J. Steadman, et al. 2012. *Drug Deliv* 19 (7):346–53.
Rassu, G., E. P. Porcu, S. Fancello, et al. 2019. *Pharmaceutics* 11 (1):8
Rauf, A., M. Imran, I. A. Khan, et al. 2018. *Phytother Res* 32 (11):2109–30.
Riaz, M. K., X. Zhang, K. H. Wong, et al. 2019. *Int J Nanomedicine* 14:2879–902.
Sajid, M., S. S. Cameotra, M. S. A. Khan, et al. 2019. In: *New Look to Phytomedicine*, ed. M. S. A. Khan, I. Ahmad and D. Chattopadhyay, 597–623. Academic Press.
Sanders, N. N., S. C. De Smedt, and J. Demeester. 2000. *J Pharm Sci* 89 (7):835–49.
Sarker, S., and L. Nahar. 2017. *Pharm Sci* 23:170–71.
Schmidt, B. M., D. M. Ribnicky, P. E. Lipsky, et al. 2007. *Nat Chem Biol* 3 (7):360–6.

Schmidt, B.M., and D.M.K. Cheng. 2017. *Ethnobotany: A Phytochemical Perspective*: Wiley-Blackwell
Schuster, B. S., J. S. Suk, G. F. Woodworth, et al. 2013. *Biomaterials* 34 (13):3439–46.
Scolari, I. R., P. L. Páez, M. E. Sánchez-Borzone, et al. 2019. *AAPS PharmSciTech* 20 (2):67.
Sedighi, M., A. Z. Bialvaei, M. R. Hamblin, et al. 2019. *Cancer Med* 8 (6):3167–81.
Shi, X., X. Luo, T. Chen, et al. 2021. *J Cell Mol Med* 25 (5):2563–71.
Shukla, S., S. Sinha, S. Khan, et al. 2016. *Sci Rep* 6:21860.
Singh, N., A. Sachdev, and P. Gopinath. 2018. *J Nanosci Nanotechnol* 18 (3):1534–41.
Singleton, P. A., and F. E. Lennon. 2011. *J Allergy Ther* Suppl 4.
Song, Z., Y. Shi, Q. Han, et al. 2018. *Biomed Pharmacother* 105:18–26.
Suk, J. S., S. K. Lai, Y.-Y. Wang, et al. 2009. *Biomaterials* 30 (13):2591–7.
Suleria, H. A. R., M. R. Goyal, and M. S. Butt. 2019. *Phytochemicals from Medicinal Plants: Scope, Applications, and Potential Health Claims*: Apple Academic Press.
Sung, H., J. Ferlay, R. L. Siegel, et al. 2021. *CA Cancer J Clin* 71 (3):209–49.
Suzuki, M., T. Nakamura, S. Iyoki, et al. 2005. *Biol Pharm Bull* 28 (8):1438–43.
Sztiller-Sikorska, M., and M. Czyz. 2020. *Pharmaceuticals (Basel)* 13 (8).
Tan, B. J., Y. Liu, K. L. Chang, et al. 2012. *Int J Nanomedicine* 7:651–61.
Thipe, V. C., K. P. Amiri, P. Bloebaum, et al. 2019. *Int J Nanomedicine* 14:4413–28.
Verma, N. K., K. Crosbie-Staunton, A. Satti, et al. 2013. *J Nanobiotechnology* 11 (1):1.
Wang, L., J. Zhang, L. Cai, et al. 2012. *Oncol Lett* 4 (1):107–12.
Wang, P., L. Zhang, H. Peng, et al. 2013. *Mater Sci Eng C* 33 (8):4802–8.
Wang, S., R. Su, S. Nie, et al.. 2014. *J Nutr Biochem* 25 (4):363–76.
Wang, W., R. Zhu, Q. Xie, et al. 2012. *Int J Nanomedicine* 7:3667–77.
Wang, X., Y. Wang, Y. Geng, et al. 2004. *J Chromatogr A* 1036 (2):171–5.
Wei, X.W., C.Y Gong, S. Shi, et al. 2009. *Int J Pharm* 369 (1):170–5.
World Health Organization. 2021a. "Asthma." accessed 21.11.2021. www.who.int/news-room/fact-sheets/detail/asthma
World Health Organization. 2021b. "Cancer." accessed 21.11.2021. www.who.int/news-room/fact-sheets/detail/cancer
Wu, X., Y. Zhu, H. Yan, et al. 2010. *BMC Cancer* 10:269–269.
Wynn, T. A. 2011. *J Exp Med* 208 (7):1339–50.
Yang, G., S. Z. F. Phua, A. K. Bindra, et al. 2019. *Adv Mater* 31 (10):1805730.
Yao, H., J. Liu, S. Xu, et al. 2017. *Expert Opin Drug Discov* 12 (2):121–40.
Yetisgin, A. A., S. Cetinel, M. Zuvin, et al. 2020. *Molecules* 25 (9).
Yu, H., L. Teng, Q Meng, et al. 2013. *Int J Pharm* 450 (1):250–8.
Yu, Z., X. Liu, H. Chen, et al. 2020. *J Aerosol Med Pulm Drug Deliv* 33 (4):194–204.
Zappa, C. and S. A. Mousa. 2016. *Transl Lung Cancer Res* 5 (3):288–300.
Zeng, Z., X. Lin, R. Zheng, et al. 2018. *Front Pharmacol* 9 (49).
Zhai, T., S. Li, W. Hu, et al. 2018. *Nutrients* 10 (7):813
Zhang, L., W. Chen, G. Tu, et al. 2020. *Int J Nanomedicine* 15:4417–29.
Zhang, L., Z. Chen, K. Yang, et al. 2015. *Mol Pharms* 12 (11):3999–4010.
Zhang, M., M. Li, L. Du et al. 2020. *Int J Pharm* 578:119177.
Zhang, T., Y. Chen, Y. Ge, et al. 2018. *Acta Pharma Sin B* 8 (3):440–8.
Zhao, L., Y. Ye, J. Li, et al. 2010. *J Pharm Pharmacol* 63 (1):80–6.

7 Phytochemical Nanoformulations for Neurological Disorders

*Madhavan Nampoothiri**
(Department of Pharmacology, Manipal College of Pharmaceutical Sciences, Manipal Academy of Higher Education, Manipal 576104, India, madhavan.ng@manipal.edu)

Prasada Chowdari Gurram
(Department of Pharmacology, Manipal College of Pharmaceutical Sciences, Manipal Academy of Higher Education, Manipal 576104, India, prasadachowdarigurram@gmail.com)

Sree Lalitha Bojja
(Department of Pharmacology, Manipal College of Pharmaceutical Sciences, Manipal Academy of Higher Education, Manipal 576104, India, lalitha.sb@manipal.edu)

Sairaj Satarker
(Department of Pharmacology, Manipal College of Pharmaceutical Sciences, Manipal Academy of Higher Education, Manipal 576104, India, satarkers@gmail.com)

7.1 INTRODUCTION

In view of a plethora of diseases emerging in humans, the need for therapeutic strategies that can effectively combat the diseases becomes essential. Neurodegenerative disorders are one of the most concerning diseases, as they affect the central nervous system (CNS), thus affecting the lives of millions of individuals. Some of the most commonly diagnosed neurological disorders, especially in the Indian population, are epilepsy, Alzheimer's disease (AD), Parkinson's disease (PD), cerebral palsy, stroke, anxiety, and depression, among others[1 2]. With an increase in disease progression, the development of effective therapeutics for neurological manifestations has also accelerated considerably, yet it has faced numerous roadblocks and challenges, particularly precise and effective targeted therapy to the brain. Two major barriers to the entry of drugs into the CNS are the blood–brain barrier (BBB) and the blood–cerebrospinal fluid barrier[3]. Because of these barriers, conventional oral or systemic formulations have lesser efficacy and exert higher toxicity because of their inability to cross the BBB in the treatment of neurological conditions[4]. Tracing back to the ancient times in terms of effective treatment strategies, especially in Asian countries such as India and China, plant-based medicines have remarkably contributed to the development of therapeutics. Medicinal plants such as *Ginkgo biloba*, *Hypericum perforatum*, *Piper methysticum*, *Valeriana officinalis*, and *Bacopa monnieri* are widely used for treating CNS disorders in India[5]. Therefore, to amplify their therapeutic effect, nanotechnology has been used to improve drug delivery and efficacy at

DOI: 10.1201/9781003231721-7

the desired level. Phytochemical nanoformulations possess a higher surface area-to-volume ratio and greater surface reactivity, which facilitate higher absorption of phytochemicals and accurate tissue-specific drug delivery. The nanoformulation delivery systems also improve systemic bioavailability of the drug together with reduced toxicity, which has made them an attractive area for drug development in neurological diseases[6]. Thus, paying more attention to the beneficial effects of these drug delivery systems is important to emphasize the role of phytochemical nanoformulations in the treatment of neurological diseases.

7.2 CHALLENGES IN DRUG DELIVERY ACROSS THE BLOOD–BRAIN BARRIER

Electrical signals are generated in neurons, and their conduction relies on the movement of ions across the neuronal membrane. The regulation of transmembrane gradients requires a balance in the ions present in the interstitial fluid of the brain[7]. Maintaining equilibrium in the cerebral environment is important to sustain normal brain function; this necessitates the precise regulation of cells, chemicals, and ion transition between the blood and the brain. A distinctive physiological barrier established in the brain, called "BBB," maintains such tight regulation. The BBB serves as a selective interface for the movement of compounds into the brain through active transport, passive diffusion, or facilitated diffusion[8]. In fact, the BBB does not function individually, but rather as a component of a larger multicellular neurovascular unit (NVU)[9]. NVU facilitates the brain's structural and functional multicellular interaction with blood vessels[10]. The distinct morphological connections among the cellular components of the NVU, which include endothelial cells, basement membrane, pericytes, astrocytes through their end feet, neurons, microglia, and oligodendrocytes, aid in generating a dynamic cerebral microenvironment (Figure 7.1)[11].

Molecules and ions are transported across the BBB through several mechanisms. Gases such as oxygen and carbon dioxide, as well as ions, diffuse easily through the BBB owing to a concentration gradient[11]. Furthermore, diffusion aids in the transfer of lipid-soluble compounds into the brain

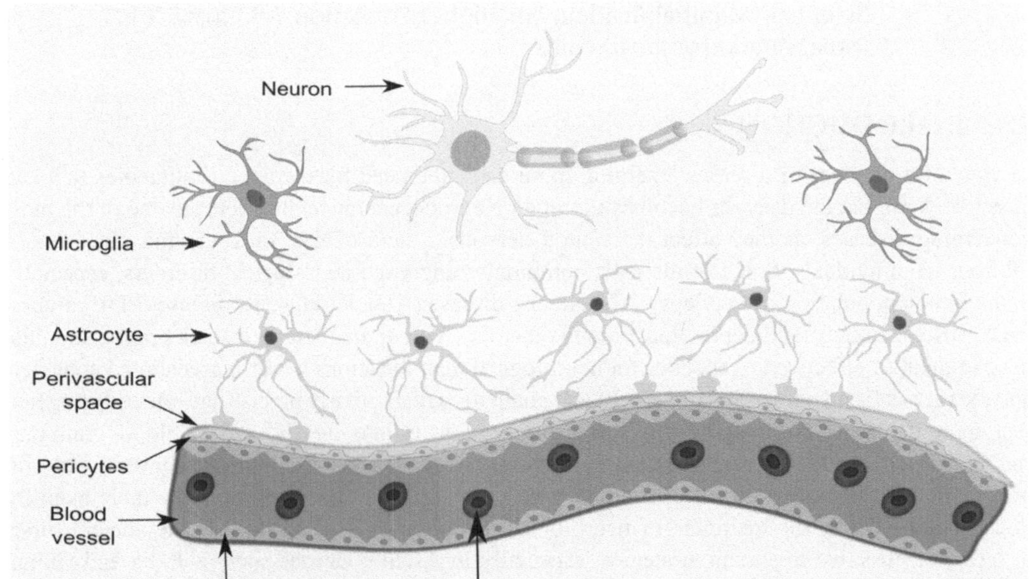

FIGURE 7.1 Components of the blood–brain barrier.

based on their ability to form hydrogen bonds (eight to ten hydrogen bonds), and their molecular weight specifically ranges from 400 to 500 Daltons[12]. These features, however, are not a complete indicator of CNS permeation, as many CNS-active medicines do not demonstrate these common features, and many solutes and medications have significantly lesser ability to enter into the brain than the expected ability depending on their lipophilic nature[13].

Various therapeutic approaches targeting specific regions of the brain across the BBB largely focus on the uneven distribution of transporters, modulation of transporters, and prevention or stimulation of the secretion of endothelial cells[4]. Furthermore, receptor-mediated transport mechanisms, mediated through the vesicular transport system, aid in transporting substrates through the BBB[14].

7.3 NANOTECHNOLOGY FOR DRUG DELIVERY TO THE BRAIN

The BBB provides resistance to the entry of therapeutic molecules into the brain, which makes pharmacological treatment for neurological disorders an unmet challenge. With advances in the field of nanotechnology, this issue can be solved. Nanoparticles (NPs), ranging in size from 1 to 100 nm, have recently emerged as an attractive tool, as they can overcome the unmet challenge of increasing drug transportation across the BBB[15]. Site-specific transport of medicines and other chemicals across the BBB using nanotechnological applications might be tailored to exert specific effects against neurological disorders. The drug forms an important component of the nano-engineered complex. The complex is designed to encapsulate the active drug, confer protection against enzymatic degradation, release the drug at a specific pH, cross the BBB, and target specific brain cells[16]. Neuroprotective phytochemicals derived from medicinal plants have beneficial effects on the nervous system owing to their antioxidant and anti-inflammatory properties. The bioavailability of herbal bioactive components in the human body is an important factor for their bioefficacy, as this aspect could be compromised by variables such as rapid metabolism, low permeability, and lack of drug stability in the CNS[17]. Liposomes, polymeric NPs, solid lipid nanoparticles (SLNs), micelles, dendrimers, and many other pharmacological nanocarriers may be used.

7.4 EPILEPSY

Epilepsy is the most common chronic neurological disorder that affects more than 70 million people globally (Thijs et al., 2019). It is a group of disorders commonly characterized by a predisposition to unprovoked seizures and cause neurobiological, cognitive, psychological, and social consequences (Fisher et al., 2017). Epileptic seizures are abnormal hypersynchronous bursts of neuronal activity. Clinical manifestations of epileptic seizures depend on the site and extent of seizures and on the mechanisms underlying the imbalance in the excitation and inhibition (Engel, 1995). Based on the etiology, the disorder can be classified into genetic, acquired (structural/metabolic), and idiopathic epilepsies. Although epilepsy is a complex disease, the pathophysiology is profoundly attributed to the imbalance between glutamate (excitation) and gamma-aminobutyric acid (GABA; inhibition) neurotransmission. Therefore, the majority of the currently used anti-seizure drugs (ASDs) rely on enhancing the GABAergic inhibitory drive (Greenfield & Jr., 2013).

Despite the availability of numerous ASDs, seizures cannot be controlled in 30% of patients with epilepsy, a phenomenon called "pharmacoresistant" or "drug-resistant" epilepsy (Löscher, 2020). Moreover, currently available drugs provide only symptomatic relief from the seizures, and no drug prevents disease progression, also called epileptogenesis. Furthermore, long-term use of ASDs causes several serious adverse effects including cognitive impairment. Hence, there exists an unmet need for developing anti-epileptogenic drugs and/or anti-epileptic drugs that have improved efficacy and can cause only fewer side effects (Bojja et al., 2021).

Traditional medicine, particularly phytomedicine, has been used for epilepsy treatment for centuries and is still being applied to date, and it remains a major fallback for such chronic diseases. Our history shows a plethora of medicinal plants such as *Cannabis sativa*, *Withania somnifera*, *Valeriana officinalis*, *Bacopa monnieri*, *Datura stramonium*, and *Uncaria rhynchophylla*, all of which are empirically used to treat epilepsy (E et al., 2019; Lin & Hsieh, 2021). These phytoconstituents have demonstrated antioxidant, neuroprotective, and GABA-promoting effects (Lin & Hsieh, 2021). Despite the superior efficacy of these phytomedicines, the clinical translation was unsuccessful. This may have been due to several pharmacological, pharmaceutical, and analytical challenges. The presence of several phytoconstituents, multi-targeted action, lack of scientific evidence, and detailed mechanism of action question their specificity. Additionally, many of these phytomedicines pose pharmaceutical challenges, as they do not fit into the conventional dosage forms, and specialized delivery strategies are needed for the intended drug bioavailability. Moreover, crossing the BBB and effectively reaching the CNS is a highly challenging task. In this regard, nanotechnology offers a robust and promising delivery approach to overcome these challenges by providing controlled drug release, enhanced surface area, protection by encapsulation, enhanced stability, and prolonged circulation time, thereby improving drug bioavailability at the target site. Various nanostructured drug delivery systems such as NPs, nanoemulsions, nanostructured lipid carriers (NLCs), liposomes, SLNs, and nano hydrogels have been successfully used for effective drug delivery (Bonifácio et al., 2014). In the following section, we discuss important studies that utilized nanotechnology-based targeted delivery of phytoconstituents for the treatment of seizures/epilepsy.

7.4.1 Polymeric Nanoparticles

Polymeric NPs are gaining increasing attention, as they provide robust stability, flexibility for surface modification, and nontoxicity. They enable prolonged duration of action, control of the drug release rate, and increased drug loading. Polymeric NPs consist of a biodegradable polymer matrix made of synthetic polymers [poly(D,L-lactide), poly(D,L-glycolide)], copolymers [poly(lactide-co-glycolide) (PLGA)], or natural biopolymers (chitosan, alginate, hyaluronic acid [HA], albumin). The polymer matrix can encapsulate hydrophilic and hydrophobic drugs, which makes it an excellent choice for drug delivery (Gagliardi et al., 2021).

7.4.1.1 Poly(lactide-co-glycolide) Nanoparticles

PLGA NPs have been widely used for the delivery of several bioactive molecules with poor aqueous solubility and instability (Cano et al., 2018). For example, chrysin derived from passion flower is a bioactive molecule with antioxidant, anti-inflammatory, and hypolipidemic properties. However, its therapeutic ability is limited by its poor aqueous solubility and low bioavailability. Chrysin-loaded PLGA NPs exhibited anticonvulsive and neuroprotective effects in a pentylenetetrazole (PTZ)-kindling mouse model. In this study, chronic treatment with the chrysin nanoformulation inhibited seizure-induced oxidative stress and neuronal apoptosis by upregulating nuclear factor erythroid 2–related factor-2 (Nrf2), heme oxygenase-1 (HO-1), and NAD(P)H quinone oxidoreductase 1 gene expression (Zhang et al., 2021).

A similar approach led to the preparation of epigallocatechin gallate (EGCG)-loaded PLGA NPs, as the physical structure of the drug is highly unstable. EGCG obtained from green tea is a well-known antioxidant that possesses antiepileptic activity, as demonstrated in several animal models. Upon PEGylation, EGCG-loaded NPs have shown enhanced stability (over 6 months upon storage at 4°C) compared with EGCG NPs and free drug. The formulation exhibited increased seizure protection compared with free EGCG by significantly decreasing the duration and severity of seizures, glial activation, and neuronal death in a kainic acid model of temporal lobe epilepsy (TLE). This finding can be attributed to the enhanced drug stability achieved by encapsulation into NPs and increased BBB permeation due to PEGylation (Cano et al., 2018).

Polymeric NPs are also advantageous in terms of achieving CNS bioavailability, as they allow intranasal delivery of drugs to the brain by avoiding the biodegradation of the drug and drug efflux by p-glycoproteins. Recently, intranasal delivery of herbal drugs such as thymoquinone (TQ) and catechin hydrate formulated with PLGA NPs has shown enhanced CNS bioavailability and seizure protection as compared with their free drugs in acute seizure models (Ahmad, Ahmad, al Qatifi, et al., 2020; Ahmad, Ahmad, Alrasheed, et al., 2020).

7.4.1.2 Natural Biopolymer Nanoparticles from Chitosan

In addition to PLGA, chitosan has received major attention owing to its mucoadhesive nature, thus emerging as a reliable drug carrier administered through the mucosal route of drug delivery. These NPs are biodegradable and biocompatible and have low toxicity. Chitosan NPs can be designed by surface modifications, thereby decreasing the burst release of drugs, enhancing the permeation and retention of NPs (Mohammed et al., 2017). Apart from chitosan, other biopolymers such as alginate, guar gum, and pectin are routinely used as mucoadhesive polymers by adhering to the mucus membrane and allowing the slow release of drugs from the formulations. Chitosan enables the transient and frequent opening of tight junctions in epithelial cells and promotes drug diffusion (Figure 7.2). In this regard, researchers have developed chitosan NPs from several herbal compounds such as madecassoside (MAD), piperine, curcumin, and quercetin and utilized them for targeting epilepsy (Anissian et al., 2018a; Firdous et al., 2021; Hashemian et al., 2017).

Centella asiatica is a phytomedicine commonly used for treating cognitive impairment and other neurological ailments including epilepsy (Visweswari et al., 2010). Among the multiple bioactive constituents of *Centella asiatica*, MAD is widely studied for its anticonvulsant, antioxidant, and anti-inflammatory activities. However, because of its poor oral bioavailability, intestinal absorption,

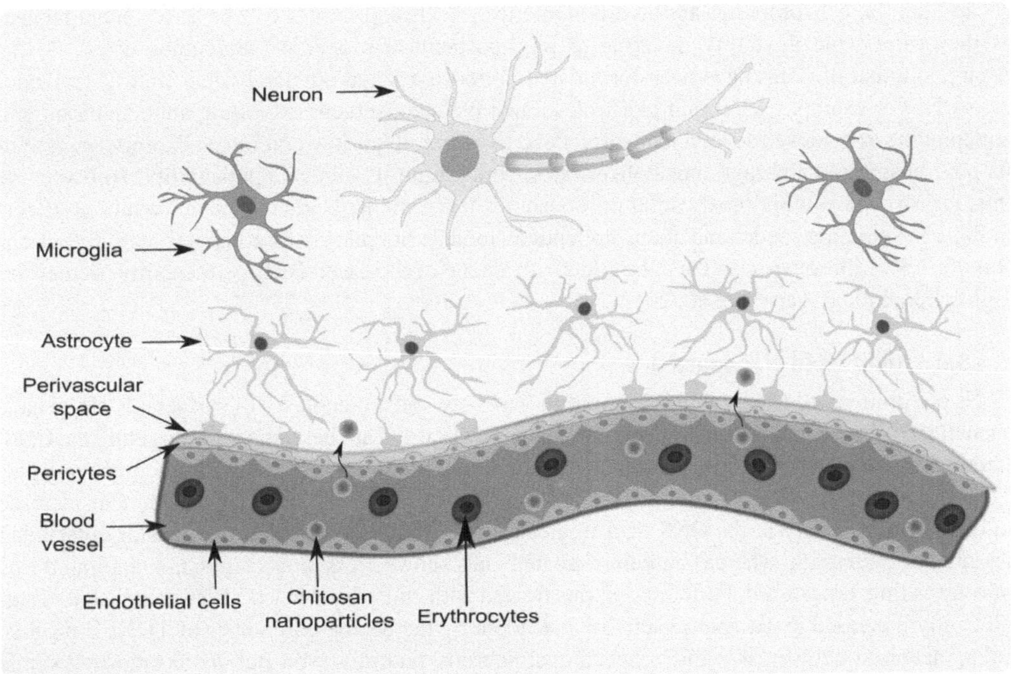

FIGURE 7.2 Entry of chitosan nanoparticles into the central nervous system by crossing the blood–brain barrier.

and cellular penetration, their clinical utilization is unsuccessful. To overcome these issues, recently, Kunjumon et al. (2021) encapsulated MAD in alginate chitosan NPs using the ionic-gelation polyelectrolyte technique. These NPs enhanced the bioavailability and cell permeation of MAD, thereby resulting in enhanced anticonvulsant, anti-apoptotic, and restoration of BBB integrity in a pilocarpine model of TLE.

7.4.2 Quantum Nanodots

Quantum nanodots have become a hotspot in nanotechnology-based drug delivery systems, particularly in tumor therapeutics. They are a robust platform for target-specific drug delivery and decrease the toxic systemic exposure of drugs. They provide an increased surface area, improve drug stability, prolong drug retention time, and improve drug bioavailability (Zhao & Zhu, 2016). Recently, Zhu et al. developed copper oxide (CuO) quantum dots-coated HA/PLGA NPs for the targeted delivery of piperine for epilepsy treatment. Piperine, a natural alkaloid obtained from black pepper, has received immense attention owing to its antiepileptic, anticancer, anti-depressive activity. However, piperine has poor solubility and bioavailability, which restrict its therapeutic exploitation. Encapsulating CuO in HA/PLGA microspheres, followed by loading them with piperine improved drug entrapment by 20-fold higher than that of the PLGA microspheres without HA and Cu-quantum dots structures. Piperine-loaded CuO-quantum dots-coated HA/PLGA (CuQDs@HA/PLGA) nanomatrices have shown an improved anticonvulsant effect in the PTZ-kindling mouse model of epilepsy when compared with HA/PLGA and control (Zhu et al., 2020).

7.4.3 Lipid-Based Nanoformulations

7.4.3.1 Liposomes

Liposomes are the most commonly used targeted drug delivery systems. They comprise one or more lamellae, comprising a phospholipid bilayer and a small volume of aqueous cavity. They can encapsulate both hydrophilic and hydrophobic drugs. The liposomes can be easily manipulated, as they offer wide flexibility in terms of physicochemical properties (Sercombe et al., 2015). A large number of drugs have been formulated in liposomes and efficiently delivered systemically or orally. For example, curcumin is a well-known phytoconstituent exhibiting anticonvulsant and antiepileptic activities in several animal models. However, it has low oral bioavailability because of its poor absorption and high metabolism rates, which limit its clinical applicability. To overcome this, Agarwal et al. developed curcumin liposomes that exhibited enhanced anticonvulsant effects in the PTZ-kindling model and status epilepticus models of epilepsy as compared with free drugs (Figure 7.3). Liposomes also have the added advantage of increased BBB permeability, as they are highly lipophilic (Agarwal et al., 2013).

7.4.3.2 Solid Lipid Nanoparticles

SLNs are lipid-based NPs with a solid lipid or wax core and a monolayer of surfactants. They have a small size, lipidic nature, and high entrapment efficiency and are best suited for targeting the CNS. Lately, SLNs have been employed to increase curcumin bioavailability in the CNS to treat epilepsy. The curcumin-loaded SLNs showed better dispersibility, stability, and more prominent increase in drug bioavailability in the CNS than free curcumin. The SLNs enabled the cellular uptake and retention of curcumin, whereas curcumin-treated cells showed less bioaccumulation and rapid loss of drug as time progressed. Furthermore, rats treated with curcumin-loaded SLNs demonstrated significantly increased antiapoptotic activity as evident by the decreased number of TUNEL-positive cells, increased number of viable hippocampal neurons, and improved Bcl-2/Bid ratio in a kainic acid model of epileptogenesis (Huang et al., 2020a).

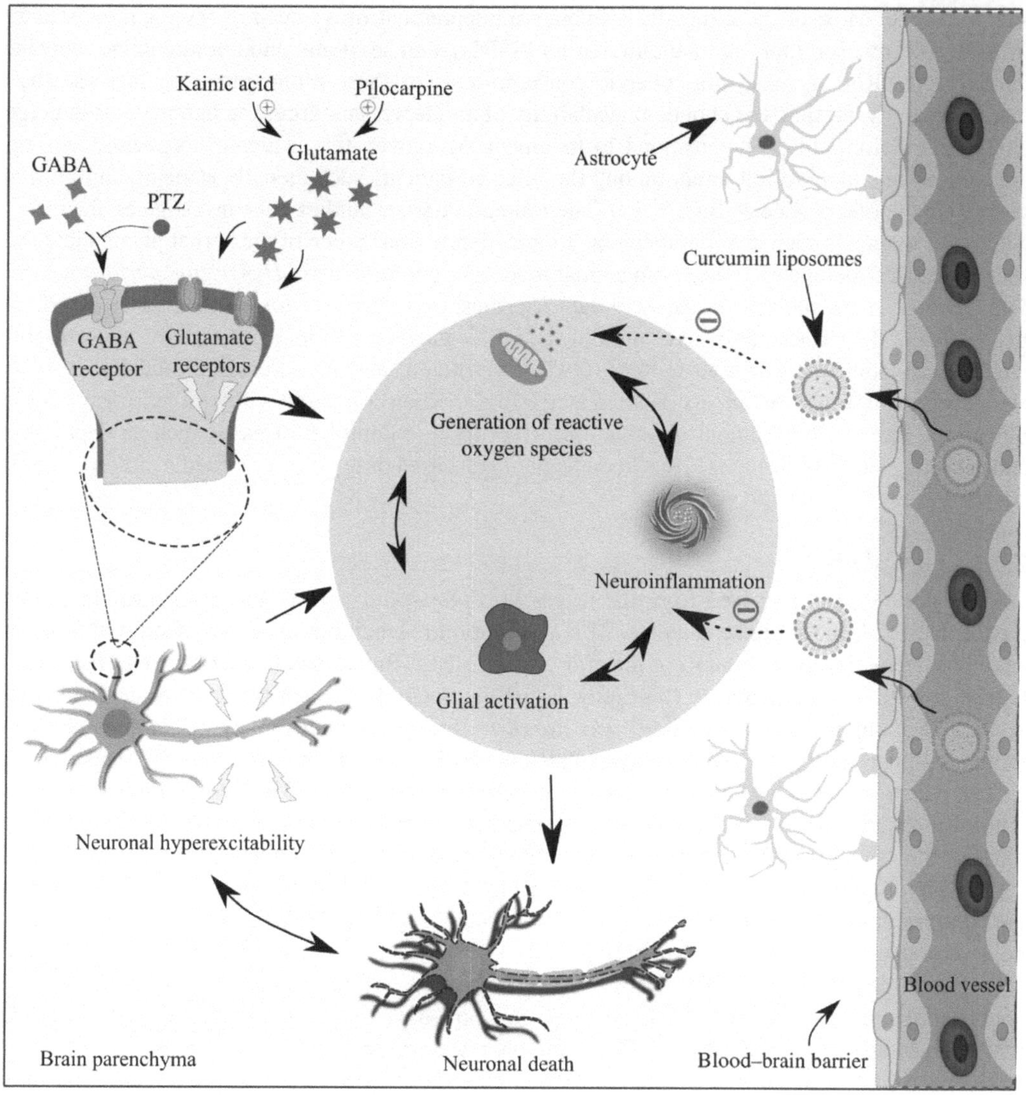

FIGURE 7.3 Therapeutic potential of curcumin liposomes in epilepsy.

7.5 DEPRESSION

The World Health Organization (WHO) describes depression as a mood disorder characterized by various symptoms including fatigue, loss of appetite, sleep, low self-esteem, anhedonia, and suicidal tendency (Jesulola et al., 2018). More than 350 million people globally are affected with depression, irrespective of age, gender, and race (Mossie et al., 2016). Depression is an emerging epidemic in the modern world and produces cognitive and socioeconomic consequences. As it is a heterogeneous and complex disorder, the exact cause and pathogenesis remain unexplored. Extensive research has indicated the contribution of genetic, environmental, and epigenetic associations to its etiology. To date, there are three hypotheses that explain the pathogenesis of depression: monoamine hypothesis, hypothalamic-pituitary-adrenal axis dysregulation, and genetic and environmental factors (Jesulola et al., 2018). In the monoamine hypothesis, the depressive symptoms are explained by the functional deficiency of the monoamine transmitters norepinephrine, serotonin, and dopamine. Most of the clinically available antidepressant drugs such as monoamine oxidase (MAO), selective serotonin reuptake inhibitors (SSRIs), and serotonin and noradrenaline reuptake inhibitors (SNRIs) increase the synaptic concentrations of these monoamines or increase their receptor activity. Despite the abundant availability of antidepressant drugs, certain patients undergo clinical remission and do not respond to treatment. Moreover, the available drugs produce several side effects that are intolerant among the affected patients such as parkinsonism, and blurred vision, among others (Lee & Bae, 2017). Therefore, numerous studies have investigated the use of alternative therapies such as phytomedicine to treat depression. Some of the herbal plants that have shown repeated therapeutic efficacy in animal models of depression are *Hypericum perforatum* (St. John's wort), *Valeriana officinalis*, *Lavandula* species, and *Curcuma longa* (Kakkar et al., 2016; Lee & Bae, 2017). These herbs have multiple activities including antioxidant activity, neurotrophic effect, modulation of neurotransmitters (GABA, serotonin, and noradrenaline) and thus elevate mood, serving as mood enhancers (Kakkar et al., 2016). When compared with synthetic drugs, these pleiotropic effects of medicinal plants are efficient to cure complex diseases such as depression and neurodegenerative diseases. Thus, nanotechnology-based drug delivery systems offer numerous advantages for targeting the CNS.

7.5.1 Phytosomes

The ethanolic extract prepared from the leaves of *Annona muricata* L. has been formulated into phytosomes and liposomes to increase BBB permeation (Mancini et al., 2018). Both phytosomes and liposomes carry high amounts of the drug and enable the functionalization of their surfaces with ligands for efficient targeting. In this study, a modified peptide derived from the receptor-binding domain of apolipoprotein E (mApoE) was linked to liposomes/phytosomes, which enabled them to mimic lipoproteins and undergo endocytosis into the BBB epithelium through low-density lipoprotein receptors, thereby resulting in their transportation across the BBB. Hydrophilic compounds with poor absorbability owing to their large size and gastrointestinal instability can be incorporated into liposomes and phytosomes. Interestingly, phytosomes are superior over liposomes in terms of carrying hydrophilic drugs. This is because, in phytosomes, covalent bonds are formed between the polar groups of the drug and the polar heads of the phospholipid bilayer, whereas in liposomes, polar compounds are physically entrapped in the aqueous compartment. *In vitro* Transwell experiments demonstrated enhanced BBB permeation by fourfold to ninefold after administering nanoformulations functionalized with mApoE. mApoE-phytosomal-AE reported maximal MAO inhibition by up to 45.1% at a dose of 60 μg/mL and increased free radical–scavenging activity. This increased activity was possible because of functionalization with mApoE, which conferred drug bioavailability in the CNS.

7.5.2 Liposomes

Low-dose piperine-encapsulated liposomes, when used for intranasal delivery of the drug, exerted antidepressant and cognitive-enhancing activities in male Wistar rats. Compared with untreated animals, both oral liposomal- and intranasal liposomal-treated rats showed low immobility time in forced swim test (FST) and enhanced learning and memory in Morris water maze test. Intranasal formulation showed a shorter time to reach the maximal concentration (T_{max}), indicating rapid onset of action (Priprem et al., 2011).

7.5.3 Solid Lipid Nanoparticles

Hibiscus rosa-sinensis is an ornamental plant widely grown in Asia and other parts of the world. It exerts antidepressant activity by modulating monoaminergic neurotransmitters. Vijayanand et al. developed *Hibiscus rosa-sinensis* extract-loaded SLNs using glycerol monostearate or beeswax as lipids and employed the emulsion-quenching technique. The formulation exerted superior antidepressant activity over free drug in FST and tail suspension test (TST) in male Swiss albino mice, which could be due to its enhanced bioavailability (Vijayanand et al., 2018). Similarly, in another study, curcumin SLNs exhibited improved bioavailability and CNS permeability and thus exerted higher antidepressant activity even at a single low dose in an FST mouse model (Kakkar & Pal Kaur, 2012).

7.5.4 Poly(Lactide-co-Glycolide) Nanoparticles

Curcumin is a medicinal herb well known for its antioxidant activity and is widely used globally for culinary purposes. It has potent MAO inhibitory activity with proven antidepressant activity[18]. However, as discussed in other sections, it has poor solubility and bioavailability and undergoes rapid metabolism. Hence, various formulations of curcumin have been verified for their therapeutic potential. Yusuf et al. prepared brain-targeted polysorbate-80–coated curcumin PLGA NPs (P-80-CUR-NP) using the solvent evaporation technique and assessed their activity in a stress-induced depression-like model. The low-dose (2.5 mg/kg) P-80-CUR-NPs showed similar protection as that of curcumin NPs (20 mg/kg) and free curcumin (20 mg/kg) in FST and TST; in fact, they improved antioxidant responses (superoxide dismutase and catalase). The study concluded that the NPs increased drug bioavailability in the CNS as evident by the higher curcumin concentration in the brain, which led to better efficacy (Yusuf et al., 2016).

7.5.5 Nanostructured Lipid Carriers

NLCs are lipid-based NPs with several advantages such as high stability, prolonged release, and targeted delivery. Silymarin NLCs were explored for their antidepressant effects in chronic mild stress-induced depressive-like behavior in mice. Silymarin NLCs showed a superior effect over fluoxetine (standard antidepressant) in terms of percentage weight gain, FST and TST results, and increasing prefrontal and hippocampal monoamine concentrations (norepinephrine and 5-hydroxytryptamine). Silymarin NPs have shown neuroprotective activities in terms of antioxidant and anti-inflammatory activities, which were not evident with fluoxetine treatment. The study demonstrated that the concentration of silibinin, the major active component of silymarin, in the brain was 12.46 times higher in silibinin nanoformulation–treated mice than in silymarin-treated mice (Ashraf et al., 2019).

7.5.6 Gold Nanoparticles

Hypericum perforatum is a widely used alternative therapy for mild-to-moderate depression. It is one of the top medicinal herbs used in the United States (US) and Europe. Prakash et al. developed gold NPs of the plant extract and investigated their strong antidepressant activity in acute immobilization stress models (6 h/d). Gold NPs of low-dose (20 mg/kg) *Hypericum perforatum* have shown superior efficacy in behavioral tests, T-maze, and mirror chamber tests over free drug (200 mg/kg). Both free drug and formulated drug exhibited similar antioxidant profiles by decreasing lipid peroxidation and increasing the activity of cellular antioxidant enzymes (Prakash et al., 2010).

7.6 ANXIETY

Anxiety is the most common psychiatric disorder that negatively affects an individual's daily life. It is characterized by a heightened emotional response to a future threat/danger or aversive conditions and manifests with symptoms such as irritability, fear, lack of concentration, palpitations, dyspnea, tremors, sweating, giddiness, and gastrointestinal disturbances. The WHO 2017 report states that approximately 260 million individuals globally were affected by anxiety (Friedrich, 2017). Most often, anxiety disorders co-exist with other disorders, predominantly with depression. The pathophysiology of anxiety largely remains unknown, as it is complex and relatively underexplored. The dysregulation of serotoninergic, dopaminergic, noradrenergic, GABAergic, and glutamatergic transmission is involved in the pathophysiology of anxiety. The pharmacological treatment available for anxiety constitutes antidepressants (SSRIs, SNRIs, tricyclic antidepressants, and MAO inhibitors) along with GABAergic drugs and benzodiazepines (Garakani et al., 2020). Caffeic acid (CA) also promoted the anxiolytic effect in a lipopolysaccharide-induced neuroinflammation mouse model[19]. However, long-term use of these drugs results in intolerance and severe side effects. Herbal drugs such as *Ginkgo biloba, Ziziphus jujuba* and *Melissa officinalis* are gaining popularity for the treatment of anxiety (Liu et al., 2015).

In this regard, Priprem et al. developed and explored the intranasal liposomal formulation of quercetin for its anxiolytic and cognitive-enhancing effects in male Wistar rats. Oral quercetin (300 mg/kg) produced a significant anxiolytic effect as evident by the increased total number of entries and time spent in the open arm. Furthermore, intranasal quercetin liposomes carrying a very minimal dose of drug (20 μg, which is 6000 times lesser than the oral dose) showed significant anxiolytic activity at par with the oral group. In addition, enhanced spatial learning and memory were observed to a similar extent with both oral quercetin and liposomal formulations. However, liposomes administered orally failed to produce the protective effect, indicating that intranasal delivery of liposomes contributed to enhanced CNS bioavailability. Additionally, the liposomal formulation effectively reduced dose loading and, thus, the possibility for toxicity (Priprem et al., 2011).

7.7 ALZHEIMER'S DISEASE

AD can cause memory loss, although the disease severity varies from one person to another. Symptoms such as spatial memory impairment, thinking difficulties, judgment problems, and impaired reasoning are signs of AD in its early stages[20]. AD is primarily caused by synaptic degeneration due to the amyloid beta (Aβ) peptide[21]. Amyloid pathogenesis begins with secretases that breakdown the amyloid precursor protein, an integral protein present on the plasma membrane, to generate Aβ oligomers. The Aβ protein oligomerizes, penetrates synaptic clefts, and disrupts neuronal signaling[22 23]. This protein polymerizes into insoluble amyloid fibrils, which clump together to form plaques. This polymerization leads to kinase activation, resulting in hyperphosphorylation of the tau protein and formation of neurofibrillary tangles (NFTs). The key players in AD progression are Aβ and NFTs. Aβ plaques and NFT aggregation promote the recruitment of microglia around the Aβ plaques, contributing to toxicity in the neuronal tissue and leading to neuroinflammation[24–26]. Phenylpropanoids can reduce NLRP3 inflammasome–mediated neuroinflammation[27]. Several receptors such as N-methyl D-aspartate receptor (NMDAR), ephrin type B receptor 2 (EphB2), Toll-like receptor 4 (TLR4), ephrin type A receptor 4 (EphA4), cellular prion protein (PrPc), and leukocyte immunoglobulin-like receptor B2 (Lilrb2) can cause synaptic toxicity, although the process driving Aβ oligomer-dependent synaptic dysfunction remains unknown. These receptors are activated when Aβ binds to them, causing neurotoxicity[28 29]. The US Food and Drug Administration (FDA) approved the acetylcholinesterase inhibitors rivastigmine, donepezil, and galantamine, and the NMDAR antagonist memantine provides only symptomatic relief and confers toxicity on long-term use[30].

Phytochemicals have anti-Alzheimer's activity. However, most of these phytochemicals have failed to treat patients with AD in clinical settings owing to their lack of selectivity, bioavailability to the brain areas, and toxicity. NPs with a particle size of <200 nm reach brain areas more efficiently and have shown remarkable results at the preclinical stages in the delivery of the phytochemicals for treating neurological disorders including AD.

7.7.1 Polymeric Nanoparticles

PLGA is a synthetic copolymer prepared from glycolic acid and lactic acid approved by the US and European regulatory authorities. PLGA NPs are nontoxic and provide a range of choices such as better drug solubility, resistance to enzyme degradation, biocompatibility, higher targeting efficiency, BBB permeability, and improved cellular internalization[31]. Curcumin nanoformulation administered orally to transgenic AD mice (Tg2576) showed a prominent effect even at low concentrations compared with disease controls in cue recognition and working memory. Curcumin encapsulated into PLGA NPs promoted the differentiation and proliferation of rat hippocampal neural cells[32 33].

As a vital component of ginseng, ginsenoside Rg3 improves memory and cognition. Ginsenoside Rg3 and thioflavin T encapsulated into biodegradable PLGA NPs improve ginsenoside administration to the brain by boosting transport across the BBB, and they exert neuroprotective effects and decrease amyloid plaque and the consequent neurodegeneration (Aalinkeel et al., 2018). For efficient intranasal delivery, Huperzine-A was loaded into lactoferrin-conjugated N-trimethylated chitosan surface-modified PLGA NPs (Lf-TMC NPs), and the NPs promoted the distribution of Huperzine-A throughout the brain, with drug targeting index values of 2.0, 1.6, 1.9, and 1.9 in the mouse olfactory bulb, cerebrum, cerebellum, and hippocampus, respectively[34]. Peptides and proteins conjugated/loaded in NPs are considered as promising therapeutic components in neurological disorders. Basic fibroblast growth factor encapsulated into *Solanum tuberosum* lectin-PLGA NPs showed enhanced brain delivery by binding to N-acetyl glucosamine on the nasal epithelial membrane. These nanoformulations improved spatial learning and memory in rat AD models[35].

7.7.2 Inorganic Nanoparticles Targeting Alzheimer's Disease

Recently, Mandal et al. reported that at very low concentrations, biogenic gold NPs suppressed cholinesterase enzymes and slowed Aβ oligomerization. Furthermore, the addition of trehalose to chemically produced gold NPs prominently increasingly prevented protein aggregation and mature fibril disintegration, suggesting that it could be used for the treatment of neurodegenerative diseases (Mandal et al. 2017). Terminalia bark is known for its anti-inflammatory activity. The neuroprotective properties of biogenic gold NPs conjugated with *Terminalia arjuna* revealed biogenic gold NPs' substantial neuroprotective properties and biocompatibility[36] (Jitta et al., 2019).

EGCG, a polyphenol present in tea, was attached to the surface of selenium NPs and coated with Tet-1 peptide. The formulation reduced the cytotoxicity of EGCG at high doses and impeded Aβ fibrillation. Furthermore, it promoted the disaggregation of toxic amyloid fibrils to nontoxic aggregates[37].

7.7.3 Lipid Nanoparticles Targeting Alzheimer's Disease

7.7.3.1 Liposomes

Targeted nanoliposomes are a practical and promising drug delivery method for AD and is yet to be tested in humans. They are biocompatible, adaptable, and capable of transporting a wide range of therapeutic chemicals across the BBB and into brain cells. There are two types of liposomes: "nontargeted" liposomes, which can direct drug transportation, and "targeted" liposomes, which are designed to conjugate with targets relevant to AD therapy. Liposomes can include either hydrophilic or hydrophobic molecules combined with several substances that can provide therapeutic benefits while also promoting BBB penetration[38–40]. They can be adjusted to lengthen their circulation in blood and focused toward single or multiple disease targets. Recently Ross et al. reviewed the importance of liposomes in AD[41].

7.7.3.2 Solid Lipid Nanoparticles

SLNs have emerged as a potential drug carrier system that can deliver drugs across the BBB to reach the target site in the brain[42]. Vakilinezhad et al. showed that oral administration of 120-nm-sized SLNs containing nicotinamide, a histone deacetylase (HDAC) inhibitor, attenuated

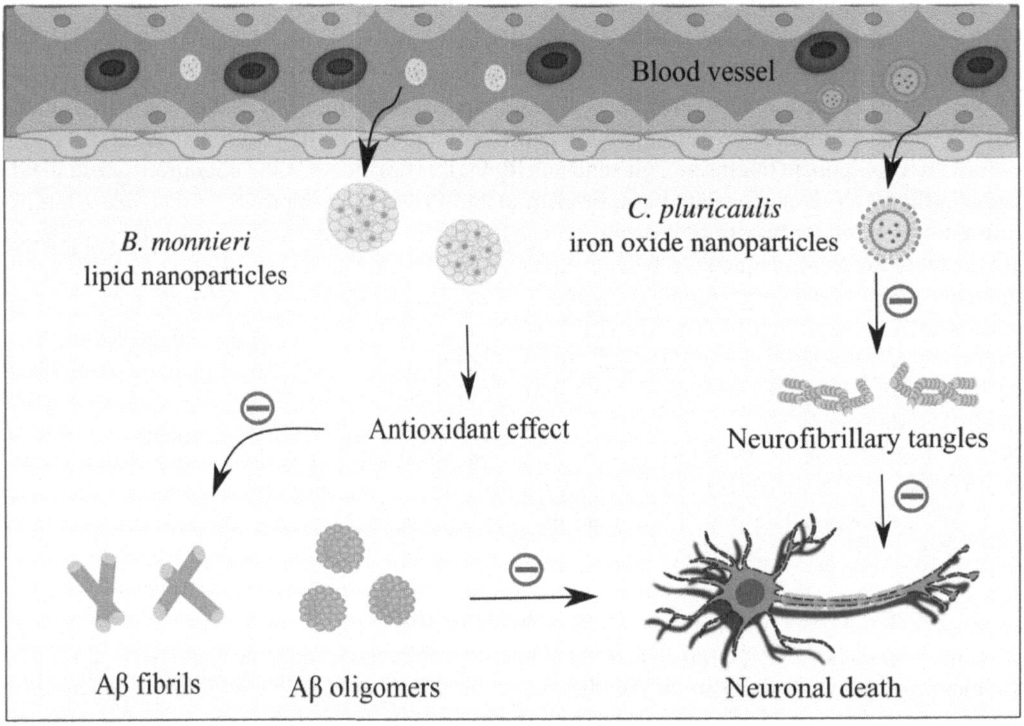

FIGURE 7.4 Use of phytochemical nanoformulations for treating the pathologies of Alzheimer's disease.

cognitive deficit in a streptozotocin (STZ)-induced AD rat model, and the SLNs showed less toxicity in SH-SY5Y cell line[43]. Chitosan-coated SLNs loaded with ferulic acid (FA) augments cognitive deficit in a STZ-induced AD rat model[44]. The crude extracts of *Bacopa monnieri* could reduce Aβ in the brain[45]. This could be the reason why, recently, the SLNs of *Bacopa monnieri* have shown to possess neuroprotective activity against AD[46]. Similarly, *Convolvulus pluricaulis* reduced tau aggregation and prevented neuronal death in AD[47]. This could be associated with improved cognition in rats that received an iron oxide nanoformulation of *Convolvulus pluricaulis* (Figure 7.4)[48].

7.7.3.3 Nanostructured Lipid Carriers

Curcumin-loaded NLCs have a greater drug entrapment, a longer release time, improved bioavailability, more stability, and improved potential in AD[49]. Sood et al. developed NLCs loaded with curcumin and donepezil for intranasal administration to the brain. When compared with intravenous treatment, intranasal administration resulted in a greater drug concentration in the brain. Similarly, administration of free medication demonstrated enhanced memory and learning in a mouse model. Nonetheless, after NLC treatment, acetylcholine levels were higher and oxidative damage was lower[50].

7.7.3.4 Self-Structured Nanoparticles Targeting Alzheimer's Disease

In an AD mouse model, quercetin NPs (NQC) attenuate Aβ formation, neurotoxicity, and improved spatial and episodic learning and memory deficit. NQC have a longer duration in systemic circulation and enhanced bioavailability in rats[51].

7.8 PARKINSON'S DISEASE

PD is one of the most common complex neurodegenerative disorders linked to the degeneration of dopaminergic neurons and occurrence of Lewy bodies[52 53]. It is the second largest neurological disorder after AD. According to the 2016 statistics report, 6.1 million people experience this disease worldwide[52]. Tremor, troubled gait, rigidity, postural instability, and akinesia are the classical motor signs of PD[54,55]. In fact, some of the nonmotor symptoms include cognitive disturbances, autonomic dysfunction, gastrointestinal and olfactory dysregulation, sleep difficulties, depression, and protein alteration, which control synaptic transmission in the substantia nigra, cingulate, and prefrontal cortex and may also accompany the disease[56]. The loss of dopaminergic neurons in the substantia nigra is a pathological feature of PD, wherein excitatory D1-type and inhibitory D2-type dopamine receptors regulate motor activity in the extrapyramidal system[54]. Accumulation of α-synuclein particles, commonly known as Lewy bodies, is a characteristic feature of PD. The build-up of misfolded α-synuclein particles in the basal ganglia causes the onset of motor symptoms in PD, followed by the involvement of motor circuits[57]. The basal ganglia, which includes the pars reticulata of the nigrostriatal pathway and globus pallidus of the ventral striatum, is linked to circuits in the thalamus and cortex. In fact, the multifunctional facet of the dopaminergic system is related to three pathways in terms of neural functions: nigrostriatal, mesolimbic, and mesocortical pathways. In the nigrostriatal pathway, dopamine exerts an important role in motor planning and execution[58]. In patients with PD, α-synuclein exists as lesser soluble aggregates that combine with other proteins[59]. α-Synuclein accumulation at presynaptic locations in the cytoplasm within the cell leads to impairment in synaptic transmission. Interestingly, this protein might have a function outside the CNS and may be responsible for nonmotor symptoms[60]. Recent data suggest that peripheral immune cells, particularly T lymphocytes, have a role in PD development. These cells invade the CNS and concentrate in the nigrostriatal pathway and produce pro-inflammatory cytokines, thereby activating the immune system and killing dopaminergic neuronal cells[61].

Currently available drugs provide only symptomatic relief, and drugs that can either restore dopamine levels in the striatum or act on striatal postsynaptic dopamine receptors are needed[62]. The US FDA-approved dopamine analogs such as levodopa (L-DOPA) and carbidopa potentiate dopamine levels to aid in the transmission of impulses across brain regions that enable smooth deliberate motions such as shaving, eating, and writing. Phytochemicals found in food, such as EGCG, CA, theaflavin, chrysin, FA, TQ, ellagic acid, vanillin, and other plant-derived antioxidants (Asiatic acid and α- and β-asarone), have been investigated for their beneficial neuroprotective effects and better treatment efficacy in PD[63]. The BBB averts the neuronal uptake of phytochemicals[12].

Several nano-delivery systems containing naturally derived phytochemicals and antioxidants are effective in reducing chronic inflammation, oxidative stress, α-synuclein aggregation, and various signaling pathways involved in most aging-related neurodegenerative disorders, particularly in PD.

7.8.1 Lipid Nanoparticles

7.8.1.1 Liposomes

Liposomes are made of one or more phospholipid bilayers that have a size range of 10–1000 nm[64]. Targeted and long-circulating stable liposomes can be used to avoid reticuloendothelial system absorption and off-target toxicities[65 59]. Liposomal formulations for various CNS-active medicines have been developed in recent years to provide a prolonged release effect by increasing L-DOPA concentration in the nigrostriatal system through more effective transportation. Liposomes are useful in PD treatment as demonstrated in a few clinical trials. Many L-DOPA dimeric prodrugs have been encapsulated in unilamellar liposomes and administered intraperitoneally. Such formulations showed an increase in baseline dopamine levels and sustained dopamine release in the CNS. Curcumin-liposomal formulations administered to transgenic rats inhibited HDAC, promoted anti-apoptotic activity, reduced α-synuclein, and ameliorated motor impairments in PD models[66]. Resveratrol liposomes for neuro-targeting are effective in PD treatment. *In vitro* studies revealed that liposomes enable sustained release of resveratrol, which is useful for the effective management of neuro-disorders including PD[67].

7.8.1.2 Solid Lipid Nanoparticles

SLNs have been used as a vehicle for targeted delivery of CNS-active medicines to treat various neurological diseases including PD in recent years[68]. Oral, ocular, topical, and parenteral delivery methods have all been investigated for SLNs. These methods are frequently used for encapsulating the agents used to treat PD owing to their particular benefits when compared with other delivery methods[69]. Resveratrol-loaded SLNs have the ability to transport resveratrol to brain cells and have an anti-PD effect[70].

7.8.1.3 Nanoemulsions

Nanoemulsions are stable formulations, wherein particles are of submicron size and are prepared from emulsions of oil and water, stabilized using emulsifiers such as surfactants and co-surfactants. These are tiny droplets with a large surface area, ranging in size from 20 to 200 nm[71].

Vitamin E is a powerful antioxidant found in natural products and can protect against PD when administered as a dietary supplement. Gaba et al. developed a naringenin nanoemulsion with vitamin E as the oil phase to transport nigercin directly to the CNS for a better effect in PD treatment[72]. Schisantherin-A nanoemulsion improved bioavailability tenfold as compared with normal Schisantherin-A suspension, and this pharmacokinetic property is attributed to the possible treatment of neurological disorders including PD[73]. Resveratrol and curcumin, two natural polyphenolic plant constituents, were combined into a mucoadhesive lipidic nanoemulsion including HA for transnasal administration in PD treatment[74].

7.8.1.4 Self-Emulsifying Drug Delivery Systems

Hydrophobic medicines have poor water solubility, which make oral administration of these medicines difficult. Self-emulsifying drug delivery systems (SEDDS) can overcome this issue[59]. Oral administration of SEDDS through gelatin capsules demonstrated a stable oil-in-water emulsion. Surfactants, and solvents are used together to improve the absorption of highly lipophilic medicines. The ability to demonstrate self-emulsification is determined by numerous parameters, including surfactant concentration, emulsion polarity, oil-to-surfactant ratio, and droplet charge, among others. To prepare effective self-emulsifying systems, it is necessary to examine particular excipient combinations[75]. In a PD model, rutein SEDDS with vitamin E reduced oxidative stress[76].

In various studies, CeO_2 NPs were synthesized using biological resources such as plants, microorganisms, and other biological products and evaluated for their neuroprotective effects in *in vitro* PD models[63 77]. Several other phytochemical-loaded NPs have been listed in Table 7.1.

TABLE 7.1
Phytochemical-Loaded Nanoparticles Used in Parkinson's Disease Models

S. No.	Nanocarrier	Loded Molecule/ Diameter/RA	Disease-Induced Model	Role of Nanocarriers in the Enhancement of Phytochemical Delivery	Therapeutic Outcome	Refs
1	Lactoferrin	Curcumin 100 nm	Rotenone-induced neurotoxicity in the dopaminergic cell line SK-N-SH	Lactoferrin enhanced BBB permeability and improved pharmacological activities	Curcumin reduces toxicity as seen in MTT and LDH assay and protects dopaminergic neurons	78 79
2	Lipid-based nanoparticles, especially glyceryl monooleate nanoparticles	Curcumin 60 nm Oral	α-Synuclein persuaded to PD in PC-12 and SH-SY5Y cell lines	Because of their lipophilic nature, biocompatibility, biodegradability, and nanoscale size, which allow them to easily penetrate the BBB's tight endothelial cells, lipid-based NPs have garnered increasing interest as a potential carrier system for CNS administration	Nanoparticles that can cross the BBB retained the rotenone-persuaded motor coordination improvement	80
3	Cerium oxide nanoparticles	CeO_2 210 nm i.p.	10 µg of 6-OHDA induced PD model in rats	CeO_2 has good catalytic properties; these are helpful for the transportation of nanomolecules	Maintains dopamine levels in the brain and decreases caspase-3 and MDA. Better motor results in OFT, Rotarod, stepping test, and initiation time. Partially retains 6-OHDA-induced toxicity	81 82
4	CeO_2 nanoparticles	CeO_2 25 nm	α-Synuclein persuaded to PD in yeast cell cultures		Restores mitochondrial damage and reduces the production of reactive oxygen species. Counteracts α-synuclein–persuaded toxicity	83
5	Liposome lipid: 1,2-dimyristoyl-sn-glycero-3-phosphocholine/ 1,2-dimyristoyl-sn-glycero-3-[phospho-rac-(1-glycerol)] (sodium salt)	Curcumin i.v.	Park 7 DJ knockout rat model of PD	1,2-dimyristoyl-sn-glycero-3-phosphocholine/ 1,2-dimyristoyl-sn-glycero-3-[phospho-rac-(1–glycerol)] conjugated with curcumin, promotes the transportation and delivery of curcumin	Liposomal-formulated curcumin inhibits histone deacetylase, attenuates apoptosis, reduces α-synuclein, and ameliorates motor impairments, with neurotrophic effects	66

(Continued)

TABLE 7.1 (Continued)
Phytochemical-Loaded Nanoparticles Used in Parkinson's Disease Models

S. No.	Nanocarrier	Loded Molecule/ Diameter/RA	Disease-Induced Model	Role of Nanocarriers in the Enhancement of Phytochemical Delivery	Therapeutic Outcome	Refs
6	Ferrosoferric oxide (Fe_3O_4)–modified liposomes	Resveratrol i.p. 124–163 nm	6-OHDA-induced PD in rats	(Fe_3O_4)–modified liposomes efficiently cross the BBB with accurate drug content for PD treatment	Sustained release of resveratrol liposomes for the effective management of PD	67
7	Nanoemulsions were prepared using peanut oil, olive oil, and other oils as carriers	Vitamin E–loaded narigenin Nasal 40 nm	6-OHDA induced PD in rats	Improved narigenin's neuro-targeted transportation and delivery	MDA levels decreased Increased GSH and SOD levels	72
8	Nanoemulsions were prepared solutol HS15 and soyabean oil	Schisantherin A Oral 230 nm	It is a comparative pharmacokinetic study in rats	Schisantherin A nanoemulsion improves bioavailability (4.3-fold to 47.3-fold) as compared to Schisantherin A	Nanoemulsion effectively alleviates PD symptoms and maintains neuronal homeostasis	73
9	Lipid nanocarrier Tween 80 as a surfactant and Transcutol P as a co-surfactant	Vitamin E–loaded resveratrol Intranasal 102 nm		Improves resveratrol transportation, delivery, and bioavailability	Scavenging efficiency estimated by the DPPH assay indicates enhanced levels of GSH and SOD, diminished levels of MDA, and attenuated PD pathology	84
10	SNEDDS Solutol HS 15–Transcutol P lipid nanocarriers	Rutein and Sefsol with Vitamin E 16 nm		Solutol HS 15–Transcutol P is a lipid-based drug delivery system that has been studied for its ability to improve the solubility and absorption of medicines that are poorly water soluble	Decreases the thiobarbituric acid reactive substances and acetyl cholinesterase and oxidative parameters. The antioxidant property of rutein attenuates PD pathogenesis	76

BBB: blood–brain barrier; LDH: lactate dehydrogenase; PD: Parkinson's disease; NP: nanoparticle; CNS: central nervous system; MDA: malondialdehyde; 6-OHDA: 6-hydroxy dopamine hydrobromide; SOD: superoxide dismutase; GSH: reduced glutathione; OFT: osmotic fragility test

REFERENCES

1 I. S.-L. D. B. I. N. D. Collaborators, *Lancet. Glob. Heal.*, 2021, **9**, e1129–e1144.
2 R. Nadig, U. S. Namapally, G. R. K. Sarma and T. Mathew, *Neurol. India*, 2019, **67**, 708.
3 J. M. Scherrmann, *Vascul. Pharmacol.*, 2002, **38**, 349–354.
4 L. A. Bors and F. Erdö, *Sci. Pharm.*, DOI:10.3390/scipharm87010006.
5 V. Kumar, *Phyther. Res.*, 2006, **20**, 1023–1035.
6 M. Dhupal and D. Chowdhury, *Int. J. Nanomedicine*, 2020, **15**, 9125–9157.
7 I. Galea, *Cell. Mol. Immunol.*, 2021, **18**, 2489–2501.
8 R. Daneman and A. Prat, *Cold Spring Harb. Perspect. Biol.*, DOI:10.1101/cshperspect.a020412.
9 H. L. McConnell, C. N. Kersch, R. L. Woltjer and E. A. Neuwelt, *J. Biol. Chem.*, 2017, **292**, 762–770.
10 A. H. Bell, S. L. Miller, M. Castillo-Melendez and A. Malhotra, *Front. Neurosci.*, 2020, **13**, 1452.
11 M. D. Sweeney, Z. Zhao, A. Montagne, A. R. Nelson and B. V. Zlokovic, *Physiol. Rev.*, 2019, **99**, 21–78.
12 W. M. Pardridge, *J. Cereb. Blood Flow Metab.*, 2012, **32**, 1959–1972.
13 N. J. Abbott, *J. Inherit. Metab. Dis.*, 2013, **36**, 437–449.
14 V. M. Pulgar, *Front. Neurosci.*, 2019, **13**, 1–9.
15 F. Re, M. Gregori, M. Masserini, et al., L. An and Y. Zhang, *Nanomedicine*, 2012, **8** Suppl 1, S51–S58.
16 G. A. Silva, *Curr. Res. Pharm. Technol.*, 2011, **9**, 81–84.
17 S. Z. Moradi, S. Momtaz, Z. Bayrami, M. H. Farzaei and M. Abdollahi, *Front. Bioeng. Biotechnol.*, 2020, **8**, 238.
18 S. K. Kulkarni, M. K. Bhutani and M. Bishnoi, *Psychopharmacology (Berl).*, 2008, **201**, 435–442.
19 J. Mudgal, S. Basu Mallik, M. Nampoothiri, M. Kinra, S. Hall, G. D. Grant, S. Anoopkumar-Dukie, A. K. Davey, C. M. Rao and D. Arora, *J. Funct. Foods*, 2020, **64**, 103638.
20 A. P. Porsteinsson, R. S. Isaacson, S. Knox, M. N. Sabbagh and I. Rubino, *J. Prev. Alzheimer's Dis.*, 2021, **8**, 371–386.
21 B. A. C. Karisetty, A. Bhatnagar, E. M. Armour, M. Beaver, H. Zhang and F. Elefant, *Front. Mol. Neurosci.*, 2020, **13**, 220.
22 J. X. Chen and S. S. Yan, *J. Alzheimer's Dis.*, DOI:10.3233/JAD-2010-100357.
23 L. Crews and E. Masliah, *Hum. Mol. Genet.*, DOI:10.1093/hmg/ddq160.
24 S. Tiwari, V. Atluri, A. Kaushik, A. Yndart and M. Nair, *Int. J. Nanomedicine*, 2019, **14**, 5541–5554.
25 J. Fernandes, J. Mudgal, C. M. Rao, D. Arora, S. Basu Mallik, K. S. R. Pai and M. Nampoothiri, *Toxicol. Mech. Methods*, 2018, **28**, 328–334.
26 B. Chowdhury, A. Sharma, S. Satarker, J. Mudgal and M. Nampoothiri, *J. Environ. Pathol. Toxicol. Oncol.*, 2021, **40**, 37–49.
27 M. Kinra, A. Joseph, M. Nampoothiri, D. Arora and J. Mudgal, *Eur. J. Pharm. Sci.*, 2021, **157**, 105637.
28 T. Guo, D. Zhang, Y. Zeng, T. Y. Huang, H. Xu and Y. Zhao, *Mol. Neurodegener.*, 2020, **15**, 1–37.
29 C. Hughes, M. L. Choi, J. H. Yi, S. C. Kim, A. Drews, P. S. George-Hyslop, C. Bryant, S. Gandhi, K. Cho and D. Klenerman, *Commun. Biol.*, 2020, **3**, 1–7.
30 K. G. Yiannopoulou and S. G. Papageorgiou, *J. Cent. Nerv. Syst. Dis.*, 2020, **12**, 117957352090739.
31 A. Cunha, A. Gaubert, L. Latxague and B. Dehay, *Pharm.*, 2021, **13**, 1042.
32 K. K. Cheng, C. F. Yeung, S. W. Ho, S. F. Chow, A. H. L. Chow and L. Baum, *AAPS J.*, 2013, **15**, 324–336.
33 S. K. Tiwari, S. Agarwal, B. Seth, A. Yadav, S. Nair, P. Bhatnagar, M. Karmakar, M. Kumari, L. K. S. Chauhan, D. K. Patel, V. Srivastava, D. Singh, S. K. Gupta, A. Tripathi, R. K. Chaturvedi and K. C. Gupta, *ACS Nano*, 2014, **8**, 76–103.
34 Q. Meng, A. Wang, H. Hua, Y. Jiang, Y. Wang, H. Mu, Z. Wu and K. Sun, *Int. J. Nanomedicine*, 2018, **13**, 705–718.
35 C. Zhang, J. Chen, C. Feng, X. Shao, Q. Liu, Q. Zhang, Z. Pang and X. Jiang, *Int. J. Pharm.*, 2014, **461**, 192–202.
36 N. Suganthy, V. Sri Ramkumar, A. Pugazhendhi, G. Benelli and G. Archunan, *Environ. Sci. Pollut. Res.*, 2018, **25**, 10418–10433.
37 J. Zhang, X. Zhou, Q. Yu, L. Yang, D. Sun, Y. Zhou and J. Liu, *ACS Appl. Mater. Interfaces*, 2014, **6**, 8475–8487.

38 S. Mancini, S. Minniti, M. Gregori, G. Sancini, A. Cagnotto, P. O. Couraud, L. Ordóñez-Gutiérrez, F. Wandosell, M. Salmona and F. Re, *Nanomedicine Nanotechnology Biol. Med.*, 2016, **12**, 43–52.

39 D. B. Vieira and L. F. Gamarra, *Int. J. Nanomedicine*, 2016, **11**, 5381–5414.

40 R. Dal Magro, F. Ornaghi, I. Cambianica, S. Beretta, F. Re, C. Musicanti, R. Rigolio, E. Donzelli, A. Canta, E. Ballarini, G. Cavaletti, P. Gasco and G. Sancini, *J. Control. Release*, 2017, **249**, 103–110.

41 C. Ross, M. Taylor, N. Fullwood and D. Allsop, *Int. J. Nanomedicine*, 2018, **13**, 8507–8522.

42 M. K. Satapathy, T. L. Yen, J. S. Jan, R. D. Tang, J. Y. Wang, R. Taliyan and C. H. Yang, *Pharmaceutics*, DOI:10.3390/pharmaceutics13081183.

43 M. A. Vakilinezhad, A. Amini, H. Akbari Javar, B. F. Baha'addini Beigi Zarandi, H. Montaseri and R. Dinarvand, *DARU, J. Pharm. Sci.*, 2018, **26**, 165–177.

44 S. Saini, T. Sharma, A. Jain, H. Kaur, O. P. Katare and B. Singh, *Colloids Surfaces B Biointerfaces*, 2021, **205**, 111838.

45 M. Dhanasekaran, B. Tharakan, L. A. Holcomb, A. R. Hitt, K. A. Young and B. V. Manyam, *Phyther. Res.*, 2007, **21**, 965–969.

46 R. Kumar, R. Garg and N. Khurana, *Curr. Bioact. Compd.*, DOI:10.2174/1573407216999201113121756.

47 S. Bihaqi, A. Singh and M. Tiwari, *Indian J. Pharmacol.*, 2012, **44**, 593.

48 L. P. Poka, K. M. G., V. R. K. and S. K., *Int. J. Pharmacogn. Phytochem. Res.*, DOI:10.25258/phyto.v9i5.8152.

49 M. Agrawal, S. Saraf, M. Pradhan, R. J. Patel, G. Singhvi, Ajazuddin and A. Alexander, *Biomed. Pharmacother.*, 2021, **141**, 111919.

50 S. Sood, K. Jain and K. Gowthamarajan, *Alzheimer's Dement.*, 2013, **9**, P299.

51 S. Palle and P. Neerati, *Bull. Fac. Pharmacy, Cairo Univ.*, 2017, **55**, 101–106.

52 B. R. Bloem, M. S. Okun and C. Klein, *Lancet*, 2021, **397**, 2284–2303.

53 A. N. MacMahon Copas, S. F. McComish, J. M. Fletcher and M. A. Caldwell, *Front. Neurol.*, 2021, **12**, 771.

54 R. J. Port, M. Rumsby, G. Brown, I. F. Harrison, A. Amjad and C. J. Bale, *J. Parkinsons. Dis.*, 2021, **11**, 715–724.

55 I. Ghosh, R. Sankhe, J. Mudgal, D. Arora and M. Nampoothiri, *Neuropeptides*, 2020, **83**, 102083.

56 J. C. Bridi and F. Hirth, *Front. Neurosci.*, DOI:10.3389/fnins.2018.00080.

57 L. Calo, M. Wegrzynowicz, J. Santivañez-Perez and M. Grazia Spillantini, *Mov. Disord.*, 2016, **31**, 169–177.

58 G. Drui, S. Carnicella, C. Carcenac, M. Favier, A. Bertrand, S. Boulet and M. Savasta, *Mol. Psychiatry*, 2014, **19**, 358–367.

59 S. Sharma, S. A. Rabbani, T. Agarwal, S. Baboota, F. H. Pottoo and R. Kadian, *Curr. Drug Metab.*, 2020, **22**, 287–298.

60 A. Bellucci, G. Collo, I. Sarnico, L. Battistin, C. Missale and P. Spano, *J. Neurochem.*, 2008, **106**, 560–577.

61 A. I. D. Thome, F. Atassi, J. Wang, A. Faridar, W. Zhao, J. R. Thonhoff, D. R. Beers, E. C. Lai and S. H. Appel, *npj Park. Dis.*, DOI:10.1038/s41531-021-00188-5.

62 A. Kouli, K. M. Torsney and W.-L. Kuan, *Park. Dis. Pathog. Clin. Asp.*, 2018, 3–26.

63 R. Balakrishnan, S. Azam, D. Y. Cho, I. Su-Kim and D. K. Choi, *Oxid. Med. Cell. Longev.*, DOI:10.1155/2021/6680935.

64 C. Zylberberg and S. Matosevic, *Drug Deliv.*, 2016, **23**, 3319–3329.

65 O. Medina, Y. Zhu and K. Kairemo, *Curr. Pharm. Des.*, 2005, **10**, 2981–2989.

66 S. Chiu, K. J. Terpstra, Y. Bureau, J. Hou, H. Raheb, Z. Cernvosky, V. Badmeav, J. Copen, M. Husni and M. Woodbury-Farina, *J. Complement. Integr. Med.*, 2013, **10**, 75–88.

67 M. Wang, L. Li, X. Zhang, Y. Liu, R. Zhu, L. Liu, Y. Fang, Z. Gao and D. Gao, *ACS Sustain. Chem. Eng.*, 2018, **6**, 17124–17133.

68 E. Esposito, M. Fantin, M. Marti, M. Drechsler, L. Paccamiccio, P. Mariani, E. Sivieri, F. Lain, E. Menegatti, M. Morari and R. Cortesi, *Pharm. Res.*, 2008, **25**, 1521–1530.

69 V. Mishra, K. K. Bansal, A. Verma, N. Yadav, S. Thakur, K. Sudhakar and M. J. Rosenholm, *Pharmaceutics*, DOI:10.3390/pharmaceutics10040191.

70 P. Ganesan, P. Ramalingam, G. Karthivashan, Y. T. Ko and D. K. Choi, *Int. J. Nanomedicine*, 2018, **13**, 1569–1583.

71 P. Savardekar and A. Bajaj, *IJRPC*, 2016, 2016, 312–322.

72 B. Gaba, T. Khan, M. F. Haider, T. Alam, S. Baboota, S. Parvez and J. Ali, *Biomed Res. Int.*, DOI:10.1155/2019/2382563.
73 F. Sa, B. J. Guo, S. Li, Z. J. Zhang, H. M. Chan, Y. Zheng and S. M. Y. Lee, *Parkinsons. Dis.*, DOI:10.1155/2015/951361.
74 M. Nasr, *Drug Deliv.*, 2016, **23**, 1444–1452.
75 J. Dubey and N. Verma, *Int. J. Pharm. Sci. Res.*, DOI: 10.13040/IJPSR.0975-8232.4(8).2893-99.
76 S. Sharma, J. K. Narang, J. Ali and S. Baboota, *Nanotechnology*, DOI:10.1088/0957-4484/27/37/375101.
77 M. Nadeem, R. Khan, K. Afridi, A. Nadhman, S. Ullah, S. Faisal, Z. U. Mabood, C. Hano and B. H. Abbasi, *Int. J. Nanomedicine*, 2020, **15**, 5951–5961.
78 S. Kumari, S. M. Ahsan, J. M. Kumar, A. K. Kondapi and N. M. Rao, *Sci. Rep.*, 2017, **7**, 1–13.
79 V. S. Bollimpelli, P. Kumar, S. Kumari and A. K. Kondapi, *Neurochem. Int.*, 2016, **95**, 37–45.
80 P. Kundu, M. Das, K. Tripathy and S. K. Sahoo, *ACS Chem. Neurosci.*, 2016, **7**, 1658–1670.
81 M. A. E. Hegazy, H. M. Maklad, D. A. Abd Elmonsif, F. Y. Elnozhy, M. A. Alqubiea, F. A. Alenezi, O. M. Al-Abbas and M. M. Al-Abbas, *Alexandria J. Med.*, 2017, **53**, 351–360.
82 C. Xu and X. Qu, *NPG Asia Mater.*, 2014, **6**, e90–e90.
83 R. Ruotolo, G. De Giorgio, I. Minato, M. G. Bianchi, O. Bussolati and N. Marmiroli, *Nanomaterials*, DOI:10.3390/nano10020235.
84 R. Pangeni, S. Sharma, G. Mustafa, J. Ali and S. Baboota, *Nanotechnology*, DOI:10.1088/0957-4484/25/48/485102.

8 Nanoparticle-Mediated Plant Bioactives for Cardiovascular Disease

Monali Kadu[a,c], *Rajesh Salve*[b,c], *Kavita Rai Gajbhiye*[d], *Ritesh Kumar Choudhary*[a,c*], *and Virendra Gajbhiye*[b,c*]

[a]Biodiversity & Palaeobiology Group, Agharkar Research Institute, Pune 411004, India

[b]Nanobioscience Group, Agharkar Research Institute, Pune 411004, India

[c]Savitribai Phule Pune University, Pune 411007, India

[d]Poona College of Pharmacy, Bharati Vidyapeeth (Deemed to be University) Pune 411038, India

8.1 INTRODUCTION

Cardiovascular diseases (CVDs) refer to a group of diseases affecting the heart and blood vessels. They comprise cerebrovascular disease, peripheral arterial disease, coronary heart disease, rheumatic heart disease, pulmonary embolism, deep vein thrombosis, congenital heart disease, and so on. More than a quarter of all deaths in low-income and middle-income countries are caused by CVDs. Nearly half of all deaths in high-income countries are due to CVDs, which could be prevented through early detection, treatment, and intervention [1]. CVDs have predominant when compared with cancer in terms of the leading cause of mortality in the 20th century [2]. According to the World Health Organization (WHO), every year, 17.9 million people die due to CVDs, which is approximately 32% of all deaths worldwide. Heart attacks and strokes cause 85% of all deaths due to CVD. More than 75% of deaths are reported in countries with low or middle income. The key mortality-related risk factors for CVDs in terms of lifestyle are unhealthy diet, tobacco use, physical inactivity, and alcohol drinking. The clinical risk factors are high blood pressure, high glucose levels, high lipid profile, overweight, and obesity [3]. These "intermediate risk factors" are associated with an elevated risk of stroke, heart attack, catastrophic cardiac events, and their related outcomes. Three primary causes of CVD-related morbidity and mortality are stroke, heart failure, and ischemic heart disease (congestive heart failure). These disorders account for at least 80% of the total burden due to CVDs across all socioeconomic areas despite the fact that they share a large number of risk factors. The obstructive nature of one of these arteries (ischemic stroke) or the rupture of another artery disrupts blood flow to a particular location (hemorrhagic stroke). Previous history of atrial fibrillation increases the risk of stroke. Additionally, blood flow disruption to the brain causes stroke [4].

After the age of 55 years, gender is a significant risk factor for stroke and CVDs. According to the US National Stroke Foundation, risk factors for the first stroke event and recurrence are comparable to those for CVD. Even though males are more likely to be susceptible to have a stroke than females, females are more likely to die due to stroke than males. According to the American Stroke Association, in both the first and recurrent stroke events, hypertension is the most important modifiable variable that may be addressed. The association between high blood pressure and stroke is stronger among East Asians than among Westerners. Other risk factors include smoking, exposure to second-hand cigarette smoke in the environment, dyslipidemia, arrhythmia (heart irregularity),

DOI: 10.1201/9781003231721-8

and reduced glucose tolerance [5–7]. Additionally, kidney dysfunction is a common complication of heart failure [8–14]. In various countries in the Middle East, North Africa, and Oceania, the incidence of obesity reached 50% in the adult population in 2013. In the general population, obesity is significantly and persistently linked to an increased risk of CVD incidence and mortality. The risk of CVD development in obese people is largely influenced by many factors [15].

According to the US National Heart, Lung and Cardiothoracic Association, CVDs, a medically fragile disease, can worsen if a person consumes too much salt in their diet. This issue can be resolved through careful treatment interventions targeting limited salt intake and thus prevent sodium retention and to halt overly active neurohumoral mechanisms simultaneously [8]. According to the WHO, giving up smoking, eating more fruits and vegetables, limiting salt intake, exercising regularly, and avoiding alcohol use may diminish CVD risk. Health policies involving healthy alternatives should be made available and affordable to promote healthy behaviors. Human diets often lack critical nutrients, which results in malnutrition. To combat this worldwide threat, crops developed for human use must be strengthened. As fortified foods include more vitamins and minerals than raw foods, they help to minimize the risk of heart disease [16-19]. However, applying phytonanotechnology to CVD treatment provides hope for imminent recovery in people affected by these diseases. Moreover, many plants have cardioprotective effects and are used for effectively treating heart diseases. For example, *Lagenaria siceraria*, known as bottle gourd and belongs to the Cucurbitaceae family, is an effective and potent cardioprotective agent, whereas *Lepechinia caulescens* (Island pitchersage) and *Pogostemon elsholtzioides* from the Lamiaceae family are used as vasorelaxant for treating CVD symptoms [20-22]. Moreover, *Ziziphora clinopodioides* (blue mint bush), *Melissa officinalis* (lemon balm), and *Sideritis raeseri* (mountain tea) have their utility in vasodilation [23–26].

The combination of plant bioactives and nanoparticles (NPs) has been largely investigated in recent years to treat life-threatening diseases. In the previous two decades, a wide range of nanotechnological applications for biomedicine have emerged, each with its distinct advantages. Nanocoatings, liposomes, dendrimers, and micelles are examples of such nanotechnological applications. Figure 8.1 depicts the diverse NPs used for diagnosing and therapeutic CVDs [27]. Plant bioactives loaded onto or inside the NPs have shown promising outcomes against CVDs. Characteristic features of NPs make them attractive vehicles for the delivery of bioactives. Nanotechnological applications can cure more than 50% of the total cases of CVDs. The ability of the nanotechnological applications to detect and treat CVDs in the early stage reduces the burden on the health care system. NP-mediated drug delivery can improve the therapeutic efficacy of a drug molecule and causes less severe side effects [28].

8.2 PHYTONANOTECHNOLOGY AND CARDIOVASCULAR DISEASES

Bioactive components derived from many plants can treat diverse diseases. Even today, scientists are investigating the possibility of employing bioactives from plants to identify new therapeutic molecules for the prevention and treatment of human diseases such as CVDs, diabetes, and cancer, among others [29]. Most nutraceuticals are derived from safe and well-tolerated natural components [30]. Numerous clinical studies are being conducted to determine the efficacy of curcumin in curing various illnesses, including cardiovascular irregularities and other health conditions [31].

Berberine, an active component isolated from some plants, enhances cardiac function in individuals with heart failure who have failed to respond to standard medical treatment with digitalis and diuretics. According to a study published in the *British Journal of Cardiology*, the improvement is likely due to peripheral vasodilation and inotropic stimulation. Berberine offers a wide range of cardiovascular advantages, including the prevention of ischemia-induced ventricular tachyarrhythmia, enhancement of cardiac contractility, and reduction of peripheral vascular resistance and blood pressure in patients with heart failure [32–34]. Lycopenes are naturally occurring

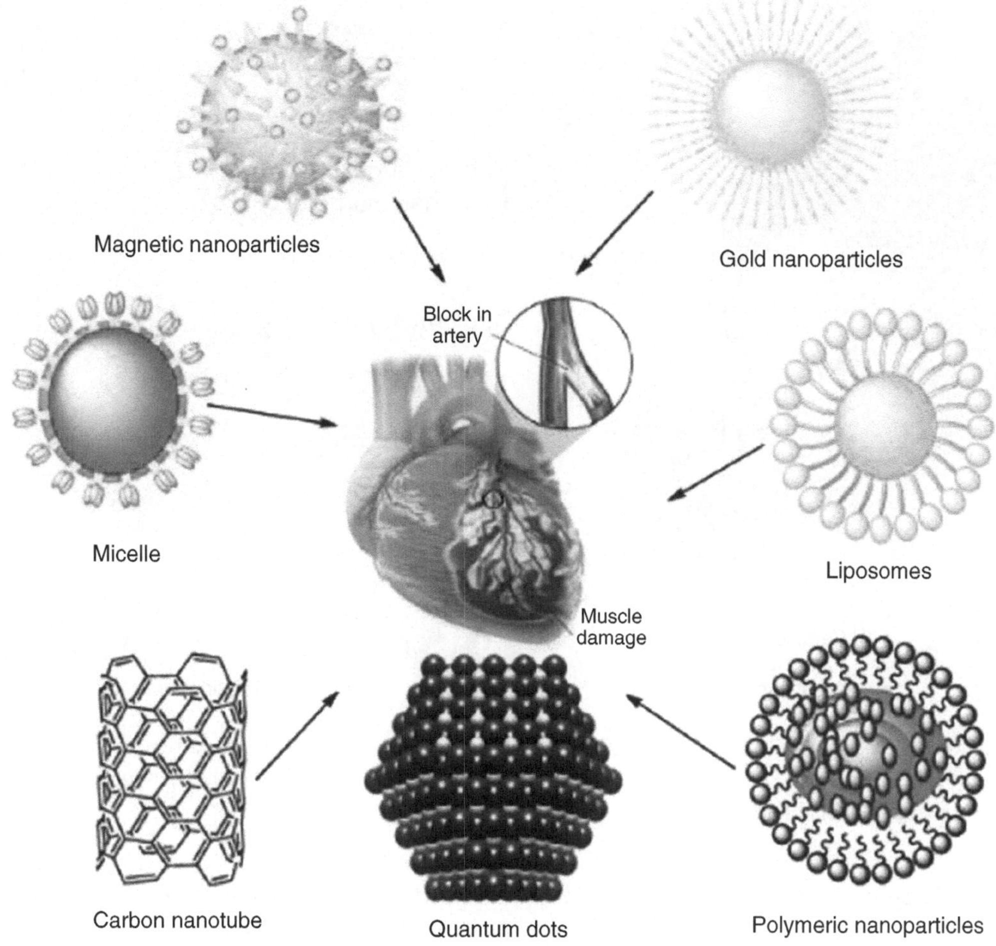

FIGURE 8.1 Nanoparticle-mediated drug delivery systems to target CVDs. The nanoparticle systems include magnetic NPs and AuNPs, self-assembling micelles, liposomes, carbon nanotubes, quantum dots, and polymeric NPs formed through the assembly of polymers containing therapeutic agents. (From Sabir, F., et al. Chemosensors. 2021 Mar 30;9(4):67. With Permission) [27]. This figure is reproduced under the terms and conditions of the Creative Commons Attribution (CC BY) license. AuNP: gold nanoparticle; CVD: cardiovascular disease; NP: nanoparticle.

compounds in plants and have a variety of therapeutic properties, including their protective effect against metabolic diseases such as diabetes and CVDs. Lycopene is derived from commonly available plants, and its use is safe and well tolerated, and it has a high concentration of antioxidants and flavonoids [30].

The pharmaceutical treatment for CVDs has spurred scientists to investigate about phytochemicals. Numerous phytochemicals are proven to have cardioprotective effects [29]. *Prunus cerasoides* and *Macrotyloma biflorum* consist of the secondary metabolite genistein, which is an effective cardioprotective agent; *Solanum tuberosum* (potato), belonging to the Solanaceae family, has bioactive compounds with cardioprotective properties [35, 36]. Grapes are particularly well suited for beverage production, most notably wine. Ripe grapes have a high concentration of phytochemicals, which have cardioprotective properties [37–40]. Numerous experimental investigations have shown

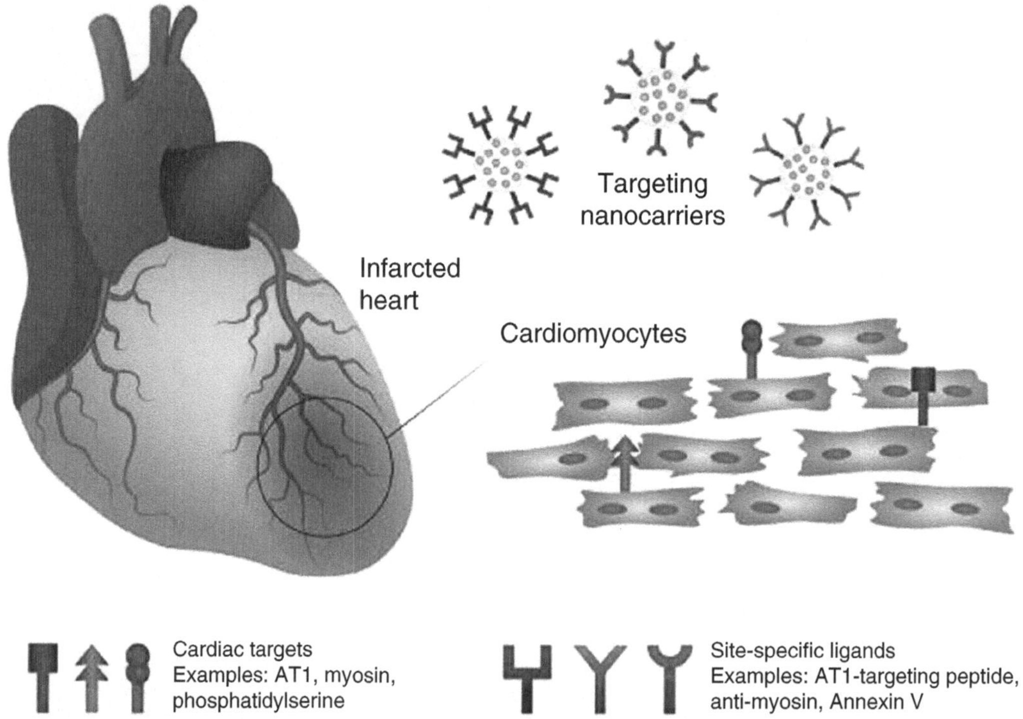

FIGURE 8.2 Schematic presenting potential cardiac targets of functionalized nanocarriers in an infarcted heart. (From Wang, D.K., et al. Nanomed. Nanotechnol. Biol. Med. 2021 June,1;34:102387. With Permission) [43]. Copyright (2021) Elsevier.

that grape polyphenols, which are mostly derived from red grapes, have cardioprotective properties and enhance endothelial function [41]. Apples, grape berries, olives, tomatoes, and mangoes all contain triterpenoids that have cardioprotective and antioxidant properties and may greatly postpone the onset of CVDs [42]. Resveratrol (3,4,5′-trihydroxy-trans-stilbene) inhibits the action of reactive oxygen species, prostaglandin E2, and cycloxygenase-2 (Cox-2) in cardiovascular cells. It is present naturally in grapes, peanuts, berries, and pine trees, as well as in red wine [29].

Plant-based NPs and bioactives-loaded synthetic NPs have promising therapeutic activity against diverse diseases. Plant-derived NPs may contain important bioactives isolated from medicinally active plants and contribute to enhanced therapeutic activity. The characteristics of NPs, such as small size, high cargo loading, biocompatibility, and so on, make them a prominent delivery vehicle for plant bioactives and other therapeutics. The field of NPs has had tremendous advances in the treatment of various cardiac diseases. Wang et al. reported that further research should be conducted on nano-risk factor therapy to obtain a better diagnosis and curative effect of CVDs. Predesigned NPs are routinely injected into the targeted artery or vasculature to avoid issues commonly encountered with traditional medication administration, such as systemic adverse effects [43]. It is well proven that targeted NPs bearing target cell-specific ligand molecules can better serve as therapeutic molecules in CVDs (Figure 8.2) [43].

8.3 PLANT BIOACTIVES IN CARDIOVASCULAR DISEASES

Numerous bioactive substances seem to have health-promoting properties. A sedentary lifestyle leads to insulin resistance, obesity, blood glucose levels, prothrombotic factors, and plasma lipids. A recent

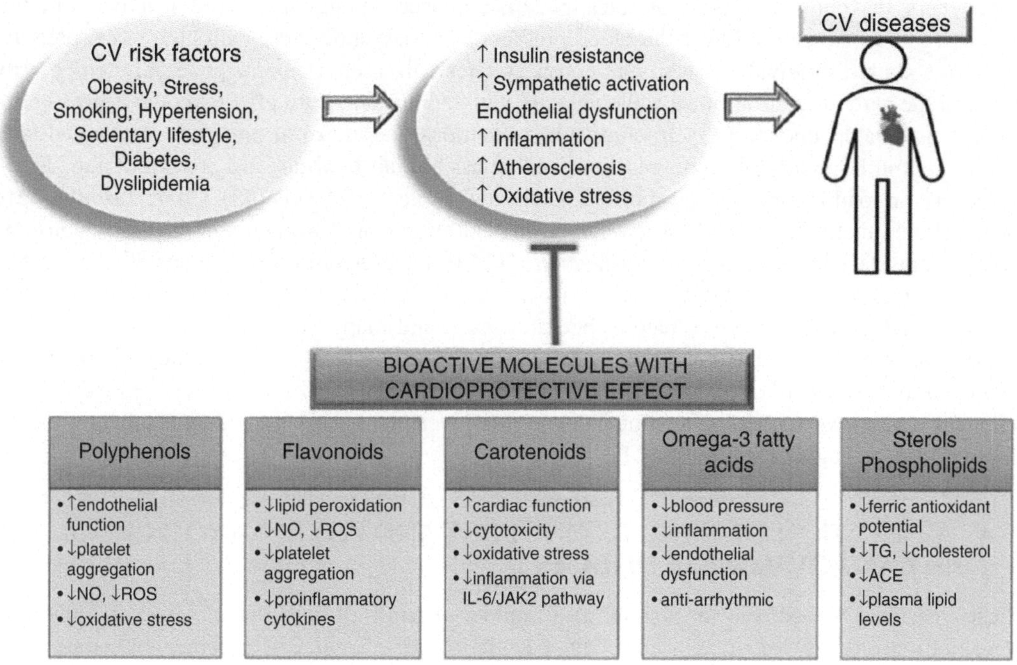

FIGURE 8.3 Schematic overview of CVD risk factors and pathophysiology, cardioprotective mechanisms, and effects of plant-based bioactive compounds. ACE: angiotensin-converting enzyme; CVD: cardiovascular disease; NO: nitric oxide; ROS: reactive oxygen species; IL-6: interleukin-6; JAK2: Janus tyrosine kinase 2; TG: triglyceride . (From Sharifi-Rad, J., et al. Int. J. Environ. Res. Public Health. 2020 Apr;17(7):2326. With Permission) [44]. The figure has been reproduced under the Creative Commons Attribution License.

study indicated that one-third of fatalities globally are attributable to CVD, and plant bioactives are active therapeutic molecules that are used to treat CVDs (Figure 8.3) [44].

Organosulfur compounds found in onion and garlic as well as monoterpenes found in citrus and cruciferous vegetables, herbs, and cherries have exhibited anticarcinogenic and cardioprotective effects in an experimental model. There is strong epidemiological evidence supporting their protective effects against numerous malignancies and CVDs, especially when these compounds are taken through diets containing high amounts of fruits, vegetables, legumes, and whole grains [45].

Digoxin, a steroid glycoside existing in high concentrations in *Digitalis lanata*, is the plant's principal active ingredient and one of the oldest and most commonly used medicinal herbs for the treatment of heart diseases. Moreover, several medicinal plants such as *Amaranthus viridis, Cichorium intybus, Daucus carota*, *Ginkgo biloba*, *Hydrocotyle asiatica*, *Mucuna pruriens*, *Nerium oleander*, *Terminalia arjuna*, and *Tinospora cordifolia* have cardioprotective properties [46]. The cardioprotective effects produced by plant bioactives are attributable to a variety of chemical compounds such as polyphenolic complexes, oligosaccharides, peptides, fatty acids, and vitamins. One of the most crucial classes of naturally occurring antioxidant, anti-inflammatory, and cardioprotective constituents is polyphenol, which exists in diverse plant parts including fruits, cereals, vegetables, seeds, and herbs. Additionally, studies have shown that foods with high levels of flavonoids can lower the risk of CVDs, including citrus fruits, products made from cocoa, black tea, and dark chocolate [44].

According to the WHO, in the United States, there is a demand for natural health products owing to their assistance in maintaining and preventing CVDs. Some of the clinical experiments associated with the cardioprotective properties of plant bioactive complexes were conducted by different researchers. In a clinical experiment, the angiotensin-converting enzyme inhibition activity of pomegranate juice was seen to lower the blood pressure of a male individual with high blood pressure [47]. In a study of individuals who had coronary artery disease, cranberry juice only temporarily reduced the carotid femoral pulse wave velocity and had no long-term effects on the heart's endothelial vasodilator function [48]. In clinical investigations, the intake of omega-3 unsaturated fatty acids, vitamin E, dietary fiber, magnesium, and potassium, all of which are present in nuts, led to a decrease in total low-density lipoprotein (LDL) and cholesterol levels [49]. Clinical studies have demonstrated that middle-aged, overweight, and healthy men and women who regularly consume whole grains have lower total cholesterol levels [50, 51]. Other natural compounds with the greatest clinical evidence of lowering the risk of possible CVDs include berberine, curcumin, green tea, garlic extract, lycopene from tomatoes, beetroot juice, and many others [52–55]. Sea buckthorn, *Hippophae rhamnoides*, has numerous beneficial therapeutic effects on the cardiovascular system and exhibits strong cardioprotective activity. Several *in vitro* and *in vivo* studies have revealed that *Hippophae rhamnoides* has therapeutic and preventive benefits against myocardial ischemia, oxidative damage, and aging [56] (Table 8.1).

8.4 APPLICATION OF NANOTECHNOLOGY AND PHYTONANOTECHNOLOGY IN CARDIOVASCULAR DISEASES

Heart disease is responsible for a significant number of fatalities worldwide; however, with early detection and effective treatment, some of these deaths may be averted. CVD treatment is hampered by its emphasis on oral medicines and invasive surgery. There is a notable gap in the research and development of medicines with the potential to enhance patient outcomes. Additionally, the regenerative nanomedicine treatment approach establishes a new paradigm for dealing with damaged heart tissue repair and the future use of nanotechnological systems [57]. Nanotechnology is becoming more applicable in cardiovascular pathology-related analytical, diagnostic, and therapeutic procedures. Continued breakthroughs in biomaterials, imaging, tissue-targeted NPs, personalized medicine, and biosensors offer the potential of nanomedicines to enhance patient care. Emerging nanotechnology-based drugs can help prevent atherosclerosis development and major adverse cardiovascular events. Nanotechnology-based treatments have the potential to help in the management of atherosclerosis progression as well as the prevention of serious cardiovascular events [58]. Some diagnostic and therapeutic applications of nanotechnology and phytonanotechnology are discussed in the following sections.

8.4.1 Diagnostic Application

Electrocardiography is a basic noninvasive technique for assessing cardiovascular health. Elevated levels of cholesterol, high-density lipoprotein (HDL), lipoproteins, and LDL are well-established indicators for CVDs. Additionally, interleukin-6, monocyte chemoattractant protein-1, fibrinogen, tumor necrosis factor-alpha, and C-reactive protein have also been used to diagnose CVDs [59]. Imaging of cardiovascular conditions provides detailed information about the state and progression of CVDs. Ultrasound, echocardiograms, positron emission tomography, computed tomography (CT), magnetic resonance imaging (MRI), coronary angiography, and single-photon emission CT, can be useful to diagnose CVDs (Figure 8.4) [60, 27]. Most of these techniques rely on a contrast agent for imaging. In the past few decades, NPs have emerged as a good contrast agent and contributed to a better understanding of the disease. Around the turn of the 21st century, the first reports of the use of NPs for molecular imaging of CVDs observed. To achieve this goal, NPs were tailored to contain a contrast-producing material in particular epitopes for diagnostic imaging [61-65]. Godin et al. examined the potential application of NPs in molecular imaging, which improved therapeutics and *in vivo* and *ex vivo* early detection. They may help in diagnosing and treating

TABLE 8.1
Cardioprotective Effects of Phytochemicals

Sr. No.	Secondary Metabolites	Class	Plants/Family of Plants	Activity	Refs
1.	Polyphenols	-	*Aristotelia chilensis*	Cardioprotective activity	[109, 110]
2.	Quercetin	Flavonoid	Onion, apple, wine, and tea	β-Hydroxy β-methylglutaryl-CoA activity was lowered, and cholesterol acyl transferase activity was increased in the liver, which reduced lipid peroxidation products	[111, 112]
3.	Catechin	Epicatechin, epicatechin-3-gallate, epigallocatechin, and epigallocatechin-3-gallate	Green tea, apple, cocoa, and berries	Multiple biological effects including antioxidative, antihypertensive, anti-inflammatory, antiproliferative, antithrombogenic, and lipid-lowering effects, contributing to its vascular-protective properties	[113, 114]
4.	Terpenoids	Monoterpenes, diterpenes, and sesquiterpenes	Lamiaceae, Rutaceae, Piperaceae, Zingiberaceae, Apiaceae, Myrtaceae, and Lauraceae. *Ginkgo biloba*	*In vitro* reperfusion of heart tissues in the presence of terpenes	[115]
5.	Resveratrol	Stilbene moiety	Dried roots of *Polygonum cuspidatum*, skin of grapes, berries, leaves of eucalyptus, and spruce	Resveratrol has the potential to stimulate autophagy and heart regeneration and may enhance a variety of health benefits, ranging from chemoprevention to cardioprotection	[116]
6.	Naringenin	Flavanone naringin	Citrus fruits such as grapefruit and orange	Improvement of the blood lipid profile	[117, 118]
7.	Flavonoids	Flavones, flavanols, flavanones, and isoflavones	*Scutellaria baicalensis*, *Vitis vinifera*, tea	Inflammatory diseases, hyperlipidemia, and arteriosclerosis	[119, 120]
8.	Carotenoids	Vitamin precursors are xanthophylls and carotenes. Lycopene and vitamin A precursor, beta-carotene, alpha- carotene, lutein, zeaxanthin, beta-cryptoxanthin	Tomatoes and carrots	Free radical scavenging, inhibition of lipid biosynthesis enzymes, chelation of transition metal ions, activation of endothelial nitric oxide, suppression of leukocyte adhesion to the endothelium, and inhibition of vascular smooth muscle cell proliferation are all mechanisms by which phytochemicals exert cardioprotective effects	[121, 122]
9.	Alkaloids	Propane alkaloids, purine alkaloids, acridone alkaloids, indole alkaloids, imidazole alkaloids, purine alkaloids, and morphine alkaloids	Solanaceae, Apocynaceae, Acanthaceae, and Papaveraceae	Cardioprotective benefits are shown by their lipid-lowering antioxidant and anti-inflammatory properties	[123-125]

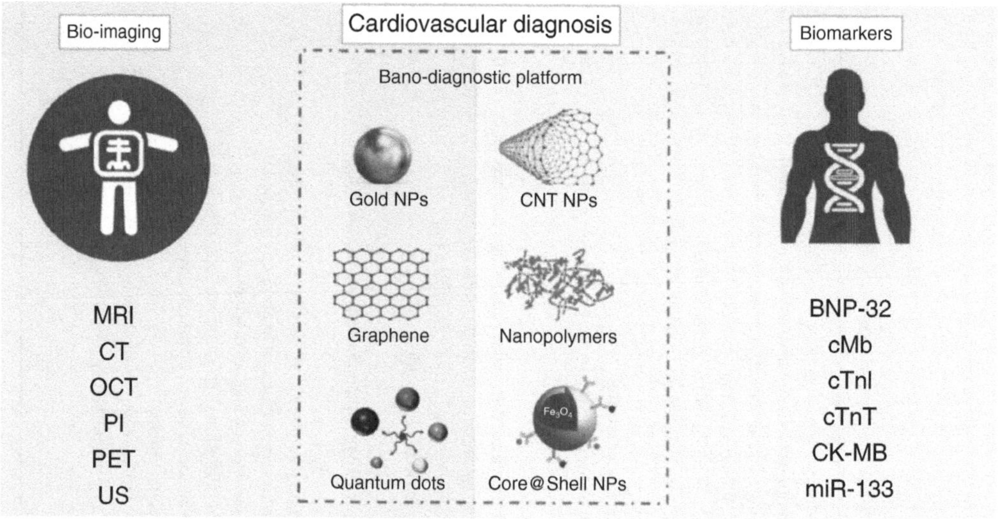

FIGURE 8.4 Diagnosis methods for CVDs (From Sabir, F., et al. Chemosensors. 2021 Mar 30;9(4):67. With Permission) [27]. This figure is reproduced under the terms and conditions of the Creative Commons Attribution (CC BY) license. CVD: cardiovascular disease.

atherosclerosis, heart attacks, and recurrent arterial disease [66]. In CVDs, gold NPs (GNPs) were used for photoacoustic imaging. When pulsed laser photons are used, GNPs induce thermal expansion of contrast chemicals or tissues. This pulsed extension produces acoustic emissions, which are subsequently detected by an ultrasonic transducer. By enabling their accumulation in diseased site-specific areas of the vasculature, atherosclerotic plaques have been visualized using GNPs, wherein they contain a gold nanoshell containing vascular cell adhesion molecule 1 [67]. For superior evaluations of the atherosclerotic plaque, gold nanorods that can be used for combined photoacoustic and ultrasound imaging have been used for *ex vivo* imaging of atherosclerotic plaques [68]. Another cardiovascular imaging method based on backscattered light from the target tissue is optical coherence tomography (OCT). GNPs have a high backscattering coefficient and are reported for OCT-based detection of individual cells in biocompatible fluids [69]. Because of its high mass absorption coefficient and atomic number, gold is a desirable candidate for CT applications and may perform better than conventional iodine-based contrast agents [70].

MRI is a prominent noninvasive technique that can effectively diagnose CVDs. To promote better atherosclerotic plaque accumulation and longer circulation, gadolinium (Gd), a contrast agent used for T1-weighted imaging, can be encapsulated in poly(lactic-co-glycolic acid) (PLGA)- or polyethylene glycol-modified polylactide NPs or HDL nanocarriers [71, 72]. Superparamagnetic iron oxide NPs have been extensively used as contrast agents for T2-weighted MRI. Dual-contrast agent-based imaging systems have been proposed to capture the effects of contrast agents for T1- and T2-weighted imaging. To reduce the possibility of producing false-positive and false-negative signals in *in vivo* experiments, iron oxide and gold core-shell (Fe3O4/Gd2O3) nanocubes have been created and reported to improve the contrast between negative and positive signals [73]. Furthermore, iron oxide NPs coated with gold have been embellished with antibodies specific to CD163, which is elevated in macrophages present near atherosclerotic plaques. These designed NPs had aggregated at the atherosclerotic lesions, as demonstrated in *in vivo* tests, which made them detectable by MRI [74].

Troponin T (cTnT)x and troponin I (cTnI) are potent biomarkers for CVDs and are released into the bloodstream when cardiomyocytes are damaged [75]. Hence, electrochemical immunosensors based on troponin detection are applied to diagnose CVDs. Accordingly, an amperometric sensor was created to detect cTnT utilizing the ionic organic compound (E)-4-[(4-decyloxyphenyl)diazenyl]-1-methylpyridinium iodide–modified GNPs stabilized with chitosan, which senses changes in electrical currents caused by oxidation or reduction in metabolic activities [76]. In this instance, the organic molecule served as a redox probe, and to detect cTnT in the sample, monoclonal anti-cTnT antibodies were mounted onto the GNPs. The analyte could be quantified because the signal decline was proportional to cTnT concentration. In another study, an artificial anti-cTnT antibody was created and bonded to the surface of multiwalled carbon nanotubes using a molecular imprinting process to serve as the recognition element. This designed sensor can precisely detect cTnT in serum samples. Conductance variations caused by analytes and electrodes can also be detected by conductometric sensors [77]. Single target-specific polyaniline nanowire biosensors have been developed using a variety of antibodies against biomarkers for CVD such as myoglobin, cTnI, B-type natriuretic peptide, and creatine kinase-MB [78]. The presence of fibrin in a developing plaque indicates that the plaque will break down and leak, which can be further detectable in the serum. Using paramagnetic or ultrasonic resonance contrast agents, NPs containing antibody fragments targeting cross-linked fibrin peptide domains can detect susceptible plaques [58, 79].

8.4.2 Therapeutic Applications

Nanotechnological applications have the potential to be used worldwide for the treatment of heart and vascular diseases. The use of NPs to deliver cardiovascular drugs might aid in the precise and targeted delivery in the body. NPs could be silica, lipids, carbon nanotubes, dendrimers, proteins, and/or nucleic acids, among others. Several novel and effective treatment approaches have greatly reduced morbidity and mortality due to cardiovascular disorders during the past half-century. Nanotechnological strategies can be used to fix, restore, and grow parts of the body, such as blood vessels, lungs, and heart. In tissue engineering, the ability to manipulate and control the microcellular environment in the laboratory is a very important feature for creating complex tissues. Recent advances in microfabrication and nanofabrication have led to progress in the field of tissue engineering in many different ways. Zhang et al.'s study focused on cardiac tissue engineering, showing how different fabrication methods are used in various types of tissue engineering research [80]. Nanotechnological approaches may be employed to treat thrombosis and nonhemostatic CVDs. Preclinical research is currently being conducted to develop novel nanotechnological strategies for treating CVD and other cardiac conditions. Mitochondrial dysfunction plays a critical part in CVD etiology and is seen as an exciting target for developing novel treatments. Among the principal unpleasant mechanisms responsive to pharmacological treatment are oxidative stress, mitochondrial permeability transition, and excessive fission and pore opening. A review article by Forini presents recent advances and lingering roadblocks in developing mitochondrial targeted therapies for CVDs. The review also describes the benefits of targeted NP-mediated delivery versus nontargeted techniques [81]. In recent years, several *in vitro* and *in vivo* approaches have assessed the health-protective characteristics of PCs. PCs, being one of the most common natural polyphenol compounds, have played critical roles owing to their antioxidant, anti-inflammatory, anticancer, and cardiovascular-protective effects [82].

Statins can be used to control a variety of CVDs by decreasing LDL and cholesterol levels, as described in earlier sections. Nabel et al. demonstrated that statins have an LDL-declining effect by blocking an enzymatic step in hepatic cholesterol production [83]. In one study, a class of statin, pitavastatin, was added to PLGA NPs, which, after being administered intravenously, were later ingested by Ly-6GCD11b+ monocytes and transported into atherosclerotic plaques [84]. Simvastatin, a derivative of statin, was encased in reconstituted HDL NPs in another study, which enabled more targeted administration to atherosclerotic plaques [85]. Simvastatin encapsulated in HDL NPs were administered to ApoE-knockout mice, and these NPs accumulated at lesion sites, slowing plaque progression at low doses and reducing inflammation at higher doses. A subsequent study demonstrated that this formulation was used to suppress macrophage growth, which was a remarkable factor in the decreased inflammation seen in the atherosclerotic plaques [86].

Triterpene glycosides or saponins are plant-derived organic chemicals that reduce LDL and cholesterol levels, thus demonstrating a hypolipidemic effect. This component from *Panax notoginseng* exerts therapeutic effects such as preventing ischemic brain damage, expanding peripheral vessels, and inhibiting thrombosis, among others [87]. Moreover, phytochemicals such as anolides, berberine, tilianin, baicalin, silymarin, curcumin, quercetin, cyclovirobuxine D, naringenin, allicin, myricitrin, and resveratrol have shown cardioprotective activities in various studies [88-91]. The chemical compound allicin, an anti-angiogenic agent abundant in garlic, can inhibit clot formation in arteries that may otherwise cause a stroke or heart attack [89]. The medicinal property of curcumin is reported for the prevention of stroke, aortic aneurysm, myocardial infarction, and atherosclerosis [90]. *In vitro* experiments conducted by Rachmawati et al. demonstrated that curcumin has antihypertensive and antihypercholesterolemic effects as analyzed using 3-hydroxy-3-methylglutaryl coenzyme A reductase and acetylcholinesterase assay kits. This nanoemulsified curcumin significantly inhibits acetylcholinesterase and 3-hydroxy-3-methylglutaryl coenzyme A reductase as compared to only curcumin treatment [92]. Another study conducted in a diabetic cardiomyopathy model showed that the multipolymer poly(gamma-benzyl-L-glutamate)-poly(ethylene glycol)-poly(gamma-benzyl-L-glutamate) NPs had increased biological efficacy, water solubility, and dispersion of curcumin in aqueous solutions. This nanoformulated curcumin enhanced serum Ca^{2+} and H_2S levels in myocardial cells and improved the expression of the calcium-sensing receptor, cystathionine-γ-lyase and calmodulin. Finally, this study revealed reduced pathological damage to myocardial cells and decreased diabetic cardiomyopathy [93]. Myricitrin, one of the plant glycosides, demonstrated antioxidant, anxiolytic, and anti-inflammatory effects [94]. Quercetin, an antioxidant, prevents CVDs by scavenging free radicals, which is crucial in atherosclerosis. In both anesthetized animals and isolated tissues, phosphatidylcholine liposomes loaded with quercetin exerted a protective effect against peroxynitrite-induced myocardial injuries by acting as a direct scavenger and destroyer of endogenously generated peroxynitrite ions, thus restoring normal myocardial contractility [95]. Another study assessed the cardioprotective effects of polymeric micelles containing a mixture of quercetin and resveratrol against doxorubicin-induced cardiotoxicity. In both *in vitro* and *in vivo* tests, this combination regimen proved efficient in reducing cardiotoxicity [96].

Diverse plants such as *Bombax ceiba*, *Centella asiatica*, *Dracocephalum moldavica*, *Ginkgo biloba*, *Nerium oleander*, *Paeonia emodi*, *Panax notoginseng*, *Pinus merkusii*, *Prunus* spp., *Sesbania grandiflora*, *Sonchus asper*, *Syzygium cumini*, *Terminalia arjuna*, *Tinospora cordifolia*, *Zingiber officinale*, and so on, belonging to various families, have shown cardioprotective activities [97]. Panax notoginsenoside (PNS), an active component of *Panax notoginseng*, plays a curative role in CVDs, but their oral administration is restricted owing to their poor absorption in the body. To increase the efficiency of PNS, methyl ether poly(ethylene glycol)-poly(lactide-co-glycolide)-based NPs were used. This encapsulation delayed the release of PNS components; significantly lowered malondialdehyde, lactate dehydrogenase, and H_2O_2 levels in the serum; and enhanced superoxide dismutase pituitrin-induced myocardial ischemia in rats [98]. Furthermore, salvianolic acid B and PNS co-delivery for acute myocardial ischemia treatment was performed using arginyl-glycyl-aspartic acid (RGD)-fabricated lipid polymer hybrid NPs. This nanosystem showed a significant infarct therapy effect and cardiac distribution *in vivo* when compared with the control group [99]. N,α-l-rhamnopyranosyl vincosamide, a bioactive chemical extracted from *Moringa oleifera* leaves, was used to generate a magnetic hydrogel nanocomposite that exhibited cardioprotective properties in rats [100, 101]. Another component, berberine, is a fluorescent isoquinoline quaternary alkaloid derived from *Berberis* spp. It shows anti-CVD effects by reducing systolic blood pressure and triglyceride levels. Its utility in medicine is terminated owing to minimal solubility in aqueous buffers, minimal absorption rate in the intestine, and short half-life in blood circulation. Meanwhile, encapsulation of berberine by liposomes could increase solubility, improve its local delivery, promote the accumulation of the liposomes in the infarcted heart tissue, enhance treatment efficacy, preserve left ventricular ejection fraction, and reduce adverse cardiac remodeling in myocardial infarction, as demonstrated in *in vivo* studies [91]. A natural flavonoid compound, 'Tilianin' has been used to treat CVDs such as coronary hypertension, heart disease, myocardial ischemia, and atherosclerosis [102]. Moreover, breviscapine is a flavone glucuronide extracted from *Erigeron breviscapus* that is used to treat cerebral infarction, angina pectoris, and coronary artery disease. It is beneficial for treating CVDs by eliminating blood clots and improving blood circulation [103]. Epigallocatechin gallate or epigallocatechin-3-gallate is a kind of catechin made of the ester of epigallocatechin and gallic acid. Epigallocatechin-3-gallate can reduce inflammatory reactions and cholesterol accumulation in macrophages, thereby inhibiting atherosclerosis development [104, 105]. Tanshinone IIA, a compound derived from the rhizome of *Salvia miltiorrhiza*, is being investigated to treat CVDs. It possesses a wide range of therapeutic properties including vasodilation, scavenging activity, endothelial dysfunction prevention, macrophage-derived foam cell delay, and inhibition of vascular smooth muscle cell proliferation and migration [106, 107]. *Allium sativum*, more generally referred to as garlic, is a member of the Amaryllidaceae family and has been utilized medicinally since ancient times. It has been used to prevent and treat CVDs and other metabolic diseases such as atherosclerosis, dementia, hyperlipidemia, diabetes, thrombosis, hypertension, and cancer. S-allyl cysteine, a key component of aged garlic extract, has undergone substantial research. S-allyl cysteine has antioxidant activity at both indirect and direct levels. In addition to reducing lipid peroxidation and DNA fragmentation, it decreases oxidation and nitration [108].

8.5 CONCLUSION

CVDs are the most prevalent diseases and disorders responsible for most of the 21st century's health outcomes and death rates. According to forecasts by the WHO, the increasing incidence of CVDs accounts for 48% of total deaths globally. When included in the diet, plant products reduce the risk of onset of CVDs. Plant supplement intake in adequate amounts will reduce the risk of CVDs. Plant bioactives include anolides, berberine, tilianin, baicalin, silymarin, curcumin, quercetin, cyclovirobuxine D, naringenin, allicin, myricitrin, and resveratrol, all of which have potent cardioprotective activity. However, the delivery of plant bioactives through NPs could enhance the therapeutic activity through longer circulation and effective accumulation inside the target organ. This delivery strategy minimizes the drawbacks of traditional drug delivery. Moreover, the emergence of theranostic NPs provides dual-responsive diagnostic and therapy techniques for pathophysiological diseases in a single platform. The revealed effects have changed our understanding of the pathophysiology of CVDs and led to the development of the most effective treatment strategies. The potential influence of nanotheranostics on the progression of atherosclerosis, ischemic heart disease, myocardial infarction, and aneurysms appears promising.

8.5.1 Future Perspective and Challenges

Nanotechnological applications may considerably increase the number of methods accessible to physicians to combat life-threatening conditions such as CVD. Nanosensors might enhance the specificity and sensitivity of the diagnostic instrument. By modifying well-established surface chemistry and by using nanomaterials, it is crucial to build highly selective and sensitive biosensor platforms to detect disease indicators and accurately diagnose heart diseases. Despite significant developments in nanotechnology for the identification of cardiovascular diseases, early-stage diagnosis remains challenging owing to the obscurity of symptoms and the poor expression of early-stage cardiac biomarkers, which make nanotechnology-based testing problematic. Concerns related to the sensitivity and specificity of the biomarker are critical. A previous study revealed that single indicators might be insufficient, which may result in the lack of sensitivity and specificity for CVD diagnosis. Owing to the unpredictability, variety, and complexity of pathophysiology in different groups, a single biomarker cannot provide reliable diagnostic results.

Furthermore, biomarkers may be directed in different ways as the disease progresses. Nanomaterials may exhibit severe adverse side effects in several experimental settings. Yet, their toxicity can be reduced or avoided by coating their surfaces with various types of natural or synthetic polymers or other chemicals. However, more study is required to demonstrate how integrating modern computational approaches and machine learning may significantly improve the diagnostic precision of CVDs.

REFERENCES

[1] Lopez AD, Mathers CD, Ezzati M, Jamison DT, Murray CJ. Lancet. 2006 May 27;367(9524):1747–57.
[2] Okwuosa IS, Lewsey SC, Adesiyun T, Blumenthal RS, Yancy CW. Int. J. Cardiol. 2016 Jan 1;202:433–40.
[3] www.who.int/health-topics/cardiovascular-diseases#tab=tab_1
[4] Gaziano T, Reddy KS, Paccaud F, Horton S, and Chaturvedi V. *Disease Control Priorities in Developing Countries*. 2nd edition (2006).
[5] Thorvaldsen P, Asplund K, Kuulasmaa K, Rajakangas AM, Schroll M. Stroke. 1995 Mar;26(3):361–7.
[6] Goldstein LB, Adams R, Becker K, Furberg CD, Gorelick PB, Hademenos G, Hill M, Howard G, Howard VJ, Jacobs B, Levine SR. Stroke. 2001 Jan;32(1):280–99.
[7] Collaborative ES. Lancet. 1998 Dec 5;352(9143):1801–7.
[8] Hall WD. Am. J. Med. Sci. 1999 Mar 1;317(3):176–82.
[9] Deferrari G, Cipriani A, La Porta E. J. Nephrol. 2021 Feb;34(1):137–53.
[10] Nakayama M, Metoki H, Terawaki H, Ohkubo T, Kikuya M, Sato T, Nakayama K, Asayama K, Inoue R, Hashimoto J, Totsune K. Nephrol. Dial. Transplant. 2007 Jul 1;22(7):1910–5.
[11] Yamamoto S, Kon V. Curr. Opin. Nephrol. Hypertens. 2009 May;18(3):181.
[12] Vanholder R, Massy Z, Argiles A, Spasovski G, Verbeke FR, Lameire N. Nephrol. Dial. Transplant. 2005 Jun 1;20(6):1048–56.
[13] Losito A, Pittavini L, Zampi I, Zampi E. Int. J. Nephrol. 2016 Nov 3;2016:7180784.
[14] Levin A, Djurdjev O, Barrett B, Burgess E, Carlisle E, Ethier J, Jindal K, Mendelssohn D, Tobe S, Singer J, Thompson C. Am. J. Kidney Dis. 2001 Dec 1;38(6):1398–407.
[15] Ortega FB, Lavie CJ, Blair SN. Circ. Res. 2016 May 27;118(11):1752–70.
[16] Bouis H, Saltzman A, Low J, Ball A, Covic N. African J. Food Agric. Nutr. Dev. 2017 Apr 21;17(2):11848–64.
[17] World Health Organization. *Nutrition for health and development: a global agenda for combating malnutrition*. No. WHO/NHD/00.6. World Health Organization, 2000.
[18] Hossain SM, Mohiuddin AK. Int. J. Agric. Res. Innov. Technol. 2012;2(2):25–35.
[19] Smith IF. Food Nutr Bull. 2000;21(4):532–7.
[20] Mali VR, Mohan V, Bodhankar SL. Pharm Biol. 2012 Nov 1;50(11):1428–35.
[21] Estrada-Soto S, Navarrete-Vázquez G, Léon-Rivera I, Rios MY, Aguilar-Guadarrama B, Castillo-España P, Ortiz-Andrade R, Aguirre-Crespo F. Phytopharmacol. 2012;2:170–8.
[22] Kumar AS, Jeyaprakash K, Chellappan DR, Murugan R. J. Ethnopharmacol. 2017 Mar 6;199:86–90.
[23] Senejoux F, Girard C, Kerram P, Aisa HA, Berthelot A, Bévalot F, Demougeot C. J. Ethnopharmacol. 2010 Oct 28;132(1):268–73.
[24] Ersoy S, Orhan I, Turan NN, Şahan G, Ark MU, Tosun F. Phytomedicine. 2008 Dec 1;15(12):1087–92.
[25] Kitic D, Brankovic S, Radenkovic M , Savikin K, Zdunic G, Kocic B, Velickovic-Radovanovic R. J Physiol Pharmacol. 2012 Oct 1;63(5):531.
[26] Michel J, Abd Rani NZ, Husain K. Front. Pharmacol. 2020 Jun 5;11:852.
[27] Sabir F, Barani M, Mukhtar M, Rahdar A, Cucchiarini M, Zafar MN, Behl T, Bungau SN. Chemosensors 2021;9:67.
[28] Pala R, Anju VT, Dyavaiah M, Busi S, Nauli SM. Int. J. Nanomed. 2020;15:3741.
[29] Swamy MK, editor. Springer Nature; 2020 May 11.

[30] Hedayati N, Oskouei Z, Tabeshpour J, Naeini MB. Eur. J. Pharmacol. 2021 Dec 15;913:174590.
[31] Avila-Carrasco L, Majano P, Sánchez-Toméro JA, Selgas R, López-Cabrera M, Aguilera A, González Mateo G. Front. Pharmacol. 2019 Jul 30;10:715.
[32] Marin-Neto JA, Maciel BC, Secches AL, Gallo L. Clin. Cardiol. 1988 Apr;11(4):253–60.
[33] Imanshahidi M, Hosseinzadeh H. Phytother. Res. 2008 Aug;22(8):999–1012.
[34] Avila-Carrasco L, Majano P, Sánchez-Toméro JA, Selgas R, López-Cabrera M, Aguilera A, González Mateo G. Front. Pharmacol. 2019 Jul 30;10:715.
[35] De Sotillo DR, Hadley M, Holm ET. J. Food Sci. 1994 May;59(3):649–51.
[36] Umadevi M, Kumar PS, Bhowmik D, Duraivel S. J. Med. Plants Stud. 2013 Mar;1(1):16–25.
[37] Xia EQ, Deng GF, Guo YJ, Li HB. Int. J. Mol. Sci. 2010 Feb 4;11(2):622–46.
[38] Muñoz-Bernal ÓA, Coria-Oliveros AJ, Laura A, Rodrigo-García J, del Rocío Martínez-Ruiz N, Sayago-Ayerdi SG, Alvarez-Parrilla E. Int. Food Res. J. 2021 Feb 1;140:110069.
[39] Tenore GC, Manfra M, Stiuso P, Coppola L, Russo M, Ritieni A, Campiglia P. Food Chem. 2013 Oct 15;140(4):803–9.
[40] Ruan Y, Jin Q, Zeng J, Ren F, Xie Z, Ji K, Wu L, Wu J, Li L. Front. Pharmacol. 2020:2040.
[41] Li SH, Tian HB, Zhao HJ, Chen LH, Cui LQ. PloS One. 2013 Jul 24;8(7):e69818.
[42] Han N, Bakovic M. J Bioanal Biomed S. 2015;12(005):1948–5.
[43] Wang DK, Rahimi M, Filgueira CS. Nanomed. Nanotechnol. Biol. Med . 2021 Jun 1;34:102387.
[44] Sharifi-Rad J, Rodrigues CF, Sharopov F, Docea AO, Can Karaca A, Sharifi-Rad M, Kahveci Karıncaoglu D, Gülseren G, Şenol E, Demircan E, Taheri Y. Int. J. Environ. Res. Public Health 2020 Apr;17(7):2326.
[45] Kris-Etherton PM, Hecker KD, Bonanome A, Coval SM, Binkoski AE, Hilpert KF, Griel AE, Etherton TD. Am. J. Med. 2002 Dec 30;113(9):71–88.
[46] Shah SM, Akram M, Riaz M, Munir N, Rasool G. Dose-response. 2019 May 24;17(2): 1559325819852243.
[47] Asgary S, Sahebkar A, Afshani MR, Keshvari M, Haghjooyjavanmard S, Rafieian-Kopaei M. Phytother. Res. 2014 Feb;28(2):193–9.
[48] Dohadwala MM, Holbrook M, Hamburg NM, Shenouda SM, Chung WB, Titas M, Kluge MA, Wang N, Palmisano J, Milbury PE, Blumberg JB. Am. J. Clin. Nutr. 2011 May 1;93(5):934–40.
[49] Griel AE, Kris-Etherton PM. Br. J. Nutr. 2006 Nov;96(S2):S68–78.
[50] Nouri F, Sarrafzadegan N, Mohammadifard N, Sadeghi M, Mansourian M. Eur. J. Clin. Nutr. 2016 Feb;70(2):217–21.
[51] Tighe P, Duthie G, Vaughan N, Brittenden J, Simpson WG, Duthie S, Mutch W, Wahle K, Horgan G, Thies F. Am. J. Clin. Nutr. 2010 Oct 1;92(4):733–40.
[52] Cicero AF, Fogacci F, Colletti A. Food Funct. 2017;8(6):2076–88.
[53] Caliceti C, Franco P, Spinozzi S, Roda A, Cicero AF. Curr. Med. Chem. 2016 Apr 1;23(14):1460–76.
[54] Thies F, Mills LM, Moir S, Masson LF. Proc. Nutr. Soc. 2017 May;76(2):122–9.
[55] Ried K. J. Nutr. 2016 Feb 1;146(2):389S–96S.
[56] Eccleston C, Baoru Y, Tahvonen R, Kallio H, Rimbach GH, Minihane AM. J. Nutr. Biochem. 2002 Jun 1;13(6):346–54.
[57] Dilnawaz F, Iqbal Z, editors. Bentham Science Publishers; 2021 Jul 27.
[58] Iverson N, Plourde N, Chnari E, Nackman GB, Moghe PV. BioDrugs. 2008 Jan;22(1):1–10.
[59] López-Mejías R, Genre F, González-Juanatey C, González-Gay MA. VASA. Zeitschrift fur Gefasskrankheiten. 2014 Mar 1;43(2):83–5.
[60] Sanz J, Fayad ZA. Nature. 2008 Feb;451(7181):953–7.
[61] Mulder WJ, Fayad ZA. Arterioscler. Thromb. Vasc. Biol. 2008 May 1;28(5):801–2.
[62] Winter PM, Morawski AM, Caruthers SD, Fuhrhop RW, Zhang H, Williams TA, Allen JS, Lacy EK, Robertson JD, Lanza GM, Wickline SA. Circulation. 2003 Nov 4;108(18):2270–4.
[63] Choudhury RP, Fuster V, Fayad ZA. Nat. Rev. Drug Discov. 2004 Nov;3(11):913–25.
[64] Sosnovik DE, Schellenberger EA, Nahrendorf M, Novikov MS, Matsui T, Dai G, Reynolds F, Grazette L, Rosenzweig A, Weissleder R, Josephson L. Magn. Reson. Med. 2005 Sep;54(3):718–24.
[65] Nahrendorf M, Jaffer FA, Kelly KA, Sosnovik DE, Aikawa E, Libby P, Weissleder R. Circulation. 2006 Oct 3;114(14):1504–11.
[66] Godin B, Sakamoto JH, Serda RE, Grattoni A, Bouamrani A, Ferrari M. Trends Pharmacol. Sci. 2010 May 1;31(5):199–205.

[67] Rouleau L, Berti R, Ng VW, Matteau-Pelletier C, Lam T, Saboural P, Kakkar AK, Lesage F, Rhéaume E, Tardif JC. Contrast Media Mol Imaging. 2013 Jan;8(1):27–39.
[68] Wang B, Su JL, Amirian J, Litovsky SH, Smalling R, Emelianov S. Opt. Express. 2010 Mar 1;18(5):4889–97.
[69] Hu J, Sanz-Rodríguez F, Rivero F, Rodríguez EM, Torres RA, Ortgies DH, Solé JG, Alfonso F, Jaque D. Gold Nanoshells: Nano Res. 2018 Feb;11(2):676–85.
[70] Cole LE, Ross RD, Tilley JM, Vargo-Gogola T, Roeder RK. Nanomedicine. 2015 Jan;10(2): 321–41.
[71] Doiron AL, Chu K, Ali A, Brannon-Peppas L. Proc. Natl. Acad. Sci. 2008 Nov 11;105(45): 17232–7.
[72] Sigalov AB. Contrast Media Mol Imaging. 2014 Sep;9(5):372–82.
[73] Li F, Zhi D, Luo Y, Zhang J, Nan X, Zhang Y, Zhou W, Qiu B, Wen L, Liang G. Nanoscale. 2016;8(25):12826–33.
[74] Tarin C, Carril M, Martin-Ventura JL, Markuerkiaga I, Padro D, Llamas-Granda P, Moreno JA, García I, Genicio N, Plaza-Garcia S, Blanco-Colio LM. Sci. Rep. 2015 Nov 30;5(1):1–9.
[75] Sharma S, Jackson PG, Makan J. J. Clin. Pathol. 2004 Oct 1;57(10):1025–6.
[76] Brondani D, Piovesan JV, Westphal E, Gallardo H, Dutra RA, Spinelli A, Vieira IC. 2014;139(20):5200–8.
[77] Mikkelsen SR, Rechnitz GA. Anal. Chem. 1989 Aug 1;61(15):1737–42.
[78] Lee I, Luo X, Huang J, Cui XT, Yun M. Biosensors. 2012 May 14;2(2):205–20
[79] Wickline SA, Neubauer AM, Winter P, Caruthers S, Lanza G. Arterioscler. Thromb. Vasc. Biol. 2006 Mar 1;26(3):435–41.
[80] Zhang B, Xiao Y, Hsieh A, Thavandiran N, Radisic M. Nanotechnology. 2011 Nov 21;22(49):494003.
[81] Forini F, Canale P, Nicolini G, Iervasi G. Pharmaceutics. 2020 Nov 20;12(11):1122.
[82] Hou K, Wang Z. Food Rev. Int. 2021 Feb 26:1–5.
[83] Nabel EG. N. Engl. J. Med. 2003 Jul 3;349(1):60–72.
[84] Katsuki S, Matoba T, Nakashiro S, Sato K, Koga JI, Nakano K, Nakano Y, Egusa S, Sunagawa K, Egashira K. Circulation. 2014 Feb 25;129(8):896–906.
[85] Duivenvoorden R, Tang J, Cormode DP, Mieszawska AJ, Izquierdo-Garcia D, Ozcan C, Otten MJ, Zaidi N. Lobatto Chapter 8 ME, van Rijs SM. Priem B, Kuan EL, Martel C, Hewing B, Sager H, Nahrendorf M, Randolph GJ, Stroes ES, Fuster V, Fisher EA, Fayad ZA and Mulder WJ. Nat. Commun. 2014;5:3065.
[86] Tang J, Lobatto ME, Hassing L, van der Staay S, van Rijs SM, Calcagno C, Braza MS, Baxter S, Fay F, Sanchez-Gaytan BL, Duivenvoorden R. Sci. Adv. 2015 Apr 3;1(3):e1400223.
[87] Dong TT, Cui XM, Song ZH, Zhao KJ, Ji ZN, Lo CK, Tsim KW. J. Agric. Food Chem. 2003 Jul 30;51(16):4617–23.
[88] Giannouli M, Karagkiozaki V, Pappa F, Moutsios I, Gravalidis C, Logothetidis S. Mater. Today: Proc. 2018 Jan 1;5(8):15998–6005.
[89] Li X, Fang Q, Tian X, Wang X, Ao Q, Hou W, Tong H, Fan J, Bai S. Mol. Med. Rep. 2017 Oct 1;16(4):4455–62.
[90] Campbell MS, Fleenor BS. Crit. Rev. Food Sci. Nutr. 2018 Nov 2;58(16):2790–9.
[91] Allijn IE, Czarny BM, Wang X, Chong SY, Weiler M, Da Silva AE, Metselaar JM, Lam CS, Pastorin G, De Kleijn DP, Storm G. J. Control. Release. 2017 Feb 10;247:127–33.
[92] Rachmawati H, Soraya IS, Kurniati NF, Rahma A. Sci. Pharm. 2016 Mar;84(1):131–40.
[93] Tong F, Chai R, Jiang H, Dong B. Int. J. Nanomedicine. 2018;13:1945.
[94] Pereira M, Siba IP, Chioca LR, Correia D, Vital MA, Pizzolatti MG, Santos AR, Andreatini R. Prog. Neuropsychopharmacol. Biol. Psychiatry. 2011 Aug 15;35(7):1636–44.
[95] Soloviev A, Stefanov A, Parshikov A, Khromov A, Moibenko A, Kvotchina L, Balavoine G, Geletti Y. Cardiovasc. Toxicol. 2002 Jun;2(2):129–39.
[96] Moradi SZ, Momtaz S, Bayrami Z, Farzaei MH, Abdollahi M. Front. Bioeng. Biotechnol. 2020 Apr 7;8:238.
[97] Shaito A, Thuan DT, Phu HT, Nguyen TH, Hasan H, Halabi S, Abdelhady S, Nasrallah GK, Eid AH, Pintus G. Front. Pharmacol. 2020 Apr 7;11:422.
[98] Zhang J, Han X, Li X, Luo Y, Zhao H, Yang M, Ni B, Liao Z. Int. J. Nanomedicine. 2012;7:4299.
[99] Qiu J, Cai G, Liu X, Ma D. Biomed. Pharmacother. 2017 Dec 1;96:1418–26.

[100] Cheraghi M, Namdari M, Daraee H, Negahdari B. J. Microencapsul. 2017 May 19;34(4):335–41.
[101] Tiloke C, Anand K, Gengan RM, Chuturgoon AA. Biomed. Pharmacother. 2018 Dec 1;108:457–66.
[102] Yuan Y, Cao W, Hong Y, Guo X, Wang Y, Wang Y, Wang X, Hu P. Phytomedicine. 2017 Oct 15;34:106–14.
[103] Fei X, Chen X, Liang G, Yue-Jian C, Hao W, Ning G, Jia-Bi Z. J. Drug. Target. 2009 Jun 1;17(5):408–14.
[104] Hong MH, Kim MH, Chang HJ, Kim NH, Shin BA, Ahn BW, Jung YD. Life Sci. 2007 May 1;80(21):1957–65.
[105] Hong J, Lu H, Meng X, Ryu JH, Hara Y, Yang CS. Cancer Res. 2002 Dec 15;62(24):7241–6.
[106] Feng J, Li S, Chen H. Eur. J. Pharmacol. 2016 Nov 15;632–9.
[107] Gao S, Liu Z, Li H, Little PJ, Liu P, Xu S. Atherosclerosis. 2012 Jan 1;220(1):3–10.
[108] Powell CR, Dillon KM, Matson JB. Biochem. Pharmacol. 2018 Mar 1;149:110–23.
[109] Céspedes CL, El-Hafidi M, Pavon N, Alarcon J. Food Chem. 2008 Mar 15;107(2):820–9.
[110] Manach C, Mazur A, Scalbert A. Curr. Opin. Lipidol. 2005 Feb 1;16(1):77–84.
[111] Aliaga C, Lissi EA. Can. J. Chem. 2004 Dec 1;82(12):1668–73.
[112] Punithavathi VR, Prince PS. Chem.-Biol. Interact. 2009 Oct 30;181(3):322–7.
[113] Velayutham P, Babu A, Liu D. Curr. Med. Chem. 2008;15(18):1840.
[114] Heim KE, Tagliaferro AR, Bobilya DJ. J. Nutr. Biochem. 2002 Oct 1;13(10):572–84.
[115] Vasanthi RH, ShriShriMal N, Das KD. Curr. Med. Chem. 2012 May 1;19(14):2242–51.
[116] Ghanim H, Sia CL, Abuaysheh S, Korzeniewski K, Patnaik P, Marumganti A, Chaudhuri A, Dandona P. J. Clin. Endocrinol. Metab. 2010 Sep 1;95(9):E1–8.
[117] O'Leary KA, de Pascual-Tereasa S, Needs PW, Bao YP, O'Brien NM, Williamson G. Mutat. Res. Fundam. Mol. Mech. Mutagen. 2004 Jul 13;551(1–2):245–54.
[118] Lekli I, Ray D, Mukherjee S, Gurusamy N, Ahsan MK, Juhasz B, Bak I, Tosaki A, Gherghiceanu M, Popescu LM, Das DK. J. Cell. Mol. Med. 2010 Oct;14(10):2506–18.
[119] Moghaddam RH, Samimi Z, Moradi SZ, Little PJ, Xu S, Farzaei MH. Eur. J. Pharmacol. 2020 Nov 15;887:173535.
[120] Wang CZ, Mehendale SR, Calway T, Yuan CS. Am. J. Chin. Med. 2011;39(04):661–71.
[121] Ross JA, Kasum CM. Annu. Rev. Nutr. 2002;22:19.
[122] Yamagata K. Carotenoids. 2017 Jun 14;25:106.
[123] Britton G. FASEB J. 1995 Dec;9(15):1551–8.
[124] Kim KJ, Lee MS, Jo K, Hwang JK. Biochem. Biophys. Res. Commun. 2011 Jul 22;411(1):219–25.
[125] Tian KM, Li JJ, Xu SW. Pharmacol. Res. 2019 Mar 1;141:541–50.

9 Treating Autoimmune Diseases Using Phytonanotechnology

*M. Bacanlı
University of Health Sciences Turkey, Turkey, mervebacanli@gmail.com

H. Erdoğan
University of Health Sciences Turkey, Turkey, hakan.erdogan@sbu.edu.tr

The immune system components participate in the defense mechanism of an organism by conferring protection against attacks by foreign bodies while maintaining control of cell proliferation in the body. Autoimmune diseases occur when a specific adaptive immune response is elicited against a self-antigen. In these diseases, the immune system fails to correctly distinguish self-antigens from nonself-antigens and attacks the self-antigens. Immune-mediated diseases can be treated in different ways such as immunomodulation. Although many immunomodulatory drugs are currently available for the treatment of autoimmune diseases, their use is restricted owing to serious side effects following prolonged duration therapy and high cost. Therefore, there is an increasing interest and need for medicinal phytochemicals with anti-inflammatory effects and fewer side effects than those with synthetic drugs. Nanoparticle systems are designed to increase drug stability, regulate drug release kinetics, deliver targeted drugs, reduce side effects, and improve biodistribution properties. As the possible benefits of treatment with herbs are known, the importance given to studies on phytonanotechnology has been increasing in recent years. This chapter presents a comprehensive description about autoimmune diseases, phytochemicals effective against autoimmune diseases, and nanotechnological approaches for autoimmune disease phytotherapy.

9.1 INTRODUCTION

The immune system, which is the most complex system in the body, comprises different cells, cell products, tissues, organs, and mediators that interact with each other. The components of the immune system participate in the defense mechanism of the organism against foreign attacks (organisms and cells) while maintaining control of cell proliferation in the body (Castelo-Branco and Soveral 2014, Roberts and Adams 2000).

The immune system cells have two functional responses against nonself, nonspecific (innate immunity), and specific (acquired or adaptive) immune reactions. Innate immunity is a nonspecific first-line defense mechanism of the body's response to foreign organisms, cells, and substances and is associated with nonimmunologic memory. Acquired immunity involves a character-specific response and is associated with immunologic memory. Prior exposure to foreign organisms/substances is required to elicit an acquired immunological response. Because of the memory of antigens following initial exposure, there is a faster and more rapid response during the next exposure. Some defects in the innate immune system are related to the predisposition to infections

DOI: 10.1201/9781003231721-9

or to autoimmune diseases. In the defense systems, antibody- and cell-mediated immunity systems interact in harmony with specific and nonspecific mechanisms. In general, an effective immunological response involves that B lymphocytes form antibodies for neutralizing foreign antigens, after which the macrophages phagocytose the neutralized antigens and present to T lymphocytes; subsequently, natural killer cells and T lymphocytes are regulated to remove cells with foreign antigens, and the required cytokines are released (Roberts and Adams 2000).

Immune-mediated diseases can be treated in different ways such as immunomodulation. Immunomodulation can be addressed through a variety of specific and nonspecific approaches. Many synthetic and natural compounds have immunomodulatory effects. A pharmacological agent shows immunomodulatory effects including immunosuppression, immunoadjuvant activity, and immunostimulation under various dose and time regimens (Brindha 2016). Figure 9.1 shows the immunomodulatory activity of drugs.

Although many drugs are currently available for the treatment of autoimmune diseases, their use is restricted owing to serious side effects following prolonged treatment duration and high cost (Bacchi et al. 2012). Therefore, there is an increasing interest and need for medicinal phytochemicals with anti-inflammatory effects and fewer side effects than synthetic drugs (Yatoo et al. 2018).

The term "phyto" is a Greek word that means "plant." Plants are the biosynthetic laboratory of phytochemicals. Phytochemicals are naturally occurring compounds with various potential bioactives. These compounds can be found in fruits, vegetables, and various medicinal plants (Brindha 2016, Behl et al. 2021). The thousands of phytochemicals that have been discovered thus far are grouped based on their functions and, sometimes, sources, and phytochemicals with potential immunostimulant activity can be classified as high- and low-molecular-weight compounds (Brindha 2016). Currently, studies on the effects of such phytochemical molecules on different chronic diseases are ongoing.

Various nanotechnological applications that can be useful in the treatment of different diseases including autoimmune diseases are available. Nanoparticle (NP) systems are designed to increase drug stability, regulate drug release kinetics, deliver targeted drugs, reduce side effects, and improve biodistribution properties. Given that the possible beneficial effects of treatment with herbs are known, the importance given to studies on phytonanotechnology has been increasing in recent years. Phytochemicals and their

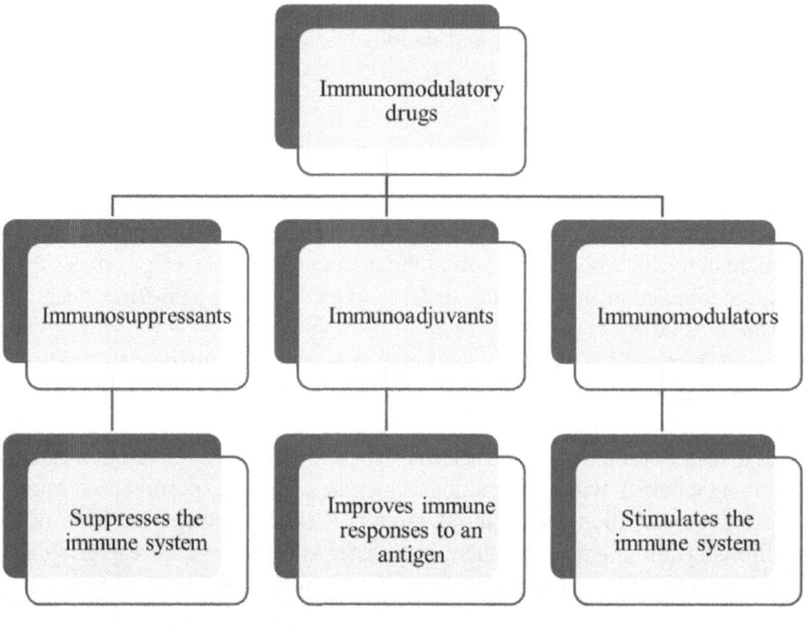

FIGURE 9.1 Types and mechanism of action of immunomodulatory drugs.

derivatives are used alone or in combination with synthetic drugs or as nanomedicine formulations for better therapy over the conventional approach (Dhupal and Chowdhury 2020).

This chapter presents a comprehensive description about autoimmune diseases, phytochemicals against autoimmune diseases, and nanotechnological approaches for phytotherapy against autoimmune diseases.

9.2 AUTOIMMUNE DISEASES

An autoimmune disease can be defined as "a disease characterized by chronic inflammatory conditions wherein the immune system attacks its body parts in an abnormal manner" (Shin et al. 2020). Autoimmune diseases are identified as a specific adaptive immune response mounted against a self-antigen. In these diseases, the immune system fails to properly distinguish self-antigens from non-self-antigens and attacks the self-antigens (Miller 1993). Autoimmune diseases are chronic, life-threatening diseases for which therapy is often complex. Although they occur relatively rare, their effects on mortality and morbidity are quite high (Chien et al. 2015, Cooper and Stroehla 2003).

Autoimmune diseases are generally uncommon, but their effects on mortality and morbidity are significant. The overall prevalence of autoimmunity is approximately 3%–5% in the general population (Eaton et al. 2007). However, they are an important health problem worldwide owing to the growing rates of morbidity (Shen et al. 2021). Autoimmune diseases are suggested to be relatively higher in developed countries than in developing countries (Theofilopoulos, Kono, and Baccala 2017). Furthermore, the incidence and prevalence of autoimmune diseases differ between geographical regions (Wang, Wang, and Gershwin 2015).

Autoimmune diseases can be divided into two distinct groups clinically: non-organ-specific autoimmune diseases and organ-specific autoimmune diseases (Mackay and Rose 2013, Fairweather 2007). At least 80 known autoimmune disorders are well known. Organ-specific autoimmune diseases include autoimmune hepatitis (liver), type 1 diabetes (pancreas), celiac disease, Crohn's disease, and ulcerative colitis (gastrointestinal tract), multiple sclerosis (brain/spinal cord), myasthenia gravis (acetylcholine nicotinic postsynaptic receptors in neuromuscular cells), and hemolytic anemia (erythrocytes). Non-organ-specific autoimmune diseases are systemic lupus erythematosus (joints, skin, brain, kidneys, heart, etc.), rheumatoid arthritis (joints, lungs, heart, eyes), scleroderma (skin, intestine, lungs, joints), juvenile idiopathic arthritis (joints, skin, lungs), and dermatomyositis (skin, muscles) (Lopomo and Berrih-Aknin 2017, Retamozo et al. 2017).

Autoimmune diseases can be affected by several etiological factors, which are listed in Table 9.1. However, the most important of these factors is the interaction between environmental factors and genetic factors (Singh, Waldron, and Hahn 2012).

TABLE 9.1
Etiological Factors Affecting Autoimmune Diseases

Genetic Variables	Gender	External Variables	Random Variables
Disease-suppressing genes	Sex hormones	Microorganisms	Immunoglobulin collection
Predisposing genes	Sex chromosomes	Drugs	T-cell receptor collection
Severity-modifying genes		Chemicals	
		Pollution	
		Sunlight	
		Diet	
		Stress	

(From [Khan et al. 2021]. With Permission.)

Younger and middle-aged groups (30–50 years) are the most sensitive ones to some autoimmune disorders such as myasthenia gravis and multiple sclerosis. However, juvenile idiopathic autoimmune diseases (e.g., arthritis) specifically starts in childhood. Rheumatoid arthritis and primary systemic vasculitis mostly occur in older adults. Type 1 diabetes is primarily seen in childhood and adolescence (Cooper and Stroehla 2003).

Generally, multiple mechanisms are involved in the immunopathogenesis of autoimmune disorders. Antibody-dependent cell-mediated cytotoxicity, complement-dependent antibody-mediated lysis, or direct/indirect effects of cytotoxic T lymphocytes can contribute to the disease pathogenesis (Lopomo and Berrih-Aknin 2017, Retamozo et al. 2017).

In patients with autoimmune diseases, the immune system is activated by B cells, T cells, monocytes/macrophages, and dendritic cells (Gabrilovich and Nagaraj 2009, Geissmann et al. 2010). T cells are the most important ones among these cells, and antigen specificity is required for T-cell activation. Significant amounts of cytokines are released from T cells in autoimmune diseases (Vivier et al. 2008, Bettelli et al. 2008). Autoantibodies are generated in autoimmune diseases, and many serum antibodies are directed against functional cellular structures including nucleic acids, nuclear molecules, receptors, or other functional cell components. They play a central role in diagnosis and classification and may be involved in tissue damage (Wang, Wang, and Gershwin 2015). The mechanisms underlying autoimmunity can be considered as complement activation, cytotoxicity, and/or direct/indirect effects of cytotoxic T lymphocytes (Ohishi et al. 1995).

Tumor necrosis factor-alpha (TNF-α) and possibly other related cytokines are important for the progression of inflammatory cascades in autoimmune disorders. The activity and expression of TNF-α are regulated by a transcription factor, named nuclear factor-kappa B (NF-κB); however, TNF-α stimulates NF-κB activity. As inflammatory mediators are also regulated by NF-κB, the upregulation and downregulation of NF-κB by gene products have been implicated in the treatment of inflammatory diseases (Lorenz 2001, Sethi, Sung, and Aggarwal 2008, Veldman, Nagel, and Hertl 2006).

Effective and safe therapeutic agents or regimens have anti-inflammatory and/or immune system–blocking properties such as selective immunosuppression agents in the management of autoimmune diseases. However, these drugs are not completely effective in the treatment of some of the autoimmune disorders. At therapy initiation, drugs show quick beneficial effects, but prolonged intake of these drugs can cause severe adverse effects and impose a high financial burden on the patient. Clinical immunosuppressants produce many side effects such as risk of infections, nephrotoxicity, neurotoxicity, headaches, nausea, intestinal discomfort, reduction in platelet count, anemia, and delay of wound healing. Patients receiving immunosuppression regimens have a significantly higher risk of developing cancer than the general population, particularly lymphomas and skin cancer (Kapoor 2008, Kaufman et al. 2004, Ponticelli et al. 2004, Wong and Chapman 2008). For this reason, these drugs may not be useful in long-term therapy of autoimmune diseases. Replacement therapy is another way to cure the diseases. Immunomodulation with medicinal plants can serve as an alternative to conventional chemotherapy. Phytochemicals have gained significant interest because they reduce the toxicity that is otherwise seen in conventional treatment, improve compliance, and reduce high drug costs associated with clinical immunosuppression (Chien et al. 2015). Plants contain numerous and diverse nutrients including polyphenols, antioxidants, vitamins, and potent chemo-protectants, all of which significantly contribute to their protective effects on health. Many investigations have proven the beneficial effects of phytochemicals including the prevention and delay of chronic degenerative diseases. Phytochemicals seem to have promising potential for the development of novel pharmaceuticals (Brindha 2016).

9.3 PHYTOCHEMICALS

Natural products contain important constituents that can prevent and treat diseases. Herbal medicines obtained from natural products are the most used resources for this purpose (Firenzuoli and Gori 2007).

Phytochemicals derived from natural plants are commonly used for the prevention and/or treatment of different diseases owing to their safety. The therapeutic importance of plants has been quoted by ancient cultures and traditions of many countries and societies, and plant-based agents are believed to be cost-effective and safe. Phytochemicals, generally known as secondary metabolites, are found in fruits and vegetables; therefore, these compounds form an important part of human diet. These phytochemicals impart color, fragrance, flavor, and taste to plants. Phytochemicals can be classified into seven main groups according to their structure and function: glycosides, alkaloids, polyphenols, polysaccharides, lectins, saponins, and terpenoids. Most of these compounds have been regarded as possible antioxidants (Bacanlı et al. 2017, Molyneux et al. 2007). When evaluating the relationship between the chemical structures of phytochemical compounds and their anti-inflammatory effects, compounds with a phenolic structure generally have more potent anti-inflammatory effects than those without a phenolic structure. In addition, depending on the differences in chemical structures, phytochemicals may have anti-inflammatory effects with different mechanisms of action in different types of inflammation (Shin et al. 2020).

With an increase in the incidence of autoimmune diseases, it has become very important to evaluate new approaches to reduce the side effects of immunomodulatory agents used in the treatment of autoimmune diseases. Among these new approaches, the most important one is the use of plants and phytochemical compounds that control the occurrence of diseases of the immune system by causing less undesirable effects (Behl et al. 2021).

Phytochemical compounds show their anti-inflammatory properties through several mechanisms including antioxidant and free radical-scavenging activities; modulation of proinflammatory enzyme activities such as cyclooxygenase (COX), lipoxygenase, and inducible nitric oxide synthase (iNOS); inhibition of cytokines (TNF-α and interleukin [IL]); modulation of inflammatory mediators and the pro-inflammatory pathway (NF-κB) (Kour et al. 2021). Phytochemical compounds exert their anti-inflammatory effects in autoimmune diseases by altering the signaling mechanisms of inflammation-related transcription factors such as NF-κB, signal transducer, and activator of transcription (STAT)1 and Nrf-2 signaling pathways (Shin et al. 2020).

Phytochemicals with immunomodulatory effects and their effects on autoimmune diseases are listed in Table 9.2.

TABLE 9.2
Phytochemicals with Immunomodulatory Effects and Their Effects on Autoimmune Diseases

Phytochemical	Effects on Autoimmune Diseases
Berberine	• Suppression of Th17 differentiation through the activation of extracellular signal-regulated kinase (Erk) and inhibition of Th1 differentiation through the inhibition of p38 MAPK and activation of the c-Jun N-terminal kinase (JNK) pathways in type 1 diabetes (Cui et al. 2009). • Reducing the permeability of blood–brain barrier, decreasing the expression and activity of MMP-9, and decreasing inflammatory cell infiltration in experimental autoimmune encephalomyelitis (Ma et al. 2010). • The increase in Th17/Th1 responses was restored by berberine. Phosphorylated (p)-STAT1, STAT3, and STAT4 proteins increased significantly in autoimmune myocarditis (Liu et al. 2016). • Suppression of lymphocyte (CD4+ T cell) proliferation, downregulation of Th1 (TNF-α) and Th2 (IL-10) cytokines, immunoglobulin (Ig) G1 and IgG2a in autoimmune neuritis (Li et al. 2014). • Reduced expression of IL-17 mRNA and proteins in the ocular tissues of animals with uveitis (Du et al. 2020). • Suppression of the severity of rheumatoid arthritis, significantly reduced IL-6 and IL-17 levels, and increased IL-10 and TGF-β levels (Wang et al. 2017).

(Continued)

TABLE 9.2 (Continued)
Phytochemicals with Immunomodulatory Effects and Their Effects on Autoimmune Diseases

Phytochemical	Effects on Autoimmune Diseases
Caffeine	• Reduced incidence of experimental autoimmune encephalomyelitis and attenuation of its pathology at the behavioral, histological, and neurochemical levels in Wistar rats (Chen et al. 2010). • Amelioration of experimental autoimmune encephalomyelitis induced neurobehavioral deficits, reduction in inflammatory cell infiltration into the spinal cord and demyelination of the spinal cord in mice (Wang, Xi, et al. 2014). • Reduction in clinical scores, inflammatory cell infiltration, demyelination level, and activation of microglia (Wang et al. 2021).
Chlorogenic acid	• Control of the total (CD3) and differentiated (CD4 and CD8) T-cell count in rheumatoid arthritis-induced Wistar rats (Chauhan et al. 2012).
Curcumin	• Inhibition of prostaglandin (PGE) 2 production in synovial fibroblasts of patients with rheumatoid arthritis (Park et al. 2007). • Decreased NF-κB expression and TNF-α and IL-1β levels on adjuvant-induced arthritis in rats (Zheng et al. 2015). • Reducing leukocyte adhesion to the endothelium and downregulating elevated malondialdehyde levels in diabetes (Wongeakin, Bhattarakosol, and Patumraj 2014).
Epigallocatechin gallate	• Suppression of inflammation by inhibiting the proliferation of and TNF-α synthesis in T cells and downregulating NF-κB in T cells (Zipp and Aktas 2006). • Regulation of serum total autoantibody levels and reduction of the development of insulin-dependent diabetes (Gillespie et al. 2008). • Reduction in IL-1β, INF-γ, and TNF-α levels in autoimmune thyroiditis–induced rats through the suppression of the NF-κB pathway (Li 2018).
Quercetin	• Reduction in the production of prostaglandin E-2, IL-1β, and leukotriene B-4 in rats (Ahmed et al. 2018). • Reduction in arthritic scores and significantly improved symptoms in rheumatoid arthritis mice by inhibiting neutrophil infiltration and decreasing the levels of pro-inflammatory cytokines (e.g., IFN-γ, TNF-α, MCP-1, IL-6, and IL-17) in rheumatoid arthritis (Yuan et al. 2020). • Reduction in the concentrations of TNF-α, CC chemokine ligand 17 (CCL17), IL-4, IL-6, and IFN-γ in the skin tissues of animals with atopic dermatitis (Hou et al. 2019). • Reduction in proteinuria levels and tissue expression of IL-6, TNF-α, TGF-β1, Bax, and thiobarbituric acid reactive substances. Increase in catalase and superoxide dismutase (SOD1) enzyme activity in pristane-induced lupus nephritis (dos Santos et al. 2018).
Resveratrol	• Reduction in cellular infiltration, fibrosis, and expression of inflammatory cytokines in the myocardium of experimental autoimmune myocarditis. Expression of antioxidant genes was upregulated in myosin-immunized hearts, and resveratrol decreased the expression in experimental autoimmune myocarditis (Yoshida, Shioi, and Izumi 2007). • Suppression of proinflammatory cytokines (IFN-γ and TNF-α) and reduction in the levels of anti-inflammatory cytokines (IL-4 and IL-10) in experimental autoimmune encephalomyelitis (Wang et al. 2016). • Suppression of CD69 and CD71 expression on CD4+ T cells as well as CD4+ T-cell proliferation, induced CD4+ T-cell apoptosis and decreased CD4 IFN-γ+ Th1 cells and the ratio of Th1/Th2 cells *in vitro* (Wang, Luo, et al. 2014).
Tetrandrine	• Inhibition of the proliferation of CD28-co-stimulated T cells and the production of cytokines in T cell lymphocytes in rheumatoid arthritis (Lai 2002). • Treatment of microglia with tetrandrine blocked the formation of NF-κB and ameliorated microglial activation in lipopolysaccharide-induced microglial activation (Xue et al. 2008). • Reducing the expression and levels of COX-2 and inflammatory markers of the symptoms of rheumatoid arthritis (Li et al. 2018).

MAPK: mitogen-activated protein kinase; MMP: matrix metalloprotein; CD: cluster of differentiation; IL: interleukin; TNF-α: tumor necrosis factor alpha; TGF-β: transforming growth factor beta; NF-κB: nuclear factor kappa B; INF-γ: interferon gamma; MCP: monocyte chemoattractant protein

9.3.1 Berberine

Berberine (BBR), is a quinoline quaternary alkaloid (Figure 9.2), isolated from various plants such as *Hydrastis canadensis*, *Berberis aristata*, *Coptis chinensis*, *Coptis rhizome*, *Coptis japonica*, *Phellodendron amurense*, and *Phellodendron chinense* Schneid (Imanshahidi and Hosseinzadeh 2008). BBR has different pharmacological effects such as antioxidant, anti-inflammatory, anticancer, antimicrobial, lipid-lowering, neuroprotective, and hepatoprotective effects (Kour et al. 2021). BBR is used in different clinical applications including hyperlipidemia, diabetes, metabolic syndrome, polycystic ovary syndrome, obesity, fatty liver disease, and coronary artery disease owing to its pharmacological activity (Cordell, Quinn-Beattie, and Farnsworth 2001).

BBR has immunomodulatory effects. BBR shows anti-inflammatory effects by inhibiting the production of TNF-α, IL-6, and monocyte chemoattractant protein 1 (MCP-1), suppressing COX-2 expression, retarding prostaglandin E2 (PGE2) production and exudate formation, and downregulating the expression of matrix metalloproteinase-2 (MMP-2) and matrix metalloproteinase-9 (MMP-9) through mitogen-activated protein kinase (MAPK) and NF-κB signaling pathways (Li et al. 2006, Ma et al. 2010).

Cui et al. (2009) reported that BBR suppresses Th17 differentiation through the activation of extracellular signal-regulated kinase (Erk) and inhibition of Th1 differentiation by suppressing the activation of p38 MAPK and c-Jun N-terminal kinase (JNK) pathways in type 1 diabetes (Cui et al. 2009).

The ameliorative effects of BBR on experimental autoimmune encephalomyelitis were studied in female C57 BL/6 mice. The results of that study demonstrated that BBR was effective in attenuating the clinical severity of experimental autoimmune encephalomyelitis in C57 BL/6 mice by reducing the permeability of blood–brain barrier, decreasing the expression and activity of MMP-9, and reducing inflammatory cell infiltration (Ma et al. 2010).

Liu et al. (2016) reported the protective effects of BBR against experimental autoimmune myocarditis in rats. BBR significantly reduced impaired cardiac function and pathophysiological severity and impeded high levels of anticardiac myosin antibody in rats with autoimmune myocarditis. Th17 and Th1 cells and the expression of IL-17 and interferon gamma (IFN-γ) were increased in autoimmune myocarditis. The increase in Th17/Th1 responses was restored by BBR. The levels of phosphorylated (p)-STAT1, STAT3, and STAT4 proteins increased significantly in autoimmune myocarditis, while BBR notably attenuated their upregulated expression (Liu et al. 2016). BBR is also effective in the regulation of Th1 and Th17 cell differentiation and amelioration in autoimmune encephalomyelitis (Qin et al. 2010). BBR treatment can ameliorate autoimmune neuritis, suppress lymphocyte (in particular CD4+ T cell) proliferation, and downregulate Th1 (TNF-α) and Th2 (IL-10) cytokines and immunoglobulin (Ig) G1 and IgG2a (Li et al. 2014).

Uveitis is an intraocular inflammatory disease that is often caused by an autoimmune response. The treatment for autoimmune uveitis primarily involves the systemic use of steroids and

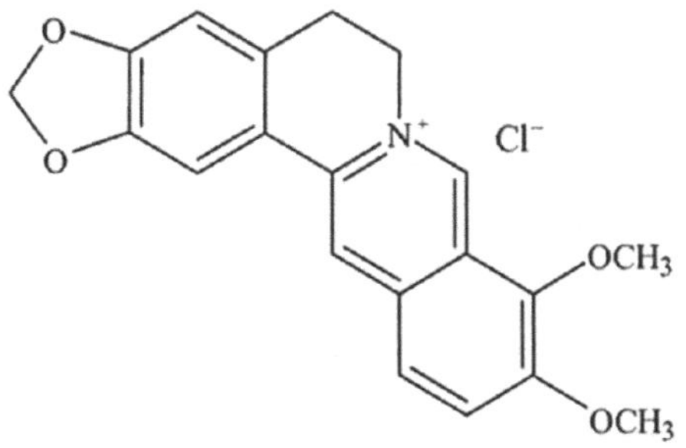

FIGURE 9.2 Berberine structure.

immunosuppressive drugs. However, BBR is also useful in autoimmune uveitis treatment. In a study of B10R.III mice, the frequency of splenic Th1 and Th17 cells had decreased, whereas that of Treg cells was increased after BBR treatment (Du et al. 2020). The effects of BBR were studied in autoimmune uveoretinitis-induced animals. IL-17 mRNA and protein expression in the ocular tissues of animals significantly reduced after BBR treatment (Lee et al. 2003).

BBR may also be effective in the treatment of rheumatoid arthritis, an autoimmune disease. Wang et al. (2017) evaluated the immunomodulatory effects of BBR in rheumatoid arthritis-induced rats. BBR suppressed the severity of rheumatoid arthritis, significantly decreased the levels of IL-6 and IL-17, and increased the levels of IL-10 and transforming growth factor (TGF)-β. Furthermore, BBR attenuated synovial hyperplasia and inflammatory cell infiltration in joint tissues (Wang et al. 2017).

9.3.2 Caffeine

Caffeine (Figure 9.3) is a natural alkaloid found in coffee beans, tea leaves, cocoa beans, cola nuts, and various plants. It is, probably, the most ingested bioactive substance through different dietary sources (tea, cola, coffee, energy drinks, and chocolate) (Nawrot et al. 2003). The beneficial effects of caffeine on health, such as neuroprotective, analgesic, and immunomodulatory activities are well known (Temple et al. 2017).

Chen et al. (2010) showed that caffeine decreased the incidence of experimental autoimmune encephalomyelitis and attenuated experimental autoimmune encephalomyelitis pathology at behavioral, histological, and neurochemical levels in Wistar rats (Chen et al. 2010). Similarly, Wang et al. (2014) reported that caffeine administration significantly ameliorated experimental autoimmune encephalomyelitis-induced neurobehavioral deficits, reduced inflammatory cell infiltration into the spinal cord and the demyelination of spinal cord in mice (Wang, Xi, et al. 2014). In a study of C57BL/6 mice, caffeine reduced the clinical score, inflammatory cell infiltration, demyelination level, and microglial activation. Furthermore, caffeine increased LC3-II/LC3-I levels and decreased NLRP3 and P62 levels (Wang et al. 2021).

9.3.3 Chlorogenic Acid

Chlorogenic acid is an ester of caffeic acid and quinic acid (Figure 9.4). It is found in peaches, prunes, potatoes, coffee, and beans. Chlorogenic acid has antitumor, antioxidant, hepatoprotective, hypoglycemic, and hypolipidemic activities (Naveed et al. 2018).

Chauhan et al. (2011) reported that chlorogenic acid effectively controls total (CD3) and differentiated (CD4 and CD8) T cell count in rheumatoid arthritis–induced Wistar rats (Chauhan et al. 2012).

In a study of lipopolysaccharide (LPS)-stimulated RAW 264.7 cells, chlorogenic acid significantly inhibited nitric oxide (NO) production, the expression of COX-2 and iNOS and the activation of NF-κB (Hwang et al. 2014). Similarly, chlorogenic acid could inhibit the inflammatory proliferation of fibroblast-like synoviocytes (RSC-364) mediated by IL-6 through apoptosis induction (Lou et al. 2016).

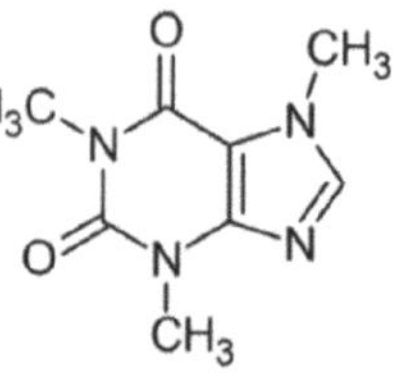

FIGURE 9.3 Caffeine structure.

9.3.4 Curcumin

Curcumin (diferuloylmethane; Figure 9.5), a phenolic compound isolated from the rhizome of turmeric (*Curcuma longa*), has been conventionally used for pain and wound healing. Curcumin, the main active compound present in turmeric, exerts antioxidant, anti-inflammatory, and anticancer activities. Curcumin can improve rheumatoid arthritis, psoriasis, multiple sclerosis, and inflammatory bowel disease in humans or *in vivo* experimental models. It can regulate various factors, enzymes, protein kinases, adhesion molecules, redox status, and gene expression of cytokines associated with inflammation (Hatcher et al. 2008).

In a clinical study, curcumin was safe, effective, and well tolerated in patients with rheumatoid arthritis (Chandran and Goel 2012). Curcumin demonstrates antiarthritic effects by decreasing the expression of NF-κB and levels of TNF-α and IL-1β on adjuvant-induced arthritis in rats (Zheng et al. 2015). In another study, curcumin alleviated synovial hyperplasia and reduced the expression levels of IL-1β, TNF-α, MMP-1, and MMP-3 in rats with collagen-induced arthritis (Dai, Di Zhou, and Song 2018). Funk et al. (2006) investigated the *in vivo* efficacy of curcumin in an animal model of rheumatoid arthritis using streptococcal cell wall-induced arthritis to determine its protective effect on arthritis. Curcumin can prevent joint inflammation when the treatment is started before the occurrence of joint inflammation (Funk et al. 2006). Park et al (2007) showed that curcumin induced apoptosis and inhibited prostaglandin (PGE) 2 production in the synovial fibroblasts of patients with rheumatoid arthritis (Park et al. 2007).

Vascular inflammation is an important clinical condition in patients with diabetes. Curcumin can regulate these undesirable effects in patients with diabetes. In studies on rats, curcumin attenuated vascular inflammation by reducing leukocyte adhesion to the endothelium and downregulating elevated malondialdehyde levels (Wongeakin, Bhattarakosol, and Patumraj 2014). Curcumin may also show therapeutic efficacy by reducing IL-1β expression and NF-κB activation in the retina of diabetic rats (Kowluru and Kanwar 2007).

9.3.5 Epigallocatechin Gallate

Green tea is the most frequently consumed beverage worldwide. Tea and its polyphenols, a product of the dried leaves of *Camellia sinensis*, have many pharmacological properties. The major

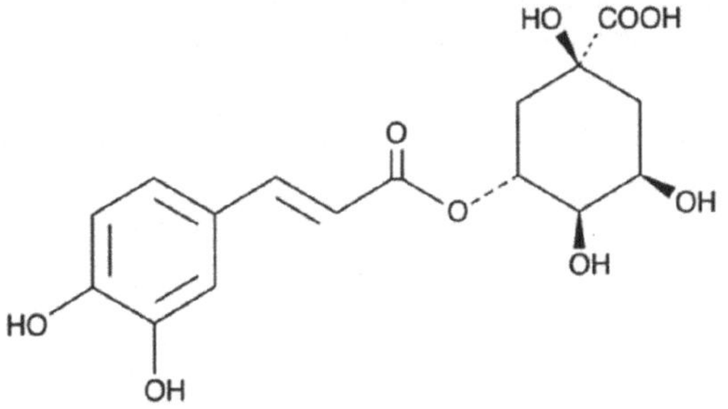

FIGURE 9.4 Chlorogenic acid structure.

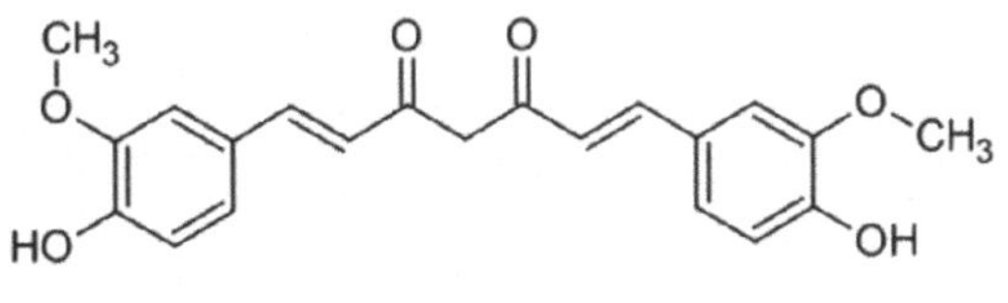

FIGURE 9.5 Curcumin structure.

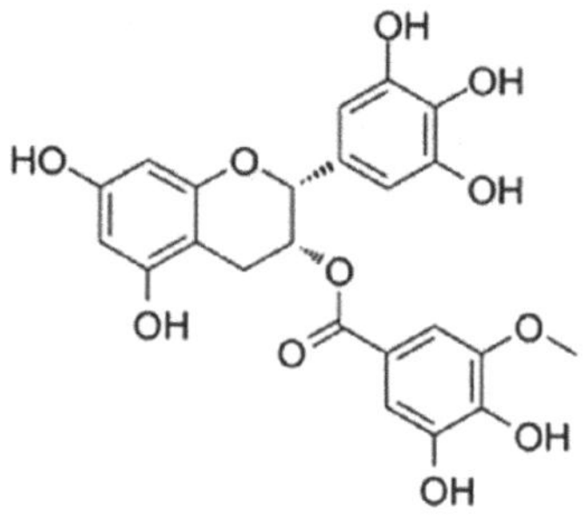

FIGURE 9.6 Epigallocatechin-3 gallate structure.

polyphenols in green tea are (-)-epicatechin, (-)-epigallocatechin, (-)-epicatechin-3-gallate, and (-)-epigallocatechin-3-gallate (EGCG).

EGCG (Figure 9.6) was shown to suppress autoimmune encephalomyelitis *in vivo* as demonstrated in an experimental animal model (Aktas et al. 2004). EGCG reduced the severity of symptoms when administered at initiation or after the onset of experimental autoimmune encephalomyelitis by limiting brain inflammation and reducing neuronal damage. Zipp et al. (2006) showed that oral treatment with EGCG suppressed inflammation by inhibiting the proliferation of and TNF-α synthesis in T cells and effectively prevented the recurrence of autoimmune diseases in the central nervous system *in vivo*. The anti-inflammatory potential of EGCG is mediated by the downregulated expression of NF-κB in T cells, thus preventing the catalytic activity of the proteasome in activated T cells, which results in the intracellular accumulation of ubiquitinated proteins including the NF-κB suppressor I-kB (Zipp and Aktas 2006).

EGCG can be used to delay or manage Sjogren's syndrome-like autoimmune disorders that affect multiple tissues or organs. Hsu et al. (2005) demonstrated that EGCG inhibited autoantigen expression in normal human keratinocytes and immortalized normal human salivary acinar cells (Hsu et al. 2005). Similarly, Gillepsie et al. (2008) investigated whether EGCG ameliorated autoimmune disorder-induced pathological changes in the salivary glands of mice with experimental nonobese diabetes-induced Sjogren's syndrome and reported that EGCG significantly retarded the onset of diabetic autoimmune disease and effectively protected the salivary gland cells from autoimmune disorder-induced damage at multiple levels (Gillespie et al. 2008).

EGCG regulates serum total autoantibody levels and reduces the development of insulin-dependent diabetes in animals. Oral administration of green tea extract reduced serum total autoantibody levels and autoimmune lymphocyte infiltration in the submandibular glands of nonobese diabetic mice (Gillespie et al. 2008).

Green tea induces changes in arthritis-related immune responses, and systematic exploration of dietary supplementation with green tea extract as an adjunct nutritional strategy for the management of rheumatoid arthritis has been proposed (Kim et al. 2008).

In experimental autoimmune thyroiditis-induced rats, ECGC supplementation exerted neuroprotective effects. EGCG treatment inhibits urinary iodine values and thyroid pathological features and significantly reduced IL-1β, INF-γ, and TNF-α levels in autoimmune thyroiditis-induced rats through suppression of the NF-κB pathway. EGCG pretreatment significantly increased B-cell lymphoma-2 protein expression and suppressed caspase-3 activity and TNF-α–related apoptosis-inducing ligand protein expression levels in rats (Li 2018).

In a recent study, EGCG treatment reduced the severity of experimental autoimmune encephalomyelitis and macrophage inflammation in the central nervous system (Cai et al. 2021).

9.3.6 Quercetin

Quercetin (Figure 9.7) is a naturally sourced flavonoid present in apples, onions, broccoli, and different beverages and has antioxidant and anti-inflammatory effects. Because of its anti-inflammatory effects, quercetin has been applied for the treatment of different autoimmune diseases (Kour et al. 2021). Quercetin exerts anti-inflammatory effects through the inhibition of cytokine production, reduction in cytokine production, and decrease in lipoxygenase expression (Carullo et al. 2017).

In one study, quercetin treatment (dose of 25 mg/kg) to diabetic rats reduced the production of prostaglandin E2, IL-1β, and leukotriene B4, thus promoting wound healing in rats (Ahmed et al. 2018).

In a randomized, double-blind, placebo-controlled clinical study, quercetin significantly alleviated morning stiffness and pain in patients with rheumatoid arthritis (Javadi et al. 2017). In an animal study, quercetin reduced arthritis scores and improved symptoms significantly in a rheumatoid arthritis mouse model by inhibiting neutrophil infiltration and decreasing the levels of pro-inflammatory cytokines (e.g., IFN-γ, TNF-α, MCP-1, IL-6, and IL-17) (Yuan et al. 2020).

Graves' disease is the most common autoimmune thyroid disorder associated with hyperthyroidism. Quercetin can reportedly inhibit DNA damage by modulating oxidative stress in the lymphocytes of patients with Graves' disease (Tang et al. 2005).

In a mouse model of atopic dermatitis, quercetin treatment reduced the concentrations of TNF-α, CC chemokine ligand 17 (CCL17), IL-4, IL-6, and IFN-γ in the skin tissues of the animals (Hou et al. 2019).

As for the ameliorative effects of quercetin against experimental autoimmune myocarditis, quercetin treatment at a dose of 10 mg/kg showed significant suppression of myocardial endothelin-1 and MAPK (Arumugam et al. 2012).

Pristane-induced lupus nephritis is characterized by inflammation and high expression of pro-fibrotic, apoptotic, and pro-oxidant markers. dos Santos et al. (2018) determined the nephroprotective effects of quercetin in pristane-induced lupus nephritis rats. Quercetin administration decreased proteinuria levels and significantly lowered tissue expression of IL-6, TNF-α, TGF-β1, Bax, and thiobarbituric acid reactive substances. Simultaneously, quercetin significantly upregulated the expression of catalase and superoxide dismutase 1 enzymes in mice (dos Santos et al. 2018).

Quercetin administration is associated with the inhibition of IL-12 expression and decreased IL-12–related T-cell proliferation and T helper (Th) 1 differentiation in SJL/J mice (Muthian and Bright 2004).

9.3.7 Resveratrol

Resveratrol (Figure 9.8), a polyphenol stilbene, found in grape skin, berries, and nuts, has important health benefits such as antioxidant, neuroprotective, and immunomodulatory effects. Resveratrol

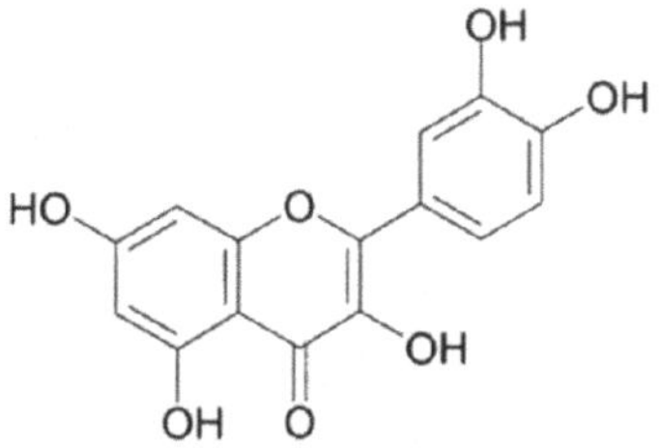

FIGURE 9.7 Quercetin structure.

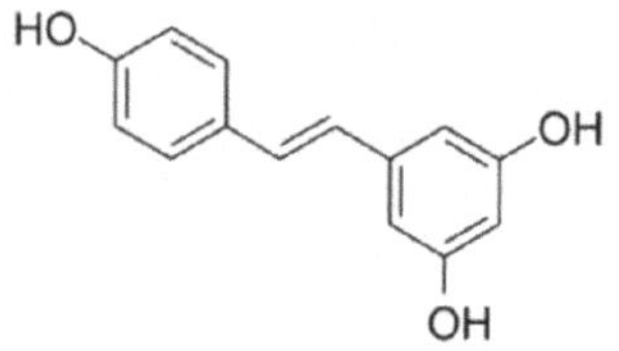

FIGURE 9.8 Resveratrol structure.

reduces the severity of autoimmune myocarditis, osteoarthritis, pancreatitis, optic neuritis, experimental autoimmune encephalomyelitis, and multiple sclerosis (Petro 2011).

In a model of autoimmune uveitis, resveratrol significantly reversed the effect of TNF-α on each target in either a time- or dose-dependent manner (Paladino et al. 2020).

Significant neuroprotective effects of resveratrol were observed in a chronic mouse model of multiple sclerosis, with a higher number of retinal ganglion cells being found in the eyes of resveratrol-treated experimental autoimmune encephalomyelitis mice with optic nerve inflammation (Fonseca-Kelly et al. 2012).

In experimental autoimmune myocarditis, resveratrol significantly decreased cellular infiltration, fibrosis, and expression of inflammatory cytokines in the myocardium. Expression of antioxidant genes was increased in myosin-immunized heart tissues, and resveratrol downregulated the expression (Yoshida, Shioi, and Izumi 2007).

In an experimental autoimmune encephalomyelitis model, efficacy studies have shown that treatment with resveratrol alone led to a significant reduction in clinical scores of the illness, and combined therapy of mouse bone marrow mesenchymal stem cells and resveratrol led to a more important reduction. Combined therapy also enhanced the immunomodulatory effects, showing suppressed production of proinflammatory cytokines (IFN-γ and TNF-α) and increased anti-inflammatory cytokines (IL-4 and IL-10) (Wang et al. 2016).

Wang et al. (2014) determined the nephroprotective effects of resveratrol in pristane-induced lupus nephritis mice. Resveratrol administration attenuated proteinuria, Ig deposition in the kidney, and glomerulonephritis as well as IgG1 and IgG2a levels in the serum of mice. Resveratrol also suppressed CD69 and CD71 expression in CD4+ T cells as well as CD4+ T-cell proliferation, induced CD4+ T-cell apoptosis, and decreased CD4 IFN-γ+ Th1 cells and the ratio of Th1 to Th2 cells *in vitro* (Wang, Luo, et al. 2014).

9.3.8 Tetrandrine

Tetrandrine (Figure 9.9), a bis-benzyl isoquinoline alkaloid used for treating autoimmune disorders in China for decades, is isolated from the root of the plant *Stephania tetrandra*. *In vivo* and *in vitro* studies have revealed that the immunosuppressive effects of tetrandrine, apart from its antioxidant and anticancer activities, are different from those of known antirheumatic drugs such as sulfasalazine, methotrexate, and cyclosporine (Lai 2002).

Tetrandrine is used effectively in rheumatoid arthritis, as it inhibits the proliferation of CD28-co-stimulated T cells and the production of cytokines in T-cell lymphocytes. Both Th1 and Th2 cytokines are susceptible to the suppressive effects of tetrandrine. Tetrandrine effectively inhibits CD28-co-stimulated NF-κB activity (Lai 2002).

Suppressive potential of tetrandrine on LPS-induced microglial activation was evaluated. According to the data, microglial treatment with tetrandrine blocked NF-κB formation and ameliorated the microglial activation (Xue et al. 2008). Tetrandrine pretreatment significantly inhibited the levels of IL-1β and TNF-α (in both transcription and translation) and suppressed LPS-induced microglial activation in a dose-dependent manner in BV2 cells (Dang et al. 2014).

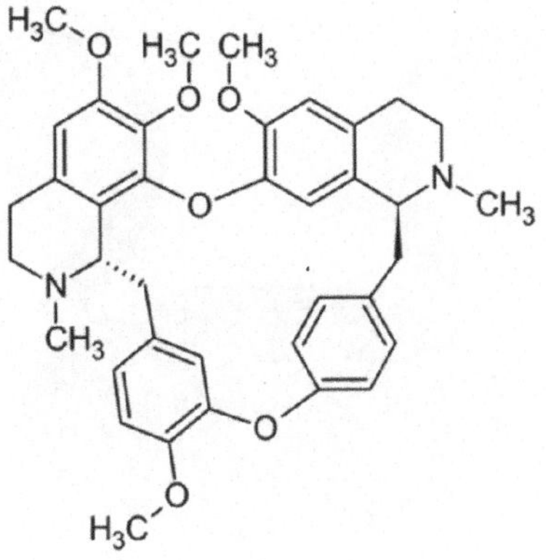

FIGURE 9.9 Tetrandrine structure.

Li et al. (2018) conducted a study in a rat model and reported that tetrandrine may be beneficial in the treatment of rheumatoid arthritis by reducing COX-2 expression and levels of inflammatory markers in the symptoms of rheumatoid arthritis (Li et al. 2018). Yuan et al. (2016) showed that tetrandrine may be beneficial in mice with collagen-induced arthritis. In that study, tetrandrine reduced the serum levels of pro-inflammatory cytokines and restored the Th17/Treg balance (Yuan et al. 2016).

9.4 PHYTONANOTECHNOLOGY

Nanotechnology includes the preparation or synthesis of nanostructures with unique physicochemical properties such as surface area, various surface structures, enhanced reactivity, and so on designed at the nanometer scale and its applications in various fields (Rizvi and Saleh 2018, Elahi, Kamali, and Baghersad 2018). The physicochemical properties of the structure stem from molecular, morphological, and structural features such as its dimensions, surface properties, chemical compositions, crystal structures, and aggregation properties. The unique properties arising from the nanostructures are increasingly used in various products including electronics, catalyst, and drug delivery (Madkour 2019, Chouhan 2018). As mentioned above, nanotechnological approaches have many applications and working abilities. One of these application areas is the nanotechnological approaches used in the primary biology approach, more specifically plant-based nanocarriers (Shaik et al. 2018, Botha et al. 2019, Rajeshkumar et al. 2019).

During the past few years, phytonanotechnology has become a research hotspot in preparing nanostructures from plant-derived substances and synthesizing nanostructures with phytonanotechnological approaches (Li and Yan 2020). In phytonanotechnological applications, plants are used as biological sensors to cope with environmental stresses (to detect nutritional deficiencies or pollution) and to improve their ability to fight or detect environmental pollutants and hazardous chemicals (Li and Yan 2020). Another application in the plant-based approach is targeted drug delivery. Phytonanotechnology has been used to explore its potential applications in human systems, particularly in the treatment of cancer and genetic diseases that cause loss of function (Figure 9.10).

Although phytonanotechnology is understood, in the first place, as the application of nanoscience in nature, specific to plant sciences and plant production systems, it includes controlled application processes involved in the synthesis of plant-derived molecules and nanostructures and the transport of different plant-derived therapeutic agents with the aid of nanosystems.

Accumulating evidence suggests the numerous benefits of herbal medicines. However, when using plant-derived molecules for the treatment of various ailments, the main difficulty in this

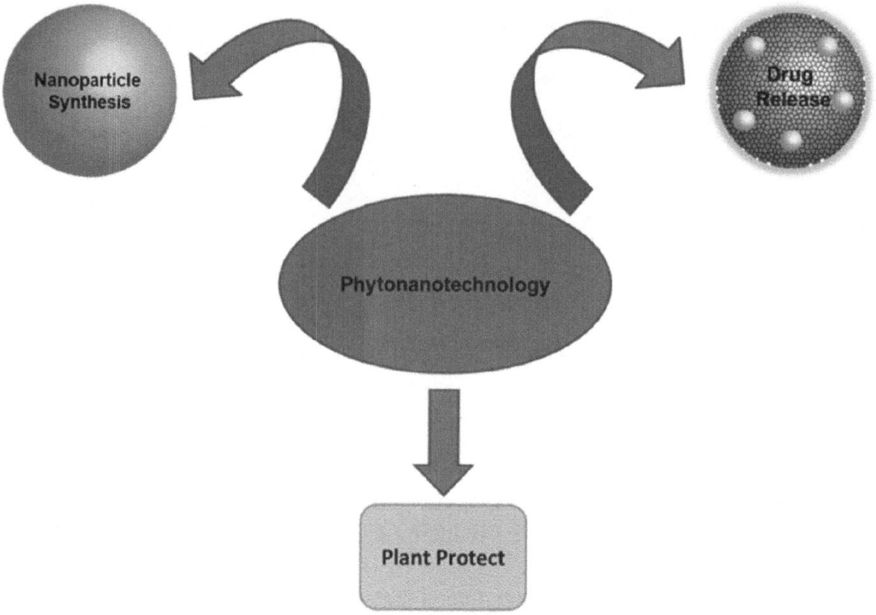

FIGURE 9.10 Areas of application of phytonanotechnology.

area is the loss of bioavailability, poor solubility, and weak stability, which arise because of their physicochemical properties when isolating the structures (Mallik, Sharangi, and Sarkar 2020). The combination of phytotherapeutics and nanotechnological applications is the most useful approach and is used to prepare a phytonanotechnological product with a high synergistic effect. In phytonanotechnological applications, carrier systems are used to carry and facilitate controlled release of herbal therapeutic agents, wherein bioactive compounds are designed by reducing the particle size and coating them in appropriate shapes as nanocarriers such as liposomes, solid lipid NPs (SLNs), niosomes, and micelles. Thus, the phytotropic charged particles designed with suitable surface properties enhance better tissue penetration of the nanocarriers, thereby increasing nanocarrier bioavailability and efficiency (Aziz and Setapar 2020). An important point here is that the release of the phytotherapeutic agent is controlled, which reduces the degree of toxicity from herbal therapeutics. Thus, a safer treatment approach can be mentioned (Yaktine and Pray 2009). The potential for developing drug delivery systems by applying phytonanotechnological approaches for treating many diseases, mainly cancer, is being investigated. Here, the most straightforward approach is to transport the herbal-derived therapeutics, which are loaded into a suitable nanocarrier, to the target site and deliver them into the tumors; this process occurs owing to their high absorbency ability, which retains the nanocarriers at the target site, thus delivering the phytotherapeutic dose at the appropriate concentration to the target area (Yaktine and Pray 2009, Lushchak et al. 2020, Khan and Gurav 2018).

9.5 PHYTONANOTECHNOLOGICAL APPROACHES IN THE TREATMENT OF AUTOIMMUNE DISEASES

One of the most critical problems in applying phytotherapeutic agents is to ensure that they reach the target tissue or cell while maintaining their effectiveness or that they reach other tissues without causing any damage to the cells. In this regard, the most useful approach may be to overcome these problems with the use of drug delivery systems. In this regard, different phytotherapeutic agents with suitable particulate formulations have been studied. One of the approaches studied frequently in recent years is the analysis of the combined effects of some phytochemicals, loaded into drug delivery systems, known to cure autoimmune diseases. In this section, phytotherapy applications based on nanoparticulate formulation have been examined.

The main disadvantage of using BBR is its poor bioavailability, which is attributed to its weak aqueous solubility. In this regard, Ahrari et al. (2021) reported that the encapsulation of BBR by poly(lactic-co-glycolic acid) (PLGA) NPs can overcome this disadvantage. In this approach, PLGA will be reduced to nontoxic substances, and the decomposed products will be rapidly metabolized in the physiological medium. In that study, BRR encapsulated by PLGA was examined for its effects on carbon tetrachloride (CCl_4)-induced liver injury in Wistar rats. Thus, to evaluate the efficacy of BBR-PLGA-NPs on damaged liver tissue, the authors tried to evaluate the α-smooth muscle actin (α-SMA) gene expression in rats. In the study, BBR-PLGA-NPs significantly reduced serum alanine transferase and aspartate aminotransferase in CCl_4-treated rats. However, the expression level of α-SMA was downregulated in CCl_4-injected rats treated with BBR, but α-SMA levels were significantly higher than those in the control group. As a result, α-SMA mRNA expression was significantly downregulated by BBR-PLGA-NPs ($p<0.05$), and in terms of hepatic histology, BBR-PLGA-NPs showed a better improvement than free BBR (Ahrari et al. 2021).

Fan et al. (2020) investigated the potential of BBR use in rheumatoid arthritis-induced inflammation. In this study, multifunctional nanoplatforms were prepared by combining BBR, an inflammation-targeted nano-drug anti-rheumatoid arthritis agent, and a reactive oxygen species (ROS)-sensitive delivery system. The conventional approach was studied in a mouse rheumatoid arthritis model. The developed inflammation-targeted nanocarriers were applied to *in vivo* therapy, wherein they showed high efficacy in rheumatoid arthritis treatment. It can be thought that the nanoplatform prepared in that study could be an application of NP-based phytonanotechnology for the treatment of advanced rheumatoid arthritis, which opens a new avenue for its potential clinical application. In the developed method, a polymeric structure of selenocystamine was prepared with BBR as a carrier nanoplatform sensitive to ROS. Thus, ROS-sensitive BBR polymeric micelles were prepared. In general, the prepared micelle structures significantly increased BBR uptake in the rheumatoid arthritis fibroblasts, and the *in vitro* and *in vivo* efficacy was improved tenfold. In that study, micelles responsive to ROS have demonstrated ten times higher efficacy than free BBR by binding to the mitochondria of the targeted inflammatory tissue. In this regard, it is thought that BBR NPs sensitive to ROS may be a practical application to eliminate the possible toxic effects of BBR accumulation in tissues damaged by rheumatoid arthritis (Fan et al. 2020). Ma et al (2021) reported that when BBR is loaded into platelet-derived extracellular vesicles (PEVs), the BBR-PEVs inhibit inflammation by acting on the phenotype of macrophages and dendritic cells. In that study, high BBR-PEV accumulation was observed in the marked joints after intravenous administration of the BBR-PEVs in the rheumatoid arthritis mouse model. This is particularly important for the rheumatoid arthritis targeted drug release approach. Furthermore, the motility of the mice was significantly increased after the administration of BBR-PEVs. Moreover, the infiltration of inflammatory cells in the arthritis part significantly reduced when compared with that in the group wherein free BBR was administered with NPs at a similar dose. The study shows that BBR-PEV particles can effectively increase the efficacy of rheumatoid arthritis treatment (Ma et al. 2021).

Another biopharmaceutical that can be used in autoimmune disorders is caffeine. It is well known that caffeine has good solubility, good adsorption ability, and lipid membrane permeability, and its plasma bioavailability ranged from 15 to 120 min when taken orally, and the bioavailability concentration was half within 3–5 hours. This situation can be overcome by shortening the caffeine application interval. As a phytonanotechnological application, different caffeine nanoformulations can be prepared and used to solve this problem. Caffeine, which has a hepatoprotective effect, has been applied through a novel delivery system, named Pickering emulsion (stabilized with solid particles). This approach was developed to reduce the possibility of developing liver disease when consumed regularly as a preventative treatment against damage due to hepatic oxidative stress. Caffeine is preferred because it alleviates inflammatory and fibrotic processes. In that study, caffeine-loaded Pickering emulsion stabilized with MgO NPs was prepared with a droplet size of less than 1μm. The prepared formulation exhibited significant hepatoprotection in rats in which liver damage was induced with CCl_4. In light of the results obtained, it was thought that caffeine could be a useful approach to produce a new, inexpensive, quickly synthesizable Pickering emulsion formula that can be applied as a prophylactic treatment in some severe liver problems (Elmotasem, Farag, and Salama 2018).

In another study, a new caffeine-based formulation was prepared to treat Alopecia areata, an idiopathic skin disease. Alopecia areata can generally occur due to stress, autoimmune disorders, and genetic predisposition. In this regard, the study focused on the preparation and application of the formulations in a two-stage process. The first stage involved the preparation of lipid-based formulations (nanostructured lipid carriers [NLCs]) with excipients suitable for drug formulation. The second stage involved the assessment of the effectiveness of drug formulation for NLCs. The results showed that caffeine-loaded, nano-sized, lipid-based carriers can be used for the topical delivery of drugs for the treatment of alopecia areata (Makky, Sadddar, and Khattab 2021).

Caffeine has many beneficial properties, one of which is the prevention of inflammation. However, as mentioned in the previous sections, despite its excellent solubility, caffeine has limited therapeutic applications because of limited bioavailability and increased metabolic clearance. Babu et al. (2018) used mesoporous silica NPs as carriers owing to their biocompatibility, biodegradability, and loading efficiency to overcome such limitations in their study. The study was conducted to prevent inflammation in macrophage cells activated by LPS. The effects of the particulate formulation on different enzymes such as myeloperoxidase, COX, lipoxygenase, and iNOS were investigated. The results showed that NP formulation was more effective than free caffeine. In addition, in the *in vitro* scratch test, which compared the caffeine- and LPS-treated groups (control), caffeine-loaded silica NPs exerted more effects on the wound area. In addition, indirect enzyme-linked immunosorbent assay results showed that caffeine-loaded silica NPs exerted COX-2 and TNF-α activity. This situation confirms the partial decrease in COX-2 and TNF-α reporter RNA levels in the reverse transcription–polymerase chain reaction assay. In this regard, the use of nanoformulations is promising owing to their anti-inflammatory effect in autoimmune disorders (Babu et al. 2018).

Hwang et al. (2015) developed a new green synthesis approach for gold NPs (CA-AuNPs) in the presence of reducing chlorogenic acid as a phytonanotechnological application. The authors investigated the *in vitro* and *in vivo* anti-inflammatory activity of free CA and CA-AuNPs in mouse RAW 264.7 macrophages and inflammatory retinal models. The results determined that chlorogenic acid attached to AuNPs showed potent anti-inflammatory effects when compared with chlorogenic acid. However, further studies may be needed to explain the detailed mechanism by which chlorogenic acid exerts greater efficiency. However, the results of *in vitro* and *in vivo* tests have confirmed that the approach is promising (Hwang et al. 2015).

Curcumin is one of the phytotherapy agents frequently used in NP-based drug release applications in recent years. Curcumin application may lead to favorable outcomes owing to its anti-inflammatory, antioxidant, and antiapoptotic effects (Rauf et al. 2018). In general, the potential for the use of curcumin has been investigated in various studies such as studies on atherosclerosis, aging, neurodegenerative disease, and autoimmune disease (Fan et al. 2013, Darvesh, Aggarwal, and Bishayee 2012). However, the most significant handicap is the hydrophobicity of curcumin, which hinders its effectiveness and causes poor *in vivo* bioavailability and poor pharmacokinetics. At this stage, curcumin should be developed into nanoformulations through different approaches, and its potential on various diseases should be examined. Lu et al. (2020) used a high-density lipoprotein–mimicking peptide-phospholipid scaffold (HPPS) particles as carriers and developed an approach for the controlled release of peripheral monocytes as immunomodulation targets, with curcumin being an anti-inflammatory agent. When a mouse model of experimental autoimmune encephalomyelitis was treated with curcumin-HPPS particles targeting the central nervous system, experimental autoimmune encephalomyelitis morbidity altered from 100% to 30%. Curcumin-HPPS inhibits monocytes passing the blood–brain barrier because of the downregulated expression of intercellular adhesion molecule 1 and macrophage-1 antigen expression by inhibiting the activation of NF-κB. In general, the study draws attention to the use of HPPS as an NP to increase the bioavailability of curcumin and demonstrates efficient and specific uptake of these NPs by monocytes through the SR-B1 receptor. For these reasons, monocytes, as different therapeutic agents, and immunomodulation targets combined with HPPS NPs are thought to have a broad application potential in monocyte-related diseases (Lu et al. 2020).

El-Missiry et al. (2020) reported the preparation method of ellagic acid-alginate (EA-ALG) NPs to evaluate anticonvulsant properties. The prepared EA-ALG NPs showed a biphasic release profile with a rapid phase for 3 hours after application. Afterward, the release was completed after 8 hours. In the study, the approach was applied for 33 days using a mouse epilepsy model, and the efficacy of EA-ALG NPs and free EA *in vivo* as an anticonvulsant and a neuroprotective agent was evaluated. The EA-ALG NPs prepared in light of the results obtained in the study prevented seizures and reduced damage due to oxidative stress during the experimental period. They also decreased cytokine levels by inhibiting apoptosis (El-Missiry et al. 2020).

EGCG, the primary anti-inflammatory compound found in green tea, can be used to treat various autoimmune diseases. However, at this point, because of the physicochemical properties of EGCG, such as low water solubility, it has poor bioavailability, which restricts its clinical application. Zhu et al. (2019) reported the preparation of spherical particles with a negative surface charge coated on EGCG AuNPs and their assessment through cellular uptake study and the investigation of intracellular ROS levels. The AuNPs reduced the formation of tartrate-resistant acid phosphatase-positive multinucleated cells and actin rings in bone marrow macrophages stimulated with LPS. The EGCG-AuNPs showed a higher antiosteoclastogenic effect than free EGCG. In *in vivo* experiments, when compared with free EGCG, EGCG-AuNPs, are more effective in improving bone density and preventing bone loss (Zhu et al. 2019). Dry eye is an ocular disease that occurs secondary to various disorders. Huang et al. (2018) prepared NPs containing anti-inflammatory agents that are effective against eye dryness. NPs were prepared by coating the gelatin-EGCG surface with hyaluronic acid (HA). Thus, a new nanoformulation was prepared in an attempt to prevent solubility-related problems of EGCG. *In vitro* biocompatibility and anti-inflammatory effects of the prepared NPs were studied in human corneal epithelial cells. In addition, *in vivo* compatibility tests were performed in an animal model of dry eye. The number of NPs in the cell cytoplasm and the number of NPs deposited on the ocular surfaces of animals indicated its advantages for ocular administration of the drug. The ocular surface of nanoformulated rabbits showed a standard corneal architecture with no appreciable change in inflammatory cytokine levels in the corneal lysate (Huang et al. 2018).

Phytonanotechnological approaches can also be used to treat or reduce the clinical effects of asthma, which can also be considered an autoimmune disease. Dhayanandamoorthy et al. (2020) reported that chitosan (CS) NPs (FACHA) decorated with aerosolized HA and loaded with ferulic acid (FA) were investigated for effective asthma control. An *in vivo* inhalation toxicity assessment showed that the as-prepared NPs are safe. FACHA prophylaxis reduces inflammation, airway hyperresponsiveness, and remodeling in ovalbumin-induced mouse models. The results showed that the NPs had outstanding features such as easy transport across the mucus barrier gained by NPs with HA modification and FA's thermal stability (Dhayanandamoorthy et al. 2020). In another study, Lima et al. (2018) carried out the preparation and *in vitro* evaluation of FA-loaded PLGA NPs coated with CS, which was administered orally. In this method, particles with an average diameter of 242 nm and positive zeta potential were prepared using the emulsion evaporation technique. The particles obtained showed an encapsulation efficiency of 50%. In the *in vitro* evaluation, similar properties with the free drug were observed to have HOCl-scavenging activity and cytotoxicity on B16-F10 and HeLa cells. Thus, the nanoparticulate formulation can be used to treat orally without decreasing treatment efficacy. In addition, CS contributed positively to the therapeutic effect of FA in antioxidant and cytotoxic tests (Lima et al. 2018).

Piperine ($C_{17}H_{19}NO_3$) is an alkaloid found in the fruits and roots of some plants in the Piperaceae family. Recently, piperine can be used in rheumatoid arthritis-like diseases. However, as with other phytotherapeutic compounds, piperine may have disadvantages in terms of drug bioavailability. Attempts have been made to solve this problem by designing various NP approaches. Bhalekar et al. (2017) reported a method to prepare piperine encapsulated by SLNs, which could be applied orally and topically. To prepare the most suitable formulation in the study, a piperine SLN dispersion was formulated using a 32 full factorial design. The prepared SLN dispersion was administered orally and topically to rats with arthritis. Piperine from the SLN gel formulation accumulates in the skin as observed in *ex vivo* studies using the Franz diffusion cell. The results of pharmacodynamic studies suggested that both topical and oral administrations of piperine have shown significant efficacy when compared with orally administered chloroquine suspension (Bhalekar et al. 2017).

Quercetin is a powerful antioxidant and shows antipsoriatic activity as a COX antagonist and COX-2 selective antagonist, exerting antiproliferative and anti-inflammatory effects. Recently, the role of quercetin in psoriasis, a chronic autoimmune skin disease characterized by thick patches of red, scaly, and itchy, has been studied. In the study, the authors tried to prepare new liposphere-based drug release systems for psoriasis treatment, prepared using antipsoriatic herbal drugs such as *Commiphora mukul* and quercetin. Separate liposphere structures were prepared for *Commiphora mukul* and quercetin using Precirol ATO 5, phosphotidylcholine, and poloxamer 338. The physical and structural properties of the liposphere structures such as particle size and drug encapsulation efficiency were investigated. Release and *in vitro* permeability tests were investigated using Franz diffusion cells. The maximum encapsulation efficiency for quercetin-loaded lipospheres was approximately 81.77% in that study. When combined with the results obtained in *in vitro* studies, the prepared liposphere gel provided more extended release than conventional creams. Formulations containing quercetin prepared with these aspects may be effective in treating psoriasis (Mestry, Rane, and Bajaj 2020). In another study, Wang et al. (2019) showed that quercetin was encapsulated by nanomicelles using amphiphilic CS as the shell material. The resulting nanomicelles released quercetin for 0–8 hours. The release rate gradually decreased after this time, and 40% of the quercetin was released at approximately 72 hours. Hemocompatibility, good water solubility, and biocompatibility were significantly increased in the quercetin-loaded nanomicelle formulations used in the study (Wang et al. 2019). In another study, Simon et al. (2019) used harmless amphiphilic polyoxazolines (POx) to release quercetin. In this method, micelles consisting of POx and phosphatidylcholine were synthesized using a high-pressure homogenization process. When the particle diameters of the micelles were examined, they were found to have a spherical shape with a diameter of approximately 20 nm. The quercetin-loading efficiency of these micelles was relatively high and was determined as 94.4%. When the micelle structures were compared with those of free quercetin in terms of *in vitro* activity, the micelle structures showed improved cell viability and antioxidant activity (Simon et al. 2019).

Caldas et al. (2021) prepared different nanolipid systems containing resveratrol and omega 3 (ω3). In the approach developed for the effective delivery of omega 3 and resveratrol from the skin layer, which finds application in dermocosmetics, resveratrol could be encapsulated only in liposomes that provided an adequate amphiphilic environment. Each formulation prepared in the study were characterized for its physical and physicochemical properties. The *in vitro* treatment performances of the prepared nanoformulations, the evaluation of their protective role against lipid peroxidation in the RAW 264.7 cell line, and their effects on the inhibition of COX and NO production were investigated. The study emphasized that the inclusion of both COX and NO lipid nanosystems in the same semi-solid form could be a promising strategy for the treatment of autoimmune disorders, inflammatory diseases, and skin cancer (Caldas et al. 2021). Jin et al. (2021) designed a CS/alginate nano-drug delivery system to treat colitis that may occur due to various autoimmune diseases. In this system, resveratrol was used as the active ingredient, which may have efficacy against colitis. The *in vivo* study examined dextran sodium sulphate-induced colitis in mice for amelioration. The results showed that resveratrol-loaded particles could target the inflammatory colon for maximum effectiveness (Jin et al. 2021).

Bisbenzylisoquinoline alkaloid tetrandrine is widely used in clinical application in the therapy of rheumatoid arthritis. Considering its properties, different nanoformulation approaches need to be developed in transport and targeting. Hu et al. (2022) synthesized an amphiphilic PLGA with PEGylation as a topical drug carrier for tetrandrine in an attempt to increase drug loading and preserve stability. To gain controlled release of the encapsulated drugs, the authors successfully developed a pH-sensitive release system by applying a controllable approach with $CaCO_3$. In addition, the prepared nanoformulation could be integrated into a soluble microneedle to increase and regulate drug uptake, thus achieving an increased synovial uptake of tetrandrine through the transdermal route. With the developed strategy, stealth escape from phagocytes and inflammatory acid-initiated release increased the efficiency. With this method, the improvement observed in arthritis includes stronger regulation of the vascular endothelial growth factor, JAK2/p-JAK2, and STAT3/p-STAT3 pathways, superior to the deletion of specific currently engineered delivery ability (Hu et al. 2022).

9.6 CONCLUSION

Although autoimmune diseases are seen in a very specific part of the population, more attention should be paid to these diseases because they reduce patients' quality of life and cause morbidity. Because of the high cost and side effects of drugs used in the long term in the treatment of autoimmune diseases, the use of phytochemicals is preferred in the treatment of these diseases.

The fact that phytochemical compounds are less harmful than synthetic chemicals increases the popularity of these compounds. However, only a few studies have assessed phytochemical compounds used for the treatment of autoimmune diseases. Moreover, the low bioavailability of these compounds creates a need for formulations prepared using different approaches.

Different nanoformulations improve the low bioavailability properties of phytotherapeutic chemicals in the free form. The bioavailability properties of the nanoformulations were assessed based on *in vitro* and *in vivo* results, which revealed that drug-NP bioavailability was considerably increased compared with that of their free form. However, future studies should examine drug release studies in terms of different diseases as an application of phytonanotechnology to target and examine their potential use in tissues in different diseases. Studies on different bioinspired NPs should identify an approach to achieve both controlled and targeted release of phytotherapeutic agents.

REFERENCES

Ahmed, Osama M, Tarek Mohamed, Hala Moustafa, Hany Hamdy, Rasha R Ahmed, and Ebtsam Aboud. 2018. *Biomedicine & Pharmacotherapy* 101:58–73.

Ahrari, Asma, Hossein Najafzadehvarzi, Ali Taravati, and Fatemeh Tohidi. 2021. *Life Sciences* 284:119884.

Aktas, Orhan, Timour Prozorovski, Alina Smorodchenko, Nicolai E Savaskan, Roland Lauster, Peter-Michael Kloetzel, Carmen Infante-Duarte, Stefan Brocke, and Frauke Zipp. 2004. *The Journal of Immunology* 173 (9):5794–5800.

Altaf S Darvesh, Aggarwal Bharat B, and Bishayee Anupam. 2012. *Current Pharmaceutical Biotechnology* 13 (1):218–228.

Arumugam, Somasundaram, Rajarajan A. Thandavarayan, Wawaimuli Arozal, Flori R. Sari, Vijayasree V. Giridharan, Vivian Soetikno, Suresh S. Palaniyandi, Meilei Harima, Kenji Suzuki, Masaki Nagata, Ritsuo Tagaki, Makoto Kodama, and Kenichi Watanabe. 2012. *Free Radical Research* 46 (2):154–163.

Aziz, Zarith, Asyikin Abdul, Siti Hamidah, and Mohd Setapar. 2020. *Functional Bionanomaterials*:47–71.

Babu, Mahesh, Chinnu Jerard, Bibin Punnoose Michael, Sini Suresh, and Rajesh Ramachandran. 2018. *Journal of Applied Pharmaceutical Science* 8 (12):124–131.

Bacanlı, Merve, Sevtap Aydın, A Ahmet Başaran, and Nurşen Başaran. 2017. *Food and Chemical Toxicology* 109:210–217.

Bacchi, Simona, Paola Palumbo, A Sponta, and Maria Francesca Coppolino. 2012. *Anti-Inflammatory & Anti-Allergy Agents in Medicinal Chemistry (Formerly Current Medicinal Chemistry-Anti-Inflammatory and Anti-Allergy Agents)* 11 (1):52–64.

Behl, Tapan, Keshav Kumar, Ciprian Brisc, Marius Rus, Delia Carmen Nistor-Cseppento, Cristiana Bustea, Raluca Anca Corb Aron, Carmen Pantis, Gokhan Zengin, and Aayush Sehgal. 2021. *Biomedicine & Pharmacotherapy* 133:110959.

Bettelli, Estelle, Thomas Korn, Mohamed Oukka, and Vijay K Kuchroo. 2008. *Nature* 453 (7198):1051–1057.

Bhalekar, Mangesh R., Ashwini R. Madgulkar, Puja S. Desale, and Gyce Marium. 2017. *Drug Development and Industrial Pharmacy* 43 (6):1003–1010.

Botha, Tarryn L, Elias E Elemike, Suranie Horn, Damian C Onwudiwe, John P Giesy, and Victor Wepener. 2019. *Scientific Reports* 9 (1):1–8.

Brindha, P. 2016. *International Journal of Green Pharmacy* 10 (1):1–18.

Cai, Feiyang, Sailiang Liu, Yunxuan Lei, Shuxin Jin, Zizhen Guo, Dehao Zhu, Xin Guo, Hanqing Zhao, Xiaoyin Niu, Yebin Xi, Zhaojun Wang, and Guangjie Chen. 2021. *Cellular Immunology* 368:104421.

Caldas, Ana R., José Catita, Raul Machado, Artur Ribeiro, Fátima Cerqueira, Bruno Horta, Rui Medeiros, Marlene Lúcio, and Carla M. Lopes. 2021. *Pharmaceutics* 13 (8):1202.

Carullo, Gabriele, Anna Rita Cappello, Luca Frattaruolo, Mariateresa Badolato, Biagio Armentano, and Francesca Aiello. 2017. *Future Medicinal Chemistry* 9 (1):79–93.

Castelo-Branco, Camil, and Iris Soveral. 2014. *Gynecological Endocrinology* 30 (1):16–22.

Chandran, Binu, and Ajay Goel. 2012. *Phytotherapy Research* 26 (11):1719–1725.

Chauhan, Prashant Singh, Naresh Kumar Satti, Punita Sharma, Vijay Kumar Sharma, Krishan Avtar Suri, and Sarang Bani. 2012. *Phytotherapy Research* 26 (8):1156–1165.

Chen, Guo Qian, Yan Yan Chen, Xin Shi Wang, Sai Zhen Wu, Hui Min Yang, Hui Qin Xu, Jin Cai He, Xiao Tong Wang, Jiang Fan Chen, and Rong Yuan Zheng. 2010. *Brain Research* 1309:116–125.

Chien, Shih-Chang, Yueh-Chen Wu, Zeng-Weng Chen, and Wen-Chin Yang. 2015. *Evidence-Based Complementary and Alternative Medicine* 2015:1–13.

Chouhan, Neelu. 2018., InTech Open, London, United Kingdom.

Cooper, Glinda S, and Berrit C Stroehla. 2003. *Autoimmunity reviews* 2 (3):119–125.

Cordell, Geoffrey A, Mary Lou Quinn-Beattie, and Norman R Farnsworth. 2001. *Phytotherapy Research: An International Journal Devoted to Pharmacological and Toxicological Evaluation of Natural Product Derivatives* 15 (3):183–205.

Cui, Guoliang, Xia Qin, Yuebo Zhang, Zhenwei Gong, Baoxue Ge, and Ying Qin Zang. 2009. *Journal of Biological Chemistry* 284 (41):28420–28429.

Dai, Qiaoding, Liping Xu, Di Zhou, and Xinwei Song. 2018. *Drug Design, Development and Therapy* 12:4095.

Dang, Yalong, Yongsheng Xu, Wentao Wu, Weiyi Li, Yanran Sun, Jing Yang, Yu Zhu, and Chun Zhang. 2014. *PloS one* 9 (8):e102522.

Dhayanandamoorthy, Yamini, M. Gover Antoniraj, Chaya A. Babu Kandregula, and Ruckmani Kandasamy. 2020. *International Journal of Pharmaceutics* 591:119958.

Dhupal, Madhusmita, and Devasish Chowdhury. 2020. *International Journal of Nanomedicine* 15:9125–9157.

dos Santos, Mariane, Priscila Tamar Poletti, Gaia Favero, Alessandra Stacchiotti, Francesca Bonomini, Carolina Caruccio Montanari, Silvia Regina Bona, Norma Possa Marroni, Rita Rezzani, and Francisco Veríssimo Veronese. 2018. *Autoimmunity* 51 (2):69–80.

Du, Ziyu, Qingfeng Wang, Xinyue Huang, Shenglan Yi, Suyin Mei, Gangxiang Yuan, Guannan Su, Qingfeng Cao, Chunjiang Zhou, Yao Wang, Aize Kijlstra, and Peizeng Yang. 2020. *International Immunopharmacology* 81:106270.

Eaton, William W, Noel R Rose, Amanda Kalaydjian, Marianne G Pedersen, and Preben Bo Mortensen. 2007. *Journal of Autoimmunity* 29 (1):1–9.

El-Missiry, Mohamed A., Azza I. Othman, Maher A. Amer, Mohammed Sedki, Sara M. Ali, and Ibrahim M El-Sherbiny. 2020. *Metabolic Brain Disease* 35 (2):385–399.

Elahi, Narges, Mehdi Kamali, and Mohammad Hadi Baghersad. 2018. *Talanta* 184:537–556.

Elmotasem, Heba, Hala K Farag, and Abeer A.A. Salama. 2018. *International Journal of Pharmaceutics* 547 (1–2):83–96.

Fairweather, DeLisa. 2007. *Encyclopedia of Life Sciences* 2007:1–6.

Fan, Xia, Chun Zhang, Dong-bo Liu, Jun Yan, and Hua-ping Liang. 2013. *Current Pharmaceutical Design* 19 (11):2011–2031.

Fan, Xing-Xing, Meng-ze Xu, Elaine Lai-Han Leung, Cai Jun, Zhen Yuan, and Liang Liu. 2020. *Nano-micro Letters* 12 (1):1–14.

Firenzuoli, Fabio, and Luigi Gori. 2007. *Evidence-Based Complementary and Alternative Medicine* 4 (S1):37–40.

Fonseca-Kelly, Zoe, Mayssa Nassrallah, Jorge Uribe, Reas Khan, Kimberly Dine, Mahasweta Dutt, and Kenneth Shindler. 2012. *Frontiers in Neurology* 3 (84):1–9.

Funk, Janet L, Jennifer B Frye, Janice N Oyarzo, Nesrin Kuscuoglu, Jonathan Wilson, Gwen McCaffrey, Gregory Stafford, Guanjie Chen, R Clark Lantz, and Shivanand D Jolad. 2006. *Arthritis & Rheumatism: Official Journal of the American College of Rheumatology* 54 (11):3452–3464.

Gabrilovich, Dmitry I, and Srinivas Nagaraj. 2009. *Nature Reviews Immunology* 9 (3):162–174.

Geissmann, Frederic, Markus G Manz, Steffen Jung, Michael H Sieweke, Miriam Merad, and Klaus Ley. 2010. *Science* 327 (5966):656–661.

Gillespie, Kevin, Isamu Kodani, Douglas P Dickinson, Kalu UE Ogbureke, Amy M Camba, Mengjie Wu, Stephen Looney, Tin-Chun Chu, Haiyan Qin, and Frederick Bisch. 2008. *Life Sciences* 83 (17–18):581–588.

Hatcher, H, R Planalp, J Cho, FM Torti, and SV Torti. 2008. *Cellular and Molecular Life Sciences* 65 (11):1631–1652.

Hou, Dian-Dong, Wei Zhang, Ya-Li Gao, Yu-zhe Sun, He-Xiao Wang, Rui-Qun Qi, Hong-Duo Chen, and Xing-Hua Gao. 2019. *International Immunopharmacology* 74:105676.

Hsu, Stephen, Douglas P Dickinson, Haiyan Qin, Carol Lapp, David Lapp, James Borke, Douglas S Walsh, Wendy B Bollag, Hubert Stöppler, and Tetsuya Yamamoto. 2005. *Journal of Pharmacology and Experimental Therapeutics* 315 (2):805–811.

Hu, Hongmei, Hang Ruan, Shuyao Ruan, Lixia Pei, Qian Jing, Tong Wu, Xiaolin Hou, Hao Xu, Youjie Wang, Nianping Feng, and Yongtai Zhang. 2022. *Chemical Engineering Journal* 431:134196.

Huang, Hsin-Yi, Ming-Chen Wang, Zhi-Yu Chen, Wen-Ying Chiu, Ko-Hua Chen, I-Chan Lin, Wei-Chung Vivian Yang, Chi-Chang Wu, and Ching-Li Tseng. 2018. *International Journal of Nanomedicine* 13:7251.

Hwang, Su Jung, Sang Hui Jun, Yohan Park, Song-Hyun Cha, Minho Yoon, Seonho Cho, Hyo-Jong Lee, and Youmie Park. 2015. *Nanomedicine: Nanotechnology, Biology and Medicine* 11 (7):1677–1688.

Hwang, Su Jung, Yong-Wan Kim, Yohan Park, Hyo-Jong Lee, and Kyu-Won Kim. 2014. *Inflammation Research* 63 (1):81–90.

Imanshahidi, Mohsen, and Hossein Hosseinzadeh. 2008. *Phytotherapy Research* 22 (8):999–1012.

Javadi, Fatemeh, Arman Ahmadzadeh, Shahryar Eghtesadi, Naheed Aryaeian, Mozhdeh Zabihiyeganeh, Abbas Rahimi Foroushani, and Shima Jazayeri. 2017. *Journal of the American College of Nutrition* 36 (1):9–15.

Jin, Mengfei, Shangyong Li, Yanhong Wu, Dandan Li, and Yantao Han. 2021. *Nanomaterials* 11 (8):1884.

Kapoor, Anil. 2008. *Drugs* 68 (1):11–19.

Kaufman, Dixon B, Ron Shapiro, Michael R Lucey, Wida S Cherikh, Rami T. Bustami, and David B Dyke. 2004. *American Journal of Transplantation* 4:38–53.

Khan, Haroon, Hammad Ullah, Sumaira Khattak, Michael Aschner, Cristobal N Aguilar, Syed MA Halimi, Omar Cauli, and Syed MM Shah. 2021. *Phytotherapy Research* 35 (1):50–62.

Khan, Tabassum, and Pranav Gurav. 2018. *Frontiers in Pharmacology* 8:1002.

Kim, Hong Ro, Rajesh Rajaiah, Qing-Li Wu, Shailesh R Satpute, Ming T Tan, James E Simon, Brian M Berman, and Kamal D Moudgil. 2008. *The Journal of Nutrition* 138 (11):2111–2116.

Kour, Gurleen, Syed Assim Haq, Bijender Kumar Bajaj, Prem N Gupta, and Zabeer Ahmed. 2021. *Pharmacological Research*:105618.

Kowluru, Renu A, and Mamta Kanwar. 2007. *Nutrition & Metabolism* 4 (1):1–8.

Lai, Jenn-Haung. 2002. *Acta Pharmacologica Sinica* 23 (12):1093–1101.

Lee, Dong-Ung, Young Jin Kang, Min Kyu Park, Young Soo Lee, Han Geuk Seo, Tae Sung Kim, Cheorl-Ho Kim, and Ki Churl Chang. 2003. *Life Sciences* 73 (11):1401–1412.

Li, Chengjun, and Bing Yan. 2020. *Environmental Science: Nano* 7 (10):2863–2874.

Li, Fei, Hua-dong Wang, Da-Xiang Lu, Yan-ping Wang, Ren-bin Qi, Yong-mei Fu, and Chu-jie Li. 2006. *Acta Pharmacologica Sinica* 27 (9):1199–1205.

Li, H, X-L Li, M Zhang, H Xu, C-C Wang, S Wang, and R-S Duan. 2014. *Scandinavian Journal of Immunology* 79 (1):12–19.

Li, Junfeng. 2018. *Experimental and Therapeutic Medicine* 15 (1):1087–1092.

Li, Xiang, Zhongxiu Wu, Bin He, and Wei Zhong. 2018. *Experimental and Therapeutic Medicine* 16 (3):2670–2676.

Lima, Isabela Angeli de, Najeh Maissar Khalil, Tania Toyomi Tominaga, Anna Lechanteur, Bruno Sarmento, and Rubiana Mara Mainardes. 2018. *Artificial Cells, Nanomedicine, and Biotechnology* 46 (Suppl 2):993–1002.

Liu, Xuefei, Xinghua Zhang, Lin Ye, and Haitao Yuan. 2016. *Biomedicine & Pharmacotherapy* 79:222–230.

Lopomo, Angela, and Sonia Berrih-Aknin. 2017. *Frontiers in Endocrinology* 8:169.

Lorenz, Hanns-Martin, Joachim R. Kalden. 2001. *Scandinavian Journal of Clinical and Laboratory Investigation* 61 (235):16–26.

Lou, Lixia, Jingwei Zhou, Yujun Liu, YI Wei, Jiuli Zhao, Jiagang Deng, Bin Dong, Lingqun Zhu, Aiming Wu, and Yingxi Yang. 2016. *Experimental and Therapeutic Medicine* 11 (5):2054–2060.

Lu, Lisen, Shuhong Qi, Yuzhou Chen, Haiming Luo, Songlin Huang, Xiang Yu, Qingming Luo, and Zhihong Zhang. 2020. *Biomaterials* 245:119987.

Lushchak, Oleh, Olha Strilbytska, Alexander Koliada, Alina Zayachkivska, Nadia Burdyliuk, Ihor Yurkevych, Kenneth B. Storey, and Alexander Vaiserman. 2020. *GeroScience* 42 (1):117–139.

Ma, Qingle, Jinyu Bai, Jialu Xu, Huaxing Dai, Qin Fan, Ziying Fei, Jiacheng Chu, Chenlu Yao, Haoliang Shi, and Xiaozhong Zhou. 2021. *Advanced NanoBiomed Research* 1 (11):2100071.

Ma, Xiaomeng, Ying Jiang, Aimin Wu, Xiaohong Chen, Rongbiao Pi, Mei Liu, and Yingying Liu. 2010. *PLoS One* 5 (10):e13489.

Mackay, Ian R, and Noel R Rose. 2013. *The Autoimmune Diseases*: Elsevier Inc.

Madkour, Loutfy H. 2019. In: *Nanoelectronic Materials*, 1–47. Springer.

Makky, Amna, Eman Sadddar, and Abeer Khattab. 2021. *International Journal of Applied Pharmaceutics* 14 (4), 235–250.

Mallik, Shreyasi, AB Sharangi, and Tapas Sarkar. 2020. *National Academy Science Letters* 43 (5):477–480.

Mestry, Madhura, Meenal Rane, and Amrita Bajaj. 2020. *Indian Journal of Pharmaceutical Education and Research* 54 (3):654–667.

Miller, JFAP. 1993. *Immunologic Research* 12 (2):115–130.

Molyneux, Russell J, Stephen T Lee, Dale R Gardner, Kip E Panter, and Lynn F James. 2007. *Phytochemistry* 68 (22–24):2973–2985.

Muthian, Gladson, and John J Bright. 2004. *Journal of Clinical Immunology* 24 (5):542–552.

Naveed, Muhammad, Veghar Hejazi, Muhammad Abbas, Asghar Ali Kamboh, Ghulam Jilany Khan, Muhammad Shumzaid, Fawwad Ahmad, Daryoush Babazadeh, Xia FangFang, and Faezeh Modarresi-Ghazani. 2018. *Biomedicine & Pharmacotherapy* 97:67–74.

Nawrot, P., S. Jordan, J. Eastwood, J. Rotstein, A. Hugenholtz, and M. Feeley. 2003. *Food Additives & Contaminants* 20 (1):1–30.

Ohishi, Kazuhito, Makoto Kanoh, Hiroto Shinomiya, Yasuo Hitsumoto, and Sayaka Utsumi. 1995. *The Journal of Immunology* 154 (7):3173–3179.

Paladino, Rachel A., Sierra N. Miller, Karlee F. Kleiber, and Donna M. Byers. 2020. *European Journal of Integrative Medicine* 36:101137.

Park, Cheol, Dong-Oh Moon, Il-Whan Choi, Byung Tae Choi, Taek-Jeong Nam, Chung-Ho Rhu, Taeg Kyu Kwon, Won Ho Lee, Gi-Young Kim, and Yung Hyun Choi. 2007. *International Journal of Molecular Medicine* 20 (3):365–372.

Petro, Thomas M. 2011. *International Immunopharmacology* 11 (3):310–318.

Ponticelli, C, A Tarantino, M Campise, G Montagnino, A Aroldi, and P Passerini. 2004. *Transplantation Proceedings* 36(2):S557–S560.

Qin, Xia, Bingshi T Guo, Bing Wan, Lei Fang, Limin Lu, Lili Wu, Ying Qin Zang, and Jingwu Z Zhang. 2010. *The Journal of Immunology* 185 (3):1855–1863.

Rajeshkumar, S, Soumya Menon, S Venkat Kumar, Murtaza M Tambuwala, Hamid A Bakshi, Meenu Mehta, Saurabh Satija, Gaurav Gupta, Dinesh Kumar Chellappan, and Lakshmi Thangavelu. 2019. *Journal of Photochemistry and Photobiology B: Biology* 197:111531.

Rauf, Abdur, Muhammad Imran, Ilkay Erdogan Orhan, and Saud Bawazeer. 2018. *Trends in Food Science & Technology* 74:33–45.

Retamozo, Soledad, Pilar Brito-Zerón, Luca Quartuccio, Salvatore De Vita, and Manuel Ramos-Casals. 2017. *Expert Review of Clinical Pharmacology* 10 (10):1085–1101.

Rizvi, Syed AA, and Ayman M Saleh. 2018. *Saudi Pharmaceutical Journal* 26 (1):64–70.

Roberts, Stephen M, and Louis Adams. 2000. *Principles of Toxicology: Environmental and Industrial Applications*:189–206.

Sethi, Gautam, Bokyung Sung, and Bharat B Aggarwal. 2008. *Experimental Biology and Medicine* 233 (1):21–31.

Shaik, Mohammed Rafi, Mujeeb Khan, Mufsir Kuniyil, Abdulrahman Al-Warthan, Hamad Z Alkhathlan, Mohammed Rafiq H Siddiqui, Jilani P Shaik, Anis Ahamed, Adeem Mahmood, and Merajuddin Khan. 2018. *Sustainability* 10 (4):913.

Shen, Pan, Weiji Lin, Xuan Deng, Xin Ba, Liang Han, Zhe Chen, Kai Qin, Ying Huang, and Shenghao Tu. 2021. *Frontiers in Immunology* 12:1991.

Shin, Seong Ah, Byeong Jun Joo, Jun Seob Lee, Gyoungah Ryu, Minjoo Han, Woe Yeon Kim, Hyun Ho Park, Jun Hyuck Lee, and Chang Sup Lee. 2020. *Molecules* 25 (24):5932.

Simon, L, M Vincent, S Le Saux, V Lapinte, N Marcotte, M Morille, C Dorandeu, JM Devoisselle, and S Bégu. 2019. *International Journal of Pharmaceutics* 570:118516.

Singh, Ram P, Richard T Waldron, and Bevra H Hahn. 2012. *Autoimmunity Reviews* 11 (9):664–669.

Tang, Xu-Lei, Xiao-Ju Liu, Wei-Ming Sun, Jin Zhao, and Rong-Liang Zheng. 2005. *Die Pharmazie-An International Journal of Pharmaceutical Sciences* 60 (9):696–700.

Temple, Jennifer L., Christophe Bernard, Steven E. Lipshultz, Jason D. Czachor, Joslyn A. Westphal, and Miriam A. Mestre. 2017. *Frontiers in Psychiatry* 8 (80):1–19.

Theofilopoulos, Argyrios N, Dwight H Kono, and Roberto Baccala. 2017. *Nature Immunology* 18 (7):716–724.

Veldman, Christian, Angela Nagel, and Michael Hertl. 2006. *International Archives of Allergy and Immunology* 140 (2):174–183.

Vivier, Eric, Elena Tomasello, Myriam Baratin, Thierry Walzer, and Sophie Ugolini. 2008. *Nature Immunology* 9 (5):503–510.

Wang, Dong, Shi-Ping Li, Jin-Sheng Fu, Lin Bai, and Li Guo. 2016. *International Journal of Developmental Neuroscience* 49:60–66.

Wang, Hui-Qi, Kai-Yi Song, Jin-Zhou Feng, Si-Yuan Huang, Xiu-Ming Guo, Lei Zhang, Gang Zhang, Ying-Chao Huo, Rong-Rong Zhang, Yue Ma, Qing-Zhe Hu, and Xin-Yue Qin. 2021. *Journal of Molecular Neuroscience*. https://doi.org/10.1007/s12031-021-01894-8.

Wang, Hui, Ziming Yang, Zuyu He, Chuang Zhou, Chao Wang, Yu Chen, Xianwu Liu, Sidong Li, and Puwang Li. 2019. *Colloids and Surfaces B: Biointerfaces* 179:519–526.

Wang, Lifeng, Fu-Sheng Wang, and M Eric Gershwin. 2015. *Journal of Internal Medicine* 278 (4):369–395.

Wang, Tan, Na-na Xi, Yu Chen, Xiao-feng Shang, Qiang Hu, Jiang-fan Chen, and Rong-yuan Zheng. 2014. *Neuropharmacology* 86:203–211.

Wang, Xue, Xin He, Chun-Feng Zhang, Chang-Run Guo, Chong-Zhi Wang, and Chun-Su Yuan. 2017. *Biomedicine & Pharmacotherapy* 89:887–893.

Wang, Zhuo-Long, Xiao-Fang Luo, Meng-Tao Li, Dong Xu, Shuang Zhou, Hou-Zao Chen, Na Gao, Zhen Chen, Ling-Ling Zhang, and Xiao-Feng Zeng. 2014. *PLoS One* 9 (12):e114792.

Wong, Germaine, and Jeremy R Chapman. 2008. *Transplantation Reviews* 22 (2):141–149.

Wongeakin, Natchaya, Parvapan Bhattarakosol, and Suthiluk Patumraj. 2014. *BioMed Research International* 2014:1–11.

Xue, Yang, Ying Wang, De-chun Feng, Bao-guo Xiao, and Ling-yun Xu. 2008. *Acta Pharmacologica Sinica* 29 (2):245–251.

Yaktine, Ann, and Leslie Pray. 2009. *Nanotechnology in Food Products: Workshop Summary*: National Academies Press.

Yatoo, Mohd, Arumugam Gopalakrishnan, Archana Saxena, Oveas R Parray, Noore A Tufani, Sandip Chakraborty, Ruchi Tiwari, Kuldeep Dhama, and Hafiz Iqbal. 2018. *Recent Patents on Inflammation & Allergy Drug Discovery* 12 (1):39–58.

Yoshida, Yuki, Tetsuo Shioi, and Tohru Izumi. 2007. *Circulation Journal* 71 (3):397–404.

Yuan, Kai, Qingqing Zhu, Qingyi Lu, Haixu Jiang, Mengmeng Zhu, Xiaohong Li, Guangrui Huang, and Anlong Xu. 2020. *The Journal of Nutritional Biochemistry* 84:108454.

Yuan, Xusheng, Bei Tong, Yannong Dou, Xin Wu, Zhifeng Wei, and Yue Dai. 2016. *Biochemical Pharmacology* 101:87–99.

Zheng, Zhaoling, YanHua Sun, Ziliang Liu, Mingqin Zhang, Chunqing Li, and Hui Cai. 2015. *Drug Design, Development and Therapy* 9:4931.

Zhu, Shenting, Lingxin Zhu, Jingjing Yu, Yanqing Wang, and Bin Peng. 2019. *International Journal of Nanomedicine* 14:5017.

Zipp, Frauke, and Orhan Aktas. 2006. *Trends in Neurosciences* 29 (9):518–527.

10 Phytochemicals as Therapeutics for Respiratory Disorders

Nitin Verma[1*], *Neha Kanojia*[1], *B. Mishra*[2], *and Kamal Dua*[3]
[1]Chitkara University, School of Pharmacy, Chitkara University, Himachal Pradesh 174103, India
[2]Department of Pharmaceutics, Institute of Technology, Banaras Hindu University, Varanasi, India
[3]Discipline of Pharmacy, Graduate School of Health, University of Technology Sydney, PO Box 123 Broadway NSW 2007, Australia

10.1 INTRODUCTION

Anatomically, the lower and upper respiratory systems are split into two segments in the throat region. Respiratory problems affect the lungs, respiratory tract, and blood vessels. Many respiratory diseases involve a combination of these organs[1,2]. The most common respiratory diseases are obstructive pulmonary disease, restrictive lung disease, pulmonary vascular disease, and other disorders[2]. Traditional medicine has a long and illustrious history of providing care to people worldwide. Phytotherapy, which is an intrinsic approach that uses extracts derived from natural sources for therapeutic effects or health-promoting characteristics, is commonly referred to as "alternative medicine." Natural sources have garnered extensive attention lately, and they have emerged as promising candidates for disease prevention because they provide a variety of health benefits with fewer side effects and less severe toxicity[3,4,5,6].

Natural components play a crucial role in the development and testing of new drugs. The use of medicinal plants has shown positive outcomes in the prevention and treatment of disorders that are difficult-to-cure conditions, such as atherosclerosis. Non-nutritive plant-based bioactive molecules that exert disease-preventive effects are known as phytochemicals. These are non-essential nutrients prepared by plants. Dietary intake of phytochemicals provides health benefits and protects the body from illness. Because of its many uses, such plant-derived bio active compounds have lately gained increasing attention. Traditional medicines are currently the greatest bio-resources for elements of ancient healthcare systems, contemporary medications, nutraceuticals, dietary supplements, folk remedies, and molecular entities for manufactured pharmaceuticals[7]. Natural antioxidants are divided into two main categories: *in vivo* and *in vitro* antioxidants. Hydrogen donors, peroxide decomposers, electron donors, enzyme inhibitors, synergists, and metal-chelating and singlet-oxygen quenchers functions as free radical scavengers. Flavonoids, polyphenols, isoflavonoids, phytoestrogens, anthocyanidins, carotenoids, terpenoids, phytosterols, glucosinolates, and limonoids are some of the phytochemicals with potential health benefits[8].

10.2 TYPES OF PHYTOCHEMICALS

The exact classification of phytochemicals is yet to be determined owing to their various activities and structures. Phytochemicals are divided into two major categories based on metabolic functions of plants: primary metabolites and secondary metabolites. Primary metabolites include amino acids,

DOI: 10.1201/9781003231721-10

proteins, nucleic acids, carbohydrates. Secondary metabolites are residual plant compounds such as terpenes, lignans, flavonoids, plant steroids, and alkaloids. A literature review showed that phenolics are the most frequent and structurally varied plant compounds.

10.2.1 Essential Oils

Several chemical components, primarily mono- and sesquiterpenoids, as well as phenylpropanoids, are present in essential oils. Because of their volatility as well as antibacterial and anti-inflammatory effects, they have excellent therapeutic effects for reducing the symptoms of respiratory diseases, particularly in the upper and lower respiratory tracts. Anise, tea tree, pepper mint, and thyme essential oils are widely used to treat respiratory disorders. Essential oils can be inhaled (e.g., eucalyptus oil), consumed orally (e.g., peppermint oil), or applied transdermally (e.g., peppermint oil) for therapeutic purposes (e.g., rosemary oil)[8,9].

10.2.2 Polyphenols

Phenolic compounds are the most abundant category of phytochemicals. Plant-based polyphenols are secondary plant metabolites present in many plant species. These compounds have molecular weight ranging from 500 to 3000–4000 D and are soluble in water owing to the presence of 12–16 phenolic groups and 57 aromatic rings per 1000 molecular mass, inter-molecular complex formation, and categorization as compressed proanthocyanins, galloyl and hexahydroxydiphenoyl ester and derivative products, or phlorotannin. Nutrition experts have long regarded polyphenols to be antinutrients because some polyphenols such as tannins have negative effects including lowering the activity of proteolytic enzymes, energy, protein and amino acid availability, and absorption of minerals and causing other harmful consequences[8,9,10]. Many edible fruits have abundant polyphenolic antioxidants and anti-inflammatory substances, and men who consume a diet rich in fruits and vegetables have a lower risk of chronic obstructive pulmonary disease (COPD).

10.2.3 Flavonoids

Flavonoids are a type of polyphenols that contain polyphenolic secondary plant metabolites of low molecular weight. Overall, approximately 8000 compounds are contained in this group. Vegetables, tea, nuts, fruits, stems, flowers, roots, seeds, wine, and coffee contain flavonoids[59,60]. They have a chromane and heterocyclic hydrocarbon structure. Flavans or isoflavans are obtained when a phenyl group (B-ring) is substituted at positions 2 and 3 of its ring C. Flavanones and isoflavanones are obtained when an oxo-group is present at position 4. Flavones and isoflavones are obtained when the carbon atoms C2 and C3 create a double bond. Such compounds are known as anthocyanidins because they have an extra double bond in the midst of C1 and C2. Flavonoids are classified into eight classes depending on their structure: isoflavanones, flavones, isoflavones, flavans, anthocyanidins, chalcones, flavanones, and flavonolignans[11,12,13,14,15].

10.2.4 Terpenes

Some species use certain substances for the synthesis of reproductive hormones and growth regulatory mechanisms (vitamin A), and the presence of such compounds in animal tissues offers protection, to some extent, against certain ailments, particularly diseases that are chronic and cause dysregulation. Terpenes do have particular antioxidant function with regard to free radicals. They respond to reactive oxygen species by migrating to the fatty membrane surface owing to their long carbon chain length. Extensive and in-depth research has focused on the two terpene antioxidants tocopherols and tocotrienols. Limonoids are a type of terpene present in citrus fruits. Several investigations in animal models of allergic asthma have revealed that some volatile terpene chemicals have anti-inflammatory effects on airway inflammation, which are exerted through anti-inflammatory responses. Inhalation of limonene can lower the levels of different proinflammation-related chemicals and alleviate airway fibrosis in a rat asthma model, suggesting that limonene has the ability to minimize airway remodeling[16,17,18].

10.2.5 Alkaloids

Alkaloids are compounds derived from plants and possess physiological activity or toxicity. Isopteropodine and pteropopine are antimicrobial alkaloids that actively stimulate white blood cells to regulate pathogenic microorganisms and cellular waste. Berberine, piperine, and harmane are highly aromatic planar quaternary alkaloids that intercalate DNA and cell walls (Cowan, 1999). Other alkaloids mimic neurotransmitters including acetylcholine, dopamine, and serotonin, which affect synapses in the central nervous system. Some of the alkaloids have therapeutic properties; for instance, narcotics; antimalarials; topical anesthetics for ophthalmology; and treatment of hypertension, respiratory disorders, neuralgia, rheumatism, motion sickness, and hormone longevity. They contain a wide range of analgesic characteristics; therefore, they are used to treat boils; infected wounds; and complaints such as headache, abdominal pain, and eye problems. They also have antineoplastic properties; for example, indole alkaloids are used in chemotherapy for leukemia and Hodgkin's disease. Indole alkaloids tend to be particularly effective against respiratory illnesses, which make them very attractive as prospective therapeutics. Several *in vivo* and *in vitro* studies have assessed the efficacy of indole alkaloids in the treatment of respiratory diseases such as asthma, emphysema, tuberculosis (TB), cancer, and pulmonary fibrosis. However, some alkaloids are hallucinogenic, addictive, and toxic and are therefore used as arrow poison in wild game hunting[19,20,21].

10.2.6 Antioxidant Vitamins

10.2.6.1 Carotenoids

Carotenoids are pigments present in plants and microorganisms. They help prevent or inhibit malignancies, atherosclerosis, immunological problems, asthma, and several-other diseases. Conjugated double bonds of carotenoids permit unpaired electrons to detach, thereby exerting antioxidant activity. To avoid cascading reactions that cause lipid peroxidation or DNA damage, singlet oxygen is quenched by carotene, without causing any damage to it, and when it comes into contact with free radicals such as peroxyl, hydroxyl, and superoxide.

Several transcription factors are regulated by carotenoids. Carotene enters cells that have undergone oxidative stress, and it inhibits the activation of oxidative stress-induced NF-κB and the subsequent generation of inflammatory cytokines such as interleukin (IL)-6 and tumor necrosis factor alpha (TNF-α), thus potentially influencing the apoptotic process in healthy cells. In normal cells, carotene can increase the production of the anti-apoptotic protein Bcl-2 but decrease the production of the pro-apoptotic protein Bax after external stimulation. In colon and leukemic cancer cells, carotene exerts a pro-apoptotic effect, which is attributed to a redox-dependent mechanism associated with NF-κB activity. Carotenoids are useful as anti-inflammatory agents for a variety of disorders towing to vitamin A's dual activities[22,23,24,25,26].

10.2.6.2 Vitamin C

Vitamin C, commonly known as ascorbic acid, is a significant and essential antioxidant that needs an aqueous physiological environment to exert its function. Vitamin C exists in two bioactive forms: ascorbic acid, and its oxidized derivative, namely, dehydro-ascorbic acid. These two forms are interconvertible. Vitamin C shields the cell membrane against oxidation by functioning as a reducing agent, thereby allowing oxidation to be reversed (i.e., hydrogen donor). Vitamin C inhibits the production of N-nitroso compounds through acid catalysis and mimics nitrosating species with secondary amines and amides. Higher doses of vitamin C (up to 2,000 mg per day) appear to have no positive effects. Vitamin C deficiency is linked to a higher incidence of asthma in children. Ascorbic acid levels in patients with asthma are significantly lower, and vitamin C treatment alleviates asthma in patients who are smokers by actively reducing cough and wheeze[27,28,29,30,31,32].

10.2.6.3 Vitamin E

Vitamin E is a fat-soluble vitamin that exists in eight distinct forms. It comprises a group of compounds that belong to families with intimate ties, namely tocopherols and tocotrienols, which exist in various isomeric forms (α, β, γ, and δ)[33]. The most active form of vitamin E in humans is α-tocopherol, which is regarded as the main antioxidant significantly utilized by membrane-bound cells[34,35,36,37]. Protection against lipid peroxidation is the main antioxidant function of α-tocopherol[38].

Lipid peroxidation alters cell membrane function and causes irreversible effects to metabolic pathways in the case of fatty acids (e.g., polyunsaturated fatty acids in cell membranes)[39,40]. Vitamin E interacts with nutrients, especially selenium and vitamin C. Ascorbic acid can reduce the α-tocopherol radical to the original α-tocopherol form[40,41,42].

10.2.6.4 Alpha-Lipoic Acid

α-Lipoic acid (LA), commonly known as thiothic acid, is a natural organosulfur chemical produced from octanoic acid. LA is taken into the body through our diet and quickly transformed to dihydrolipoic acid (DHLA), a dithiol form. LA and DHLA both have potent antioxidant properties[44]. Lipoyllysine is abundant in plants such as broccoli, tomatoes, and spinach. LA is a "non-vitamin" because it is an essential nutrient that is generated by the body. LA is involved in the oxidative decarboxylation of keto acids and is used as a growth factor by some species. LA is a fat- and water-soluble chemical that may easily cross cell membranes and be absorbed in the body. LA suppresses the function of transcription factors (e.g., NF-κB), and following the removal of free radicals, LA can replenish antioxidants such as vitamin C, vitamin E, and glutathione (GSH) for future use[43,44].

10.2.6.5 Phytoncides

Phytoncides are naturally occurring volatile substances produced by trees and plants to protect themselves from insects, animals, and microorganisms. The main components of phytoncides are terpenoids including carene, pinene, and myrcene, all of which are very volatile in nature[45]. Terpenoids obtained from plant sources are present in high quantities in essential oils. Monoterpenoids, which are particularly valuable in perfumery and flavoring, are separated from oils derived from higher plants.

Insect defense secretions, pheromones, and a variety of marine organisms all contain phytoncides, which are usually halogenated compounds. Japanese researchers have developed four different types of phytoncide solutions: A-type plants include chemical components generated by trees, AB-type plants include very bactericidal plants, D-type plants are flowering grass, and G-type plants are nonallergic plants[46]. An injection of a phytoncide solution reduces asthmatic symptoms and restores GSH levels, as proven in an animal model of asthma (unpublished information). To fully comprehend the mechanism of action of phytoncides in asthma, more research is needed. Volatile chemicals such as phytoncides are used as an additional method for treating asthma in the future[47,48,49,50,51,52,53,54,55].

TABLE 10.1
Different Phytochemicals Used for Treating Respiratory Disorders

Phytochemical	Disease	Mechanism of Action	Refs.
Isoliensinine	Pulmonary fibrosis	MDA, hydroxyproline, TGF-β, and TNF-α levels increase with decrease in SOD activity	[56]
β-Carboline	Pulmonary fibrosis	Decrease in the NF-κB/p65 mechanism, with reduction in vimentin, EMT, E-cadherin, and α-SMA expression	[57]
Matrine	Pulmonary fibrosis	Decrease in the JAK-STAT pathway	[58]
Rutecarpine	Pulmonary fibrosis	Decrease in eukaryotic initiation factor 3a signaling pathway and also decreases collagen I, EMT procedure, α-SMA and vimentin	[59]
Aloperine	Pulmonary fibrosis	Decrease in fibroblast differentiation and proliferation, with reduction in PI3K/AKT/mTOR signaling and TGF-β/Smad	[60]
Glaucocalyxin A	Pulmonary fibrosis	Deposition of hydroxyproline and collagen content reduces and decrease in infiltration of neutrophils and macrophages into the lungs	[61]
Triptolide	Pulmonary fibrosis	NOX4 and NOX2 levels in alveolar macrophages decrease	[61]
Ginkgolides meglumine	Pulmonary fibrosis	Decrease in levels of TNF-α and IL-1β	[63]
Aucubin	Pulmonary fibrosis	Disposition of collagen fibrils and reduction in inflammation	[62]
Nimbolide	Pulmonary fibrosis	Decrease in the infiltration of monocytes, lymphocytes, neutrophils, and leukocytes	[63]
Tanshinone IIA	Pulmonary fibrosis	Disposition of collagen fibrils and reduction in inflammation	[64]
Ginsenoside	Pulmonary fibrosis	Decrease in hydroxyproline, α-SMA, and TGF-β1 levels	[65]
Gentiopicroside	Pulmonary fibrosis	Decrease in the levels of TNF-α, IL-1β, CTGF, TGF-β1, and hydroxyproline	[66]
Lettuce glycoside B	Pulmonary fibrosis	Decrease in the levels of TGF-b1, TNF-α, and IL-6	[67]
Trans-Anethole	COPD	Reduction in the serum concentrations of the pro-inflammatory cytokines IL-6 and TNF-α	[68]
Andrographolide	COPD	Reduction in the concentration of IL-1β, MCP-1, and IP-10	[69]
Naringin	Chronic bronchitis	Reduction in the level of LTB4, IL-8, and TNF-α	[70]
Silymarin	COPD	Decrease in the levels of neutrophils and macrophages	[71]
Quercetin	COPD	Reduction in the inflammation of lungs and expression of MMP-12 and MMP-9	[72]
Mangiferin	Pulmonary fibrosis	Hydroxyproline, α-SMA, TGF-β1, and cytokine decreases	[73]
Isorhamnetin	Pulmonary fibrosis	Deposition of type I collagen and α-SMA levels decrease	[74]
Kaempferol	Pulmonary fibrosis	MMP-9 and MMP-2 levels reduce	[75]
Juglanin	Pulmonary fibrosis	Type I and type III collagen levels decrease	[76]

COPD: chronic obstructive pulmonary disease; MMP: matrix metalloprotein; α-SMA: alpha-smooth muscle actin; TGF-β1: transforming growth factor-beta 1; IL: interleukin; MCP-1: monocyte chemoattractant protein; TNF-α: tumor necrosis factor alpha; CTGF: connective tissue growth factor; IP-10: interferon-γ inducible protein 10; mTOR: mammalian target of rapamycin; EMT: epithelial-mesenchymal transition; NF-κB: nuclear factor kappa beta; MDA: malondialdehyde; SOD: superoxide dismutase

10.3 OVERVIEW OF RESPIRATORY DISORDERS

Sedentary lifestyle and physical inactivity over the years have opened a broad path toward a wide range of diseases that pose a great threat to humans. The most common diseases are respiratory disorders, which affect a wide spectrum of people in both middle- and low-income countries worldwide. Despite their considerable risk, respiratory disorders are frequently overlooked. Human lungs are regularly exposed to different environmental contaminants, including organic/inorganic/

biological agents that occur from a variety of natural and manmade sources. These harmful chemicals increase a person's risk of acquiring basic to complex pulmonary diseases, which can lower the person's quality of life, eventually leading to death. The following sections describe the primary respiratory disorders that currently pose a threat to people's lives worldwide.

10.3.1 Asthma

Asthma is the most prevalent chronic disease globally, and it is categorized as an inflammatory disease of the lower airways. Asthma is associated with T helper cell type 2 immunological responses; therefore, it is currently incurable[82]. Asthma-associated symptoms include cough, breathing problem, dyspnea (breathlessness), and wheezing[80]. The risk factors for infection include exposure to irritants, allergens, bacterial or viral infections in the respiratory tract, bacterial sinusitis, physical exercise, thunderstorm, and cold weather, all of which can aggravate the symptoms[80]. If a person does have a record of recurring dry cough, wheezing, tightness of the chest, or breathlessness, then treatment might be considered[81].

10.3.2 Allergic Rhinitis (Hay Fever)

Allergic rhinitis is a nasal disorder caused by exposure to an allergen that triggers an IgE-mediated inflammatory response[11]. Sneezing, rhinorrhea, nasal congestion, and nasal itching are the most common symptoms of allergic rhinitis, and the disease worsens when asthma is present[83].

10.3.3 Sinusitis

Sinusitis is a common chronic condition that causes inflammation of one or more paranasal sinuses. Allergic people, particularly those with IgE-mediated allergic rhinitis, are more likely to develop rhinosinusitis[84,85].

10.3.4 Bronchitis

Bronchitis is a pulmonary ailment characterized by marked features of augmented mucous formation and respiratory inflammation[86]. It is also difficult to ascertain from other infections that typically induce coughing, such as flu, common cold symptoms, and so on[87]. The phrase "common cold" refers to a moderate self-limiting viral illness of the upper airways[88].

10.3.5 Influenza

Influenza is a typical viral illness of the respiratory system that causes annual epidemics[89]. Acute respiratory distress syndrome[90] is a potentially fatal chronic lung disease characterized by acute respiratory collapse of clearly defined nonpulmonary and pulmonary exposures[91].

10.3.6 Emphysema

Emphysema is characterized by persistent breathlessness and difficulty to withstand physical exercise owing to the destruction to the respiratory membrane. In emphysema, the increase in the number of macrophages, T-lymphocytes, and eosinophils within the pulmonary parenchyma and distal air gaps inside lungs is indicated[92]. COPD is a highly reversible lung disease characterized by airway constriction[93]. Chronic bronchitis is an inflammatory condition in the upper airways, while emphysema is an inflammatory condition in the microscopic air sacs[94].

10.3.7 Cystic Fibrosis

Cystic fibrosis is a progressive, hereditary disease characterized by mucus accumulation and loss of elasticity in the lung tissue. This condition occurs as a result of a mutation in a gene that prevents the cystic fibrosis transmembrane conductance regulator (CFTR) from functioning properly. CFTR regulates the transport of salt into and out of cells, which is required for proper mucus formation. In cystic fibrosis, lung problems such as viral infection, inflammation, and deadly growth occur[95].

10.3.8 Pneumonia

Pneumonia is a severe illness of lower airways (lungs) that has a high fatality rate. This disease could be acquired in the public or even in hospitals and can be transmitted by inhalation or absorption of a harmful pathogenic microorganism. The most prevalent community-acquired bacterial pneumonia is *Streptococcus pneumoniae* (pneumococcus)[96].

10.3.9 Pulmonary Fibrosis

Pulmonary fibrosis is a lung disorder characterized by recognized or, most usually, undiscovered pulmonary irritation[97]. Evidence suggests that contact with metal, wood dust, pollution, stomach aspiration, tobacco, and illness as factors that can cause pulmonary fibrosis[98].

10.3.10 Tuberculosis

TB is the world's leading cause of death from a single infectious pathogen. In 2018, a total of 10 million people had TB and 1.5 million died of the disease[99]. The acid-fast bacterial species *Mycobacterium tuberculosis* is the main species causing human TB cases, with the animal-adapted strain *Mycobacterium bovis* accounting for a lower number of zoonotic infections (143,000 cases in 2018)[99]. The structural damage to the lung caused by pulmonary TB includes fibrosis, traction bronchiectasis, cavitation, bronchostenosis, and parenchymal loss.

10.3.11 Lung Cancer

Because of delayed diagnoses and deficiency in medical interventions, lung cancer-related mortality is a major threat, and 70% of patients who have lung cancer are diagnosed at the advanced stage, that is, stage III or IV. Lung cancer is characterized by severe protrusion, which undergoes rapid metastases and the prevalence is high in both genders. The death rate for lung cancer in the United States is higher than the combined fatality rate for the other four primary forms of cancer, namely, prostate, breast, colon, and pancreas[100] (Table 10.2).

10.4 PHYTOCHEMICALS AS THERAPEUTICS IN DISEASE PROGRESSION

The human lungs undergo repeated exposure to environmental oxidants, regardless of whether the oxidants are created endogenously by metabolic activities or acquired from external sources such as air pollution and cigarette smoke[3,4,5,6]. Accumulating evidence suggests that oxidative stress plays a major role in the etiology of lung diseases such asthma, lung cancer, acute lung damage, pulmonary fibrosis, and COPD[7,8,9,10,11].

Devereux et al. explored that reduced intake of fruits and vegetables in the western society has declined pulmonary antioxidant defenses, which makes the affected people highly vulnerable to the allergens and irritants that are inhaled from the environment[52]. Our diet is the major source of supplementation of micronutrient antioxidants such as vitamins A, C, and E; polyphenols; and

TABLE 10.2
Different Types of Respiratory Ailments

Respiratory Ailments	Organ Affected	Indications	Causes
Asthma	Lungs	Intensity variable recurrent episodes of wheezing, breathlessness, chest tightness, and/or coughing	Dust, smoke, pollen, and volatile organic compounds
Bronchitis (acute)	Trachea, bronchi	Persistent cough; occasionally, sputum, dyspnea, and wheeze	Pathogen, biomass fuel, dust, and chemical fumes
Chronic obstructive pulmonary disease	Lungs	Persistent cough with mucus production, dyspnea, wheezing, and chest tightness	Smoking, chemical fumes, dust, other lung irritants, and α-1 antitrypsin deficiency
Common cold	Nose, sinuses, pharynx, larynx	Sore throat, rhinitis, rhinorrhea, cough, and malaise	Rhinovirus, bacteria
Cystic fibrosis	Lungs	Cough, wheeze, repeated lung infections, shortness of breath, and labored breathing	–
Emphysema	Lungs	Shortness of breath	Active proteolytic enzyme
Hay fever (allergic rhinitis)	Nose	Sneezing, sore and runny eyes, nasal itching, congestion, sinus pain, and tickly or itchy throat	Pollen, mites, domestic animals, molds, tobacco smoke, and automobile exhaust
Influenza	Lungs, nose, and throat	Weakness, myalgia, cough, and nasal congestion	Viruses
Lung cancer	Lungs	Worsening cough, hoarseness, tiredness, and chest pain	Smoking, genetic predisposition, and repetitive respiratory infections
Pulmonary fibrosis	Lungs	Cough, shortness of breath, fatigue, and low saturation levels	–
Pneumonia	Lungs	Fever, cough, dyspnea, and chest pain	Virus, bacteria, fungi, and parasites
Respiratory distress syndrome	Lungs	Dyspnea, tachypnea, and hypoxemia	Viruses, fungi, parasites, and bacteria
Sinusitis	Sinuses	Facial pain/pressure, facial congestion/fullness, and nasal obstruction	Allergens, viruses, fungi, bacteria, animal dander, polluted air, smoke, and dust

carotenoids. Therefore, attention must be paid toward this area and the significant role it plays in protecting against airway diseases.

These bioactive substances aid in the reduction of oxidative stress and the regulation of inflammation in patients with asthma. Their activities are linked to their competence to regulate the enzymatic breakdown process through endogenous antioxidative enzymes, antioxidant synthesis, and quenchers, thus preventing the production of cytokines, chemokines, and adhesion molecules, as well as their functions. Only a few plant-derived substances have undergone testing in clinical trials to assess whether they have antioxidative and anti-inflammatory characteristics.

10.5 PLANTS USED FOR THE TREATMENT OF RESPIRATORY DISORDERS

10.5.1 Ivy Leaf (*Hedera helix*)

Antibiotic overuse for severe respiratory infections seems to be a serious issue, which results in the development of resistant strains and high healthcare costs without improving cough treatment.

Therefore, nonantibiotic health alternatives are required. Mucolytic medications and antitussives are often used as over-the-counter medicines for the treatment of severe congestion in both adults and children and are regularly recommended for use in general clinical practice. Cough fluids contributed to 102 million pounds in revenue in the United Kingdom in 2008. Herbal treatments comprising ivy (*Hedera helix* L.) leaf extract are popular among nonantibiotic cough cures in several European nations. In 2007, more than 80% of natural expectorants recommended in Germany contained ivy extracts, resulting in approximately 2 million applications and revenues surpassing 13 million Euros. Saponins present in the ivy leaf were thought to have spasmolytic, mucolytic, antibacterial, and bronchodilatory properties.

Randomized and nonrandomized controlled clinical studies, as well as observational research, were conducted to explore the efficacy of ivy formulations for acute upper respiratory tract infection. Ten appropriate studies with a total of 17,463 subjects were identified. One randomized controlled trial revealed that a combination of thyme and ivy was useful; hence, these data need to be validated further. The ivy extract is very well accepted; no major side effects have been documented[101].

10.5.2 Tavipec® (*Spicae aetheroleum*)

Acute rhinosinusitis (ARS) is a relatively common pulmonary ailment, which makes it the most likely reason for people to visit their primary care physician. Adult ARS is defined as inflammation of the paranasal sinuses and nasopharynx with two or even more signs, some of which might be nasal obstruction or nasal secretion, both with and without facial distress and/or diminution or lack of odor, persisting more than 12 weeks. The anti-inflammatory characteristics and method of action of plant-based monoterpenes as well as their usefulness in controlling clinical symptoms of respiratory tract infections, particularly infectious ARS, have already been carefully explored. Monoterpenes are absorbed from the stomach within the intestinal system, after which they enter into the bloodstream and attain the mucosal glands, wherein they have secretolytic activity by increasing saliva production and ciliary beat frequency and enhancing ciliary cellular proliferation, thereby assisting cleanup and mucosal retrieval[102,103].

Various preclinical investigations have explored the anti-inflammatory activity of monoterpenes, such as cineol or linalool, whereas other randomized controlled trials have proved its medicinal benefits in inflammatory respiratory diseases. *Spicae aetheroleum*, which has high levels of 1,8-cineoleand linalool, is widely used for its mucolytic, anti-inflammatory, and spasmolytic properties. In a placebo-controlled double-blind randomized controlled trial, the effectiveness and safety of *Spicae aetheroleum* (Tavipec®) in alleviating clinical symptoms was proved in people with symptomatic bronchitis.

A clinical trial determined the efficacy and tolerability of Tavipec®, which was prepared with the flower extract of *Lavandula latifolia*, in comparison with placebo in adult people with acute viral rhinosinusitis. People received a 150-mg dose of Tavipec® or placebo three times a day for one week. Overall, 288 individuals were registered and randomly assigned to one of two therapy groups. The Tavipec® sample considerably reduced MSS versus placebo, as well as the effect of rhinosinusitis signs on QoL was dramatically decreased on the eighth day[102].

10.5.3 Sinupret® Syrup

A typical therapeutic approach for ARS aims to lower disease intensity, decrease the frequency of illness, avoid transition to post-viral and bacterial rhinosinusitis, and prevent progression into a chronic condition in the future. Antibiotics, antihistamines, nasal decongestants, mucolytic drugs, and homoeopathic medications should not be used in acute viral rhinosinusitis because their usefulness is yet to be proven. Herbal medications, which inhibit a variety of pathogenic processes, provide an alternative option. Sinupret®, a complicated herbal treatment containing gentian root, primula, elder, verbena, and sorrel, is one example. The herbal remedy increases ciliary properties *in vitro* and exerts anti-inflammatory properties as demonstrated in animal experiments. It possesses a diverse set of pharmacological activities, comprising mucolytic, antiviral, secretomotor, immunomodulatory, and anti-inflammatory activities.

In a study, 386 adults who had acute viral rhinosinusitis were selected for a double-blind, placebo-controlled randomized experimentation. Following 15 days, the initial treatment group received a daily dosage of 3 × 160 mg of plant preparation. According to sinonasal evaluation, encompassing total index, nasal signs, rhinogenous clinical signs, and quality of life, the effective therapeutic group outperformed the control group. This review intended to unravel the efficacy of the complex herbal medication Sinupret® syrup for the management of acute viral rhinosinusitis in children. A randomized control trial of two parallel groups was conducted on patients aged 6–11 years. Overall, 184 individuals were involved in the study. The variations in the degree of face discomfort and headache among groups were negligible. The Sinupret® group also showed a shift toward antibiotic sparing (2.17% vs. 5.26% in the control group).

10.5.4 Oscillococcinum®

Oscillococcinum® is used to prevent or treat seasonal cold and respiratory inflammatory diseases as well as to enhance symptom management and minimize the number of respiratory tract infection episodes. The investigation focused on 455 people who had respiratory problems. The Anas barbariae hepatis et cordis extractum 200K was administered to 246 individuals, whereas the remaining 209 people did not receive it (random sample). The incidence of respiratory tract infections had reduced in treated patients independent of diagnosis; the drop in the number of events that occurred was significant in all the classes of individuals ($p<0.001$). The expenditures incurred by the National Health Service have been considerably lower in the patient groups treated ($p<0.001$).

10.6 PHYTOCHEMICALS USED FOR THE TREATMENT OF RESPIRATORY DISORDERS

10.6.1 Curcumin

Curcumin is derived from natural sources, such as from turmeric, and exerts significant antioxidant effects and anti-inflammatory properties. A study discovered a substantial favorable association as well as dose-response relationship among individuals who consumed curry that had high levels of curcumin and forced expiratory volume in 1 second (FEV1)/forced vital capacity and FEV1 in older Chinese people. The respiratory behavior of smokers who consumed curry rich in curcumin was nearly equivalent to that of nonsmokers, implying that curcumin has a specific advantage for smokers[104].

10.6.2 Kaempferol

It can be found in tomatoes, apples, grapes, broccoli, and other foods. The crucial effect of kaempferol has received extensive attention. It functions as an anti-apoptotic, antioxidant, and anti-inflammatory agent. Kaempferol administration reduced eosinophil accumulation and degranulation in lung cells of mice with ovalbumin (OVA)-induced asthma by downregulating the NF-κB pathway.

The effects of kaempferol-3-O-rhamnoside and kaempferol on OVA-challenged mice have been explored; the analysis indicated that kaempferol decreased Th2-related cytokine levels (IL-13, IL-4, and IL-5) in kaempferol-treated mice by exerting an antioxidant effect. When compared with primary kaempferol, kaempferol-3-O-rhamnoside exhibited a weaker antioxidant effect but a larger inhibitory activity on Th2 cytokines, TNF-α levels, and IgE formation[105, 106,107].

10.6.3 Fisetin

It is a flavonoid present in foods such as strawberries and apples, as well as onions and cucumbers. Prophylactic fisetin therapy in mice challenged with OVA can reduce airway hyperresponsiveness, mucus hypersecretion, and Th2 cytokine levels (IL-5, IL-13, and IL-4) in lavage fluid, which were elevated following OVA treatment. Fisetin significantly decreased nuclear translocation of p65, thus suppressing the NF-κB pathway. The findings were supported by *in vitro* research on human lung cell cultures wherein fisetin reduced the activation of the NF-κB product[108,109].

10.6.4 Resveratrol

A polyphenolic chemical found in the Dipterocarpaceae, Vitaceae, Cyperaceae, and Gnetaceae families. It may be found in different foods and food items such as wine, grapes, mulberries, and grape juice. In an animal model study, the effect of resveratrol on house dust mite (HDM)-induced asthma was investigated. Resveratrol therapy decreased the prevalence of IL-17, IL-6, transforming growth factor-beta (TGF-β), and TNF-α in bronchoalveolar lavage fluid, which were augmented in HDM-treated animals. It also inhibited IgE-induced Syk production in RBL-2H3 cells[110,111].

10.6.5 Galangin

It belongs to the chemical family of flavanols and are mostly found in plant species such as *Helichrysum aureonitens* and *Alpinia officinarum* as well as foods such as honey.

10.6.6 Geraniol

Geraniol is a monoterpene alcohol found in essential oils such as geranium, lemon, and other therapeutic plants. It exerts a protective effect against OVA-induced asthma. Geraniol boosted Nrf-2 levels and superoxide dismutase and glutathione S-transferase activity in OVA-challenged mice, demonstrating its antioxidant effect by reducing oxidative stress through the Nrf2/ARE network and enhancing Th1/Th2 equilibrium in the lungs[112,113,114].

10.6.7 Polydatin

Polydatin is a glucoside derived from the plant *Polygonum cuspidatum*, a natural antioxidant with several therapeutic benefits. This compound was tested in an animal model of OVA-induced asthma; OVA-treated animals demonstrated elevated TGF-β and ROS levels and reduced Nrf-2 activity. Moreover, the therapy increased the antioxidant NQO1 enzymatic activity as well as the production of Nrf2 and HO-1. The findings suggest that polydatin effectively reduces ROS generation and fibrosis by boosting Nrf2 activity[115,116].

10.6.8 Icosahydropicenic Acid

Icosahydropicenic acid (IHPA), a novel pentacyclic triterpenoid saponin isolated from the roots of *Clerodendrum serratum* L. Moon (Verbenaceae) for the first time, is used to treat allergic asthma. When compared with normal sodium cromoglycate, IHPA demonstrated a considerable protective effect against mast cell degeneration (59.62%) at a dose of 100 mg/kg (64.48%). The drug also showed a strong inhibitory effect in gout tracheal chain preparations produced by histamine[117].

10.6.9 Lobeline

Lobelia is used as an adjuvant in the treatment of asthma and other respiratory conditions. Lobelia has high levels of an alkaloid, called lobeline, which has significant effects on the respiratory tract function, which includes stimulation of breathing, aiding the cough reflexes, and improvising vascular tone. *Lobelia* is an important alternative option in treating patients with asthma because of its properties as a respiratory stimulant and an expectorant and because it can reduce the usage of other pharmaceuticals that can cause side effects[118,119,120].

10.6.10 Diallyl Disulfide

The activation of Nrf-2 and the expression of HO-1 are both induced by diallyl disulfide (DADS). DADS inhibits the activation of NF-κB, inducible nitric oxide synthase, and matrix metalloprotein-9. DADS decreases airway inflammation through regimentation of Nrf-2/HO-1 and NF-κB, and the findings suggest that DADS could be an effective novel oral medication for allergic asthma[121,122,123].

10.6.11 Quercetin

Quercetin is an active component of *Pergularia daemia* Linn. that exerts antihistamine, anti-inflammatory, and antioxidant effects, which make it useful in the treatment of allergies and respiratory disorders such as allergic rhinitis and bronchitis. Furthermore, 6-(4,7-dihydroxy-heptyl)quinine, a new bioactive molecule with antibacterial activity, has been discovered[124,125].

10.6.12 Ursolic Acid

Ursolic acid is a pentacyclic triterpenoid carboxylic chemical found in a variety of plant extracts in the Lamiaceae family. Ursolic acid confers protection from OVA-induced asthma, as it lowers the inflow of inflammatory cells in BAL as well as the levels of IgE and Th2 cytokines[126,127].

Different phytochemicals possessing respiratory functions are shown in Table 10.1.

10.7 CHALLENGES IN CURRENT THERAPEUTIC STRATEGIES

Different infectious and noninfectious respiratory ailments are being explored in detail, although the pathophysiology and epidemiology of the diseases are still being explored, and the development of more effective medicines is still in the early stage. Asthma and COPD are two of the most common noninfectious diseases that afflict huge populations in both high- and low-income countries. The pathophysiology and disease course vary greatly among populations and is mostly determined by one's personal lifestyle. Bronchodilators, as well as newly discovered long-acting β2 agnostics and long-acting muscarinic antagonists, can be used to treat patients with mild-to-severe symptoms of chronic respiratory diseases. However, their efficacy was insufficient in patients with chronic asthmatic conditions and COPD. Furthermore, personal lifestyle factors such as smoking or continuous industrial dust exposure have hampered the efficacy of existing therapy. Furthermore, other harsh conditions are characterized by inevitable lung tissue damage that leads to scarring, which is unaffected by commercially available treatments. Infectious agent-induced respiratory ailments are a major problem because most of these diseases are seasonal and can be found both within and outside the human body. Antibiotic abuse, natural selection, and mutations have led to the emergence of new virulent mutants with enhanced resistance toward treatment, resulting in global epidemic outbreaks. A recent study revealed the cross-transfer of infectious organisms between species (animals to humans and vice versa), which has a global impact. COVID-19, a recent viral outbreak, caused by a rare coronavirus, is a good example of how newborn mutants lacking natural immunity can represent a severe threat to global health[128].

10.8 CONCLUSION AND FUTURE PROSPECTS

Plant-based molecular farming and nanomedicines using conventional aids are being heavily emphasized in the development of plant-derived medications. New biologically active chemicals have been explored in plants by applying phytochemical and pharmacological investigations as analytical chemistry and biotechnology continue to progress. The concept of using plants in traditional medicine approaches is quite an old technique. Natural compounds have been identified for their therapeutic potential since ancient times, which has resulted in a paradigm change toward phytochemicals for the discovery of new therapies. Beyond their antioxidant properties, phytochemicals of distinct classes such as terpenoids, alkaloids, and polyphenols have been used for the treatment of different respiratory problems, and all compounds have already shown profound anti-inflammatory properties in preclinical animal experiments. These phytoconstituents are used either alone or in combination with different anti-inflammatory drugs to treat a variety of ailments, which results in fewer negative side effects and lower drug costs. Organic bioactive compounds could be utilized along with other anti-inflammatory medications, which will help lower drug costs and/or reduce adverse effects. The concern is whether these findings are applicable to human illness outcomes because exposure to natural biological substances occurs in the long term and at small concentrations depending on the extent of absorption and metabolism. Clinical research involving antioxidant and biologic therapy is challenging owing to the complexity of each reaction and the enormous variability in physiological factors. Therefore, future studies should include a variety of compounding elements, such as adequate dose adjustments and the identification of more potent natural bioactive constituents will need to continue. According to those studies, long-term adjunction and large population-established randomized controlled trials with placebo are required to clarify the role and benefits of antioxidants in therapeutic contexts.

REFERENCES

[1] Quiera, J. (1993) *Basic Human Histology*. 1st ed. Tehran: Argmand publication; pp. 124–126, 459–485.

[2] Guyton, AC. (1985) *Anatomy and Physiology*. Philadelphia: Saunders College Publishing; 1985, pp. 589–604.

[3] Asgary, S., Sahebkar, A., Afshani, M. R., Keshvari, M., Haghjooyjavanmard, S., and Rafieian-Kopaei, M. (2014). *Phytotherapy Research*, 28(2), 193–199.

[4] Ghasemi Pirbalouti, A., Momeni, M., and Bahmani, M. (2012). *African Journal of Traditional, Complementary, and Alternative Medicines*. 10(2), 368–385. https://doi.org/10.4314/ajtcam.v10i2.24

[5] Bahmani, M., and Eftekhari, Z. (2012). *Comparative Clinical Pathology*, 22(3), 403–407. https://doi.org/10.1007/s00580-012-1423-8

[6] Bahmani, M., Banihabib, E., Rafieian-Kopaei, M., GholamiAhangaran, M. (2014). *KafkasUniv Vet FakDerg*; doi: 10.9775/kvfd.2014.11223

[7] Ncube, N.S., Afolayan, A.J., Okoh, A.I. (2008). *African Journal of Biotechnology* 7(12), 1797e1806

[8] Scalbert, A., Manach, C., Morand, C., Remesy, C. (2005). *Critical Reviews in Food Science and Nutrition* 45, 287–306.

[9] Balasundram, N., Sundram, K., and Saman, S. (2006) *Food Chem*. 99, 191–203.

[10] Fauconneau, B., Waffo-Teguo, P., Huguet, F., Barrier, L., Decendit, A., and Merillon, J. M. (1997). *Life Sciences*, 61(21), 2103–2110. https://doi.org/10.1016/s0024-3205(97)00883-7

[11] Goh, F. Y., Upton, N., Guan, S., Cheng, C., Shanmugam, M. K., Sethi, G., Leung, B. P., and Wong, W. S. (2012). *European Journal of Pharmacology*, 679(1–3), 109–116. https://doi.org/10.1016/j.ejphar.2012.01.002

[12] Pretorius, J. C. (2003) *Current Medicinal Chemistry–Anti Infective Agents*. 2: 335–353.

[13] Darbour, N., Bayet, C., Rodin-Bercion, S., Elkhomsi, Z., Lurel, F., Chaboud, A., and Guilet, D. (2007). *Natural Product Research*, 21(5), 461–464. https://doi.org/10.1080/14786410601086871
[14] Garcia-Larsen, V., Thawer, N., Charles, D., Cassidy, A., van Zele, T., Thilsing, T., Ahlström, M., Haahtela, T., Keil, T., Matricardi, P. M., Brożek, G., Kowalski, M. L., Makowska, J., Niżankowska-Mogilnicka, E., Rymarczyk, B., Loureiro, C., Todo Bom, A., Bachert, C., Forsberg, B., Janson, C., … Burney, P. G. (2018). *Nutrients*, 10(1), 95. https://doi.org/10.3390/nu10010095
[15] Lago, J. H., Toledo-Arruda, A. C., Mernak, M., Barrosa, K. H., Martins, M. A., Tibério, I. F., and Prado, C. M. (2014). *Molecules* (Basel, Switzerland), 19(3), 3570–3595. https://doi.org/10.3390/molecules19033570
[16] Cox-Georgian, D., Ramadoss, N., Dona, C., and Basu, C. (2019). *Medicinal Plants: From Farm to Pharmacy*, 333–359.
[17] Li, G. P., Yang, H., Zong, S. B., Liu, Q., Li, L., Xu, Z. L., Zhou, J., Wang, Z. Z., and Xiao, W. (2018). *Biomedicine & Pharmacotherapy*, 99, 746–754. https://doi.org/10.1016/j.biopha.2018.01.135
[18] Chiang, Y. M., and Kuo, Y. H. (2001). *Journal of Natural Products*, 64(4), 436–439. https://doi.org/10.1021/np0004808
[19] Mishra, S. N. (1989) *Proc. All India Workshop on M & AP*, Faizabad, 492–495.
[20] Wink, M., Schmeller, T., and Latz-briining, B. (1998) *Journal of Chemical Ecology,* 24, 1888–1937.
[21] Cui, Y., Jiang, L., Yu, R., Shao, Y., Mei, L., and Tao, Y. (2019). *Journal of Ethnopharmacology*, 243, 112096. https://doi.org/10.1016/j.jep.2019.112096
[22] Guénégou, A., Leynaert, B., Pin, I., Le Moël, G., Zureik, M., and Neukirch, F. (2006). *Thorax*, 61(4), 320–326. https://doi.org/10.1136/thx.2005.047373
[23] Schünemann, H. J., Grant, B. J., Freudenheim, J. L., Muti, P., Browne, R. W., Drake, J. A., Klocke, R. A., & Trevisan, M. (2001). *American Journal of Respiratory and Critical Care Medicine*, 163(5), 1246–1255. https://doi.org/10.1164/ajrccm.163.5.2007135
[24] Grievink, L., de Waart, F. G., Schouten, E. G., and Kok, F. J. (2000). *American Journal of Respiratory and Critical Care Medicine*, 161(3 Pt 1), 790–795. https://doi.org/10.1164/ajrccm.161.3.9904040
[25] Schünemann, H. J., McCann, S., Grant, B. J., Trevisan, M., Muti, P., and Freudenheim, J. L. (2002). *American Journal of Epidemiology*, 155(5), 463–471.
[26] Park, H. S., Kim, S. R., Kim, J. O., and Lee, Y. C. (2010). *Molecules* (Basel, Switzerland), 15(10), 6810–6834. https://doi.org/10.3390/molecules15106810
[27] Riccioni, G., Barbara, M., Bucciarelli, T., di Ilio, C., and D'Orazio, N. (2007). *Annals of Clinical and Laboratory Science*, 37(1), 96–101.
[28] Retsky, K. L., Chen, K., Zeind, J., and Frei, B. (1999). *Free Radical Biology & Medicine*, 26(1–2), 90–98. https://doi.org/10.1016/s0891-5849(98)00151-8
[29] Cameron, E., and Pauling, L. (1976). *Proceedings of the National Academy of Sciences of the United States of America*, 73(10), 3685–3689. https://doi.org/10.1073/pnas.73.10.3685
[30] You, W. C., Zhang, L., Gail, M. H., Chang, Y. S., Liu, W. D., Ma, J. L., Li, J. Y., Jin, M. L., Hu, Y. R., Yang, C. S., Blaser, M. J., Correa, P., Blot, W. J., Fraumeni, J. F., Jr, and Xu, G. W. (2000). *Journal of the National Cancer Institute*, 92(19), 1607–1612. https://doi.org/10.1093/jnci/92.19.1607
[31] Kaur, B., Rowe, B. H., and Arnold, E. (2009). 2009(1), CD000993. https://doi.org/10.1002/14651858.CD000993.pub3
[32] Shorey-Kendrick, L. E., McEvoy, C. T., Ferguson, B., Burchard, J., Park, B. S., Gao, L., Vuylsteke, B. H., Milner, K. F., Morris, C. D., and Spindel, E. R. (2017). *American Journal of Respiratory and Critical Care Medicine*, 196(6), 745–755. https://doi.org/10.1164/rccm.201610-2141OC
[33] Riccioni, G., and D'Orazio, N. (2005). *Expert Opinion on Investigational Drugs*, 14(9), 1145–1155. https://doi.org/10.1517/13543784.14.9.1145
[34] Burton, G. W., and Ingold, K. U. (1989). *Annals of the New York Academy of Sciences*, 570, 7–22. https://doi.org/10.1111/j.1749-6632.1989.tb14904.x
[35] Atkinson, J., Epand, R. F., and Epand, R. M. (2008). *Free Radical Biology & Medicine*, 44(5), 739–764. https://doi.org/10.1016/j.freeradbiomed.2007.11.010
[36] DeLuca, H. F., and Zierold, C. (1998). *Nutrition Reviews*, 56(2 Pt 2), S4–75. https://doi.org/10.1111/j.1753-4887.1998.tb01686.x
[37] Moriguchi, S., and Muraga, M. (2000). *Vitamins and Hormones*, 59, 305–336. https://doi.org/10.1016/s0083-6729(00)59011-6

[38] Pryor W. A. (2000). *Free Radical Biology & Medicine*, 28(1), 141–164. https://doi.org/10.1016/s0891-5849(99)00224-5
[39] Devaraj, S., Harris, A., and Jialal, I. (2002). *Nutrition Reviews*, 60(1), 8–14. https://doi.org/10.1301/002966402760240381
[40] Yoshikawa, T., and Yoshida, N. (2000). *Antioxidants & Redox Signaling*, 2(4), 821–825. https://doi.org/10.1089/ars.2000.2.4-821
[41] Burton, G. W., Joyce, A., and Ingold, K. U. (1982). *Lancet* (London, England), 2(8293), 327. https://doi.org/10.1016/s0140-6736(82)90293-8
[42] Hanson, C., Lyden, E., Furtado, J., Campos, H., Sparrow, D., Vokonas, P., and Litonjua, A. A. (2016). *Clinical Nutrition* (Edinburgh, Scotland), 35(1), 169–174. https://doi.org/10.1016/j.clnu.2015.01.020
[43] Balsano, C., and Alisi, A. (2009). *Current Pharmaceutical Design*, 15(26), 3063–3073. https://doi.org/10.2174/138161209789058084
[44] Packer, L., and Suzuki, Y. J. (1993). *Molecular Aspects of Medicine*, 14(3), 229–239. https://doi.org/10.1016/0098-2997(93)90009-3
[45] Hisama, M., Matsuda, S., Tanaka, T., Shibayama, H., Nomura, M., and Iwaki, M. (2008). *Journal of Oleo Science*, 57(7), 381–390. https://doi.org/10.5650/jos.57.381
[46] Aoshima, H., and Hamamoto, K. (1999). *Bioscience, Biotechnology, and Biochemistry*, 63(4), 743–748. https://doi.org/10.1271/bbb.63.743
[47] Hossain, S. J., Aoshima, H., Koda, H., and Kiso, Y. (2002). *Journal of Agricultural and Food Chemistry*, 50(23), 6828–6834. https://doi.org/10.1021/jf020448e
[48] Li, Q., Morimoto, K., Kobayashi, M., Inagaki, H., Katsumata, M., Hirata, Y., Hirata, K., Shimizu, T., Li, Y. J., Wakayama, Y., Kawada, T., Ohira, T., Takayama, N., Kagawa, T., and Miyazaki, Y. (2008). *Journal of Biological Regulators and Homeostatic Agents*, 22(1), 45–55.
[49] Sadlon, A. E., & Lamson, D. W. (2010). *Alternative Medicine Review: A Journal of Clinical Therapeutic*, 15(1), 33–47.
[50] Nam, J. H., Jung, H. J., Choi, J., Lee, K. T., & Park, H. J. (2006). *Biological & Pharmaceutical Bulletin*, 29(5), 967–970. https://doi.org/10.1248/bpb.29.967
[51] Oikawa, S. (2005). *Environmental Health and Preventive Medicine*, 10(2), 65–71. https://doi.org/10.1007/BF02897995
[52] Nose K. (2000). *Biological & Pharmaceutical Bulletin*, 23(8), 897–903. https://doi.org/10.1248/bpb.23.897
[53] Li, Q., Nakadai, A., Matsushima, H., Miyazaki, Y., Krensky, A. M., Kawada, T., and Morimoto, K. (2006). *Immunopharmacology and Immunotoxicology*, 28(2), 319–333. https://doi.org/10.1080/08923970600809439
[54] Komori, T., Fujiwara, R., Tanida, M., Nomura, J., and Yokoyama, M. M. (1995). *Neuroimmunomodulation*, 2(3), 174–180. https://doi.org/10.1159/000096889
[55] Shibata, H., Fujiwara, R., Iwamoto, M., Matsuoka, H., and Yokoyama, M. M. (1991). *The International Journal of Neuroscience*, 57(1–2), 151–159. https://doi.org/10.3109/00207459109150355
[56] Xiao, J. H., Zhang, J. H., Chen, H. L., Feng, X. L., and Wang, J. L. (2005). *Planta Medica*, 71(3), 225–230. https://doi.org/10.1055/s-2005-837821
[57] Ma, X., Chen, R., Liu, X., Xie, J., Si, K., and Duan, L. (2013). *African Journal of Traditional, Complementary, and Alternative Medicines*, 10(3), 442–448. https://doi.org/10.4314/ajtcam.v10i3.10
[58] Gao, Y. X., Jiang, L. L., Zhang, Q., Zuo, D. Z., and Li, X. W. (2018). *China Journal of Chinese Materia Medica*, 43(17), 3530–3538. https://doi.org/10.19540/j.cnki.cjcmm.20180530.005
[59] Yin, W., Han, J., Zhang, Z., Han, Z., and Wang, S. (2018). *Scientific Reports*, 8(1), 6265. https://doi.org/10.1038/s41598-018-24565-y
[60] Yang, F., Cao, Y., Zhang, J., You, T., and Zhu, L. (2017). *Biochemical and Biophysical Research Communications*, 482(1), 147–153. https://doi.org/10.1016/j.bbrc.2016.11.003
[61] Chen, C., Yang, S., Zhang, M., Zhang, Z., Hong, J., Han, D., Ma, J., Zhang, S. B., Okunieff, P., and Zhang, L. (2016). *Cancer Biology & Therapy*, 17(4), 381–389. https://doi.org/10.1080/15384047.2016.1139229
[62] Zhou, Y., Li, P., Duan, J. X., Liu, T., Guan, X. X., Mei, W. X., Liu, Y. P., Sun, G. Y., Wan, L., Zhong, W. J., Ouyang, D. S., and Guan, C. X. (2017). *Inflammation*, 40(6), 2062–2073. https://doi.org/10.1007/s10753-017-0646-x

[63] Goud, M. P., Bale, S., Pulivendala, G., and Godugu, C. (2019). *International Immunopharmacology*, 75, 105755. https://doi.org/10.1016/j.intimp.2019.105755

[64] Tang, H., He, H., Ji, H., Gao, L., Mao, J., Liu, J., Lin, H., and Wu, T. (2015). *The Journal of Surgical Research*, 197(1), 167–175. https://doi.org/10.1016/j.jss.2015.02.062

[65] Zhan, H., Huang, F., Ma, W., Zhao, Z., Zhang, H., and Zhang, C. (2016). *Biological & Pharmaceutical Bulletin*, 39(8), 1284–1292. https://doi.org/10.1248/bpb.b16-00046

[66] Chen, C., Wang, Y. Y., Wang, Y. X., Cheng, M. Q., Yin, J. B., Zhang, X., and Hong, Z. P. (2018). *Biochemical and Biophysical Research Communications*, 495(4), 2396–2403. https://doi.org/10.1016/j.bbrc.2017.12.112

[67] Zhou, Y., Gao, Y., Chen, Y., Zheng, R., Zhang, W., and Tan, M. (2015). *Cell Biochemistry and Biophysics*, 71(2), 971–976. https://doi.org/10.1007/s12013-014-0295-8

[68] Kim, K. Y., Lee, H. S., and Seol, G. H. (2017). *Biomedicine & Pharmacotherapy Biomedecine & Pharmacotherapie*, 91, 925–930. https://doi.org/10.1016/j.biopha.2017.05.032

[69] Guan, S. P., Tee, W., Ng, D. S., Chan, T. K., Peh, H. Y., Ho, W. E., Cheng, C., Mak, J. C., and Wong, W. S. (2013). *British Journal of Pharmacology*, 168(7), 1707–1718. https://doi.org/10.1111/bph.12054

[70] Luo, Y. L., Zhang, C. C., Li, P. B., Nie, Y. C., Wu, H., Shen, J. G., and Su, W. W. (2012). *International Immunopharmacology*, 13(3), 301–307. https://doi.org/10.1016/j.intimp.2012.04.019

[71] Luo, Y. L., Zhang, C. C., Li, P. B., Nie, Y. C., Wu, H., Shen, J. G., and Su, W. W. (2012). *International Immunopharmacology*, 13(3), 301–307. https://doi.org/10.1016/j.intimp.2012.04.019

[72] Ganesan, S., Faris, A. N., Comstock, A. T., Chattoraj, S. S., Chattoraj, A., Burgess, J. R., Curtis, J. L., Martinez, F. J., Zick, S., Hershenson, M. B., and Sajjan, U. S. (2010). *Respiratory Research*, 11(1), 131. https://doi.org/10.1186/1465-9921-11-131

[73] Jia, L., Sun, P., Gao, H., Shen, J., Gao, Y., Meng, C., Fu, S., Yao, H., and Zhang, G. (2019). *The Journal of Pharmacy and Pharmacology*, 71(6), 1017–1028. https://doi.org/10.1111/jphp.13077

[74] Zheng, Q., Tong, M., Ou, B., Liu, C., Hu, C., and Yang, Y. (2019). *International Journal of Molecular Medicine*, 43(1), 117–126. https://doi.org/10.3892/ijmm.2018.3965

[75] Liu, H., Yu, H., Cao, Z., Gu, J., Pei, L., Jia, M., and Su, M. (2019). *DNA and Cell Biology*, 38(12), 1418–1426. https://doi.org/10.1089/dna.2019.4941

[76] Dong, Z. W., and Yuan, Y. F. (2018). *International Journal of Molecular Medicine*, 41(6), 3353–3365. https://doi.org/10.3892/ijmm.2018.3554

[77] Singh, SK., Patel, JR., Debey, PK., and Thakur, S. (2014). *International Journal of Pharmaceutical Sciences and Research,* 5(10):4109–4116.

[78] Ray, A., Oriss, T. B., and Wenzel, S. E. (2015). *Lung Cellular and Molecular Physiology*, 308(2), L130–L140. https://doi.org/10.1152/ajplung.00070.2014

[79] Pavord, I. D., Beasley, R., Agusti, A., Anderson, G. P., Bel, E., Brusselle, G., Cullinan, P., Custovic, A., Ducharme, F. M., Fahy, J. V., Frey, U., Gibson, P., Heaney, L. G., Holt, P. G., Humbert, M., Lloyd, C. M., Marks, G., Martinez, F. D., Sly, P. D., von Mutius, E., … Bush, A. (2018). *Lancet (London, England)*, 391(10118), 350–400. https://doi.org/10.1016/S0140-6736(17)30879-6

[80] Lutfi, M. F. (2017). *Multidisciplinary Respiratory Medicine*, 12, 3. https://doi.org/10.1186/s40248-017-0084-5

[81] Franks, T. J., Colby, T. V., Travis, W. D., Tuder, R. M., Reynolds, H. Y., Brody, A. R., Cardoso, W. V., Crystal, R. G., Drake, C. J., Engelhardt, J., Frid, M., Herzog, E., Mason, R., Phan, S. H., Randell, S. H., Rose, M. C., Stevens, T., Serge, J., Sunday, M. E., Voynow, J. A., … Williams, M. C. (2008). *Proceedings of the American Thoracic Society*, 5(7), 763–766. https://doi.org/10.1513/pats.200803-025HR

[82] Quirt, J., Hildebrand, K. J., Mazza, J., Noya, F., and Kim, H. (2018). *Allergy, Asthma, and Clinical Immunology: Official Journal of the Canadian Society of Allergy and Clinical Immunology*, 14(Suppl 2), 50. https://doi.org/10.1186/s13223-018-0279-0

[83] Whitsett, J. A., Wert, S. E., and Weaver, T. E. (2010). *Annual Review of Medicine*, 61, 105–119. https://doi.org/10.1146/annurev.med.60.041807.123500

[84] Hiemstra, P. S., McCray, P. B., Jr, and Bals, R. (2015). *The European Respiratory Journal*, 45(4), 1150–1162. https://doi.org/10.1183/09031936.00141514

[85] Harkness, L. M., Kanabar, V., Sharma, H. S., Westergren-Thorsson, G., and Larsson-Callerfelt, A. K. (2014). *Pulmonary Pharmacology & Therapeutics*, 29(2), 144–155. https://doi.org/10.1016/j.pupt.2014.09.003

[86] Rock, J. R., Randell, S. H., and Hogan, B. L. (2010). *Disease Models & Mechanisms*, 3(9–10), 545–556. https://doi.org/10.1242/dmm.0060
[87] Barnes, P. J., Bonini, S., Seeger, W., Belvisi, M. G., Ward, B., and Holmes, A. (2015). *The European Respiratory Journal*, 45(5), 1197–1207. https://doi.org/10.1183/09031936.00007915
[88] Dua, K., Malyla, V., Singhvi, G., Wadhwa, R., Krishna, R. V., Shukla, S. D., Shastri, M. D., Chellappan, D. K., Maurya, P. K., Satija, S., Mehta, M., Gulati, M., Hansbro, N., Collet, T., Awasthi, R., Gupta, G., Hsu, A., and Hansbro, P. M. (2019). *Chemico-Biological Interactions*, 299, 168–178. https://doi.org/10.1016/j.cbi.2018.12.009
[89] Dua, K., Shukla, S. D., de Jesus Andreoli Pinto, T., and Hansbro, P. M. (2017). *Interventional Medicine & Applied Science*, 9(1), 39–41. https://doi.org/10.1556/1646.9.2017.1.02
[90] McCracken, J. L., Veeranki, S. P., Ameredes, B. T., and Calhoun, W. J. (2017). *JAMA*, *318*(3), 279–290. https://doi.org/10.1001/jama.2017.8372
[91] Fergeson, J. E., Patel, S. S., and Lockey, R. F. (2017). *The Journal of Allergy and Clinical Immunology*, 139(2), 438–447. https://doi.org/10.1016/j.jaci.2016.06.054
[92] Ravindra, K., Sokhi, R., and Van Grieken, R. (2008). *Atmospheric Environment* 42(13), 2895–2921.
[93] Nunes, C., Pereira, A. M., and Morais-Almeida, M. (2017). *Asthma Research and Practice*, 3, 1. https://doi.org/10.1186/s40733-016-0029-3
[94] Ray, A., Oriss, T. B., and Wenzel, S. E. (2015). *Lung Cellular and Molecular Physiology*, 308(2), L130–L140. https://doi.org/10.1152/ajplung.00070.2014
[95] Pavord, I. D., Beasley, R., Agusti, A., Anderson, G. P., Bel, E., Brusselle, G., Cullinan, P., Custovic, A., Ducharme, F. M., Fahy, J. V., Frey, U., Gibson, P., Heaney, L. G., Holt, P. G., Humbert, M., Lloyd, C. M., Marks, G., Martinez, F. D., Sly, P. D., von Mutius, E., … Bush, A. (2018). *Lancet (London, England)*, 391(10118), 350–400. https://doi.org/10.1016/S0140-6736(17)30879-6
[96] Quirt, J., Hildebrand, K. J., Mazza, J., Noya, F., & Kim, H. (2018). *Allergy, Asthma, and Clinical Immunology: Official Journal of the Canadian Society of Allergy and Clinical Immunology*, 14(Suppl 2), 50. https://doi.org/10.1186/s13223-018-0279-0
[97] Fahy J. V. (2015). *Nature Reviews Immunology*, 15(1), 57–65. https://doi.org/10.1038/nri3786
[98] Robinson, D., Humbert, M., Buhl, R., Cruz, A. A., Inoue, H., Korom, S., Hanania, N. A., and Nair, P. (2017). *Clinical and Experimental Allergy: Journal of the British Society for Allergy and Clinical Immunology*, 47(2), 161–175. https://doi.org/10.1111/cea.12880
[99] Hastie, A. T., Steele, C., Dunaway, C. W., Moore, W. C., Rector, B. M., Ampleford, E., Li, H., Denlinger, L. C., Jarjour, N., Meyers, D. A., Bleecker, E. R., and NHLBI Severe Asthma Research Program (SARP) (2018). *Clinical and Experimental Allergy: Journal of the British Society for Allergy and Clinical Immunology*, 48(7), 787–797. https://doi.org/10.1111/cea.13129
[100] Holzinger, F., and Chenot, J. F. (2011). *Evidence-Based Complementary and Alternative Medicine*, 2011, 382789. https://doi.org/10.1155/2011/382789
[101] Dejaco, D., Bocian-Sobkowska, J., Szymanski, W., Zacke, G., Hockner, V., and Riechelmann, H. (2019). *Rhinology*, 57(5), 367–374. https://doi.org/10.4193/Rhin19.089
[102] Young, J., De Sutter, A., Merenstein, D., van Essen, G. A., Kaiser, L., Varonen, H., Williamson, I., and Bucher, H. C. (2008). *Lancet (London, England)*, 371(9616), 908–914. https://doi.org/10.1016/S0140-6736(08)60416-X
[103] Ng, T. P., Niti, M., Yap, K. B., & Tan, W. C. (2012). *PloS One*, 7(12), e51753. https://doi.org/10.1371/journal.pone.0051753
[104] Gong, J. H., Shin, D., Han, S. Y., Kim, J. L., and Kang, Y. H. (2012). *The Journal of Nutrition*, 142(1), 47–56. https://doi.org/10.3945/jn.111.150748
[105] Imran, M., Salehi, B., Sharifi-Rad, J., Aslam Gondal, T., Saeed, F., Imran, A., Shahbaz, M., Tsouh Fokou, P. V., Umair Arshad, M., Khan, H., Guerreiro, S. G., Martins, N., and Estevinho, L. M. (2019). *Molecules (Basel, Switzerland)*, 24(12), 2277. https://doi.org/10.3390/molecules24122277
[106] Calderón-Montaño, J. M., Burgos-Morón, E., Pérez-Guerrero, C., and López-Lázaro, M. (2011). *Mini Reviews in Medicinal Chemistry*, 11(4), 298–344. https://doi.org/10.2174/138955711795305335
[107] Kashyap, D., Garg, V. K., Tuli, H. S., Yerer, M. B., Sak, K., Sharma, A. K., Kumar, M., Aggarwal, V., and Sandhu, S. S. (2019). *Biomolecules*, 9(5), 174. https://doi.org/10.3390/biom9050174
[108] Lee, S., Ro, H., In, H. J., Choi, J. H., Kim, M. O., Lee, J., Hong, S. T., and Lee, S. U. (2018). *Cytokine*, 108, 247–254. https://doi.org/10.1016/j.cyto.2018.01.004

[109] Filardo, S., Di Pietro, M., Mastromarino, P., and Sessa, R. (2020). *Pharmacology & Therapeutics*, 214, 107613.
[110] Ramdani, L. H., and Bachari, K. (2020). *Acta Virologica*, 64(3), 276–280. https://doi.org/10.4149/av_2020_309
[111] Lei, Y., Fu, P., Jun, X., and Cheng, P. (2019). *Planta Medica*, 85(1), 48–55. https://doi.org/10.1055/a-0750-6907
[112] Xue, Z., Zhang, X. G., Wu, J., Xu, W. C., Li, L. Q., Liu, F., and Yu, J. E. (2016). *Annals of Allergy, Asthma & Immunology: Official Publication of the American College of Allergy, Asthma, & Immunology*, 116(6), 506–513. https://doi.org/10.1016/j.anai.2016.03.029
[113] Fatima, K., Wani, Z. A., Meena, A., and Luqman, S. (2021). *Phytotherapy Research*, 35(7), 3861–3874. https://doi.org/10.1002/ptr.7094
[114] Yan, X. D., Wang, Q. M., Tie, C., Jin, H. T., Han, Y. X., Zhang, J. L., Yu, X. M., Hou, Q., Zhang, P. P., Wang, A. P., Zhang, P. C., Gao, Z., and Jiang, J. D. (2017). *Scientific Reports*, 7, 40030. https://doi.org/10.1038/srep40030
[115] Tang, J., Li, Y., Wang, J., Wu, Q., and Yan, H. (2019). *Journal of Cellular Biochemistry*, 120(6), 10137–10144.
[116] Patel, J. J., Acharya, S. R., and Acharya, N. S. (2014). *Journal of Ethnopharmacology*, 154(2), 268–285. https://doi.org/10.1016/j.jep.2014.03.071
[117] Stead, L. F., and Hughes, J. R. (2012). 2012(2), CD000124. https://doi.org/10.1002/14651858.CD000124.pub2
[118] Li, K. C., Ho, Y. L., Chen, C. Y., Hsieh, W. T., Chang, Y. S., and Huang, G. J. (2016). *Respiratory Physiology & Neurobiology*, 225, 19–30.
[119] Marlin, D. J., Roberts, C. A., Schroter, R. C., and Lekeux, P. (2000). *Equine Veterinary Journal*, 32(3), 200–207. https://doi.org/10.2746/042516400776563644
[120] Adaki, S., Adaki, R., Shah, K., and Karagir, A. (2014). *Indian Journal of Cancer*, 51(4), 577–581. https://doi.org/10.4103/0019-509X.175383
[121] Suman, S., and Shukla, Y. (2016). *Advances in Experimental Medicine and Biology*, 929, 127–144. https://doi.org/10.1007/978-3-319-41342-6_6
[122] Puccinelli, M. T., and Stan, S. D. (2017). *International Journal of Molecular Sciences*, 18(8), 1645. https://doi.org/10.3390/ijms18081645
[123] Tiboc-Schnell, C. N., Filip, G. A., Man, S. C., Decea, N., Moldovan, R., Opris, R., Sas, V., and Tabaran, F. (2020). *Journal of Physiology and Pharmacology: An Official Journal of the Polish Physiological Society*, 71(4). https://doi.org/10.26402/jpp.2020.4.03
[124] Tripathi, A., Kumar, B., and Sagi, S. (2019). *PloS One*, 14(6), e0219075. https://doi.org/10.1371/journal.pone.0219075
[125] Mlala, S., Oyedeji, A. O., Gondwe, M., and Oyedeji, O. O. (2019). *Molecules* (Basel, Switzerland), 24(15), 2751. https://doi.org/10.3390/molecules24152751
[126] Huang, Y. F., Bai, C., He, F., Xie, Y., and Zhou, H. (2020). *Pharmacological Research*, 158, 104939. https://doi.org/10.1016/j.phrs.2020.104939
[127] Paz-Ares L., Balint B., de Boer R. H., van Meerbeeck J. P., Wierzbicki R., De Souza P., et al. (2013). *Journal of Thoracic Oncology*, 8(3), 329–337, https://doi.org/10.1097/JTO.0b013e31827ce554
[128] Kelly R. J., Thomas A., Rajan A., Chun G., Lopez-Chavez A., Szabo E., et al. (2013) *Annals of Oncology* 24(10), 2601–2606, https://doi.org/10.1093/annonc/mdt249

11 Nanoparticles in Anti-inflammatory Phytoconstituent Delivery

*Zahid Hussain[1,2] (zhussain@sharjah.ac.ae)
[1]Department of Pharmaceutics and Pharmaceutical Technology, College of Pharmacy, University of Sharjah, Sharjah 27272, United Arab Emirates
[2]Research Institute for Medical and Health Sciences, University of Sharjah, Sharjah 27272, United Arab Emirates

Hnin Ei Thu[3,4] (eithuu287@gmail.com)
[3]Research Institute for Medical and Health Sciences, University of Sharjah, Sharjah 27272, United Arab Emirates
[4]Department of Pharmacology, Faculty of Dentistry, Universiti Teknologi MARA, Sungai Buloh Campus, Selangor Branch, 47000, Sungai Buloh, Selangor, Malaysia

Shahzeb Khan[5,6,7] (shahzeb_333@hotmail.com)
[5]Department of Pharmacy, University of Malakand, Chakdara, Dir Lower 18800, Khyber Pakhtunkhwa, Pakistan
[6]Center for Pharmaceutical Engineering Science, Faculty of Life Sciences, School of Pharmacy and Medical Sciences, University of Bradford, Bradford BD7 1DP, UK
[7]Discipline of Pharmaceutical Sciences, School of Health Sciences, University of KwaZulu-Natal, Private Bag X54001, Westville 3631, Durban 4000, South Africa

Rai Muhammad Sarfraz[8] (sarfrazrai85@yahoo.com)
[8]College of Pharmacy, University of Sargodha, Sargodha, Punjab, Pakistan

Asif Mahmood[9] (asif_1591985@yahoo.com)
[9]Department of Pharmacy, University of Chakwal, 48800, Chakwal, Pakistan

Nor Amlizan Ramli[10] (nor.amlizan@uitm.edu.my)
[10]Department of Pharmaceutics, Faculty of Pharmacy, Universiti Teknologi MARA, Puncak Alam Campus, Bandar Puncak Alam 42300, Selangor, Malaysia

DOI: 10.1201/9781003231721-11

11.1 INTRODUCTION

Inflammation is a natural protective physiological response of the body elicited against tissue injury (i.e., wound, cuts, burns, radiation, irritants, and physical or chemical stress), microbial infection (i.e., bacterial, viral, or fungal), or autoimmune disorders. Depending on the cause of injury and the time span to heal that injury and/or remove the causative factor, inflammation can be classified as either acute or chronic. Acute inflammation is a nonspecific early response of the body against an insult, and this inflammation normally lasts for a few days to a few weeks. It is characterized by the release of local inflammatory mediators that cause vasodilation, increase in capillary permeability, increase the accumulation of fluid plasma proteins in the interstitial spaces, chemotaxis of inflammatory cells (i.e., macrophages and neutrophils), and release of pro-inflammatory mediators (i.e., histamine, cytokines, and lymphokines). The infiltrated inflammatory cells first phagocytose the offending agent (bacteria, virus, fungi, or toxic chemicals) and then undergo apoptosis, after which these cells are cleared out from the lymphatic system and tissue function is reinstated. Generally, acute inflammation is characterized by five signs: 1) redness, 2) increased heat generation, 3) swelling, 4) pain, and 5) loss of function. By contrast, chronic inflammation is a specific hyper-immune response of the body against a particular injury, insult, or autoimmune disorder that may persist long, usually from months to years. It may also occur when a condition causing the acute inflammation is not resolved or due to over-infiltration of inflammatory cells, overexpression of pro-inflammatory cytokines, or dysregulation of cellular signaling and loss of barrier function. If remain unnoticed, then chronic inflammation may predispose to several secondary complications such as complete failure/removal of the inflamed tissue/organ (i.e., diabetic wound or foot ulcer, leading to amputation), neurodegenerative disorders (neuro-inflammation, causing Alzheimer's diseases, Parkinson's diseases), diabetes-induced liver inflammation, rheumatoid arthritis (RA), sepsis, cancer, asthma, and many other inflammatory diseases. Among the most important pro-inflammatory cytokines responsible for inducing chronic inflammation, interleukin-1β (IL-1β), IL-6, IL-10, and tumor necrosis factor-α (TNF-α) are the predominant ones [1-5].

Many synthetic compounds (e.g., steroidal and nonsteroidal agents) are being clinically used as anti-inflammatory agents to effectively suppress both acute and chronic inflammatory diseases; however, these synthetic anti-inflammatory agents can cause several side effects including, but not limited to, allergic response (i.e., skin rashes, asthma, hypotensive shock, pruritus, and angioedema), gastric or intestinal ulceration, bleeding, hepatic failure, renal failure, hemolytic anemia, and many others [6,7].

Phytochemicals, also referred to as phytocompounds or phytoconstituents, are plant-derived natural compounds that exert several vital functions to maintain homeostasis in the life cycle of plants. For example, they protect plants from infection due to various microorganisms (fungal, viral, or bacterial infections) and maintain normal growth and functions of the plants. Phytochemicals, also referred to as phytonutrients, are found in various parts of the plant, such as roots, leaves, stem, beans, fruits, grains, legumes, beans, herbs, spices, nuts, and seeds and are used as food sources for humans. Amongst various phytochemicals, polyphenols, flavonoids, flavones, isoflavones, terpenoids, alkaloid, and essential oils have gained remarkable recognition owing to their promising pharmacological properties. For example, it is well established that polyphenols have the ability to interact with various proteins, enzymes, and membrane receptors, resulting in modulating various biological processes. Among these biological properties, strong antioxidant and anti-inflammatory properties have been extensively studied for many polyphenols such as curcumin (Cur), resveratrol (RSV), quercetin (QUE), and tannins in the treatment of various inflammatory disorders [8,9], obesity-mediated inflammation [10,11], inflammatory bowel disease [12-14], Alzheimer's disease [15,16], diabetes [17], and skin inflammation [18,19].

Despite their promising potency, abundance, and cost-effectiveness, clinical translation of most phytochemicals has been restricted owing to several shortcomings such as poor aqueous solubility, low stability, presystemic metabolism, first-pass effect, poor permeability across biological membranes, short plasma half-life, and rapid elimination from the body. To alleviate these limitations as well as to improve their therapeutic efficacy, many researchers have attempted to use a variety of nanodelivery systems for the effective delivery of phytochemicals. In this chapter, we summarize and critically discuss the biopharmaceutical viability and therapeutic significance of various types of nanodelivery systems (i.e., polymeric nanospheres [NSPs], nanocapsules [NCPs], solid lipid nanoparticles [SLNs], polymeric micelles [PMs], nanoemulsions [NEs] or microemulsions, liposomes, ethosomes, niosomes [NIOs], cubosomes [CBMs], transfersomes [TFMs], nanocomposites, nanodiamonds [NDs], and nanohybrid scaffolds) that have been employed for the effective delivery of anti-inflammatory phytochemicals for the treatment of various inflammatory disorders.

11.2 NANOPARTICLES FOR THE DELIVERY OF ANTI-INFLAMMATORY PHYTOCHEMICALS

11.2.1 Polymeric Nanoparticles

Polymeric nanoparticles (NPs) are nanoscale particles (1–1000 nm) made of natural or synthetic and biodegradable or nonbiodegradable polymers. Based on their structure, polymeric NPs can be classified as polymeric NSPs or polymeric NCPs. NSPs have the characteristic feature of a dense polymeric matrix formed of synthetic, biodegradable, and biocompatible polymer(s) wherein drug(s) are entrapped within the polymeric matrix or adsorbed onto the surface of the polymeric matrix. However, NCPs are nanoscale polymeric shells containing a solid or liquid core (inner cavity) wherein drug(s), proteins, genes, or other bioactive(s) are entrapped and surrounded by a distinct polymeric membrane. Drug(s) entrapment within the inner cavity of NCPs is an important feature to protect the drug(s) from undergoing premature degradation (e.g., enzymatic or pH-induced inactivation), burst release, or presystemic metabolism. The characteristic features of NSPs and NCPs can be modified by constructing them using multiple polymers having different properties (natural or synthetic and biodegradable or non-biodegradable) or by varying their polymeric composition, synthesis method, and process conditions.

Considering their salient features, many researchers have utilized polymeric NPs for improving the physicochemical properties (i.e., aqueous solubility, dissolution profile, and permeation coefficient) and chemical stability of anti-inflammatory phytochemicals in biological fluids (e.g., plasma) [20,21]. For example, Cur is one of the most potent anti-inflammatory phytochemicals (polyphenol) that possess a wide range of pharmacological activities; however, its poor aqueous solubility, photodegradation, low absorption, poor permeability, low bioavailability, and short plasma half-life restrict its biopharmaceutical application. To alleviate these problems, Xie et al. [22] developed poly(lactide-co-glycolide) (PLGA)-NPs that exhibited nanoscale particle size (<200 nm), excellent encapsulation efficiency (%EE; >90%), and 640-fold increase in the aqueous solubility of Cur. Furthermore, a significant improvement in its oral bioavailability (5.6-fold) and plasma half-life was observed when compared with that of free Cur, which was attributed to the enhanced aqueous solubility of Cur-PLGA-NPs in the intestinal fluid and inhibition of P-glycoprotein efflux. These findings suggested that Cur-PLGA-NPs could be a promising delivery vehicle for the oral administration of Cur-like potent phytochemicals [22]. Likewise, Doggui et al. [23] developed PLGA-NPs for improving the physicochemical properties, cell uptake, bioavailability, and anti-inflammatory efficacy of Cur for neuro-inflammation treatment. The fabricated Cur-PLGA-NPs exhibited ultrafine particle size (80–120 nm), good %EE (31%), smooth spherical morphology, and biphasic release pattern with a burst release initially and a relatively slow release finally. Incubation of SK-N-SH cells (human neuronal cells) with Cur-PLGA-NPs displayed a significant improvement in antioxidant effectiveness (glutathione) to counteract reactive oxygen species (ROS; H_2O_2)-induced oxidative stress. These findings concluded that Cur-PLGA-NPs could be a promising nanophytomedicine for improving the anti-inflammatory and antioxidant potential of Cur to protect neuronal tissues against ROS-induced neuro-inflammation in neurodegenerative diseases (e.g., Alzheimer's disease, Parkinson's disease) [23]. Many studies have reported the biopharmaceutical significance of polymeric NPs for the successful delivery of various anti-inflammatory phytochemicals such as QUE [24,25], epigallocatechin-3-gallate (EGCG) [26], Cur (isolated from the root extract of *Curcuma longa*) [27-29], morin (flavonoid isolated from *Maclura pomifera*) [30], naringenin (a flavonoid isolated from the Chinese herbal medicine) [31], piperine (PIP; isolated from *Piper nigrum*) [32], and genistein (GNS) (an isoflavone isolated from the root extract of *Genista tinctoria*) [33] for the treatment of various inflammatory diseases.

Psoriasis is a chronic skin inflammatory disorder characterized by the aberrant activation of immune cells (i.e., macrophages, dendritic cells [DCs], and so on) against keratinocytes. One of the challenges encountered in the treatment of such topical diseases is the poor penetration of drugs across the stratum corneum (SC) (major barrier), which is the outermost cutaneous layer. Therefore, to improve Cur penetration through the SC and for the effective management of psoriasis, Sun et al. [34] developed Cur-PLGA-NPs. The fabricated Cur-PLGA-NPs exhibited nanoscale particle size (150 nm), good %EE, and smooth spherical morphology. The incubation of Cur-PLGA-NPs with human keratinocytes (HaCaT cells) demonstrated a significant suppression of the keratinocyte proliferation. Franz diffusion experiment showed superior permeation and higher accumulation of Cur-PLGA-NPs into various skin layers (epidermis and dermis) than those of free Cur. Additionally, the therapeutic efficacy of Cur-PLGA-NPs was evaluated in a psoriasis mouse model wherein psoriasis was induced using imiquimod (IMQ). The results of that study suggested a superior antipsoriasis efficacy of Cur-PLGA-NPs after topical application compared with that of free Cur and other formulations. The authors demonstrated that the superior antipsoriasis efficacy of the Cur-PLGA-NPs was attributed to the enhanced anti-inflammatory, antioxidant, and immunomodulatory efficacy of Cur-PLGA NPs [34]. To counteract the barrier properties of the SC as well as to improve the localization of Cur into deeper skin layers, Mao et al. [35] engineered novel polymeric NPs assembled with the RRR-α-tocopheryl succinate-grafted-ε-polylysine conjugate (VES-g-ε-PLL). The fabricated Cur-VES-g-ε-PLL-NPs exhibited ultra-fine particle size (~25 nm), cationic zeta potential (+20 mV), smooth spherical morphology, and excellent %EE (~80%). To further prolong the retention time on the skin, the developed polymeric NPs (Cur-VES-g-ε-PLL-NPs) were matrixed over a silk fibroin (SF)-based hydrogel. The results suggested sustained release, prolonged retention, and improved penetrability of Cur from the Cur-NP-composite hydrogel compared with Cur-gel and free Cur-suspension. Moreover, this hybrid polymeric nanosystem of Cur also displayed superior suppression of the expression of pro-inflammatory cytokines (e.g., TNF-α, interferon [IFN]-γ, IL-2, IL-6, and IL-12). These findings suggest that encapsulation of Cur into a polymeric NP-composited hydrogel could improve skin retention time, skin penetrability, and antipsoriasis efficacy of Cur [35].

Polymeric (chitosan [CS]) NPs have been fabricated for improving the physicochemical properties and biomedical efficacy of other phytochemicals (embelin [EMB]) (Cui et al. [36]. EMB is a naturally occurring para-benzoquinone isolated from dried berries of *Embelia ribes*. It exerts a wide range of pharmacological properties such as antioxidant, anti-inflammatory, antirheumatic, antimicrobial, anthelmintic, and anticancer activities, consequently garnering increasing attention; however, poor aqueous solubility, low permeability, and degradation limit its therapeutic application. Therefore, Cui et al. [36] recently proposed the encapsulation of EMB into polymeric NPs and evaluated its anti-inflammatory effects in RA treatment. RA was induced in the experimental animals (rats) by administering a subcutaneous injection of 0.1 mL of complete Freund's adjuvant (CFA) into the footpad of the left hind paw. CFA-induced arthritic rats were then treated with different concentrations of EMB-CS-NPs (25 and 50 mg/kg); the results suggested a significant dose-dependent decrease in oxidative stress mediators such as malondialdehyde (MDA) and nitric oxide (NO), pro-inflammatory cytokines (e.g., IL-1β, IL-6, and TNF-α), and RA symptoms such as paw swelling. These results indicated that encapsulation of EMB by CS-NPs is a promising technique to improve the antioxidant and anti-inflammatory effects of EMB for RA treatment [36]. Interestingly, Das et al. [37] proposed a novel tri-polymeric composite comprising CS, alginate (ALG), and pluronic to further improve cell uptake efficiency and penetrability of phytoconstituents using Cur as the model drug. The developed Cur-loaded tri-polymeric NPs exhibited ultra-fine particle size (~100 nm), positive zeta potential, good %EE, and smooth spherical morphology. A cell uptake (HeLa cells) study revealed a significantly higher internalization of Cur-encapsulated NPs into HeLa cells than that of free Cur, which suggested the great penetrability of Cur after encapsulation into the tri-polymeric–composited NPs [37].

PEGylation is a promising approach to prolong the plasma half-life and elimination half-life of NPs. Anand et al. [38] developed Cur-PLGA-NPs stabilized with polyethylene glycol (PEG)-5000. The fabricated PEGylated Cur-PLGA-NPs exhibited ultra-fine particle size (~80 nm), good %EE, and excellent cell uptake efficiency (*in vitro*). A prolonged plasma half-life of PEG-Cur-PLGA-NPs was evident in animals when compared with that of free Cur and un-PEGylated Cur-PLGA-NPs. Additionally, a pronounced increase in cell apoptosis was observed in cells treated with PEG-Cur-PLGA-NPs when compared with free Cur, which was attributed to the pronounced inhibition of TNF-α–induced suppression of cell proliferation (cyclin D1), invasion (matrix metalloproteinase [MMP]-9), and angiogenesis (vascular endothelial growth factor). This study concluded that encapsulation of Cur into PLGA-NPs and subsequent PEGylation significantly improves physicochemical properties, cell uptake, plasma half-life, and anti-inflammatory effect of Cur [38]. Another recent adaptation to achieve the selective targeting of NPs to specific cells/tissues/organs is the conjugation of the targeting moiety to the surface of polymeric NPs. Kuo et al. [39] engineered monoclonal antibody (MAb)-conjugated polyacrylamide–cardiolipin-PLGA-NPs for the simultaneous delivery of Cur and rosmarinic acid and evaluated their use for the treatment of amyloid-β (Aβ)-induced neuro-inflammation. The results suggested that MAb-conjugated NPs resulted in significant improvement in drug permeation across the blood–brain barrier (BBB), cell uptake into neuronal cells (SK-N-MC cells), and anti-inflammatory and antioxidative potential against Aβ-induced neuro-inflammation for the treatment of Alzheimer's diseases [39]. The same research group had previously established the biopharmaceutical viability of PLGA-NPs functionalized with cross-reacting material 197 (CRM197) and apolipoprotein E (ApoE) to enhance the permeation of phytoconstituents across the BBB for rescuing neuronal tissues from Aβ-induced neuro-inflammation for the treatment of Alzheimer's diseases [40].

NCPs are polymeric vesicular delivery systems comprising a distinct core (inner cavity) surrounded by a shell (polymeric membrane). Bioactive(s) can be confined into the core or can be dispersed onto the shell of NCPs. Many studies investigated the potency of NCPs for the successful delivery of various phytochemicals owing to their unique architectural design [41-43]. Considering the unique properties of NCPs, Frozza et al. [44] fabricated NCPs for improving the physicochemical properties and biomedical efficacy (antioxidant and anti-inflammatory) of RSV for the treatment of Aβ-induced neuro-inflammation. One of the hallmarks for inducing neuro-inflammation is oxidative stress, followed by the inflammatory cascade, which result in damage to neuronal functioning (memory and cognition). RSV has been extensively investigated for the treatment of Aβ-induced neuro-inflammation owing to its powerful antioxidant and anti-inflammatory effects [45-49]; however, its poor pharmacokinetic profile and vulnerability to photo-oxidation limit its therapeutic viability. Therefore, Frozza et al. [44] incorporated RSV into NCPs and evaluated their neuroprotective effects. The results suggested that, when compared with free RSV, RSV-NCPs significantly suppressed Aβ-induced neuro-inflammation by downregulating oxidative stress (ROS, NO, and H_2O_2) and the release of pro-inflammatory cytokines (e.g., IL-1β, IL-6, and TNF-α), which finally blocked neuro-inflammation [44]. The same research group has validated these findings by conducting another study in rats and reported that animals treated with RSV-NCPs expressed a significantly higher anti-amyloidogenic effect than partial neuroprotection achieved with free RSV. The authors further demonstrated that NCPs enabled higher accumulation of RSV into intracerebral tissues, which is attributed to the superior neuroprotective effects of RSV [50]. The biopharmaceutical significance of NCPs has been validated for the intracerebral delivery of Cur [51]. Owing to its excellent antioxidant and anti-inflammatory properties, Cur is widely used for the treatment of Aβ-induced neuro-inflammation [52-54]; however, its poor aqueous solubility and poor pharmacokinetic profile limit its clinical usability. Therefore, Hoppe et al. [51] developed lipid-core NCPs for improving the pharmacokinetic profile and intracerebral delivery of Cur for the treatment of Aβ-induced neuro-inflammation. Following the induction of neuro-inflammation with a single intracerebroventricular injection of Aβ(1-42), animals were treated with an intraperitoneal (IP) injection of either Cur-NCPs or free Cur for 10 days. In contrast to the significant impairment of memory and cognitive functioning in Aβ(1-42)-infused animals (nontreated animals), animals treated with free Cur displayed good efficacy in terms of protecting against memory impairment and cognitive dysfunction. Of high interest, animals treated with an IP injection of Cur-NCPs demonstrated a 20-fold higher neuroprotection against Aβ-induced neuro-inflammation than animals treated with free Cur [51].

Owing to their great flexibility as drug cargo, NCPs have been employed for the simultaneous delivery of multiple phytoconstituents. Coradini et al. [55] designed lipid-core NCPs comprising polycaprolactone (PCL; 1.0% w/v), grape seed oil (1.65% v/v), sorbitan monostearate (0.038% w/v), and polysorbate 80 (0.038% w/v) for the co-delivery of Cur (0.05% w/v) and RSV (0.05% w/v). The developed Cur-RSV-NCPs exhibited ultra-fine particle size (<200 nm), narrow polydispersity index (PDI), anionic surface charge (−12 mV), good %EE, and spherical morphology. Experimental animals treated with Cur-RSV-NCPs showed greater control over CFA-induced RA symptoms and markers such as paw swelling, paw edema, cellular infiltration, fibrosis, cartilage loss, and bone loss as well as liver enzymes such as alkaline phosphatase, alanine aminotransferase, and aspartate aminotransferase. These results clearly indicate the superiority of Cur-RSV-NCPs over Cur-NCPs, RSV-NCPs and the vehicle for RA treatment [55]. The same research group also previously validated the potential of Cur-RSV-NCPs for preserving the photostability and antioxidant efficacy of Cur and RSV [56]. Friedrich et al. [57] developed Cur-RSV-NCPs and validated their potential in terms of skin penetration, localization efficiency in various skin layers, and treatment of skin inflammatory diseases. RSV has gained good recognition owing to its powerful antioxidant and anti-inflammatory effects; however, its poor aqueous solubility, photodegradation, and poor pharmacokinetic profile restrict its therapeutic applicability [57].

Osteoporosis is a medical condition characterized by weak and fragile bones particularly the hip bone, wrist bone, and backbone (spine). It is a silent disease that remains progressive through the years (during which bones keep losing strength), and patients usually remain asymptomatic until the bones become highly brittle, very fragile, and easily breakable accompanied with mild trauma. To treat this chronic bone disease, Dong et al. [58] developed polymeric (CS) NPs for the combined delivery of alendronate (ALN; bone antiresorptive agent) and Cur. In addition to other biomedical uses [59,60], Cur exhibits good bone density-enhancing property by improving the proliferation of osteoblasts (bone-forming cells) [61]. The fabricated Cur-ALN-CS-NPs were functionalized with hyaluronic acid (HA). HA is a potent biological macromolecule that has numerous biomedical applications including bone regeneration, wound healing, anticancer agent, skin rejuvenation, tissue remodeling, and ligand-targeting activities for the specific targetability of CD44 receptors bearing cells/tissues. To evaluate their antiosteoporotic efficacy, HA-Cur-ALN-CS-NPs were incubated with human preosteoblasts cells (MC3T3-E1 cells), and the time-dependent proliferation, differentiation, and maturation were estimated. MC3T3-E1 cells incubated with HA-Cur-ALN-CS-NPs displayed time-dependent upregulation in the protein expression of bone-forming markers (bone morphogenetic protein, RUX-2, and osteocalcin) compared with other treatments which provides evidence for the improved bone regeneration ability of HA-Cur-ALN-CS-NPs [58]. Similarly, Hussain et al. [62] also recently confirmed the antiosteoporotic efficacy of HA-functionalized raloxifene (RLX) and Cur co-encapsulated CS-NPs. A remarkably higher bone-forming ability of HA-RLX-Cur-CS-NPs was evidenced through improved proliferation, differentiation, and maturation; collagen (COL) deposition; calcium deposition; and phosphate upregulation in MC3T3-E1 cells [62].

Wound healing is a complex process of replacing the damaged/injured skin tissues with new healthy tissues. This process involves four distinct phases; 1) homeostasis, 2) inflammation, 3) proliferation, and 4) remodeling. To accomplish this process, a wide range of molecules including macrophages, neutrophils, lymphocytes, fibroblasts, keratinocytes, endothelial cells, and others play an important role. A wide range of phytoconstituents have been investigated for wound potential, and among them Cur has shown remarkable wound healing efficacy. Although substantial evidence exists regarding the effect of Cur on various phases of the wound healing process, it mainly modulates the inflammation phase. Encapsulation of Cur into polymeric NPs results in improved wound healing efficacy. Lin et al. [63] developed polymeric NPs comprising three polymers such as CS, poly-γ-glutamic acid, and pluronic using the ionic gelation method. The fabricated Cur-loaded NPs exhibited ultra-fine particle size (~195 nm), cationic zeta potential (+20 mV), good %EE, smooth spherical morphology, biphasic release profile, and good skin permeation efficiency. Furthermore, wounds treated with Cur-loaded NPs showed higher wound closure rate, neocollagen deposition, and neovascularization than those treated with free Cur and other control substances [63]. Likewise, Krausz et al. [64] developed polymeric NPs for improving physicochemical properties, antimicrobial effect, and wound healing effectiveness of Cur. The fabricated NPs exhibited ultra-fine particle size, positive zeta potential, spherical morphology, extended release profile, and a dose-dependent antimicrobial effect against methicillin-resistant *Staphylococcus aureus* and *Pseudomonas aeruginosa*. The superior wound healing efficacy of Cur-loaded polymeric NPs was also validated in animals with infected burn wounds [64]. Other studies have demonstrated that encapsulation of Cur into PLGA-NPs [65] and lignin NPs [66] resulted in significant improvement in aqueous solubility, stability against photodegradation, antimicrobial activity, and wound healing efficacy. To achieve synergistic antioxidant and anti-inflammatory efficacy, Hussain et al. [67] recently developed CS-NPs for the combo-delivery of Cur and RSV for the treatment of chronic diabetic wounds. To further improve skin repair and tissue regeneration potential, fabricated Cur-RSV-NPs were functionalized with HA. The developed HA-Cur-RSV-NPs exhibited nano-scale particle size (<200 nm), positive zeta potential (>30 mV), good %EE, narrow particle size distribution (PDI), and smooth spherical morphology. The functionalized NPs displayed non-Fickian diffusion, sustained release mechanism, and good colloidal stability. The authors suggested that HA functionalization could be a promising approach to maximize their retention and localization into the skin and improved wound healing effectiveness [67]. Utilizing the concept of HA functionalization, the same research group also developed another polymeric nanosystem for the co-delivery of Cur and QUE for the management of burn wounds [68]. The developed polymeric NPs (HA-Cur-QUE-NPs) exhibited ultra-fine particle size (~180 nm), cationic surface charge (~40 mV), high %EE (~65% for Cur and ~85% for QUE), good loading capacity (~43% for Cur and ~38% for QUE), spherical morphology, triphasic release kinetics, and good cytocompatibility when incubated with MC3T3-E1 cells. Franz diffusion experiment indicated significantly higher permeation across the SC and higher localization of both phytochemicals (Cur and RSV) into the epidermis and the dermis when applied with HA-Cur-QUE-NPs than those of free drugs (Cur and RSV). Moreover, superior wound healing efficacy was observed in burn wounds treated with HA-Cur-QUE-NPs (98 cases of wound closure within 28 days) with marked signs of greater granulation tissue formation, re-epithelization, angiogenesis, anti-inflammation, antioxidant, and tissue regeneration [68].

11.2.2 Polymeric Micelles

PMs are nanoconstructs with a core-shell architecture that are obtained as a result of the self-assembly of amphiphilic block co-polymers. The architectural design (hydrophobic inner core surrounded by hydrophilic shell) and flexible modifiable properties make PMs a promising delivery vehicle for hydrophobic drugs. Many studies have reported the biopharmaceutical viability of PMs for the successful delivery of anti-inflammatory phytochemicals for the treatment of various diseases. Fan et al. [69] developed PMs composed of Cur and HA for the purpose of improving joint lubrication while exerting anti-inflammatory and immunomodulatory effects for RA treatment. Physicochemical characterization revealed that Cur-HA-PMs exhibited ultra-fine particle size (164 nm), good zeta potential, good %EE, and spherical morphology. Fabricated Cur-HA-PMs showed good biocompatibility, which facilitated the proliferation of chondrocytes. Parenteral administration of Cur-HA-PMs into rats resulted in good control over arthritic scores, edema, and a remarkable decrease in friction between the surfaces of the cartilage around the joints, which protects the joints from RA-induced damage. Moreover, assessment of the blood biochemistry of experimental RA animals revealed a remarkable downregulation in the expression of RA-associated pro-inflammatory cytokines in the blood and synovial fluid. These results provide evidence that co-administration of Cur and HA using PMs is a promising technique to maximize the anti-RA efficacy [69]. Likewise, Li et al. [70] reported the significance of encapsulation of Cur into PMs for the treatment of corneal inflammation.

Duan et al. [71] developed methoxypoly(ethylene glycol)-poly(lactide) (mPEG-PLA)/D-α-tocopherol PEG 1000 succinate (TPGS) mixed micelles (MMs) for the successful oral delivery of Cur. The fabricated Cur-MPP-TPGS-MMs exhibited ultra-small particle size (<50 nm), good %EE, smooth spherical morphology, and sustained release behavior in simulated gastrointestinal fluids. Pharmacokinetic studies revealed a significant increase in C_{max}, mean retention time, area under the curve (AUC), and relative bioavailability (~930%), were obtained with the fabricated Cur-MPP-TPGS-MMs compared with those of the Cur-suspension. These results provide evidence that MMs could be a promising delivery vehicle for improving intestinal absorption and oral bioavailability of hydrophobic phytochemicals such as Cur [71]. Similarly, Akbar et al. [72] developed MMs composed of CS/ALG/maltodextrin/pluronic F127/pluronic P123/Tween 80 using the thin-film hydration method for the delivery of Cur and evaluated their potential for the treatment of chronic diabetic wounds. An evident decrease in the blood glucose level and lipid profile (total cholesterol, triglycerides, and high-density lipoprotein level) was observed with accelerated wound healing efficacy in animals treated with Cur-MMs when compared with that in animals treated with free Cur. Cur-MMs also displayed an excellent inhibitory effect on histopathologic changes in the liver, kidney, and pancreas [72].

Surface functionalization is an extensively adopted and well-established technique that aims to improve the biopharmaceutical properties, specific targetability, and therapeutic efficacy of PMs. Sun et al. [73] recently developed PEG-PLA-TPGS-PMs for improving the physicochemical properties and oral bioavailability of Cur, and the developed PMs were further functionalized with galactosamine (GAL) to improve the intestinal absorption of Cur. The fabricated GAL-Cur-PEG-PLA-TPGS PMs exhibited nanoscale size (~100 nm), natural zeta potential, good loading efficiency, and sustained release kinetics. A cell uptake study demonstrated a significantly higher uptake of Cur from GAL-modified micelles than the uptake of free Cur, and an intestinal absorption study indicated jejunum and ileum as the best absorption regions for the intestinal absorption of Cur from GAL-modified micelles. Pharmacokinetic studies established significantly improved AUC, C_{max}, and relative bioavailability (~260%) of Cur from GAL-modified micelles when compared with those of Cur suspension. These results concluded that GAL functionalization of PMs is a promising approach to maximize the intestinal absorption and oral bioavailability of phytoconstituents [73].

Surface functionalization of PMs has been performed with folic acid (FA) [74]. The developed FA-functionalized PMs (FA-Cur-PLA-PEG-PMs) exhibited ultra-small size (<90 nm), good %EE, smooth spherical morphology, and controlled release profile. A cell uptake study showed a significant improvement in Cur uptake FA-functionalized PMs into HepG2 cells compared with that of free Cur. These results concluded that FA-functionalization could be a promising technique to improve cell uptake and specific targetability of Cur [74]. These results were in agreement with the findings of other studies [75–78].

The therapeutic efficacy of Cur-PMs has been validated in a randomized double-blind controlled clinical trial of 65 patients with RA [79]. Patients with RA were divided into two groups: Cur-PM group (30 patients) and placebo group (35 patients). Patients with RA received either Cur-PMs (40 mg) or placebo capsules thrice daily for 12 weeks. The therapeutic efficacy of Cur-PMs, in comparison with placebo, was evaluated by estimating the "disease activity scores of 28 joints (DAS-28)" and erythrocyte sedimentation rate (ESR) at baseline and at treatment completion (after 12 weeks). At baseline, no significant difference was observed between patients with RA receiving placebo and those receiving Cur-PMs in terms of DAS-28, swollen joint count (SJC), and tender joint count (TJC); however, an evident downregulation in all study parameters such as DAS-28, SJC and TJC was observed in patients with RA treated with Cur-PMs compared with those in the placebo group. By contrast, no significant difference in ESR was observed between the study groups. These results suggested that encapsulation of Cur into PMs could improve anti-RA (anti-inflammatory) efficacy [79].

11.2.3 Nanoemulsions

NEs are thermodynamically stable nano-sized (1–1000 nm) emulsion formulations obtained by shearing a mixture of two immiscible liquids (oil/water), surfactant(s), and co-surfactant(s). NEs comprise two immiscible phases and are therefore a promising delivery vehicle for simultaneously carrying hydrophilic and hydrophobic phytochemical(s) and improving the physicochemical and therapeutic efficacy of such phytochemicals. Chen et al. [80] recently developed a novel NE for the successful delivery of tetrandrine (TET), a potent phytochemical isolated from Chinese herb (*Stephania tetrandra S.* Moor) for the treatment of psoriasis (inflammatory disease of the skin). Topical application of TET-NEs in IMQ-induced psoriasis animals showed significant improvement in TET penetration across the SC, localization into the skin, and plasma concentration. Antipsoriasis efficacy of TET-NEs was established by estimating the downregulation in the expression of various psoriasis-linked protein markers such as IL-2, IL-6, IL-12, TNF-α, and IFN-γ. These results provide evidence that NEs hold great promise as a nanovehicle for the delivery of phytochemicals for the treatment of skin inflammatory diseases (Figure 11.1) [80]. Likewise, Wang et al. [81] developed Cur-loaded NEs for improving physicochemical properties and anti-inflammatory effects and reported a significant decrease in 12-O-tetradecanoylphorbol-13-acetate (TPA)-induced ear inflammation in mice treated with Cur-NEs compared with that in mice treated with topical Cur suspension [81]. Ahmad et al. [82] also prepared Cur-NEs by ultra-sonication to improve the physicochemical properties, anti-inflammatory effect, and wound healing efficacy of Cur. For optimization of the physicochemical properties of Cur-NEs, they used various formulations (percentage of clove oil, percentage of surfactants, and PEG-400) and process parameters (ultrasonication time, ultrasonication intensity, and temperature). The developed Cur-NEs exhibited nanoscale globule size (<100 nm), anionic zeta potential (−12 mV), good %EE (~98%), and smooth spherical morphology. Cur-NEs showed a significant improvement in skin permeation, which was expected to be attributed to the use of clove oil, Tween-80, and PEG-400. The developed Cur-NEs showed greater anti-inflammatory and wound healing effects than free Cur, which provided evidence for the significant effect of NEs as a drug delivery vehicle for phytochemicals [82].

Zheng et al. [83] conducted a comparative study and evaluated the effect of encapsulation of Cur into NEs on the oral bioavailability of Cur as well as its anti-inflammatory and anti-RA efficacy in comparison with free Cur. The developed Cur-NEs exhibited ultra-fine globule size (~150 nm), smooth spherical morphology, and good protective effect against acidic or enzymatic degradation of Cur in the simulated gastric fluid. Interestingly, AUC and C_{max} were approximately threefold greater for the oral administration of Cur-NEs than for the intravenous (IV) administration of Cur

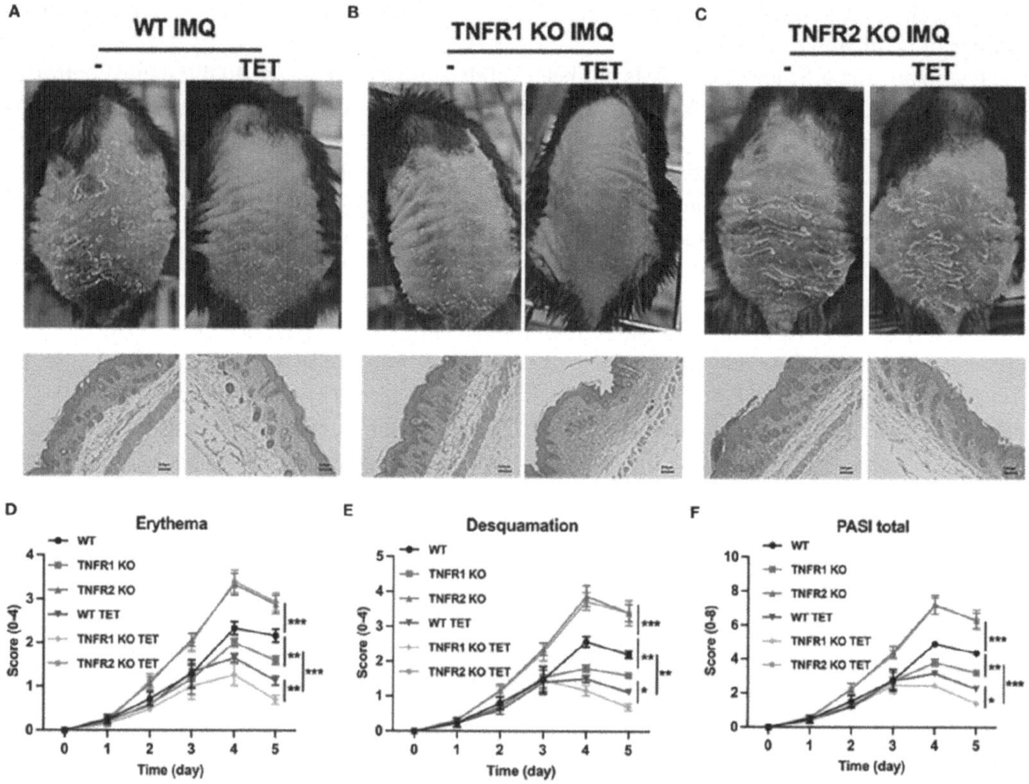

FIGURE 11.1 Anti-psoriatic efficacy of tetrandrine-nanoemulsion. (From Chen et al. [80]. With Permission.)

suspension. Moreover, a comparable downregulated expression of pro-inflammatory cytokines (e.g., IL-1β and TNF-α) was observed in the serum and synovial fluid of animals after oral administration of Cur-NEs or IV administration of the Cur suspension. These results clearly indicated that nanoencapsulation of Cur in NEs is a promising technique that allow changing the route of administration from IV to oral without alteration in efficacy [83]. Similarly, Gokhale et al. [84] developed QUE-loaded NE and incorporated it into a carbopol 940-based gel (pH 5.0–6.0) with semisolid consistency for the topical application of RA treatment. The developed QUE-NEs exhibited a mean globule size of <150 nm with narrow PDI, zeta potential of −25 mV, high %EE (~95%), and smooth spherical morphology. QUE-NEs also exhibited good storage stability with no significant changes in physicochemical properties up to 180 days. A swift release of ~85% QUE within only 4 hours was observed in QUE-NEs when compared with only 25% from QUE suspension, attributed to the larger surface area (nanoscale globule size) of QUE-NEs. Lipopolysaccharide (LPS)-induced TNF-α secretion from RAW 264.7 cells was significantly reduced when treated with QUE-NEs compared with that when treated with free QUE. TNF-α is a key inducer of synovitis through the upregulated expression of fibroblast proliferation and overexpression of pro-inflammatory cytokines (e.g., IL-1β and prostaglandin E2). Topical application of the QUE-NE gel to CFA-induced animals demonstrated a significant control over arthritis index, erythema intensity, stiffness scores, and paw circumferences compared with CFA-induced animals (nontreated) and QUE gel. These results conclude that QUE incorporation into NEs resulted in significant improvement in skin permeation efficiency and anti-RA (anti-inflammatory) efficacy of QUE [84]. Similarly, Naz et al. [85] developed Cur-NEs for RA treatment. The developed Cur-NEs exhibited nanoscale globule size (~70 nm), anionic zeta potential (−33 mV), good %EE, and smooth spherical morphology. To achieve

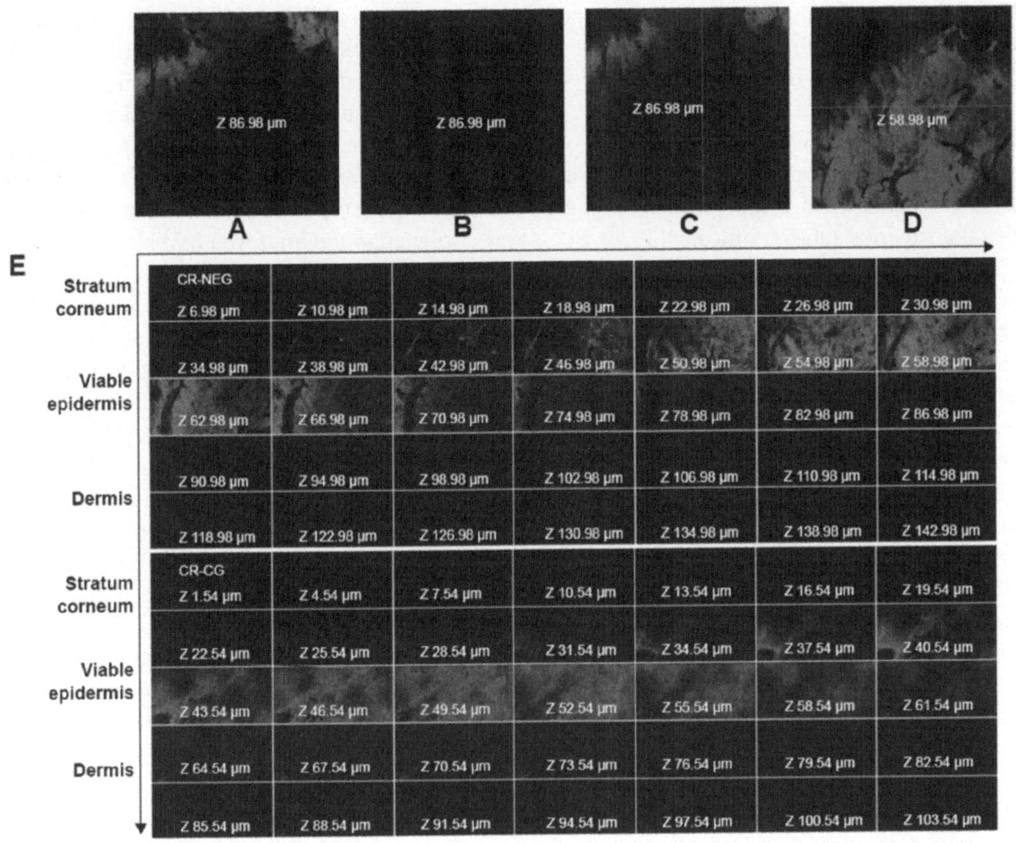

FIGURE 11.2 Skin permeation efficiency of nanoemulsion gel after topical application on Wistar rat skin, observed under a laser confocal microscope: Rhodamine B (red fluorescence) and Cur (blue fluorescence) at different depths of the skin. (From Naz et al. [85]. With Permission.)

semisolid consistency for topical application, the developed Cu-NEs were incorporated into a gel prepared from carbopol 980 (1% w/v). Skin permeation study revealed that the Cur-NE gel formulation displayed the highest permeation and skin retention, which was fourfold higher than the Cur nonaqueous suspension as well as greatest depth (~87 µm) in skin; these results were validated by laser scanning fluorescence microscopy (Figure 11.2) [85]. An *in vivo* study also revealed that FCA-induced RA animals (Wistar rats) treated (28 days) with the Cur-NE gel displayed significantly higher control over RA symptoms (e.g., body weight, paw edema, and tibiotarsal joint thickness) and downregulation of pro-inflammatory cytokines (e.g., IL-1β and TNF-α) in blood and synovial fluid. These results conclude that Cur-NEs in the form of gel is a promising formulation for the topical treatment of RA [85].

11.2.4 Solid Lipid Nanoparticles

SLNs are specialized nanoscale (1–1000 nm) colloidal dispersions formed of lipids, which remain solid at room temperature and liquid at body temperature. Initially introduced in 1991, SLNs consist of a core of solid lipid(s) with surfactants. The lipid core of SLNs plays a very important role in protecting susceptible drug(s) from undergoing premature enzymatic or pH-mediated degradation and regulating the release behavior of encapsulated drug(s). Moreover, because of their lipoidal nature and modifiable properties, SLNs are considered promising delivery vehicles to improve oral bioavailability, plasma circulation time, and biodistribution of drug(s). Many researchers have designed SLNs for the successful delivery of a wide range of phytoconstituents. Arora et al. [86] developed SLNs for improving the physicochemical properties, pharmacokinetic profile, and anti-rheumatic efficacy of Cur. Animals treated with Cur-SLNs displayed a dose-dependent (highest efficacy obtained at 30 mg/kg) significant control over RA signs and symptoms such as paw volume, mobility scores, hyperalgesia, joint stiffness, expression of RA biomarkers, and radiological alterations in joints when compared with animals treated with Cur suspension. The authors suggested that the superior anti-RA efficacy of Cur-SLNs was attributed to the improved oral bioavailability; plasma half-life; and antioxidant, anti-inflammatory, and immunomodulatory effects of Cur [86]. Wang et al. [87] reported a significant improvement in cellular uptake, skin localization, and higher anti-apoptotic activity of Cur-SLNs in RAW 264.7 cells than free Cur. The authors also observed significantly downregulated expression of pro-inflammatory cytokines (e.g., IL-1β, IL-6, and TNF-α) and NO production by LPS-activated RAW 264.7 cells treated with Cur-SLNs when compared with free Cur [87]. The biopharmaceutical utility of SLNs for the successful delivery of Cur has been reported by Maiti et al. [88]. Because of its powerful antioxidant, anti-inflammatory, and immunomodulatory properties, Cur has been extensively investigated for its effects in the treatment of Aβ-protein–induced neuro-inflammation [52-54]; however, its poor aqueous solubility, photodegradation, instability in physiological fluid, and poor pharmacokinetic profile limit its therapeutic utility. Therefore, to mitigate these challenges as well as to improve its therapeutic efficacy for the treatment of neuro-inflammation, Maiti et al. [88] proposed the encapsulation of Cur into SLNs. The results suggested that 5-day treatment to experimental animals (1-year-old 5×FAD-and age-matched wild-type mice) by administering an IP injection of Cur-SLNs (50 mg/kg) displayed better accumulation of Cur in cerebral tissues (i.e., hippocampus, striatum, and entorhinal cortex) than the accumulation of free Cur. Animals treated with Cur-SLNs displayed significantly lesser Aβ-plaque formation in the prefrontal cortex and dentate gyrus regions of the hippocampus than animals treated with free Cur. Moreover, 5-day treatment showed more pronounced effects than 2-day treatment, with an evident decrease in the size of plaques and disaggregation. Owing to the highest accumulation of Cur-SLNs into different brain tissues and potent antioxidant and anti-inflammatory effects, a significant reduction in microglial activation was seen in different brain regions, which resulted in minimal morphological changes and aggregation of microglial cells (Figure 11.3). These results conclude that encapsulation of Cur into SLNs resulted in significant improvement in the neuroprotective effects of Cur [88]. The same research group also had previously validated the medical utility of SLNs for the targeted delivery of Cur as a fluorescent dye (imaging agent) for the identification and quantification of Aβ-plaques, which is a key pathogenetic feature of neurodegeneration in Alzheimer's diseases [89]. The research group also proposed that structural similarity and preferential binding of Cur with Aβ-plaques makes it a suitable candidate for the diagnosis of neurodegeneration. The results suggested that, compared with classical Aβ-binding dyes such as Congo red and Thioflavin-S, Cur-SLNs showed a better fluorescence intensity after an IP injection at nanomolar concentration. Moreover, Cur-SLNs displayed superior labeling efficiency against different types of Aβ-plaques (i.e., neuritis, diffuse, and core) over other amyloid-binding dyes. These results conclude that Cur-SLNs are a cost-effective method for detecting Aβ-plaques (neurodegeneration) in Alzheimer's disease compared with Aβ-specific antibody or other Aβ-labeling dyes [89]. Later on, the authors developed Cur-SLNs for preserving synaptic failure, which is one of the key hallmarks of cognitive dysfunction in patients with Alzheimer's disease [90]. The results suggested that oral administration of Cur-SLNs (100 mg/kg of body weight) into 5×FAD mice displayed promising control over neurodegeneration (Aβ-plaque deposition) and neuronal morphology in different brain regions when compared with free Cur. In addition, Cur-SLNs showed superior effects in restoring various pre- and postsynaptic protein markers (e.g., synaptophysin, PSD95, Shank, Homer, Drebrin, Kalirin-7, CREB, and phosphorylated CREB [pCREB]), which finally improved cognitive functioning by preserving synaptic failure and preventing neuronal death

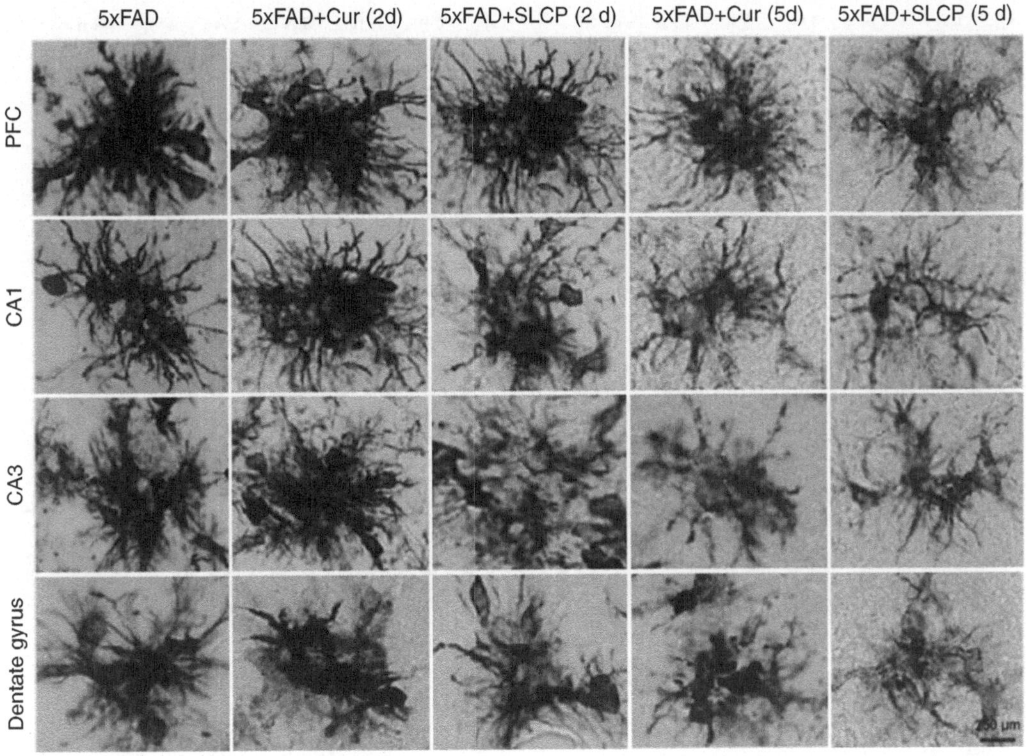

FIGURE 11.3 Neuroprotective efficacy of curcumin–solid lipid nanoparticles (2-day and 5-day treatment periods) compared with free Cur in terms of microglial aggregation in different regions of the brain of experimental animals (5×FAD mice). At the end of the treatment period (2 or 5 days), animals were killed and sections from different parts of the brain were collected for aggregation (activation) of microglia. (From Maiti et al. [88]. With Permission.)

[90]. Biopharmaceutical significance of SLNs as a drug delivery vehicle for Cur has been validated for sepsis treatment [91]. Sepsis is a notable public healthcare problem, affecting millions of people worldwide each year and killing one in four people, with increasing incidence every year. The fabricated Cur-SLNs exhibited ultra-fine particle size (~60 nm), narrow particle size distribution, anionic zeta potential (~27 mV), low %EE (36%), and smooth spherical morphology. To evaluate the antisepsis efficacy of the developed Cur-SLNs, an acute sepsis shock model was established by administering an IP injection of LPS into IL-1β-induced transgenic mice. Antiluciferase activity was monitored (in terms of luciferase signal) in all the experimental animals (sepsis shock mouse model) treated with phosphate-buffered saline (PBS), Cur, Cur-SLNs, and dexamethasone at different time points (0, 1, 3, 5, 24, and 48 h) following the LPS injection. The results suggested superior antiluciferase activity of Cur-SLNs (lowest luciferase signal) when compared with the PBS and free Cur groups; however, they had a slightly lower antiluciferase activity than dexamethasone (Figure 11.4) [91]. These results were validated by estimating the expression of various pro-inflammatory cytokines (i.e., IL-1β, IL-6, IL-10, and TNF-α) in the serum of LPS (3.0 mg in PBS/kg body weight)-induced sepsis shock animals. The results demonstrated a significantly lower expression of all pro-inflammatory cytokines in the serum of both male and female mice. The superior antisepsis efficacy of Cur-SLNs was validated through histopathological (hematoxylin-eosin staining) examination, and the results suggested no significant alterations in the histological sections of the liver, kidney, and heart when compared with marked signs of inflammation (i.e.,

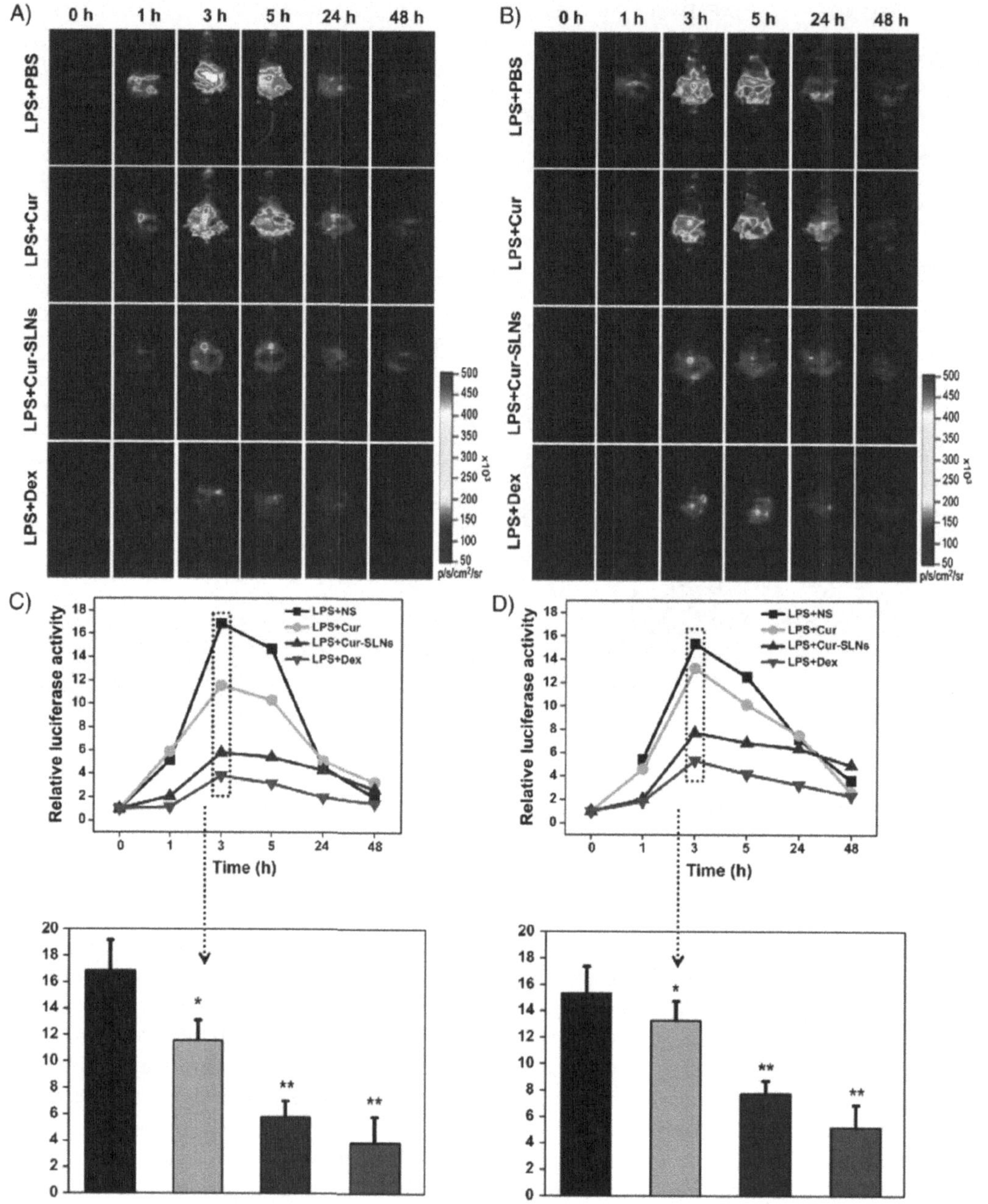

FIGURE 11.4 Anti-luciferase activity of curcumin–solid-lipid nanoparticles in a transgenic lipopolysaccharide-induced sepsis mouse model: Imaging monographs of A) male mice and B) female mice. Relative luciferase activity in C) male mice and D) female mice. (From Wang et al. [91]. With Permission.)

infiltration of inflammatory cells) in the organs of animals treated with PBS and free Cur. These results indicate comparable anti-inflammatory efficacy of Cur-SLNs with dexamethasone (a Food and Drug Administration-approved anti-inflammatory agent) [91]. Similarly, the biopharmaceutical viability of SLNs has been reported in many other studies for the delivery of various anti-inflammatory phytochemicals such as oral delivery of QUE [92,93], puerarin (PUR; isolated from the root extract of *Pueraria lobate*, a Chinese herbal medicine) [94], andrographolide (AND) (a diterpenoid extracted from *Andrographis paniculata*) [95], PIP (isolated from *Piper nigrum*) [96], and *Dracocephalum moldavica* L. (total flavonoid extract from *Dracocephalum moldavica* L.) [97] for the treatment of various inflammatory diseases.

Surface functionalization of SLNs is a promising adaptation for improving various biopharmaceutical properties such as absorption, distribution, plasma half-life, cell uptake, specific targeting, and therapeutic efficacy. Ji et al. [98] designed SLNs for improving the physicochemical properties and bioavailability of Cur. To improve the intestinal absorption, SLNs were modified with P-glycoprotein-silencing ingredients such as TPGS and Brij 78. TPGS and Brij 78 are well-recognized pharmaceutical ingredients that can improve the intestinal absorption of drugs by silencing the P-glycoprotein (efflux pump) in the intestinal brush border, which, in turn, suppresses the efflux of absorbed drugs back to the intestinal lumen. These results suggested that TPGS and Brij 78-modified Cur-SLNs exhibited nanometric size (~135 nm), anionic zeta potential (−25 mV), good %EE (~90%), spherical morphology, and sustained release profile. An *in vivo* pharmacokinetic study revealed that the AUC and relative bioavailability of Cur from Cur-SLNs were approximately 12-fold and 9-fold higher than those of Cur-suspension, respectively. Moreover, the permeability coefficient of Cur significantly improved when delivered as Cur-SLNs when compared with that of Cur suspension. These results suggest that modulation of SLNs with TPGS and Brij 78 can significantly improve the intestinal absorption and oral bioavailability of phytoconstituents [98]. Similarly, Yusuf et al. [99] developed polysorbate-80 (P80)-modified PIP-SLNs for intranasal delivery of phytochemicals for the treatment of neuro-inflammation. Biochemical analysis of the cortex (brain tissues) isolated from animals (male Wistar rats) treated with P80-PIP-SLNs revealed a significant increase in antioxidant enzymes (e.g., superoxide dismutase [SOD] and total antioxidant capacity [TAC]) and a decrease in acetylcholinesterase activity, which indicates the suppression of neuronal damage and improved neuroprotective efficacy. Compared with free PIP, a significantly lower caspase 3 activity (apoptosis) and TNF-α (neuro-inflammation) was estimated from the biochemical analysis of the cortex treated with P80-PIP-SLNs, and this indicates the highest neuroprotective efficacy of PIP when delivered as P80-PIP-SLNs [99]. Surface functionalization of SLNs with CS and its derivatives has been reported for the delivery of other phytochemicals such as QUE [100], Cur [101,102], and RSV [103], and the results suggested a significant improvement in the physicochemical properties, oral bioavailability, cell uptake efficiency, and anti-inflammatory efficacy of these phytochemicals.

Surface functionalization of SLNs with specific receptor antibodies has been well regarded for improving the specific uptake of phytochemicals into various cells/tissues/organs. Loureiro et al. [104] developed antitransferrin receptor MAb (OX26 mAb)-modified RSV-SLNs for the treatment of neuro-inflammation. OX26mAb-RSV-SLNs exhibited ultra-fine particle size (176 nm), neutral zeta potential, narrow PDI, excellent %EE (95%), smooth spherical morphology, and good storage stability with no significant variations in their physicochemical properties for 2 months. Aβ peptide aggregation is responsible for neuritic plaque formation; however, incubation of the Aβ-peptide with OX26mAb-RSV-SLNs resulted in a significant decline in Aβ-peptide fibrillation (characterized by a decrease in fluorescence intensity) compared with free RSV. OX26mAb-RSV-SLNs also exhibited superior cell uptake efficiency in human bronchial epithelial cells (BBB cell model) over free RSV, which indicates the selectivity of SLNs to permeate and translocate into the BBB [104]. Similar findings were reported in other studies [105,106]. Pinheiro et al. [105] developed RVG29 peptide-modified QUE-SLNs for the successful delivery of QUE across the BBB for the effective treatment of neuro-inflammation. RVG29-QUE-SLNs displayed significantly higher efficacy in terms of inhibiting Aβ-peptide fibrillation than unfunctionalized RSV-SLNs and free RSV [105]. Dong et al. [106] developed RGD (αvβ3 integrin receptor-specific targeting moiety) -modified PEG-PUR-SLNs for the treatment of myocardial infarction (MI). The results suggested a significant improvement in plasma half-life and biodistribution of PUR in cardiomyocytes when delivered as

RGD-PEG-PUR-SLNs when compared with those of free PUR. These results were validated *in vivo* where a significant decrease in infarct size was evident in animals treated with an intravenous injection of RGD-PEG-PUR-SLNs when compared with free PUR [106].

11.2.5 Nanovesicles

Lipid-based nanovesicles including liposomes, NIOs, CBMs, and TFMs are widely used for the delivery of a wide range of anti-inflammatory phytochemicals to improve their physicochemical properties and for the treatment of various inflammatory diseases. Shen et al. [107] reported that co-therapy with sinomenine (SIN)-loaded thermo-sensitive LPMs (TSL) and microwave-induced hyperthermia (HT) significantly improved the anti-inflammatory and anti-RA effects of SIN. The results suggested that the prepared SIN-LPMs exhibited ultra-fine particle size (100 nm), good %EE, and smooth spherical morphology. SIN-TSL exhibited thermo-sensitive release behavior, showing a predominantly faster release at 43°C and a relatively slower release at 37°C. A cell uptake study showed a greater uptake of SIN-TSL into LPS-activated human umbilical vein endothelial cells (HUVECs) than into normal HUVECs, which indicates their specificity and higher permeability into inflammatory HUVECs. Following an IV injection, SIN-TSL showed promising fluorescence in the hind limb within 4 hours, which indicated that microwave HT can be applied 4 hours after the injection of SIN-TSL. Co-therapy of microwave HT and SIN-loaded TSL showed a significantly higher reduction in paw thickness, arthritic scores, and release of pro-inflammatory cytokines (e.g., IL-6 and TNF-α) than treatment with PBS, free SIN, and SIN-TSL without HT (Figure 11.5) [107].

One of the hallmarks of psoriasis is over-stimulatory DCs which trigger the pathological interplay between immune cells and keratinocytes. Xi et al. [108] investigated a suitable alternative phytonanomedicine for the treatment of psoriasis and recently developed liposomes as a drug delivery device for celastrol (CLR), a triterpenoid isolated from the root extract of traditional Chinese medicine herbs (*Tripterygium wilfordii* and *Tripterygium regelii*). CLR possesses a wide range of pharmacological activities including anti-inflammatory, antioxidant, and anticancer effects; however, its poor aqueous solubility limits its further development and clinical translation. To improve its physicochemical properties and anti-inflammatory effect, CLR was encapsulated into LPMs. Moreover, to specifically target and optimize the uptake of CRL into DCs by utilizing receptor-mediated transcytosis through mannose (MAN) receptors on DCs, the developed CRL-LPMs were decorated with MAN. The mannosylated liposomes (MAN-CRL-LPMs) were evaluated for their physicochemical characteristics, uptake into DCs, expression of psoriatic markers, and antipsoriasis efficacy. The results suggested that MAN-CRL-LPMs inhibited expression of CD80, CD86, and major histocompatibility-II on DCs to a greater extent than unfunctionalized LPMs. An *in vivo* evaluation also validated that MAN-CRL-LPMs were more potent in suppressing the pathological symptoms of psoriasis such as severity index, histopathology, and expression of inflammatory cytokines than unfunctionalized LPMs. These results conclude that MAN functionalization of phytonanomedicine is a promising technique to treat DC-induced skin inflammatory diseases [108]. Many studies have established the significance of mannosylation of nanophytomedicine for the targeted delivery of various anti-inflammatory phytocompounds [109-114].

Qiu et al. [115] developed CRL-loaded NIOs in the gel form with semisolid consistency for topical application in psoriasis treatment (skin inflammation). The developed CRL-NIOs exhibited ultra-fine particle size (~135 nm), good %EE (~85%), smooth spherical morphology, and sustained release profile. The liquid chromatography–mass spectrometry drug quantification experiment showed that CRL predominantly accumulated in the skin layers with minimal bioavailability in the plasma and lymphatic vessels. A cell uptake study showed the prominent uptake of CRL-NIOs into human keratinocytes (HaCaT cells), demonstrating greater suppression of mRNA expression of various pro-inflammatory cytokines (e.g., IL-2, IL-6, IL-12, INF-γ, and TNF-α) than that demonstrated by free CRL. These results conclude that CRL encapsulation into NIOs resulted in a significant decline

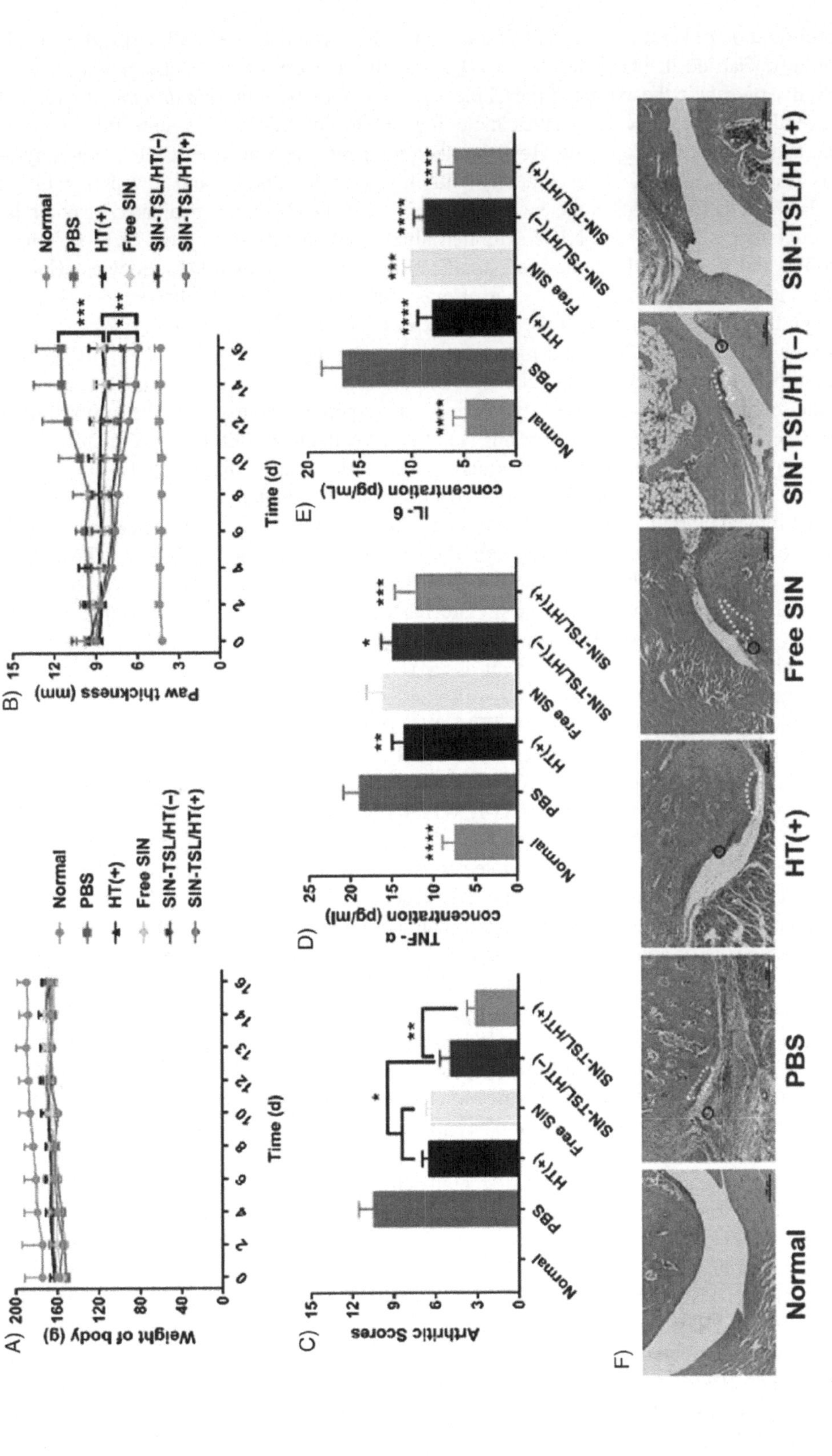

FIGURE 11.5 Anti–rheumatoid arthritis efficacy of sinomenine (SIN)-thermo-sensitive LPM (TSL)+hyperthermia (HT) compared with phosphate-buffered saline, HT (+), free SIN, SIN-TSL/HT (–), or SIN-TSL/HT (+). A) Body weight of rats. B) Hind paw thickness. C) Arthritis scores. Expression levels of pro-inflammatory cytokines. D) Tumor necrosis factor-α and E) interleukin-6 in the joint tissues measured by enzyme-linked immunosorbent assay. F) Hematoxylin-eosin staining of ankle joints; the labeled black areas indicate inflammatory cell infiltration and labeled yellow areas indicate bone erosion. (From Shen et al. [107]. With Permission.)

in hyper-proliferation of keratinocytes and skin inflammation, which are key hallmarks of psoriasis [115]. Similarly, Gao et al. [116] developed NIOs for the delivery of daphnetin (DPH), a coumarin derivative present in the extract of the Chinese medicinal herb *Euphorbia dracunculoides* that exhibits several pharmacological activities including potent antiproliferative, anti-inflammatory, antioxidant, and antiallergic properties. However, the therapeutic significance of this potent phytochemical is restricted owing to poor aqueous solubility, low dissolution, and low permeability; therefore, Gao et al. [116] proposed the encapsulation of DPH into NIOs and evaluated its potential in psoriasis treatment. DPH-loaded NIOs exhibited ultra-small particle size, good %EE, sustained release profile, and significantly higher cell uptake efficiency into human keratinocytes (HaCaT cells). DPH-NIOs also displayed greater control over IMQ-induced hyperproliferation in HaCaT cells and significant downregulation in the mRNA expression of pro-inflammatory cytokines (e.g., IL-1β, IL-6, IL-8, TNF-α, IL-23A, and monocyte chemoattractant protein 1 [MCP-1]) than free DPH. The antipsoriasis efficacy of DPH-NIOs was established *in vivo* where a pronounced amelioration in the severity of erythema intensity, scaling, and hyperperakeratosis and the infiltration of inflammatory cells was observed in the psoriatic skin treated with DPH-NIOs [116]. Similar findings were reported by other researchers in the treatment of skin inflammatory diseases [117,118].

Avadhani et al. [119] developed a specialized deformable vesicular delivery system, that is, TFMs, for the topical co-delivery of EGCG (a catechin derivative) and HA for protecting the skin against ultraviolet (UV) irradiation-induced skin damage and photoaging. EGCG-HA-TFMs were prepared from soy phosphatidylcholine and sodium cholate through methods such as thin-film hydration and high-pressure homogenization. The fabricated EGCG-HA-TFMs exhibited nanovesicular size (~100 nm), narrow PDI, anionic zeta potential (−45 mV), good %EE, and smooth spherical morphology. An *in vitro* antioxidant assay (DPPH [2,2-diphenyl-1-picrylhydrazyl] and ABTS [2,2′-azino-bis-3-ethylbenzthiazoline-6-sulfonic acid] assays) indicated significantly higher free radical-scavenging ability of EGCG-HA-TFMs than that of free EGCG, HA, and Cur (reference control). An *in vitro* cell toxicity experiment identified the lowest toxicity and highest proliferation of HaCaT cells after treatment with EGCG-HA-TFMs, which indicates the superior biocompatibility and cell regenerative potential of EGCG-HA-TFMs over those of free EGCG, EGCG-TFMs, and Cur [119]. ROS generation usually occurs in skin cells (keratinocytes) after exposure to UV irradiation (UV-A and UV-B), and the generated ROS are physiologically neutralized by cellular antioxidant mechanisms. However, in case of overproduction of ROS or downregulation of cellular antioxidant mechanisms, keratinocytes undergo inflammation and apoptosis, resulting in irreversible damage to the skin or aging. To evaluate the anti-inflammatory and antiaging effects, HaCaT cells (human keratinocytes) were treated with EGCG-HA-TFMs and other test formulation, and the results suggested that ROS generation and release of pro-inflammatory cytokines (MMP-2 and MMP-9) were significantly suppressed due to EGCG-HA-TFMs when compared with those of free EGCG, EGCG-TFMs, and Cur. In addition, encapsulation of EGCG and HA into TFMs also caused significant improvement in terms of skin permeation and accumulation in various skin layers. These results concluded that EGCG-HA-TFMs could be a promising nanodelivery system for the treatment of UV irradiation-induced skin inflammation and aging [119].

Nanovesicles are also used for the specific targeted delivery of anti-inflammatory phytochemicals such as PIP (isolated from *Piper nigrum*) [120] and GNS (an isoflavone isolated from the root extract of *Genista tinctoria*) [121] for the treatment of neuro-inflammation. Elnaggar et al. [120] developed Tween (T)-modified PIP–loaded CBMs for the targeted delivery of PIP to the brain tissues through the intranasal route. T-PIP-CBMs exhibited ultra-fine particle size (~170 nm), anionic zeta potential (−34 mV), good %EE (86%), cubic morphology, and good storage stability for up to 3 months with insignificant changes to phytochemical properties. T-PIP-CBMs exhibited a biphasic release, with an initial release of 21% of the contents in the first hour, followed by a relatively sustained release, with a total release of 45% of the PIP contents within 6 hours when compared with an 80% release of PIP from the PIP-suspension within only 2 hours. Biochemical analysis of the hippocampus (brain tissues) extracted from Wistar rats treated with T-PIP-CBMs revealed a significant increase in antioxidant enzymes (e.g., SOD and TAC) and a decrease in acetylcholinesterase activity and MDA, which indicates the suppression of neuronal damage and improvement of neuroprotective efficacy. Compared with free PIP and blank CS-NPs, a significantly lower caspase 3 activity (apoptosis) and TNF-α (neuro-inflammation) was estimated in the biochemical analysis of the hippocampus of Wistar rats treated with intranasal T-PIP-CBMs, and this indicates the highest neuroprotective efficacy of PIP when delivered as T-PIP-CBMs [120]. Similarly, Langasco et al. [121] developed GNS-TFMs for the targeted delivery of GNS to the brain tissues for neuro-inflammation treatment. Encapsulation of GNS into TFMs significantly reduced H_2O_2-induced oxidative stress (LHD release) and apoptosis in PC12 cells compared with free GNS, which indicates the improved cytoprotective efficacy of GNS-TFMs. The authors suggested that GNS-TFMs can be a promising delivery system for the treatment of oxidative stress-induced neurodegenerative diseases [121].

11.2.6 Nanofibers

Nanofibers (NFs) are fiber-shaped one-directional nanoconstructs (10–1000 nm) with versatile physicochemical properties including large surface area-to-volume ratio, high porosity (depending on the crosslinking degree), great mechanical strength, flexibility of functionalization, and high loading efficiency. The characteristics of NFs largely depend on the nature of the building polymer(s), which can be either natural (e.g., COL, cellulose, SF, gelatin [GLT], keratin, ALG, CS, and polysaccharides) or synthetic (e.g., polyurethane [PU], PLGA, PLA, PCL, or others). Moreover, the utilization of more than one polymer with different properties can produce more versatile NFs with better physicochemical properties based on the intended purpose. Owing to their unique modifiable properties, NFs have been extensively used for the delivery of a wide range of anti-inflammatory phytochemicals.

Merrell et al. [122] used the electrospinning technique to develop PCL-NFs as a delivery vehicle for Cur for improving its physicochemical properties and wound healing efficacy. The developed Cur-PCL-NFs exhibited extended release of the drug over 72 hours and good biocompatibility against human fibroblasts (HFF-1 cells) compared with free Cur. The antioxidant potential of Cur-PCL-NFs was determined using the oxygen radical absorbance capacity (ORAC) assay in terms of the ability of Cur-PCL-NFs to preserve the viability of HFF-1 cells under oxidative stress conditions. Moreover, Cur-PCL-NFs exhibited significant improvement in anti-inflammatory effects by downregulating IL-6 expression in mouse monocyte-macrophages when compared with free Cur. Cur-PCL-NFs demonstrated superior wound healing efficacy *in vivo*, which was attributed to their excellent anti-inflammatory and antioxidant potential [122]. Similarly, Dai et al. [123] developed GLT-based nanofibrous mats (NMs) through electrospinning with an aim to improve the physicochemical properties, absorption, oral bioavailability, biocompatibility, and wound healing efficacy of Cur (*in vitro* and *in vivo*). The results suggested that human fibroblasts (HS-27 cells) treated with Cur-GLT-NMs displayed the highest proliferation when compared with human fibroblasts treated with blank GLT-NMs (without Cur) and free Cur, which provide evidence for the superior biocompatibility and wound healing potential of Cur when encapsulated into Cur-GLT-NMs [123]. Furthermore, the anti-inflammatory potential of Cur-GLT-NMs was evaluated by estimating the expression of MCP-1 and thrombospondin-1 in HS-27 cells. MCP-1 and thrombospondin-1 play a very important role during the inflammatory phase of the wound healing process and induce chemoattraction of inflammatory cells (e.g., macrophages) at the wounded or injured site. A significant downregulation in the migration and proliferation of HS-27 cells and a noticeable decline in the expression of inflammatory mediators were observed in human fibroblasts treated with Cur-GLT-NMs. These results were validated in animals (Sprague Dawley rats), and the results revealed a significantly higher wound closure rate in animals treated with Cur-GLT-NMs, particularly at the late stages of treatment (after day 7) than the animals treated with blank GLT-NMs and free Cur [123].

The therapeutic potential of NF-scaffold has been investigated for the delivery of asiaticoside for the treatment of chronic diabetic wounds [124]. AT is a potent phytochemical isolated from *Centella asiatica* and is widely used owing to its powerful antioxidant, anti-inflammatory, antirheumatic, moisturizing, and tissue-regenerating properties. The developed NFs exhibited ultra-fine dimensions (100–200 nm), high surface area-to-volume ratio, good mechanical strength, extended release profile, and good biocompatibility against human keratinocytes (HaCaT cells). The AT-NF-scaffold displayed excellent antimicrobial efficacy against gram-positive and gram-negative bacteria and superior wound closure rate in diabetic wounds when compared with blank NFs and free Cur [124].

For optimizing the physicochemical properties and drug delivery efficiency, Zahiri et al. [125] recently proposed constructing/blending of natural (GLT) and synthetic (PCL) polymers. The developed NMs were successfully loaded with Cur (model drug) and evaluated for wound healing and tissue regeneration efficacy. The results suggested that the developed NMs (Cur-PCL-GLT-NMs) exhibited ultra-fine fiber size, good tensile strength, high surface area-to-volume ratio, good wettability and swellability, high porosity, biodegradability, and extended release profile. Adhesion, proliferation, and maturation of human endometrial stem cells were superior when incubated with Cur-PCL-GLT-NMs over those of cells incubated with blank PCL-GLT-NMs and free Cur. These results demonstrated the superior skin repair and tissue-regenerating potential of Cur-PCL-GLT-NMs [125]. Similarly, Agarwal et al. [126] constructed NMs by blending three polymers, namely, SF, PCL and polyvinyl alcohol (PVA). The developed Cur-PCL-PVA-SF-NMs exhibited nano-scaled fibers (200–350 nm), high surface area-to-volume ratio, and extended release profile. The wound healing efficacy of Cur-PCL-PVA-SF-NMs was evaluated in a streptozotocin-induced diabetic wound model, and the results suggested faster wound closure rate, higher granulation tissue formation, and re-epithelization with Cur-PCL-PVA-SF-NMs than with blank SF-NMs and free Cur. These findings were attributed to the excellent anti-inflammatory and antioxidant properties of Cur as well as an ideal wound environment (i.e., oxygen supply) provided by the composited PCL-PVA-SF-NMs [126]. The biomedical efficacy of the composited NMs has been verified for the successful delivery of anemoside B4 (ANE), a potent phytochemical isolated from the Chinese plant *Pulsatilla* [127]. ANE is a traditional phytomedicine commonly used for the treatment of various ailments owing to its excellent anti-inflammatory and antioxidant properties. Zhang et al. [127] developed NMs by electrospinning of CS and PVA and incorporated ANE into the NMs to improve its physicochemical properties and wound healing efficacy. The developed ANE-CS-PVA-NMs exhibited nanoscale particle size, high surface area-to-volume ratio, good mechanical strength, and extended release profile. ANE-CS-PVA-NMs also displayed promising ability to suppress LPS-induced activation of macrophages and release of pro-inflammatory cytokines (e.g., IL-1β, IL-6, and TNF-α). An *in vivo* wound healing study revealed a faster wound closure rate in diabetic wounds treated with ANE-CS-PVA-NMs, which was attributed to higher granulation tissue formation, COL deposition, re-epithelization, higher angiogenesis, and excellent antimicrobial efficacy [127]. Another constructed NM was developed by Anand et al. [128] by electrospinning two synthetic polymers, namely, PCL and cellulose acetate (CA), for the successful delivery of ferulic acid (FA). FA is a natural phytomedicine derived from the genus *Ferula*, which encompasses the giant fennel, and it is recognized for its powerful anti-inflammatory and antioxidant (free radical scavenger) properties. The developed FA-PCL-CA-NMs exhibited ultra-small fiber size (100–250 nm), high surface area-to-volume ratio, good biodegradability, extended release profile, and excellent biocompatibility against human keratinocytes (HaCaT cells). An *in vivo* wound healing study confirmed a greater wound closure rate in diabetic wounds treated with FA-PCL-CA-NMs than in wounds treated with blank NMs and free FA. The superior wound healing efficacy of FA-PCL-CA-NMs was attributed to excellent COL deposition, granulation tissue formation, re-epithelization, and angiogenesis [128].

Chrysin (CHN) is a naturally occurring flavone that has been well recognized for its wide range of pharmacological activities including potent anti-inflammatory and antioxidant properties. Despite its potent pharmacological activities, clinical significance of CHN is limited owing to poor aqueous solubility and susceptibility to chemical degradation. To mitigate these problems and improve its biomedical efficacy, Deldar et al. [129] fabricated NMs by electrospinning of PCL and PEG and loading of CHN and evaluated for their anti-inflammatory, antioxidant, and wound healing efficacies. The developed CHN-PCL-PEG-NMs exhibited ultra-fine dimensions, high surface area-to-volume ratio, extended release profile, and good biocompatibility against human fibroblasts (HFF-1 cells). As for the anti-inflammatory effects, CHN-PCL-PEG-NMs displayed promising ability in suppressing LPS-induced activation of macrophages (J774A1 cells) and releasing pro-inflammatory cytokines (IL-1β, IL-6, and TNF-α). The ORAC assay showed that CHN-PCL-PEG-NMs displayed powerful free radical-scavenging ability to protect HFF-1 from oxidative stress (NO). The excellent wound closure rate observed in diabetic wounds treated with CHN-PCL-PEG-NMs was expected to be due to the powerful anti-inflammatory and antioxidant activities of CHN-PCL-PEG-NMs [129]. The same research group had evaluated CHN-PCL-PEG-NMs for regulating the proliferation and differentiation of adipose-derived stem cells (ADSCs) [130]. The results suggested that ADSCs incubated with CHN-PCL-PEG-NMs showed an exponential increase in adhesion, viability, and maturation when compared with those incubated with blank PCL-PEG-NMs and tissue culture polystyrene [130]. These findings were consistent with those of another study [131]. Interestingly, for synergizing the anti-inflammatory and wound healing efficacies of phytochemicals, Mohammadi et al. [132] proposed the co-delivery of Cur and CHN using PCL-PEG-NMs and demonstrated that the superior wound healing efficacy of Cur-CHN-PCL-PEG-NMs was attributed to the remarkable efficacy of regulating the gene expression of various inflammatory mediators such as MMP-2, inducible NO synthase, IL-6, tissue inhibitor of metalloproteinase-1 (TIMP-1), and TIMP-2 [132]. Similar findings were reported by another research group on the biopharmaceutical viability of PCL-PEG-NMs as a drug delivery vehicle for Cur [133].

Owing to the exceptional physicochemical properties of NFs or NMs, many researchers have constructed NFs or NMs by using a variety of polymers such as PCL and CS [134]; HA, CS, and GLT [135]; PCL and tragacanth [136]; poly(3-hydroxy butyric acid-co-3-hydroxy valeric acid) [137]; PU and PCL [138,139]; PCL and quaternary ammonium salt-modified montmorillonite [140]; PCL and PVA [141,142]; CA [143,144]; CS, PVA, and GLT [145]; PV and CS [146]; CA and GLT [147]; CA and HA [148]; PCL, PVA, and COL [149]; PCL and polystyrene [150]; and PCL carboxyethyl CS and PVA [151] for improving the physicochemical properties and antioxidant, anti-inflammatory, and wound healing efficacy of various phytochemicals.

Surface functionalization with PEG (PEGylation) has been utilized to improve the physicochemical properties and biomedical efficacy of phytonanomedicines [152]. In this study, PCL-PVA based NMs were fabricated as a drug delivery device for Cur, and finally, NMs were PEGylated to sustain the release profile as well as to improve the biomedical efficacy of Cur. The results showed a sustained release profile and significant improvement in antibacterial and wound healing efficacies of PEG-Cur-NMs [152]. Recently, Farahani et al. [153] developed a novel multifunctional NM as a drug delivery device for *Zataria multiflora* (ZM), a potent phytochemical with strong anti-inflammatory and antioxidant properties, and evaluated its wound healing efficacy. In this study, ZM-loaded NE was prepared and then incorporated into CA-GLT-NMs. The fabricated multifunctional ZM-NE-CA-GLT-NMs exhibited nanoscale fiber size, good tensile strength, high loading efficiency, greater cell adhesion (L929 fibroblast cells), and accelerated wound healing potential [153].

11.2.7 Nanodiamonds

NDs are state-of-the-art carbon-based NPs that have received considerable attention recently owing to their inherent truncated octahedral (diamond-like cubic lattice) architecture, versatility of delivering a wide range of therapeutics, amazing stability, large surface area, optical properties, and flexibility of functionalization. Because of their unique properties, NDs have been extensively used in various medical applications such as bioimaging, biosensing, and drug delivery [154–156].

Many researchers have used NDs as a drug delivery system for a variety of anti-inflammatory phytochemicals for the treatment of various diseases. Considering their unique properties, Cheng et al. [157] synthesized NDs for the successful delivery of Cur. Cur-conjugated NDs were then functionalized with TPGS, which yielded a unique amorphous nanostructure. The developed TPGS-Cur-NDs exhibited ultra-small particle size (<200 nm), core-shell structure, good %EE (>80%), and sustained release kinetics. Dissolution studies displayed a significantly higher absorptive concentration of Cur from TPGS-Cur-NDs than that of Cur-suspension. Pharmacokinetic studies revealed an approximately fivefold increase in C_{max} and a tenfold increase in AUC of TPGS-Cur-NDs when compared with those of Cur suspension. These results suggest that TPGS-functionalized Cur-NDs could be a promising delivery system for improving the intestinal absorption and oral bioavailability of hydrophobic phytoconstituents [157]. Likewise, Liu et al. [158] developed TPGS-functionalized Cur-conjugated NDs for improving dissolution, oral bioavailability, and specific cell uptake of Cur. The developed TPGS-Cur-NDs exhibited ultra-small particle size (<200 nm), core-shell structure, good %EE (81%), and good biocompatibility against Caco-2 cells. A cell uptake study revealed that TPGS-functionalization of Cur-NDs significantly improved their uptake into Caco-2 cells through clathrin-mediated and caveolin-mediated endocytosis, and these results were validated in animals wherein a greater enhancement in intestinal permeation was evident. These results conclude that TPGS-modified Cur-NDs is a promising adaptation to improve intestinal absorption, oral bioavailability, and specific cell uptake efficiency of Cur [158]. TPGS and other surface functionalization approaches have been utilized and validated by other researchers to improve the oral delivery of phytoconstituents [159–162].

11.2.8 Nanocomposites/Nanohybrid Scaffolds

Nanocomposites are hybrid nanomaterials that are usually synthesized by incorporating the nanoscale particles into a matrix of standard material. Nanocomposite preparation is an innovative strategy to modify the physicochemical (i.e., mechanical strength, toughness, porosity, flexibility, biodegradability, biocompatibility, and others) and functional (i.e., release kinetics, stability, and others) properties from component materials. Because of their modifiable and versatile properties, many researchers have used nanocomposites as a drug delivery device for anti-inflammatory phytochemicals in the treatment of various diseases. Karri et al. [163] developed a nanohybrid scaffold for improving the physicochemical properties and anti-inflammatory, antioxidant, antimicrobial, and wound healing efficacy of Cur. Briefly, the authors encapsulated Cur into CS-NPs and impregnated the synthesized Cur-CS-NPs into COL scaffolds. The fabricated nanohybrid scaffold displayed good stability, sustained release behavior, good mechanical properties, biocompatibility, and water uptake capacity. Wounds treated with the nanohybrid scaffolds exhibited faster wound closure, absence of inflammatory cells, thick granulation tissue formation, COL deposition, and superior re-epithelization when compared with wounds treated with free Cur. These findings suggest that a combination of Cur, CS, and COL caused a synergistic increase in anti-inflammatory, wound healing, and tissue regenerating efficacy [163]. Similarly, Guadarrama-Acevedo et al. [164] developed a hybrid ALG membrane embedded with Cur-loaded PCL-NPs. The developed hybrid membrane displayed a good release profile, mechanical strength, biocompatibility, and antimicrobial and wound healing efficacy [164].

Mobaraki et al. [165] developed Cur-PEG-NPs through sonication and the subsequent vigorous magnetic stirring overnight. The fabricated Cur-PEG-NPs (particle size of ~45 nm with a spherical morphology) were then embedded into a COL-ALG scaffold and evaluated for wound healing efficacy. The nanohybrid scaffold exhibited superior wound healing efficacy over free Cur and blank COL-ALG scaffold, which clearly provide evidence for the significance of embedding Cur-PEG-NPs into the COL-ALG scaffold [165]. Another nanohybrid composite (hydrogel) was fabricated by Gong et al. [166]. In this study, they encapsulated Cur into PMs to improve their physicochemical properties (i.e., aqueous solubility, dissolution profile, and sensitivity to light). When the fabricated Cur-PMs were embedded into a thermosensitive hydrogel, they remained a free-flowing sol at an ambient temperature and transformed to a nonflowing gel at physiological temperature. Physicochemical characterization of the Cur-PM-embedded hydrogel displayed good adhesiveness and a sustained release profile for Cur. Wounds treated with the Cur-PM-embedded hydrogel showed faster wound closure rate, higher COL deposition and granulation tissue formation, and significant increase in antioxidant enzymes (SOD and catalase) at the wound bed [166]. Similar findings were reported in other studies, wherein encapsulation of Cur into hybrid nanocomposites or nanohybrid scaffolds improved physicochemical properties (aqueous solubility, dissolution, permeability coefficient, and others), provided protection against photodegradation, improved bioavailability, prolonged localization at the application site, and demonstrated superior wound healing efficacy of Cur [167–171].

11.2.9 Other Nanoparticles

In addition to incorporating phytoconstituent(s) into various nanodelivery systems, particle size reduction by physical methods (i.e., milling) can improve physicochemical properties (i.e., aqueous solubility), bioavailability, and biomedical efficacy of the phytoconstituent(s). Utilizing physical methods, Bhawana et al. [172] reduced the particle size of Cur and formed nano-Cur (particle size of 2–40 nm) using the wet-milling technique. Compared with traditional Cur particles (poor aqueous solubility), nano-Cur had better water solubility without the addition of surfactant(s). Moreover, the nano-Cur-suspension displayed more potent antimicrobial efficacy against a wide range of gram-positive and gram-negative bacteria and fungi than the traditional Cur suspension. Transmission electron microscopy observations revealed that the superior antimicrobial efficacy of nano-Cur was attributed to its exceptional ability to enter into microbial cells and cause death of bacteria [172]. Similarly, many studies have reported that wet-milling or dry-milling techniques have promising outcomes in terms of improving physicochemical properties and biomedical efficacy (antimicrobial, antioxidant, anticancer, and wound healing) of Cur [173–180].

11.3 CONCLUSIONS

Despite the potency of phytochemicals, clinical translation of phytomedicines (phytocompounds used as medicine) is constrained owing to their macromolecular size, poor aqueous solubility, susceptibility to enzymatic or pH-mediated degradation, chemical instability, low permeation, poor bioavailability, short plasma half-life, and sub-therapeutic efficacy. To mitigate these shortcomings, nanotechnology has been utilized in the past few decades. The application of nanotechnology-aided interventions resulted in significant improvement in physicochemical properties, stability, bioavailability, and therapeutic efficacy of a wide range of phytochemicals used for the treatment of various diseases. However, this chapter mainly focuses on the biopharmaceutical and therapeutic significance of various types of nanodelivery systems used for the delivery of anti-inflammatory phytochemicals for the treatment of various inflammatory diseases.

To the best of our knowledge, a wide variety of nanodelivery systems such as polymeric NSPs, polymeric NCPs, SLNs, PMs, NEs, micro-emulsions, liposomes, ethosomes, NIOs, CBMs, TFMs, nanocomposites, NDs, and nanohybrid scaffolds have been explored for their use in the successful delivery of a wide range of anti-inflammatory phytochemicals. The efficiency of phytonanomedicines for the successful delivery of phytochemicals has been investigated through different routes including, but not limited to, oral, intravenous, ophthalmic, IP, nasal, respiratory, subcutaneous, and intra-articular routes as well as in *in vitro* settings. Generally, encapsulation of anti-inflammatory phytochemicals by various nanodelivery systems has resulted in significant improvements in the physicochemical properties (i.e., aqueous solubility, permeation coefficient, and chemical stability), biocompatibility, bioavailability, pharmacokinetic profile (absorption, metabolism, distribution, and elimination) and therapeutic efficacy of the phytochemicals for the treatment of various inflammatory diseases. These outcomes were mainly attributed to the intrinsic physicochemical properties of nanodelivery systems such as nanoscale dimensions, high %EE, protection of susceptible phytochemical(s) against premature degradation, higher permeation efficiency and transcellular diffusion (biodistribution), and molecular interaction with subcellular components.

Interestingly, many recent adaptations to phytonanomedicines have been utilized in an attempt to further improve their pharmacokinetic profile, cell uptake efficiency, specific targeting, and therapeutic efficacy. For example, PEGylation (surface coating of NPs with PEG) is one of the adaptations applied for different phytonanomedicines to prolong plasma circulation time to upregulate their passive diffusion into different tissues of the body. Surface functionalization of phytonanomedicines have been carried out with different functional moieties such as penetration enhancers, surfactants/emulsifiers, transporter proteins, and targeting ligands. Functionalization with TPGS, which is a multirole pharmaceutical excipient, has been carried out to improve the intestinal, corneal, topical, and nasal absorption of phytochemicals. Similarly, functionalization with various targeting moieties (i.e., FA, transferrin, RVG29 peptide, and others) has been carried out to target specific cells/tissues/organs with distinct receptors for optimum accumulation and superior therapeutic efficacy while preventing the off-target effects. Conclusively, nanotechnology-aided interventions have significantly improved physicochemical properties, pharmacokinetic profile, and therapeutic efficacy of anti-inflammatory phytochemicals for the treatment of various inflammatory disorders (e.g., UV-induced skin inflammation, psoriasis, RA, liver inflammation, MI, corneal inflammation, acute and chronic diabetic wound healing, sepsis, and neuro-inflammation).

ACKNOWLEDGMENT

The authors would like to express their gratitude to the College of Pharmacy at the University of Sharjah (Sharjah, United Arab Emirates) for providing financial support (SEED grant) to accomplish this chapter.

CONFLICT OF INTEREST

The authors reported no conflict of interest associated with the present work.

REFERENCES

1. Golia, E.; Limongelli, G.; Natale, F.; Fimiani, F.; Maddaloni, V.; Pariggiano, I.; Bianchi, R.; Crisci, M.; D'Acierno, L.; Giordano, R.; et al. Curr. Atheroscler. Rep. 2014, 16, 435.
2. Montecucco, F.; Liberale, L.; Bonaventura, A.; Vecchiè, A.; Dallegri, F.; Carbone, F. Curr. Atheroscler. Rep. 2017, 19, 11.
3. Amor, S.; Peferoen, L.A.; Vogel, D.Y.; Breur, M.; van der Valk, P.; Baker, D.; van Noort, J.M. Immunology 2014, 142, 151–166.
4. Chen, W.W.; Zhang, X.; Huang, W.J. Mol. Med. Rep. 2016, 13, 3391–3396.
5. Perretti, M.; Cooper, D.; Dalli, J.; Norling, L.V. Nat. Rev. Rheumatol. 2017, 13, 87–99.
6. Bjarnason, I.; Hayllar, J.; MacPherson, A.J.; Russell, A.S. Gastroenterology 1993, 104, 1832–1847.
7. Schacke, H.; Docke, W.D.; Asadullah, K. Pharmacol. Ther. 2002, 96, 23–43. Bellik, Y.; Boukraâ, L.; Alzahrani, H.; Bakhotmah, B.; Abdellah, F.; Hammoudi, S.; Iguer-Ouada, M. Molecules 2013, 18, 322–353.
8. Zhang, Y.J.; Gan, R.Y.; Li, S.; Zhou, Y.; Li, A.N.; Xu, D.P.; Li, H.B. Molecules 2015, 20, 21138–21156.
9. Pan, M.H.; Lai, C.S.; Ho, C.T. Food Funct 2010, 1, 15–31.
10. Chuang, C.C.; McIntosh, M.K. Annu. Rev. Nutr. 2011, 31, 155–176. Dong, J.; Zhang, X.; Zhang, L.; Bian, H.X.; Xu, N.; Bao, B.; Liu, J. J. Lipid Res. 2014, 55, 363–374.
11. Talero, E.; Avila-Roman, J.; Motilva, V. Curr. Pharm. Des. 2012, 18, 3939–3965.

12. Andlujar, I.; Recio, M.C.; Giner, R.M.; Cienfuegos-Jovellanos, E.; Laghi, S.; Muguerza, B.; Rios, J.L. J. Agric. Food Chem. 2011, 59, 6474–6483.
13. Fruet, A.C.; Seito, L.N.; Rall, V.L.M.; Di Stasi, L.C. BMC Complement. Altern. Med. 2012, 12, 62.
14. Orsi, P.R.; Seito, L.N.; Di Stasi, L.C. J. Ethnopharmacol. 2014, 151, 380–385.
15. Poulose, S.M.; Miller, M.G.; Shukitt-Hale, B. J. Nutr. 2014, 144, 561S–566S.
16. Davinelli, S.; Sapere, N.; Zella, D.; Bracale, R.; Intrieri, M.; Scapagnini, G. Longev. 2012, 2012, 386527.
17. Meng, B.; Li, J.; Cao, H. Curr. Pharm. Des. 2013, 19, 2101–2113.
18. Nichols, J.A.; Katiyar, S.K. Arch. Dermatol. Res. 2010, 302, 71–83.
19. Neves, A.R.; Lucio, M.; Lima, J.L.; Reis, S. Curr. Med. Chem. 2012, 19, 1663–1681.
20. Barras A, Mezzetti A, Richard A, Lazzaroni S, Roux S, Melnyk P, Betbeder D, Monfilliette-Dupont N. Int. J. Pharm. 2009;379(2):270–77.
21. Zhan S, Paik A, Onyeabor F, Ding B, Prabhu S, Wang J. Molecules. 2020;25(15):3422.
22. Xie X, Tao Q, Zou Y, Zhang F, Guo M, Wang Y, Wang H, Zhou Q, Yu S. J. Agric. Food Chem. 2011;59(17):9280–9289.
23. Doggui S, Sahni JK, Arseneault M, Dao L, Ramassamy C. J. Alzheimers Dis. 2012;30(2):377–92.
24. Derman S, Uzunoglu D, Acar T, Arasoglu T, Ucak S, Ozalp VC, Mansuroglu B. Iran J. Pharm. Res. 2020;19(1):424–435.
25. Pool, H.; Quintanar, D.; Figueroa, J.D.D.; Marinho Mano, C.; Bechara, J.E.H.; Godinez, L.A.; Mendoza, S. J. Nanomater. 2012, 2012, 12.
26. Wu YR, Choi HJ, Kang YG, Kim JK, Shin JW. Int. J. Nanomedicine. 2017;12:7007–7013.
27. Ameruoso A, Palomba R, Palange AL, Cervadoro A, Lee A, Di Mascolo D, Decuzzi P. Front. Immunol. 2017;8:1411.
28. El-Naggar ME, Al-Joufi F, Anwar M, Attia MF, El-Bana MA. Colloids Surf. B Biointerfaces. 2019;177:389–398.
29. Djiokeng Paka G, Doggui S, Zaghmi A, Safar R, Dao L, Reisch A, Klymchenko A, Roullin VG, Joubert O, Ramassamy C. Mol. Pharm. 2016;13(2):391–403.
30. Shetty PK, Venuvanka V, Jagani HV, Chethan GH, Ligade VS, Musmade PB, Nayak UY, Reddy MS, Kalthur G, Udupa N, Rao CM, Mutalik S. Int. J. Nanomedicine. 2015;10:6477–6491.
31. Joshi H, Hegde AR, Shetty PK, Gollavilli H, Managuli RS, Kalthur G, Mutalik S. Photodermatol. Photoimmunol. Photomed. 2018;34(1):69–81.
32. Elnaggar YS, Etman SM, Abdelmonsif DA, Abdallah OY. J. Pharm. Sci. 2015;104(10):3544–3556.
33. Rassu G, Porcu EP, Fancello S, Obinu A, Senes N, Galleri G, Migheli R, Gavini E, Giunchedi P. Pharmaceutics. 2018;11(1):8.
34. Sun L, Liu Z, Wang L, Cun D, Tong HHY, Yan R, Chen X, Wang R, Zheng Y. J. Control Release. 2017;254:44–54.
35. Mao KL, Fan ZL, Yuan JD, Chen PP, Yang JJ, Xu J, ZhuGe DL, Jin BH, Zhu QY, Shen BX, Sohawon Y, Zhao YZ, Xu HL. Colloids Surf. B Biointerfaces. 2017;160:704–714.
36. Cui P, Qu F, Sreeharsha N, Sharma S, Mishra A, Gubbiyappa SK. IUBMB Life. 2020;72(5):1054–1064.
37. Das RK, Kasoju N, Bora U. Nanomedicine. 2010;6(1):153–160.
38. Anand P, Nair HB, Sung B, Kunnumakkara AB, Yadav VR, Tekmal RR, Aggarwal BB. Biochem. Pharmacol. 2010;79(3):330–338.
39. Kuo YC, Tsai HC. Mater. Sci. Eng. C Mater. Biol. Appl. 2018;91:445–457.
40. Kuo YC, Rajesh RE. Int. J. Pharm. 2017;528(1–2):228–241.
41. Zanotto-Filho A, Coradini K, Braganhol E, Schröder R, de Oliveira CM, Simões-Pires A, Battastini AM, Pohlmann AR, Guterres SS, Forcelini CM, Beck RC, Moreira JC. Eur. J. Pharm. Biopharm. 2013;83(2):156–167.
42. Figueiró F, Bernardi A, Frozza RL, Terroso T, Zanotto-Filho A, Jandrey EH, Moreira JC, Salbego CG, Edelweiss MI, Pohlmann AR, Guterres SS, Battastini AM. J. Biomed. Nanotechnol. 2013;9(3):516–526.
43. Lollo G, Ullio-Gamboa G, Fuentes E, Matha K, Lautram N, Benoit JP. Mater. Sci. Eng. C Mater. Biol. Appl. 2018;91:859–867.
44. Frozza RL, Bernardi A, Hoppe JB, Meneghetti AB, Battastini AM, Pohlmann AR, Guterres SS, Salbego C. J. Biomed. Nanotechnol. 2013;9(12):2086–2104.
45. Simão F, Matté A, Pagnussat AS, Netto CA, Salbego CG. Eur. J. Neurosci. 2012;36(7):2899–2905.

46. Li F, Gong Q, Dong H, Shi J. Curr. Pharm. Des. 2012;18(1):27–33.
47. Braidy N, Jugder BE, Poljak A, Jayasena T, Mansour H, Nabavi SM, Sachdev P, Grant R. Curr. Top. Med. Chem. 2016;16(17):1951–1960.
48. Jia Y, Wang N, Liu X. Nutrients. 2017;9(10):1122.
49. Albani D, Polito L, Signorini A, Forloni G. Biofactors. 2010;36(5):370–376.
50. Frozza RL, Bernardi A, Hoppe JB, Meneghetti AB, Matté A, Battastini AM, Pohlmann AR, Guterres SS, Salbego C. Mol. Neurobiol. 2013;47(3):1066–1080.
51. Hoppe JB, Coradini K, Frozza RL, Oliveira CM, Meneghetti AB, Bernardi A, Pires ES, Beck RC, Salbego CG. Neurobiol. Learn. Mem. 2013;106:134–144.
52. Reddy PH, Manczak M, Yin X, Grady MC, Mitchell A, Tonk S, Kuruva CS, Bhatti JS, Kandimalla R, Vijayan M, Kumar S, Wang R, Pradeepkiran JA, Ogunmokun G, Thamarai K, Quesada K, Boles A, Reddy AP. J. Alzheimers Dis. 2018;61(3):843–866.
53. Reddy PH, Manczak M, Yin X, Grady MC, Mitchell A, Kandimalla R, Kuruva CS. J. Investig. Med. 2016;64(8):1220–1234.
54. Su IJ, Chang HY, Wang HC, Tsai KJ. Int. J. Mol. Sci. 2020;21(15):5459.
55. Coradini K, Friedrich RB, Fonseca FN, Vencato MS, Andrade DF, Oliveira CM, Battistel AP, Guterres SS, da Rocha MI, Pohlmann AR, Beck RC. Eur. J. Pharm. Sci. 2015;78:163–170.
56. Coradini K, Lima FO, Oliveira CM, Chaves PS, Athayde ML, Carvalho LM, Beck RC. Eur. J. Pharm. Biopharm. 2014;88(1):178–185.
57. Friedrich RB, Kann B, Coradini K, Offerhaus HL, Beck RC, Windbergs M. Eur. J. Pharm. Sci. 2015;78:204–213.
58. Dong J, Tao L, Abourehab MAS, Hussain Z. Int. J. Biol. Macromol. 2018;116:1268–1281.
59. Hussain Z, Thu HE, Amjad MW, Hussain F, Ahmed TA, Khan S. Mater. Sci. Eng. C Mater. Biol. Appl. 2017;77:1316–1326.
60. Hussain Z, Thu HE, Ng SF, Khan S, Katas H. Colloids Surf. B Biointerfaces. 2017;150:223–241.
61. Bukhari SNA, Hussain F, Thu HE, Hussain Z. J. Integr. Med. 2019;17(1):38–45.
62. Hussain Z. Int. J. Polym. Mater. Polym. Biomaterials 2021;70(1):1–13.
63. Lin YH, Lin JH, Hong YS. J. Biomed. Mater. Res. B Appl. Biomater. 2017; 105(1):81–90.
64. Krausz AE, Adler BL, Cabral V, Navati M, Doerner J, Charafeddine RA, Chandra D, Liang H, Gunther L, Clendaniel A, Harper S, Friedman JM, Nosanchuk JD, Friedman AJ. Nanomedicine. 2015;11(1):195–206.
65. Chereddy KK, Coco R, Memvanga PB, Ucakar B, des Rieux A, Vandermeulen G, Préat V. J. Control. Release. 2013;171(2):208–215.
66. Alqahtani MS, Alqahtani A, Kazi M, Ahmad MZ, Alahmari A, Alsenaidy MA, Syed R. J. Drug Deliv. Sci. Technol. 2020;60:102020.
67. Hussain Z, Pandey M, Choudhury H, Ying PC, Xian TM, Kaur T, Jia GW, Gorain B. J. Drug Deliv. Sci. Technol. 2020;57:101747.
68. Hussain Z, Pandey M, Thu HE, Kaur T, Jia GW, Ying PC, Xian TM, Mohammad AS. Biomed. Pharmacother. 2022;150:112992.
69. Fan Z, Li J, Liu J, Jiao H, Liu B. ACS Appl. Mater. Interfaces. 2018;10(28):23595–23604.
70. Li M, Xin M, Guo C, Lin G, Wu X. Drug Dev. Ind. Pharm. 2017;43(11):1846–1857.
71. Duan Y, Zhang B, Chu L, Tong HH, Liu W, Zhai G. Colloids Surf. B Biointerfaces. 2016;141:345–354.
72. Akbar MU, Zia KM, Akash MSH, Nazir A, Zuber M, Ibrahim M. Int. J. Biol. Macromol. 2018;120(Pt B):2418–2430.
73. Sun S, Du X, Fu M, Khan AR, Ji J, Liu W, Zhai G. Int. J. Pharm. 2021;595:120227.
74. Phan QT, Le MH, Le TT, Tran TH, Xuan PN, Ha PT. Int. J. Pharm. 2016;507(1–2):32–40.
75. Yang C, Chen H, Zhao J, Pang X, Xi Y, Zhai G. Colloids Surf. B Biointerfaces. 2014;121:206–213.
76. Kheiri Manjili H, Ghasemi P, Malvandi H, Mousavi MS, Attari E, Danafar H. Eur. J. Pharm. Biopharm. 2017;116:17–30.
77. Lin YH, Chen CY. Polymers (Basel). 2020;12(10):2280.
78. Zhang Y, Zhou J, Yang C, Wang W, Chu L, Huang F, Liu Q, Deng L, Kong D, Liu J, Liu J. Int. J. Nanomedicine. 2016;11:1119–1130.
79. Javadi M, Khadem Haghighian H, Goodarzy S, Abbasi M, Nassiri-Asl M. Int. J. Rheum. Dis. 2019;22(10):1857–1862.

80. Chen S, Lin Z, He T, Islam MS, Xi L, Liao P, Yang Y, Zheng Y, Chen X. Front. Immunol. 2022;13:800283.
81. Wang X, Jiang Y, Wang YW, Huang MT, Ho CT, Huang Q. Food Chem. 2008;108(2):419–424.
82. Ahmad N, Ahmad R, Al-Qudaihi A, Alaseel SE, Fita IZ, Khalid MS, Pottoo FH. RSC Adv. 2019;9(35):20192–20206.
83. Zheng Z, Sun Y, Liu Z, Zhang M, Li C, Cai H. Drug. Des. Devel. Ther. 2015;9:4931–4942.
84. Gokhale JP, Mahajan HS, Surana SJ. Biomed. Pharmacother. 2019;112:108622.
85. Naz Z, Ahmad FJ. Int. J. Nanomedicine. 2015;10:4293–4307.
86. Arora R, Kuhad A, Kaur IP, Chopra K. Eur. J. Pain. 2015;19(7):940–952.
87. Wang J, Zhu R, Sun D, Sun X, Geng Z, Liu H, Wang SL. J. Biomed. Nanotechnol. 2015;11(3):403–415.
88. Maiti P, Paladugu L, Dunbar GL. BMC Neurosci. 2018;19(1):7.
89. Maiti P, Hall TC, Paladugu L, Kolli N, Learman C, Rossignol J, Dunbar GL. Histochem Cell Biol. 2016;146(5):609–625.
90. Maiti P, Bowers Z, Bourcier-Schultz A, Morse J, Dunbar GL. Alzheimers. Res. Ther. 2021;13(1):37.
91. Wang J, Wang H, Zhu R, Liu Q, Fei J, Wang S. Biomaterials. 2015;53:475–483.
92. Li H, Zhao X, Ma Y, Zhai G, Li L, Lou H. J. Control Release. 2009;133(3):238–244.
93. Dhawan S, Kapil R, Singh B. J. Pharm. Pharmacol. 2011;63(3):342–351.
94. Luo CF, Yuan M, Chen MS, Liu SM, Zhu L, Huang BY, Liu XW, Xiong W. Int. J. Pharm. 2011;410(1–2):138–144.
95. Yang T, Sheng HH, Feng NP, Wei H, Wang ZT, Wang CH. J. Pharm. Sci. 2013;102(12):4414–4425.
96. Bhalekar MR, Madgulkar AR, Desale PS, Marium G. Drug Dev. Ind. Pharm. 2017;43(6):1003–1010.
97. Tan ME, He CH, Jiang W, Zeng C, Yu N, Huang W, Gao ZG, Xing JG. J. Nanomedicine. 2017;12:3253–3265.
98. Ji H, Tang J, Li M, Ren J, Zheng N, Wu L. Drug Deliv. 2016;23(2):459–470.
99. Yusuf M, Khan M, Khan RA, Ahmed B. J. Drug Target. 2013;21(3):300–311.
100. Vijayakumar A, Baskaran R, Jang YS, Oh SH, Yoo BK. AAPS PharmSciTech. 2017;18(3):875–883.
101. Ramalingam P, Ko YT. Pharm Res. 2015;32(2):389–402.
102. Baek JS, Cho CW. Eur. J. Pharm. Biopharm. 2017;117:132–140.
103. Ramalingam P, Ko YT. Colloids Surf. B Biointerfaces. 2016;139:52–61.
104. Loureiro JA, Andrade S, Duarte A, Neves AR, Queiroz JF, Nunes C, Sevin E, Fenart L, Gosselet F, Coelho MA, Pereira MC. Molecules. 2017;22(2):277.
105. Pinheiro RGR, Granja A, Loureiro JA, Pereira MC, Pinheiro M, Neves AR, Reis S. Pharm. Res. 2020;37(7):139.
106. Dong Z, Guo J, Xing X, Zhang X, Du Y, Lu Q. Biomed Pharmacother. 2017 May;89:297–304.
107. Shen Q, Zhang X, Qi J, Shu G, Du Y, Ying X. Int. J. Pharm. 2020;576:119001.
108. Xi L, Lin Z, Qiu F, Chen S, Li P, Chen X, Wang Z, Zheng Y. Acta Pharm Sin B. 2022;12(1):339–352.
109. White KL, Rades T, Furneaux RH, Tyler PC, Hook S. J. Pharm. Pharmacol. 2006;58(6):729–737.
110. Hattori Y, Kawakami S, Suzuki S, Yamashita F, Hashida M. Biochem. Biophys. Res. Commun. 2004;317(4):992–999.
111. Hattori Y, Kawakami S, Nakamura K, Yamashita F, Hashida M. J. Pharmacol. Exp. Ther. 2006;318(2):828–834.
112. Lu Y, Kawakami S, Yamashita F, Hashida M. Biomaterials. 2007;28(21):3255–3262.
113. Espuelas S, Thumann C, Heurtault B, Schuber F, Frisch B. Bioconjug. Chem. 2008;19(12):2385–2393.
114. Pichon C, Midoux P. Methods Mol. Biol. 2013;969:247–274.
115. Qiu F, Xi L, Chen S, Zhao Y, Wang Z, Zheng Y. Celastrol. Niosome. Int. J. Nanomedicine. 2021;16:6171–6182.
116. Gao J, Chen F, Fang H, Mi J, Qi Q, Yang M. Biol Res. 2020;53(1):48.
117. Meng S, Sun L, Wang L, Lin Z, Liu Z, Xi L, Wang Z, Zheng Y. Colloids Surf. B Biointerfaces. 2019;182:110352.
118. Gollavilli H, Hegde AR, Managuli RS, Bhaskar KV, Dengale SJ, Reddy MS, Kalthur G, Mutalik S. Colloids Surf. B Biointerfaces. 2020;193:111122.
119. Avadhani KS, Manikkath J, Tiwari M, Chandrasekhar M, Godavarthi A, Vidya SM, Hariharapura RC, Kalthur G, Udupa N, Mutalik S. Drug Deliv. 2017;24(1):61–74.
120. Elnaggar YS, Etman SM, Abdelmonsif DA, Abdallah OY. Int. J. Nanomedicine. 2015;10:5459–5473.

121. Langasco R, Fancello S, Rassu G, Cossu M, Cavalli R, Galleri G, Giunchedi P, Migheli R, Gavini E. Phytomedicine. 2019;52:23–31.
122. Merrell JG, McLaughlin SW, Tie L, Laurencin CT, Chen AF, Nair LS. Clin. Exp. Pharmacol. Physiol. 2009;36(12):1149–1156.
123. Dai X, Liu J, Zheng H, et al. NPG Asia Mater. 2017;9:e368.
124. Anand S, Rajinikanth PS, Arya DK, Pandey P, Gupta RK, Sankhwar R, Chidambaram K. Pharmaceutics. 2022;14(2):273.
125. Zahiri M, Khanmohammadi M, Goodarzi A, Ababzadeh S, Sagharjoghi Farahani M, Mohandesnezhad S, Bahrami N, Nabipour I, Ai J. Int. J. Biol. Macromol. 2020;153:1241–1250.
126. Agarwal Y, Rajinikanth PS, Ranjan S, Tiwari U, Balasubramnaiam J, Pandey P, Arya DK, Anand S, Deepak P. Int. J. Biol. Macromol. 2021;176:376–386.
127. Zhang H, Zhang M, Wang X, Zhang M, Wang X, Li Y, Cui Z, Chen X, Han Y, Zhao W. Drug Deliv. 2022;29(1):174–185.
128. Anand S, Pandey P, Begum MY, Chidambaram K, Arya DK, Gupta RK, Sankhwar R, Jaiswal S, Thakur S, Rajinikanth PS. Pharmaceuticals (Basel). 2022;15(3):302.
129. Deldar Y, Pilehvar-Soltanahmadi Y, Dadashpour M, Montazer Saheb S, Rahmati-Yamchi M, Zarghami N. Artif. Cells Nanomed. Biotechnol. 2018;46(4):706–716.
130. Deldar Y, Zarghami F, Pilehvar-Soltanahmadi Y, Dadashpour M, Zarghami N. Cell Tissue Bank. 2017;18(4):475–487.
131. Mohammadi Z, Sharif Zak M, Majdi H, Seidi K, Barati M, Akbarzadeh A, Latifi AM. Chem. Biol. Drug Des. 2017;90(6):1106–1114.
132. Mohammadi Z, Sharif Zak M, Majdi H, Mostafavi E, Barati M, Lotfimehr H, Ghaseminasab K, Pazoki-Toroudi H, Webster TJ, Akbarzadeh A. Artif. Cells Nanomed. Biotechnol. 2019;47(1):1642–1652.
133. Fu SZ, Meng XH, Fan J, Yang LL, Wen QL, Ye SJ, Lin S, Wang BQ, Chen LL, Wu JB, Chen Y, Fan JM, Li Z. J. Biomed. Mater. Res. B. Appl. Biomater. 2014;102(3):533–542.
134. Fahimirad S, Abtahi H, Satei P, Ghaznavi-Rad E, Moslehi M, Ganji A. Carbohydr. Polym. 2021;259:117640.
135. Prasathkumar M, Sadhasivam S. Int J Biol Macromol. 2021 Sep 1;186:656–685.
136. Ranjbar-Mohammadi M, Rabbani S, Bahrami SH, Joghataei MT, Moayer F. Mater. Sci. Eng. C Mater. Biol. Appl. 2016;69:1183–1191.
137. Mutlua G, Calamakab S, Ulubayram K, Guvena E. J. Drug Deliv. Sci. Tech. 2018;43:185–193.
138. Shababdoust A, Ehsani M, Shokrollahi P, Zandi M. Prog. Biomater. 2018;7(1):23–33.
139. Shababdoust A, Zandi M, Ehsani M, Shokrollahi P, Foudazi R. Int. J. Pharm. 2020;575:118947.
140. Sadeghianmaryan A, Yazdanpanah Z, Soltani YA, Sardroud HA, Nasirtabrizi MH, Chen X. J. Biomater. Sci. Polym. Ed. 2020;31(2):169–187.
141. Saeed SM, Mirzadeh H, Zandi M, Barzin J. Prog. Biomater. 2017;6(1–2):39–48.
142. Akrami-Hasan-Kohal M, Tayebib L, Ghorbani M. New. J. Chem. 2020;44:10343–10351.
143. Ullah A, Ullah S, Khan MQ, Hashmi M, Nam PD, Kato Y, Tamada Y, Kim IS. Int. J. Biol. Macromol. 2020;155:479–489.
144. Ullah A, Saito Y, Ullah S, Haider MK, Nawaz H, Duy-Nam P, Kharaghani D, Kim IS. Int. J. Biol. Macromol. 2021;166:1009–1021.
145. Ardekani NT, Khorram M, Zomorodian K, Yazdanpanah S, Veisi H, Veisi H. Int. J. Biol. Macromol. 2019;125:743–750.
146. Mouro C, Dunne CP, Gouveia IC. Molecules. 2020;26(1):83.
147. Samadian H, Zamiri S, Ehterami A, Farzamfar S, Vaez A, Khastar H, Alam M, Ai A, Derakhshankhah H, Allahyari Z, Goodarzi A, Salehi M. Sci. Rep. 2020;10(1):8312.
148. Wang F, Wang X, Hu N, Qin G, Ye B, He JS. J. Biomed. Nanotechnol. 2022;18(1):77–86.
149. Salami MS, Bahrami G, Arkan E, Izadi Z, Miraghaee S, Samadian H. BMC Complement. Med. Ther. 2021;21(1):111.
150. Motealleh B, Zahedi P, Rezaeian I, Moghimi M, Abdolghaffari AH, Zarandi MA. J. Biomed. Mater. Res. B Appl. Biomater. 2014;102(5):977–987.
151. Shokrollahi M, Bahrami SH, Nazarpak MH, Solouk A. Int. J. Biol. Macromol. 2020;147:547–559.
152. Saeed M, Mirzadeh H, Zandi M, Barzin J. Prog. Biomater. 2020;9(4):175–185.
153. Farahani H, Barati A, Arjomandzadegan M, Vatankhah E. Int J Biol Macromol. 2020;162:762–773.

154. Kaur R, Badea I. Int. J. Nanomedicine. 2013;8:203–220.
155. Whitlow J, Pacelli S, Paul A. J Control Release. 2017;261:62–86.
156. Bondon N, Raehm L, Charnay C, Boukherroub R, Durand JO. J. Mater. Chem. B. 2020;8(48):10878–10896.
157. Cheng B, Pan H, Liu D, Li D, Li J, Yu S, Tan G, Pan W. Int. J. Pharm. 2018;540(1–2):162–170.
158. Liu D, Cheng B, Li D, Li J, Wu Q, Pan H. Int. J. Pharm. 2018;551(1–2):177–183.
159. Liu D, Qiao S, Cheng B, Li D, Chen J, Wu Q, Pan H, Pan W. AAPS PharmSciTech. 2020;21(5):187.
160. Jariwala DH, Patel D, Wairkar S. Mater. Sci. Eng. C Mater. Biol. Appl. 2020;113:110996.
161. Tinwala H, Wairkar S. Mater. Sci. Eng. C Mater. Biol. Appl. 2019;97:913–931.
162. Huang H, Liu M, Jiang R, Chen J, Mao L, Wen Y, Tian J, Zhou N, Zhang X, Wei Y. J. Colloid Interface. Sci. 2018;513:198–204.
163. Karri VV, Kuppusamy G, Talluri SV, Mannemala SS, Kollipara R, Wadhwani AD, Mulukutla S, Raju KR, Malayandi R. Int. J. Biol. Macromol. 2016;93(Pt B):1519–1529.
164. Guadarrama-Acevedo MC, Mendoza-Flores RA, Del Prado-Audelo ML, Urbán-Morlán Z, Giraldo-Gomez DM, Magaña JJ, González-Torres M, Reyes-Hernández OD, Figueroa-González G, Caballero-Florán IH, Florán-Hernández CD, Florán B, Cortés H, Leyva-Gómez G. Pharmaceutics. 2019;11(8):389.
165. Mobaraki M, Bizari D, Soltani M, Khshmohabat H, Raahemifar K, Akbarzade Amirdehi M. Polymers (Basel). 2021;13(24):4291.
166. Gong C, Wu Q, Wang Y, Zhang D, Luo F, Zhao X, Wei Y, Qian Z. Biomaterials 2013;34(27):6377–6387.
167. Li X, Chen S, Zhang B, Li M, Diao K, Zhang Z, Li J, Xu Y, Wang X, Chen H. Int. J. Pharm. 2012;437(1–2):110–119.
168. Dang LH, Nguyen TH, Tran HLB, Doan VN, Tran NQ. J. Healthcare Eng. 2018;2018:5754890.
169. Xu H, Li J, Yang X. et al. A J Clust Sci 31, 823–830 (2020).
170. Li X, Wang C, Yang S, Liu P, Zhang B. Int. J. Nanomedicine. 2018;13:5287–5299.
171. Xu, H., Li, J., Yang, X. et al. J. Clust. Sci. 31, 823–830 (2020).
172. Bhawana, Basniwal RK, Buttar HS, Jain VK, Jain N. J. Agric. Food Chem. 2011;59(5):2056–2061.
173. Adahoun MA, Al-Akhras MH, Jaafar MS, Bououdina M. Nanomed. Biotechnol. 2017;45(1):98–107.
174. Basniwal RK, Khosla R, Jain N. Nutr Cancer. 2014;66(6):1015–1022.
175. Rai M, Ingle AP, Pandit R, Paralikar P, Anasane N, Santos CAD. Expert. Rev. Anti. Infect Ther. 2020;18(4):367–379.
176. Tagde P, Tagde P, Islam F, Tagde S, Shah M, Hussain ZD, Rahman MH, Najda A, Alanazi IS, Germoush MO, Mohamed HRH, Algandaby MM, Nasrullah MZ, Kot N, Abdel-Daim MM. Molecules. 2021;26(23):7109.
177. Ghalandarlaki N, Alizadeh AM, Ashkani-Esfahani S. Biomed. Res. Int. 2014;2014:394264.
178. Shome S, Talukdar AD, Tewari S, Choudhury S, Bhattacharya MK, Upadhyaya H. Biotechnol. Appl. Biochem. 2021;68(3):603–615.
179. Flora G, Gupta D, Tiwari A. Crit. Rev. Ther. Drug. Carrier. Syst. 2013;30(4):331–368.
180. Karthikeyan A, Senthil N, Min T. Front. Pharmacol. 2020;11:487.

12 Delivery of Antimalarial Phytochemicals by Nanoscale Approaches

Niladri Haldar[a,b], *Rajesh Salve*[a,b], *Kavita R. Gajbhiye*[c], *and Virendra Gajbhiye*[a,b,*]
[a]Nanobioscience Group, Agharkar Research Institute, Pune 411038, India
[b]Savitribai Phule Pune University, Pune 411007, India
[c]Poona College of Pharmacy, Bharati Vidyapeeth (Deemed to be University) Pune 411038, India

12.1 INTRODUCTION

Malaria is one of the most severe and life-threatening mosquito-borne diseases caused by *Plasmodium* species, and for decades it remains a major health concern worldwide, particularly affecting children and pregnant women in tropical countries [1-4]. Malaria is diagnosed in more than 200 million people and is associated with 600,000 deaths each year, with 241 million cases and 627,000 associated deaths being recorded in 2020 [5]. Five different *Plasmodium* species cause malaria in humans: *Plasmodium vivax*, *Plasmodium malariae*, *Plasmodium ovale*, *Plasmodium falciparum*, and *Plasmodium knowlesi*. Among them, *Plasmodium falciparum* and *Plasmodium knowlesi* predominantly cause most infections, manifesting as acute and deadly forms of this disease [2, 6, 7]. Malaria is transmitted through the sting of a female *Anopheles* mosquito infected with the *Plasmodium* parasite. Approximately 60 of the 400 *Anopheles* species play a role as malaria vectors, 30 of which are the most significant ones [8]. *Plasmodium* life cycle starts when an infected female *Anopheles* mosquito stings a human skin and sucks blood while injecting *Plasmodium* in the form of sporozoites; these sporozoites subsequently enter liver cells and undergo division inside the cell [9, 10]. In the intraerythrocytic cycle, merozoites are discharged into the blood, wherein they invade red blood cells (RBCs), and after the invasion, the merozoites grow and reinvade new RBCs. The merozoites then develop gametocytes, which are released from the RBCs. An *Anopheles* mosquito then stings the infected person and takes up these gametocytes, generating gametes that unite to form the zygote. The zygote matures as an ookinete, which travels to the mosquito's midgut to become an oocyst, which grows there until the release of infective sporozoites (Figure 12.1) [9]. Because of increased global warming, geographic regions covering *Anopheles* distribution are expected to expand, resulting in greater transmission of malarial parasites [11]. Thus, efforts have been made to discover innovative malaria prevention and treatment strategies.

Plants are a reliable source of antimalarial drugs, such as quinine (QN), quinidine, and artemisinin (ART), all of which are derived from plants and therapeutically useful for alleviating severe malaria [12]. Various phytochemicals such as alkaloids, terpenes, and flavonoids have shown an antimalarial effect. Alkaloids are a notable category of phytochemicals with varied biological functions and comprised antimalarial activity. They comprise a family of structurally diverse compounds produced from amino acids with a heterocyclic ring possessing nitrogen atoms [13]. Terpenoids, also known as isoprenoids, are isoprene-derived compounds. They are produced through a three-step process: C5 formation, terpenoid skeleton formation by condensing subunits, and prenyl diphosphate

DOI: 10.1201/9781003231721-12

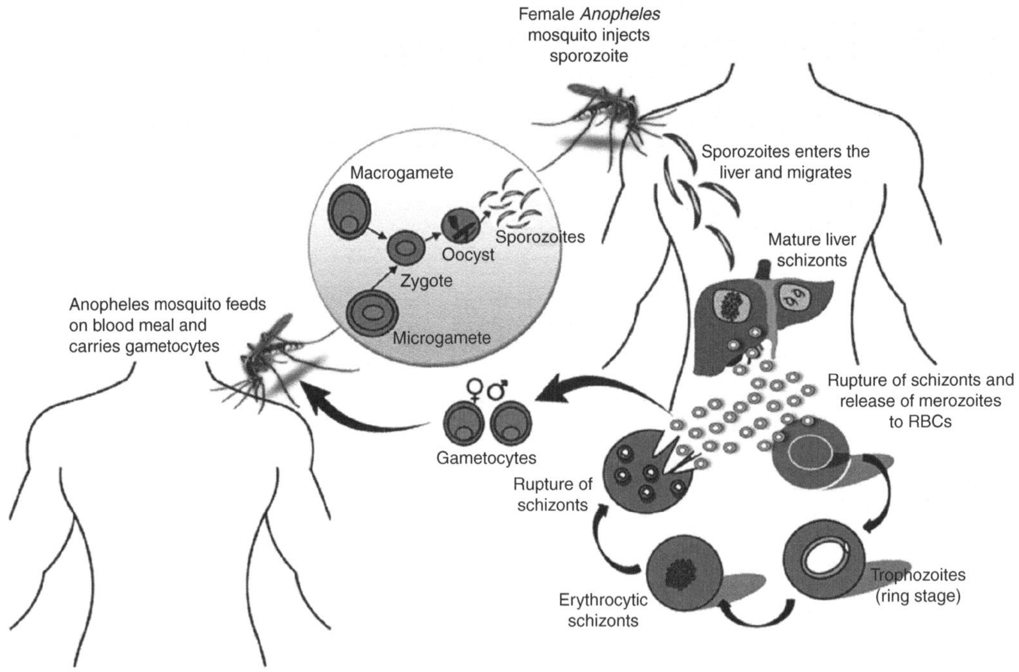

FIGURE 12.1 Life cycle of the malarial parasite. (From Puttappa et al., 2019. *Acta Trop.*, vol. 195. April, pp. 103–114 [9]. With Permission.) Copyright (2019) Elsevier.

conversion. The mevalonate or methylerythritol 4-phosphate process is associated with terpenoid synthesis. Isopentenylpyrophosphate (known as "activated isoprene") is the common intermediate from which all terpenoids are produced [14]. Flavonoids are a major subgroup of secondary metabolites classified as phenolic chemicals that are extensively distributed in plants. Flavonoids are produced from phenylalanine and malonyl-coenzyme A. The term "flavonoid" refers to a wide range of natural compounds that comprise the carbon framework C6-C3-C6. This group of natural products may be divided into three categories based on the position of the connection between the aromatic ring and the benzopyrano moiety: flavonoids (2-phenylbenzopyrans), neoflavonoids (4-benzopyrans), and isoflavonoids (3-benzopyrans) [15]. Alkaloids, terpenes, and flavonoids are responsible for the plant's antimalarial activity.

Moreover, nonspecific toxicities; limited availability; and undesirable physicochemical or pharmacokinetic characteristics such as poor lipid–water partitioning behavior, short plasma half-life, and inadequate bioavailability are some of the major drawbacks of phytochemical-based anti-malarial treatment. Therefore, treating multidrug resistance malaria has become more difficult and challenging in most malaria-endemic regions of the world, necessitating the urgent development of novel and more effective antimalarial medications to combat drug-resistant malaria [16]. In recent years, nanoparticle (NP)-based delivery of therapeutics is receiving more attention in curing life-threatening diseases. Compared with chemically synthesized NPs and free phytochemicals, plant-based NPs have higher drug solubility and some additional benefits. The modification of phytochemical pharmacokinetics and biodistribution resulting from NP formation might increase the therapeutic index of the phytochemicals by decreasing toxicity and boosting effectiveness [17]. Nanocarriers (NCs), which have a prolonged blood circulation time, have been used to transport anti-malarial medications to lengthen their resident time in the body, especially in stages at which drugs are likely to interact with the infected RBCs and parasites. Furthermore, nano-drug delivery systems can shield unstable medicines, facilitate cell adhesion, and bind to couple with particular ligand molecules on their surface [18]. Thus, a nanotechnological approach for delivering phytochemicals possesses great potential in malarial treatment.

12.2 CHALLENGES IN MALARIAL TREATMENT

Despite having achieved remarkable decreases in malaria-associated death and morbidity rates globally through the use of several antimalarial approaches, malaria not only is a potentially fatal disease but also remains a serious public health concern [19, 20]. The World Health Organization defined "drug resistance" in 1967 as the ability of a strain of a parasite to live or reproduce independently of the application and absorption of medication when administered in dosages equivalent to or greater than that usually recommended but within the tolerance range of the particular person. Drug resistance typically results in a delay in or failure of discharge of the asexual parasites from peripheral circulation, thus allowing the gametocytes to be produced, transmitting the resistant genotype. According to the recommendations of the WHO in 2001 for the usage of ART-mediated combination treatments as the primary treatment strategy for *P. falciparum* infections, a significant decrease was noted in outbreaks of this illness after 2005. However, drug-resistant parasites, specifically resistant to ART and its derivatives, have lately evolved in several regions of Southeast Asia, posing a threat to the previous success in malaria control, treatment, and eradication efforts. Currently available antimalarial medicines act on a limited number of biological targets [21]. Furthermore, because malaria is a vector-borne parasitic disease, it could be detected at any phase in the disease process, from the vector (mosquito) phase to the infected RBC phase, posing significant obstacles to the development of innovative antimalarial therapies. The majority of efforts are directed at drug delivery systems, which generally carry recognized antimalarial medicines. Two critical goals must be met when designing anti-infective treatments: administering sufficiently high local doses to promptly kill the pathogen without enabling it to develop resistance, and attaining it with overall dosages that are not harmful to the human host. Both objectives could be met simultaneously by using tailored drug delivery systems such as NCs [10].

QN has been used to treat malaria and remains the most popular and successful treatment choice even today. This medicine is effective against all schizonts of *Plasmodium* spp. as well as gametocytes of *Plasmodium malariae*, *Plasmodium vivax*, and *Plasmodium ovale* except *Plasmodium falciparum*. QN is no longer used as the primary therapy owing to drug resistance, although it is still used for the treatment of severe malaria. After oral dosing, chloroquine (CQ) is efficiently taken up by the body through the gastrointestinal system. This medication is used for the uncomplicated malarial treatment triggered by *Plasmodium vivax*, *Plasmodium malariae*, and *Plasmodium ovale*, but for CQ-resistant (CQR) strains, this drug is not effective. Lumefantrine is another medicine extensively used in conjunction with an ART derivative (artemether) in repeated dosages for the treatment of malaria caused by *Plasmodium falciparum*. However, these antimalarial drugs have some disadvantages such as low bioavailability, unsatisfactory pharmacokinetic profile, and limited RBC penetration. Some medications such as amodiaquine and all derivatives of ART except artesunate (ARTSN) have shown low bioavailability when administered orally, and less than 35% of the consumed medicines can reach the bloodstream, necessitating large doses, which may result in hazardous side effects. Because ART derivatives have a shorter half-life ($t_{1/2}$), the companion medicine must have a long bioavailability ($t_{1/2}$ >4 days) to completely eliminate the parasite. To improve penetrability, hydrophobic drugs should be used to target RBCs (this requirement is fulfilled by ART and its derivatives, which is the frontline medicine). However, hydrophobicity results in off-target delivery and deficient pharmacokinetics, resulting in aggregation at nontargeted tissues and interaction with plasma proteins [10].

Less effective single-drug therapies increase the likelihood of the emergence of parasites and their resistance to therapy; combining antimalarial medications with distinct mechanisms of action can prevent the development of resistance to each individual component of the combination regimen. There are several obstacles ahead in generating next-generation antimalarial therapy [22].

12.3 PLANT BIO ACTIVES FOR MALARIAL TREATMENT

Since ancient times, plant-based therapies have had essential functions in malarial treatment. The discovery of QN is the beginning of contemporary chemotherapeutic malarial therapy, and for a long time, quinoline-based synthetic antimalarial medications such as CQ, mefloquine (MQ), and others have been used for malarial treatment [16]. QN was first isolated from *Cinchona* bark in 1820 [23]. *Cinchona* species have long been known for their antimalarial potential, and the component alkaloid QN is still used as a medicine. Quinidine is a less well-known stereoisomer and may be more potent than QN [24]. ART, an antimalarial component extracted from *Artemisia annua*, is the most significant medicine, identified and defined by Chinese scientists, and it has been used to treat malaria for thousands of years [23]. The reduced form of ART, called dihydro ART, is a semisynthetic derivative of ART and has strong antimalarial potential for multidrug-resistant malarial treatment, notably the disease caused by *Plasmodium falciparum* [16]. Plant antimalarial activity could be attributed to a variety of phytochemicals comprising alkaloids, terpenes, steroids, and flavonoids [13].

12.3.1 Alkaloids as Antimalarial Agents

Alkaloids are naturally produced organic compounds containing at least one nitrogen atom in their structure. Several classes of alkaloids have shown antimalarial activity, such as indole alkaloids, naphthoisoquinolines, and furoquinolines [25]. Indole alkaloids are widely distributed secondary metabolites found in plant roots, leaves, seeds, fruits, flowers, bark, sponges, fungi, and algae. They are a family of substances with fascinating biological activity, comprising antimalarial efficacy against drug-resistant strains. Furthermore, several indole alkaloids such as vinblastine and vincristine have been investigated in clinical studies or are assessed in clinical trials, and the results suggest that indole alkaloids represent a substantial source of active pharmacological chemicals [26]. A novel active indole alkaloid, named decursivine, was extracted from the stems and leaves of *Rhaphidophora decursiva*; this compound exerts antimalarial activity against D6 and W2 of *Plasmodium falciparum* clones, with half-maximal inhibitory concentration rates amounting to 3.93 and 4.41 mg/mL, respectively [27]. Monoterpenoid indole alkaloids (MIAs) such as N-methyl-tetrahydrolivacine, uleine, aspidoscarpine, and apparicine are found in *Aspidosperma olivaceum*. The antimalarial potentiality of *Aspidosperma Olivaceum*, assessed using a biomonitoring technique, was attributed to aspidoscarpine and, to a smaller degree, N-methyl-tetrahydrolivacine. Other MIA compounds identified were active but had a lesser selectivity index (SI) because of their elevated level of toxicities [28].

There are four kinds of bisindole skeleton alkaloids with powerful and specific anti-*Plasmodium* activity [29]. A bisindole alkaloid (IC_{50}: 2.0 μM) was extracted from *Aristolochia cordigera*, which showed significant efficacy against the CQ-sensitive (CQS) 3D7 strain of *Plasmodiun falciparum*. Furthermore, this alkaloid (IC_{50}: >250 μM) was found to be nontoxic to HepG2 cells. The SI value was greater than 125. Bisindole alkaloids (IC_{50}: 190–670 nM) were extracted from the leaves of *Flindersia pimenteliana*; although they demonstrated significant efficacy against *Plasmodium falciparum* CQS 3D7 and CQR Dd2 strains, the efficacy was not superior to that of CQ (IC_{50}: 23 and 130 nM) [26]. Using mass-directed purification, a novel tribrominated indole alkaloid, named kororamide A, was extracted from the bryozoan *Amathia tortuosa* from northern part of New South Wales in Australia. Kororamide A was slightly effective against CQS and *Plasmodium falciparum*-resistant strains but ineffective against human cell lines (normal human embryonic kidney [HEK], pancreatic, and breast cancer cell lines) [30]. Naphthylisoquinolines are a group of alkaloids consisting of naphthalene and isoquinoline moieties. Dioncopeltine A, dioncophylline B, and dioncophylline C are the most active secondary metabolites that have shown superior antiplasmodial activity in both *in vitro* and animal studies. Habropetaline A, a new naphthylisoquinoline alkaloid extracted from *Triphyophyllum peltatum*, has excellent antiplasmodial activity. Habropetaline A has IC_{50} values of 5.0 and 2.3 ng mL^{-1} for K1 strains that are resistant against CQ and pyrimethamine [31]. Jozipeltine A is a naphthylisoquinoline alkaloid and the constitutionally 6′-coupled symmetrical dimer of dioncopeltine A, which has extreme antimalarial activity. The first dimer of naphthylisoquinoline called jozipeltine A has oxygen functionalities in the naphthalene side chain. Interestingly, although jozipeltine A has considerable antiplasmodial activity against *Plasmodium falciparum in vitro* against K1 strain (875 ng mL^{-1}) and NF54 strain (2530 ng mL^{-1}), it exerts considerably lesser activity than dioncopeltine A (K1 strain 4.8 ng mL^{-1}, NF54 strain 3.3 ng mL^{-1}), which is its monomeric precursor [32]. Four novel naphthylisoquinoline alkaloids were identified from a liana plant named *Ancistrocladus robertsoniorum*. Biological testing demonstrated that these alkaloids exerted only mild antimalarial properties [33]. The 7,3′-coupled ancistrotanzanine C; 5,1′-coupled O-methylancistrocladinine; 5,1′-coupled O,N-dimethylancistrocladine naphthylisoquinoline; and 7,3′-coupled naphthylisoquinoline alkaloids ancistrocladidine and ancistrotectorine alkaloids were isolated from *Ancistrocladus tanzaniensis*. The compounds ancistrocladidine and ancistrotectorine had modest antiplasmodial activity against K1 strain of *Plasmodium falciparum*, which is CQR, but these compounds were significantly less effective against *Plasmodium falciparum* CQS 3D7 strain [34].

Quinoline alkaloids are major groups of N-based heterocyclic aromatic chemicals, which have pooled the interest of researchers worldwide during the last 200 years owing to their diverse biological activities and uses. Most compounds and their derivatives have antimalarial, anti-inflammatory, antiparasitic, insecticidal, antiviral, antibacterial, anticancer, and antioxidant, and other biological activities. QN, CQ, furoquinolines, quinidine, and MQ are quinoline alkaloids with antimalarial activity [35]. A quinoline, named heliparvifoline (6-methoxy-7-hydroxydictamnine), was isolated from *Melicope madagascariensis*, which demonstrated limited antimalarial activity (IC_{50}: 35 μM) against Dd2, the CQR strain of *Plasmodium falciparum* [36]. Alkaloids such as fagarine, skimmianine, kokusaginine, evoxine, 7-O-isopentenyl-fagarine, 4-methoxy-1-methyl-quinolin-2-one, and anhydro-evoxine were isolated from the wood and leaf extracts of *Peltostigma guatemalense*. These furoquinolines showed antimalarial activity against *Plasmodium falciparum* strains by inhibiting 80% invasion at a concentration of 140 ppm [37]. A novel quinoline alkaloid, named 2′R-2′,3′-epoxy-N-methylatanine, and potent known products such as maculosidine, skimmianin, pteleine, fagarine, evolitrine, werneria chromene, N-methylatanine, and dictamnine were purified from the Australian plant *Drummondita calida*. 2′R-2′,3′-epoxy-N-methylatanine inhibited the Dd2 strain of *Plasmodium falciparum* by 74% at a dose of 80 μM and was nontoxic against HEK293 cells at the same concentration [38]. Four furoquinoline alkaloids, namely, tecleaverdoornine, montrifoline, maculine, and kokusaginine, have been extracted from the stem bark of *Teclea afzelii* (Rutaceae). An *in vitro* study showed that the NF54 strain was partially inhibited by all four compounds at a dose of 3 μM, whereas kokusaginine showed the best antiplasmodial activity. Other *in vitro studies* showed that treatment with kokusaginine at doses ranging from 2 to 8 μM resulted in partial inhibition of the NF54 strain at a dose of 4 μM and an even greater effect at a dose of 8 μM [39].

12.3.2 Terpenoids as Antimalarial Agents

Terpenes (also known as terpenoids or isoprenoids) are the largest category of natural compounds and are particularly understudied [40]. Terpenes are natural compounds and encompass approximately 30,000 different secondary metabolites with varied functions such as defense compounds in plants against parasitic, viral, and bacterial infections. Terpenes have a vast variation in compound structure and may be found in plants (particularly in resins, latex essential oils, and waxes), algae, bacteria, fungi, and certain mammals (insects and marine organisms) [41]. Terpenoids are components present in plant-based essential oils and are derived from five-carbon terpene hydrocarbons. Various types of terpenoids include monoterpenoids, diterpenoids, triterpenoids, tetraterpenoids, sesquiterpenoids, sesterterpenoids, hemiterpenoids, and polyterpenoids. Some of these terpenoids may be detected in plant-based essential oils, which are utilized in conventional medicine for the treatment of malaria and several other disorders [42]. Monoterpenes, also known as C10 isoprenoids, are a broad family of natural components that are best recognized as ingredients of aromatic plant-based essential oils and protective oleoresins. Monoterpenes are widely used in food, pharmaceutical, and cosmetic industries, in addition to their ecological functions in pollinator attraction, allelopathy, and plant defense mechanisms [43]. 1,8-Cineole is a monoterpene commonly found in nature and is detected in many plant-based essential oils, and it has antimalarial properties. 1,8-Cineole shows an irreversible inhibitory effect at a dose of 972 μM, independent of parasitemia. Furthermore, it can slow down intracellular growth across two life cycles of the parasite, causing significant morphological alterations. The monoterpene significantly lowered the adhesion index of infected RBCs to brain microvascular endothelial cells. A brain malarial mouse model experiment revealed that treatment with 1,8-cineole for 6 days at a dose of 100 mg/kg per day led to a 50%decrease in cerebral edema and parasitemia. These findings suggest a possible inhibitory activity of 1,8-cineole on the erythrocyte cycle of the malarial parasite and cure for acute disease [44]. Cuminaldehyde, an aromatic monoterpenoid isolated from the fruits of *Bunium persicum*, was tested for its antimalarial activity using a cell-free hematin production assay. The purified chemical had no inhibitory activity against hematin production. Cuminaldehyde is distinguished from other monoterpenoids by its aromatic ring, which resembles phenyl propanoids. Moreover, the antimalarial activity of phenyl propanoids can be seen when monoterpenoid glycosides are linked to a phenyl propanoid moiety [45]. 1,8-Cineole, terpinen-4-ol (T4O), p-cymene, α-terpinene, and limonene monoterpene compounds were extracted from *Melaleuca alternifolia*, the Australian tea tree oil (TTO), and their antimalarial activity against *Plasmodium falciparum* indicated promising therapeutic potential. TTO compounds have antimalarial potential, and the IC_{50} values ranged from 70.2 to 1528.8 μM for α-terpinene, p-cymene, limonene, and 1,8-cineole, which confirmed that the monoterpenes of TTO can kill the FCR-3 CQR strain of *Plasmodium falciparum* [46].

Sesquiterpenoids are made of three isoprene units (C_{15}) in various shapes and sizes, including linear, monocyclic, bicyclic, and tricyclic frameworks. They are the most varied class of terpenoids [47, 48]. Sesquiterpenes are colorless lipophilic molecules containing a 15-carbon backbone. In plants, biosynthesis of sesquiterpenes begins with three isoprene units and occurs in the endoplasmic reticulum, with farnesyl pyrophosphate being the primary compound [49]. While terpenoids in aerial tissues are often explored in research, unique activities of sesquiterpenoids as signaling molecules in the below-ground environment have recently been discovered [48]. Soft corals found in the Red Sea coastlines contain high levels of natural bioactive compounds. The soft coral *Litophyton arboreum* was chemically probed; it is used for the isolation and structural characterization of litoarbolide A, an unknown sesquiterpenoid, and 14 previously reported metabolites. Litoarbolide A, a biosynthetic precursor of other sesquiterpenoids, is generated by further post-translational modifications. However, a few isolated compounds such as leptocladols B and nephthetetraol have shown limited antimalarial activity against *Plasmodium falciparum* strains that are CQR (W2). Furthermore, only the acyclic sesquiterpenoid of a sec-germacrane nucleus showed antimalarial activity against the CQR (W2) and CQS (D6) strains of *Plasmodium falciparum*, with IC_{50} values of 2.2 and 3.7 mg mL^{-1}, respectively. Moreover, all identified bioactives were nontoxic to the Vero cell line [50]. Recently, lindenane sesquiterpenoid dimers, named chlorajaponilide C, were isolated from *Chloranthecae* spp., which showed antiplasmodial activity against Ugandan *Plasmodium falciparum*, with the half-maximal effective concentration (EC_{50}) being 1.6 nM. The EC_{50} of chlorajaponilide C increased by more than 250-fold in all chosen resistant clones, and whole-genome sequencing indicated changes in the recently identified *Plasmodium falciparum* prodrug activation and resistance esterase. Thus, when assessing *in vitro* efficacy, chlorajaponilide C assessed parasite esterase activity, and these

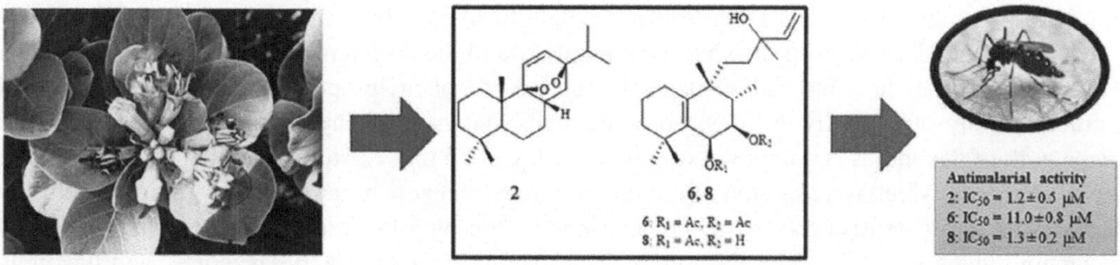

FIGURE 12.2 Graphical representation of the antimalarial activity of *Vitex rotundifolia* against *Plasmodium falciparum*. (From Kim et al., 2020. *Bioorg. Chem.*, vol. 100. May, p. 103925 [56]. With Permission.) Copyright (2020) Elsevier.

compounds served as the basis for the generation of powerful and selective antimalarials [51]. The fruit extract of *Reneilmia cincinnata*, containing high levels of 6,7,10-trihydoxyisodaucane, showed high antimalarial activity against *Plasmodium falciparum* strains [52].

Diterpenoids are a broad and structurally varied family of natural compounds that are formed by joining four C5 isoprene units in a head–tail configuration. They are mostly found in plants and fungi but have also been discovered in marine species and insects. They are produced through biosynthesis from mevalonic acid with 2E, 6E, and 10E-geranylgeranyl pyrophosphate [53]. The chloroform extract of the whole plant of *Andrographis paniculata* (IGNTU_06) demonstrated antimalarial efficacy, with the half-maximal inhibitory concentration being 6.36 μg mL^{-1} and 5.24 μg mL^{-1} against the 3D7 strain and K1 strain, respectively, without any indication of considerable cytotoxic properties against VERO cell line (CC_{50} > 100 μg/mL). *In vitro* findings clearly confirm the traditional usage of *Andrographis paniculata* for malarial therapy. The findings have revealed that the IGNTU 06 extract contains diterpenoids as prospective antimalarial chemicals, which need to be isolated and investigated for further study of the antimalarial properties and for drug development [54]. Traditional Ethiopian medicine utilizes the leaf extract of *Otostegia integrifolia* to cure a variety of ailments including malaria. Ongoing research on efficient, safe, and inexpensive plant-based antimalarial medicines investigated an 80% methanolic leaf extract of *Otostegia integrifolia* in an antimalarial animal model against *Plasmodium berghei* in a 4-day suppression experiment. This extract contained otostegindiol, a labdane diterpenoid that exhibited substantial antiplasmodial action. At dosages of 25, 50, and 100 mg/kg, otostegindiol demonstrated considerable ($p<0.001$) antimalarial activity *in vivo*, with respective chemo-suppression rates of 50.13%, 65.58%, and 73.16%. This study reported that otostegindiol exerts antimalarial activity, which validates the use of *Otostegia integrifolia* leaves for malarial treatment as traditional medicine [55]. *Vitex rotundifolia* is an essential medicinal plant commonly utilized in conventional medicine to treat malaria. Abieta-11(12)-ene-9β,13β-endoperoxide, vitetrifolin D, and vitetrifolin E diterpenoids were isolated and purified with other diterpenoids, and these three compounds were identified as promising antimalarials (Figure 12.2). These compounds showed strong *in vitro* activity against *Plasmodium falciparum*, with the half-maximal inhibitory concentration being 1.2, 1.3, and 11.0 μM for abieta-11(12)-ene-9β,13β-endoperoxide, vitetrifolin D, and vitetrifolin E, respectively. The existence of the β-peroxy bridge in abieta-11(12)-ene-9β,13β-endoperoxide proved to have more powerful antimalarial activity [56].

12.3.3 Flavonoids as Antimalarial Agents

Flavonoids are secondary metabolites including polyphenolic compounds (benzo-γ-pyrone structure) and comprise 10,000 structural compounds that are distributed in almost all plants worldwide [16,

57]. Flavonoids are biosynthesized in plants as metabolic hybrids by combining the phenylpropanoid (→C6-C3) route derived from the shikimate pathway and the acetatemevalonatepolyketide (→C6) pathway. Thus, they have the carbon skeletal structure of a phenylpropanoid (C6-C3 unit) and form a crucial family of naturally synthesized compounds known as phenylpropanoids. The molecular framework of flavonoids is composed of a parent C6-C3-C6 unit existing as a nucleus of chromone (benzo-pyrone), which is a skeleton of flavonoid (phenyl-benzopyrone) [16]. Flavonoids have a 15-carbon structure comprising two rings of benzene linked by a heterocyclic pyrene ring. Flavones (apigenin, luteolin, and flavone), flavanones (naringenin, hesperetin, and flavanone), and flavonols (quercetin, fisetin, myricetin, and kaempferol) are some classes of flavonoids [57]. Flavonoids exert antimalarial activity against malarial parasites, notably *Plasmodium falciparum* [16]. A flavone, named artemetin, has been isolated from *Artemisia gorgonum* Webb. from the Asteraceae family, an endemic shrub in Cape Verdean that was tested for its antimalarial activity. An *in vitro* study revealed that artemetin has antimalarial potential against CQR *Plasmodium falciparum* (FcB1 strain), with the IC_{50} value being 3.50 μg/mL. Artemetin showed no significant cytotoxicity in normal murine cells (colony-forming unit granulocyte-macrophage). The antimalarial efficacy and low cytotoxicity of artemetin suggest an amalgamation of certain metabolites as drug templates to treat malaria [58]. A variety of commonly consumed flavonoids such as the isoflavone genistein, the flavanone naringenin, flavones (baicalein, acacetin, chrysin, luteolin, and apigenin), and flavonols (quercetin, isoquercitrin myricetin, and kaempferol) were examined for their antimalarial activity by treating them against CQR (7G8) and CQS (3D7) strains of *Plasmodium falciparum*. Among the 11 flavonoids, naringenin, genistein, and acacetin did not show any antiplasmodial activity against the 3D7 strain, whereas the other flavanoids showed antiplasmodial activity, with IC_{50} values ranging from 11 to 66 μM. Luteolin was the most active chemical in both strains, with the half-maximal inhibitory concentration rates being 11±1 μM and 12±1 μM for 3D7 and 7G8 strains, respectively [59]. Using conventional and microwave irradiation-based methods, a variety of flavone-stilbene-centered hybrid compounds were synthesized in high-to-outstanding yield (65%–95%), with Heck coupling as the essential step. Using the SYBR Green I assay, these hybrid compounds were examined for their antiplasmodial efficacy against the *Plasmodium falciparum* strains CQS Pf3D7 and CQR PfINDO. The compound hybridized flavone-stilbene scaffold (11 g) showed excellent antimalarial activity, with the half-maximal inhibitory concentration being 6.01 and 5.67 μM for CQS Pf3D7 and CQR PfINDO strains, respectively [60].

A novel flavone-centred tetradentate Schiff base ligand was designed by combining o-phenylenediamine and 2-(4-chlorophenyl)-5-hydroxy-6-nitrochroman-4-one in a 1:2 ratio, as well as its copper (II), nickel (II), and cobalt (II) complexes. These complexes were tested for their antiplasmodial efficacy against W2 and D6 strains. The result showed that the copper complex had the greatest antiplasmodial activity, with the IC_{50} values being 0.11±0.09 and 0.07±0.03 μg mL^{-1} for clones W2 and D6, respectively. β-Hematin test likewise demonstrated highest activity of the copper complexes, with the half-maximal inhibitory concentration being 0.13±0.04 mg/mL. The parasitic lactate dehydrogenase technique was performed to assess antiplasmodial activity, and the findings revealed that complexes containing copper and nickel had high activity against both W2 and D6 strains. By contrast, the complex containing cobalt was relatively active [61]. In another study, *Artemisia annua* cell suspension cultures demonstrated effective *in vitro* antiplasmodial activity against *Plasmodium falciparum*. The therapeutic activity of the chloroform extract of the plant (*Artemisia annua*) was attributed to the methoxy group present in the flavonoids artemetin, chrysosplenol-D, chrysoplenetin, and cirsilineol. An *in vitro* study showed that the pure flavonoids had a half-maximal inhibitory concentration of 2.4–6.5×10^{-5} M against *Plasmodium falciparum*, whereas purified artimisinin had an IC_{50} value of nearly 3×10^{-8} M [62].

The research group working in the same field isolated three novel flavonol arabinosides from the immature leaves of *Calycolpus warszewiczianus*. The isolated compounds were tested for antimalarial activity against *Plasmodium falciparum* under *in vitro* conditions. Among these compounds, 5-galloylquercetin-3-O-α-l-arabinofuranoside showed poor efficacy against a CQR strain of *Plasmodium falciparum*, with the half-maximal inhibitory concentration being 14.5 μM, but at a concentration below 100 μg/mL, no significant cytotoxicity was observed against Vero cells [63]. For screening antimalarial activity, three flavonol glycosides, namely, myricitrin, afzelin, and quercitrin, were isolated using the bioassay-guided fractionation technique of the methanolic extracts of *Euphorbia hirta*, belonging to the family Euphorbiaceae, and their structures were confirmed by nuclear magnetic resonance and mass spectrometry analyses. The three compounds, namely, afzelin, quercitrin, and myricitrin, showed antimalarial activity by inhibiting the proliferation of the CDC1 strain of *Plasmodium falciparum*, with the respective IC_{50} values being 1.1, 4.1, and 5.4 μg/mL. Moreover, these flavonol glycosides have a weak effect on the proliferation of human epidermoid carcinoma KB 3-1 cells, which exhibit cytotoxic potential, with the respective IC_{50} values being 276.1, 88.2, and 156.4 μg/mL [64]. In the study conducted by Ganesh et al., quercetin and its derivatives showed antiplasmodial activity. Quercetin-3β-glucoside, quercetin, and the quercetin analog rutin caused inhibition of both clones at lesser micromolar doses and were the most active analogs at both K1 and 3D7. Quercetin had an IC_{50} value equivalent to that of rutin but showed tenfold greater activity against the 3D7 strain and sevenfold higher activity against the K1 strain when compared with previously reported data. Interestingly, rutin showed an attractive result against *Plasmodium falciparum* parasites, with an IC_{50} value of 7.10 μM [65]. Thus, different plant species produce different substances that act as antimalarial agents. These functions of various plant secondary metabolites such as alkaloids, terpenoids, and flavonoids suggest great potential against the malarial parasites and vectors for malarial treatment.

12.4 NANO-BASED TREATMENT FOR MALARIA

Nanotechnology appears to be a potential solution to overcome these restrictions, and it can help design more powerful and less toxic antimalarial medications in various ways [66]. Because of low solubility, stability, and bioavailability, various plant-based antimalarial agents have failed to attract therapeutic importance. To overcome these disadvantages, rather than spending all efforts on identifying new medications, the production of efficient delivery carriers is a vital option. Nanomaterials have been studied for their ability to transport antimalarial medications at sufficiently high local dosages to kill parasites and minimize drug resistance while keeping a low total dosage to avoid unwanted toxicity and side effects [67].

12.4.1 Phytonanoparticles for Malarial Treatment

The overall purpose of novel therapies for malarial treatment can be divided into two types. First, new medications that are effective against the parasite's evolving resistance forms must be made accessible [68]. Second, vector control is the primary preventative measure for limiting parasite spread [69]. Researchers have used various types of plant-based NPs to treat malaria, which are discussed in this section (Figure 12.3) [7].

For parasite control, Ojemaye et al. developed potential silver NPs (AgNPs) from the leaf and fruit extracts of *Crataegus ambigua*, which showed antagonistic activity against free radicals, bacterial strains, and malarial parasites. Spectrophotometric methods, micro-dilution, and parasite viability were used to assess their antioxidant, antibacterial, and antimalarial activities, respectively. The NPs generated from the fruit and leaf components had outstanding antimalarial activities, showing 100% inhibition at a dose of 20 μg mL^{-1}, whereas the extract of the whole plant showed 48.5% inhibition [70]. Similarly, AgNPs were prepared from the aqueous plant extract of *Launaea sarmentosa*, and the phytoconstituents indicated the presence of terpenoids, alkaloids, phenolics, and tannins. Highly stable spherical AgNPs with a particle size of approximately 30 nm, on average, were successfully synthesized. Strong antiplasmodial activity with an IC_{50} value of 17.21±1.18 μg mL^{-1} and

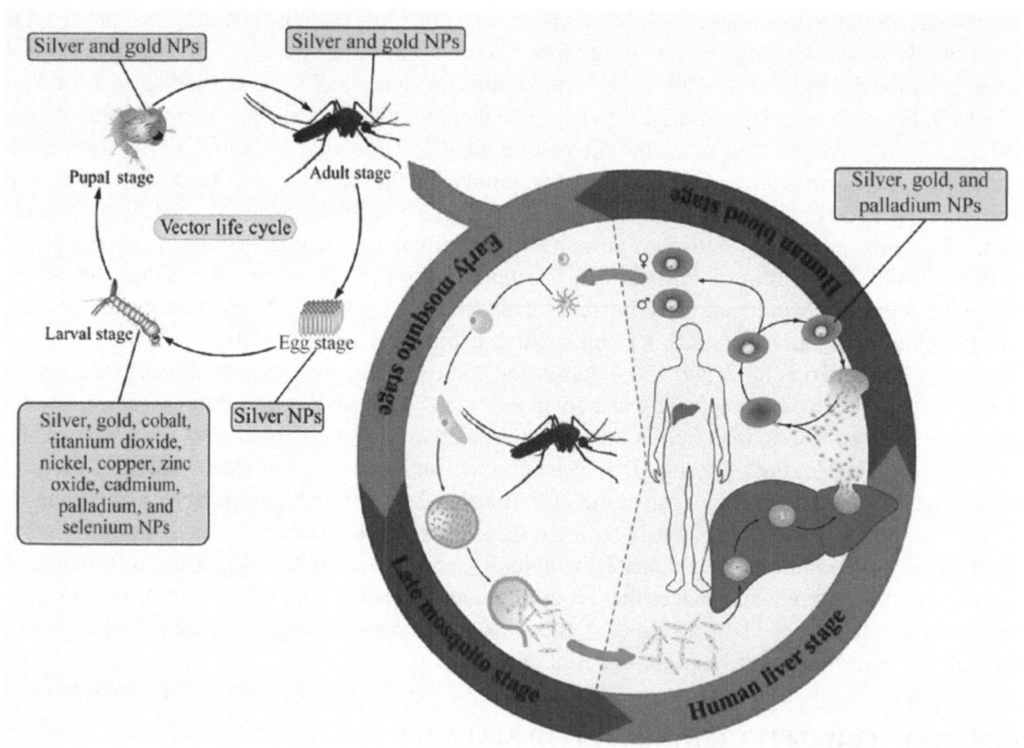

FIGURE 12.3 Detailed scheme of the nanoparticles used against malaria (*Plasmodium* parasite and mosquito vector). (From Barabadi et al., 2019. *Nanomedicine Nanotechnology Biol. Med.*, vol. 18, pp. 221–233 [7]. With Permission.) Copyright (2019) Elsevier.

100% growth inhibition at a dose of 312.5 μg mL^{-1} was demonstrated by AgNPs generated from the *Launaea sarmentosa* extract against the *Plasmodium falciparum* K1 strain. Using the suggested technology, silver NPs are prepared at a low cost, sustainable, nontoxic, quick, and reliable manner, which might prove as an important alternative treatment against malaria [71]. In the same vein, AgNPs were synthesized from the aqueous leaf extract of *Salvia officinalis*, and the transmission electron microscopy (TEM) analysis revealed that the AgNPs were spherical and 41 nm in size. The X-ray diffraction data confirmed the crystalline nature of AgNPs, and the Fourier transform infrared method identified certain essential phytochemical components that acted as capping and stabilizing agents. The antiplasmodial activities were seen in both the aqueous extract of *Salvia officinalis* and AgNPs. High parasite viability was seen in cells treated with 50 μg/mL of *Salvia officinalis* aqueous extract, indicating limited antiplasmodial activity as compared with that in AgNPs, which entirely killed the parasites at this dose. The synthesized AgNPs were less toxic to the HeLa cell line and had strong anti-plasmodial potential (IC_{50}: 3.6 μg/mL) [72].

For vector control, fern-mediated production of AgNPs from *Pteridium aquilinum* plant extract as capping and reducing agent was reported. Liquid chromatography–mass spectrometry analysis discovered 19 compounds from *Pteridium aquilinum* plant extract. These extracted plant metabolites showed strong antioxidant activity and have been recommended for single-step, efficient production of AgNPs with mosquitocidal properties. Biosynthesized AgNPs are particularly toxic to the juvenile instars of *Anopheles stephensi* malaria vector and reduce mosquito adult fertility and lifespan. AgNPs produced from *Pteridium aquilinum* were extremely poisonous to *Anopheles stephensi* larval and pupa, with LC_{50} values of 7.48, 10.68, 13.77, 18.45, and 31.51 ppm for larval stages I, II,

III, IV, and pupae, respectively. Both the extract from fern and AgNPs caused inhibition of the development of *Plasmodium falciparum* in CQR and CQS strains, with IC_{50} values being lower than those of CQ [73]. In another study, AgNPs were green synthesized from the flower of *Chrysanthemum indicum* and examined for activities against the malarial vector *Anopheles stephensi* mosquito larvae and pupae. For 24 hours, larvae and pupae were treated with varying concentrations of the aqueous extract of *Chrysanthemum indicum* and AgNPs synthesized from the plant, and the AgNPs caused the greatest mortality against the vector *Anopheles stephensi*, with LC_{50} values of 5.07, 10.35, 14.19, 22.81, and 35.05 ppm and LC_{90} values of 29.18, 47.15, 65.53, 87.96, and 115.05 ppm, respectively, for larval stages I, II, III, IV, and pupae. This larva- and pupa-killing effect of the produced AgNPs might be attributed to their relatively small size, which allows them to cross the membrane and attach to sulfur-carrying proteins or phosphorus-carrying elements such as DNA, causing cellular malfunction and, eventually, cell death [74]. Furthermore, AgNPs were synthesized from the extracts of *Holarrhena antidysenterica* bark by combining 0.090 L of $AgNO_3$ with 10 mL of aqueous extracts of *Holarrhena antidysenterica* bark, which had excellent activity against *Anopheles stephensi* larvae. AgNPs and extracts from bark prepared in various solvents such as water, methanol, chloroform, hexane, and ethyl acetate were tested for 24 hours against the *Anopheles stephensi* larvae. Highest mortality of larval stages was observed with AgNPs synthesized from the bark extract, which had LC_{90} and LC_{50} values of 4.482 ppm and 2.672 ppm, respectively, whereas acetone, water, methanol, ethyl acetate, hexane, and chloroform bark extracts had LC_{50} values of 1.91E3, 121.53, 96.40, 41.92, 31.56, and 3.0 ppm, respectively [75]. In one study, AgNPs synthesized from *Aloe vera* leaf extract had mosquitocidal and antibacterial characteristics. The mosquitocidal abilities of *Anopheles stephensi* larvae (stages I–IV instar) and pupae were tested with green-synthesized AgNPs. Even at modest doses, the *Aloe vera* extract proved harmful to *Anopheles stephensi* larvae and pupae in the laboratory. The LC_{50} values were 48.79, 59.09, 70.88, 83.58, and 152.55 ppm for stage I, II, III, IV, and V instar larvae, respectively. Moreover, the application of freshly produced AgNPs ($10 \times LC_{50}$) resulted in a 74.5%, 86.6%, and 97.7% decrease in *Anopheles stephensi* larvae after 24, 48, and 72 hours, respectively [76]. AgNPs synthesized from the leaf extract of *Belosynapsis kewensis* showed *in vitro* larvicidal ability against *Aedes aegypti* and *Anopheles stephensi*. TEM analysis showed that AgNPs synthesized from *Belosynapsis kewensis* had a mean size of 24±1.6 nm. These AgNPs were investigated for larvicidal effectiveness against stage IV instars *of Aedes aegypti and Anopheles stephensi*. An *in vitro* study confirmed that the AgNPs had the highest LC_{50} and LC_{90} values (78.4 and 144.7 ppm, respectively) against *Anopheles stephensi*, followed by *Aedes aegypti* (LC_{50}: 84.2 and LC_{90}: 117.3 ppm) [77]. Thus, several plant-based NPs have shown activity against malarial parasites and vectors, which confirms the outstanding potency of phyto nanomaterials for malarial treatment.

12.4.2 Nanoparticles for the Delivery of Antimalarial Phytochemicals

Phytochemicals have been encapsulated or conjugated with NCs for their transport to specific target sites. Phytochemicals are secondary plant metabolites that are beneficial to human health for the treatment of diseases. However, these phytochemicals are huge in size and have polarity, which makes them difficult to penetrate the blood–brain barrier, gastrointestinal system, endothelial lining of blood vessels, and mucosa. Furthermore, they are destroyed enzymatically in the gastrointestinal system. Thus, encapsulating or conjugating these chemicals with NCs may be another strategy to improve their bio-efficiency by improving their gastrointestinal stability, rate of absorption, and dispersion ability [78]. Various biological components such as proteins, DNA, drug molecules, phytochemicals, and other nanoscale entities can be delivered to the target site through NCs [66]. Novel structures with superior thermal, optical, electrical, or mechanical abilities of nano-structured components can be created by atomic-level manipulations in organic and inorganic materials. Chemical procedures can be adapted to a specific use, such as enhancing the biocompatibility of implanted materials. Nanomaterials can have stronger reactivity, improved mechanical strength, or electrical or magnetic characteristics owing to a larger surface area-to-volume ratio [79]. These features of nanomaterials can increase the effectiveness of the phytochemicals and cause enhanced functional ability.

Curcumin (CUR) exerts antiplasmodial effect; nevertheless, its low bioavailability, quick metabolic activity, and poor chemical stability cause hindrance to its therapeutic potential. Cetyl palmitate and coconut oil are used as lipids in respective solid and liquid forms to produce nanostructured lipid carriers (NLCs) through which CUR can be delivered into the cells. For the *in vivo* investigation of antiplasmodial ability of CUR-loaded NLCs (CUR-NLCs), *Plasmodium berghei*-infected mice were used. In this study, pseudo-spherical CUR-NLCs were prepared with a size of 145 nm and a zeta potential of −25 mV. CUR-NLCs had a loading ability of 3.1±0.015% and an encapsulation efficiency of 74±3.32%. CUR release profiles *in vitro* revealed a gradual release pattern lasting for approximately 5 days from the synthesized CUR-NLCs. In the animal experiment, antiplasmodial activity of CUR-NLCs against *Plasmodium berghei* was demonstrated, wherein considerably stronger antimalarial activity of CUR-NLCs than that of free CUR at a dosage of 40 mg/kg per day was observed [80]. In another study, delivery of CUR using a NC was investigated to examine antimalarial activity. For this, nanocapsules were developed using various polymers such as polysorbate 80 (P80), polyethylene glycol (PEG), chitosan, and Eudragit RS100® for CUR delivery. *In vitro* release experiments revealed that nanoencapsulation was an effective technique for controlling CUR release, with P80-coated nanocapsules exhibiting the slowest CUR release among all nanoformulations examined. The antiplasmodial activity of CUR-loaded nanocapsules was determined by treating *Plasmodium falciparum* CQR (W2) and QN strains with the formulation. Furthermore, regardless of coating (anionic or cationic), all CUR-loaded nanocapsules demonstrated detectable antimalarial activity, with more significant outcomes for chitosan-coated nanocapsules and an IC_{50} value of 0.17±0.03 μM [81]. In one study, lipid-based drug delivery systems (LBDDSs) were used to transport CUR (Figure 12.4). In a biorelevant gastric medium, CUR-LBDDSs produced particles with diameters ranging from 30 to 40 nm. Throughout *in vitro* experiments of lipid NC lysis, 90%–95% of the CUR remained solubilized, although 5%–10% of the CUR precipitated in the form of amorphous solid, which redissolved rapidly in a biorelevant intestinal medium. Because of the improved CUR solubility, the movement of CUR-LBDDS across monolayers of the Caco-2 cell line was increased when compared with that of free drug. Orally administration of CUR-LBDDSs to mice infected with *Plasmodium berghei* demonstrated mild antimalarial activity. However, combining CUR-LBDDS with a sub-therapeutic dosage of β-arteether-LBDDS offered an upsurge in survival and protection rates, as well as a considerable delay in recurrence. These data imply that

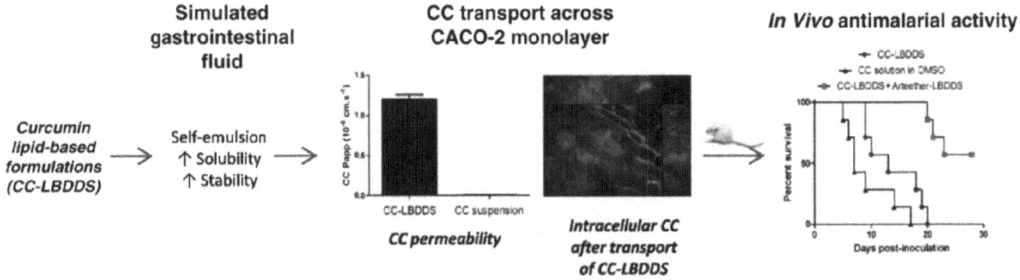

FIGURE 12.4 Graphical representation demonstrating the stability and solubility of Cur–lipid-based drug delivery systems (LBDDS) in simulated gastrointestinal fluid, transport of Cur-LBDDS across monolayers of Caco-2 cell line, and *in vivo* activity in mice. (From Memvanga et al., 2013. *J. Control. Release*, vol. 172, no. 3, pp. 904–913 [82]. With Permission.) Copyright (2013) Elsevier.

combining oral CUR with β-arteether lipid-mediated preparations may be a suitable method for malarial therapy [82].

In another study, curcuminoid-encapsulated lipid NPs (LNPs) were synthesized using a nanoemulsion approach. Tristearin, trimyristin, and glyceryl monostearate were used as solid lipids and medium-chain triglyceride as liquid lipids for NP synthesis. The prepared NPs were spherical and had size varying from 120 to 250 nm and zeta potential ranging from −28 and −45 mV. These curcuminoid-loaded LNPs have showed an entrapment efficiency ranging from 80% to 94% and drug-loading capacity (LC) ranging from 1.62% to 3.27%. The *in vivo* pharmacodynamic activity showed that the curcuminoid-encapsulated LNPs exhibited a twofold enhanced antimalarial activity when compared with free curcuminoids at the verified dose range. Furthermore, new preparations may be taken parenterally, avoiding the gastrointestinal path, where the majority of molecules are degraded; this is the primary reason for the limited bioavailability of free curcuminoids. This leads to greater levels of drug use in the blood, spleen, and liver, wherein parasites are found. Additionally, LNPs may boost pharmacological concentrations at the affected site, which promotes the treatment of cerebral malaria [83].

Owing to some disadvantages of curcuminoids, such as water insolubility, liposomes have been used for the transport of curcuminoids in a malaria mouse model. Thin-film hydration has been used to create curcuminoid-loaded liposomes made of phosphatidylcholine. Liposomes have shown an encapsulation efficiency of 83.63%–94.36%, with the highest entrapment at a drug-to-lipid ratio of 1:10. An intravenous injection of sole curcuminoid-encapsulated liposomes and conjugated with α/β arteether was administered to *Plasmodium berghei*-infected mice for determining the antimalarial efficiency of curcuminoid-loaded liposomes. Compared with the control group, the curcuminoid-encapsulated liposome treatment group had lower parasitemia and greater survival in animals. Collaborated treatment with curcuminoid-encapsulated liposomes (40 mg/kg body weight) and α/β arteether (30 mg/kg body weight) could alleviate and prevent disease recurrence in infected mice [84]. In another study, CUR was delivered using hydrogel NPs fabricated with polyvinyl pyrrolidone and hydroxyl propyl methyl cellulose for better intake and sustained the elimination of CUR for possible reticuloendothelial system evasion. Through solvent emulsion-evaporation, hydrogel NPs were generated, which had a size of 98.33±0.665 nm and an encapsulation efficiency of 72.34±0.34%. Animal experiments conducted in *Plasmodium berghei*-infected murine model demonstrated that NPs at a dosage comparable to CUR 25 mg/kg body weight resulted in a substantial ($p<0.05$) enhancement in percent decrease of parasitemia. Moreover, a higher percentage of activity than that of free CUR was noted. Thus, the formulation may be used as an adjuvant therapy for malarial treatment in addition to conventional therapy. Oral and cellular safety of the synthesized NPs was demonstrated by several toxicity experiments. These findings may be useful for the continued use of the formulation as an additional antiplasmodial treatment to avoid recurrence and/or lower the dose of conventional malarial medications [85]. In another study, chitosan NPs were established for the transport of CUR for malarial treatment. *In vitro* experiments confirmed that CUR can suppress parasite-generated β-hematin production in a dosage-dependent manner, with a lesser IC_{50} value of 122±2.7 μM than that of CQ, which had an IC_{50} value of 198±3.7 μM. Thus, this result indicated that CUR can suppress parasite lysate-generated heme polymerization *in vitro* in a dosage-dependent manner like CQ, with a lesser IC_{50} rate than that of CQ. When the same formulation was injected into mice, mouse RBC absorption was significantly greater with CUR-attached chitosan NPs than with the free form of CUR. CUR-attached chitosan NPs delivered orally to regular mice demonstrated penetration of the mucosal barrier, and confocal imaging revealed the presence of NPs in the blood. CUR-attached chitosan NPs increased CUR bioavailability in plasma and RBC when administered orally. Mice treated with a deadly *Plasmodium yoelii* (N-67) strain deceased within 8 and 9 days after infection. Chitosan NPs alone allowed them to live for an additional 5 days. Diseased mice that fed 1 mg of native CUR each day for 7 days caused one-third of the mouse population in the study to survive. However, when 1 mg of CUR conjugated with chitosan NP was administered, all animals survived [86].

In another study, poly (lactic-co-glycolic) acid (PLGA) NPs were used for the transport of CUR for malarial treatment, and CUR-encapsulated PLGA NPs were formed through solvent evaporation. CUR-encapsulated PLGA NPs (CUR-PLGA) toxicity in RAW 264.7 cell line increased at higher doses (1000 and 500 µg/mL) in contrast to the free drug. At lower doses, cell viability was greater with CUR-PLGA NPs than with free CUR ($p>0.05$). Concentrations of 5 and 10 mg/kg were used in *in vivo* antimalarial tests and toxicity assessments for free and loaded medication. After treatment, on the fourth day, the percentage of parasite inhibition at 5 mg/kg of CUR-PLGA NPs was substantially greater by 56.8% than that with an equal concentration of free CUR (by 40.5%) ($p<0.05$). [87]. Solid LNPs (SLNs) were developed for the transport of luteolin, isolated from *Allium fistulosum* for the treatment of malaria or severe malaria anemia (SMA). SMA occurs due to the phagocytosis of PfHz (*Plasmodium falciparum*-derived hemozoin) by neutrophils, macrophages, and monocytes, which finally cause a reduction in cyclooxygenase 2-prostaglandin E2 (COX-2-PGE2). PfHz is a crystalline substance generated through heme aggregation in the host. Because Fe^{2+} is a component of heme, luteolin may bind to Fe^{2+} such that the association between luteolin and Fe^{2+} in heme inhibits the formation of PfHz, thus preventing acute anemia. Luteolin was loaded using SLN and PEG because of poor bioavailability. SLN was effective in enhancing luteolin bioavailability in the body, while PEG was useful in preventing the reticuloendothelial system from destroying luteolin-SLN. The study showed that PEG-modified SLN-loaded luteolin could be used as a new treatment for acute anemia in children with *Plasmodium falciparum*-associated malaria [88].

Researchers are putting in efforts to identify a novel chemical with high antimalarial properties. The development of such a chemical entails significant obstacles, such as decreasing treatment duration, enhancing compliance, and primarily blocking the transmission of the parasite back into the insect vector. Most pharmacological compounds produced from natural sources have a slow therapeutic potential. Thus, there is a need to create a new molecule that is more effective and provides quicker therapy that is also cost-effective. Each year, the Food and Drug Administration approves novel potential therapeutic substances, yet malaria incidence continues to be a serious concern. The complexity of the life cycle, as well as dispersed sites of malarial parasites, is a significant obstacle for scientists in developing a novel chemical. Nanotechnology offers various approaches that are advantageous for delivering a substance in an incredibly feasible manner. NCs can be used to upgrade target significance characteristics and shield the drug or immunizing agent from extracellular degradation, reduce the frequency of doses, and change the pharmacological property of the drug and biopharmaceutics, potentially overpowering side effects and thus enhancing medication efficiency. The primary goal of NP-mediated therapy for malaria is to transport the delivered medicine to the intracellular parasite vacuoles and to extend the retention time of the drug in the blood. It also prolongs the residence duration of medicine in circulation, facilitating contact with afflicted RBCs and the parasite's membrane. This situation promotes the sensible and successful use of current agents with the support of some innovative medication delivery methods [9].

The World Health Organization currently recommends ART derivatives as the first-line therapy for malaria in combination with additional medications such as lumefantrine, amodiaquine, MQ, and sulfadoxinepyrimethamine. *Artemisia annua*-generated ART, namely, sesquiterpene lactone, consists of a ring structure containing 1,2,4-tiroxane. ART exerts antimalarial, antiamyloid, anti-inflammatory, antibacterial, and antioxidant properties, but its use is hindered owing to its poor oral bioavailability and low water solubility. Researchers used a one-step nano-precipitation approach to encapsulate ART into micelles, resulting in the creation of ART-loaded poly(ε-caprolactone)-poly(ethylene glycol)-poly (ε-caprolactone) (PCL-PEG-PCL) micelles with a zeta potential of −8.37 mV and an average particle size of 91.87 nm. The antimalarial efficacy of ART-encapsulated micelles was tested in *Plasmodium berghei*-infected Swiss albino mice. Injection of ART-encapsulated micelles extended circulation time and boosted ART efficacy. These findings indicate that PCL-PEG-PCL micelles might be used as a vehicle for ART in malarial treatment [89]. ARTSN is a derivative of ART, which is derived from the herb *Artemisia annua* [90]. This ARTSN was encapsulated in PLGA (ARTSN-PLGA) through the solvent evaporation process. The formulated medication had an average particle size of 329.3±21.7 nm and an encapsulation efficiency of 38.4±10.1%. The cellular toxicity of free ARTSN and ARTSN-PLGA to the RAW 264.7 cell line was measured *in vitro* at doses ranging from 7.8 to 1000 µg/mL. ARTSN-PLGA NPs suppressed parasites more effectively, showing 62.6% suppression when compared with that of free ARTSN, which had a 58.2% suppression ability. The *in vitro* cellular toxicity study revealed that the IC_{50} of ARTSN-PLGA was substantially greater than that of free ARTSN. In mice, platelet counts in controls were considerably higher than those in mice injected with free ARTSN or ARTSN-PLGA. There was no evidence of liver toxicity after using the formulations. Thus, the data confirmed that treatment with ARTSN-encapsulated PLGA NPs had minimal toxicity and improved antiplasmodial activity [91].

Because of effective antimalarial ability, ARTSN is commonly recommended for multidrug-resistant malarial treatment. In another study, iron oxide NPs were prepared for the delivery of ARTSN in a *Plasmodium falciparum*-infected malaria mouse model. This iron oxide NP-loaded ARTSN can increase the specificity and toxic effect of ARTSN against the parasite. Iron oxide NP-loaded ARTSN was tested for its antimalarial activity against wild-type and ARTSN-resistant *Plasmodium falciparum* (R539T)-cultured human blood with O^+ blood group and swiss albino mice infected with *Plasmodium berghei*. The NP-enriched ARTSN was very effective at the one-eighth concentration of ARTSN IC_{50} and resulted in *Plasmodium falciparum* growth retardation, with considerable damage to cellular macromolecules facilitated by increased reactive oxygen species generation. Likewise, preclinical animal investigations revealed a significant decrease in parasitemia with an eightfold to tenfold reduction in the ARTSN dose when supplemented with iron oxide NPs. Surface-coated iron-oxide NPs loaded with ARTSN have the potential to be a powerful therapeutic mediator against multidrug-resistant and ARTSN-resistant malaria (Figure 12.5) [92].

Another ART derivative, named artemether, was loaded in zein NPs for malarial treatment. Zein is a plant protein obtained from corn. Artemether-encapsulated zein NPs were prepared using sodium caseinate as a stabilizer. Surprisingly, encapsulating artemether in zein NPs significantly reduced hemolysis, a frequent clinical occurrence following ART-based antimalarial medication. The drug release experiment showed that the free artemether disintegrated slowly in the first 2 hours, most likely because of poor water solubility. Although most of the free artemether was released within 5 hours, only 54±2% of the artemether discharged from the NPs within the same period. Finally, intravenous administration of artemether in the form of artemether-encapsulated zein NPs in animal models appears to be a potential treatment alternative for acute malaria [93]. In one study, MCM-41 mesoporous silica NPs were used for the delivery of QN. QN-loaded MCM-41 (QN-MCM-41) was treated against the *Plasmodium falciparium* W2 strain for *in vitro* study, whereas for the *in vivo* study, *Plasmodium berghei* NK65 strain was chosen. At dosages greater than 0.0625 mg/kg body weight, QN-MCM-41 had the strongest antiplasmodial potential in the animal model against *Plasmodium berghei* NK65 (effective dose 50 of 0.0625 mg/kg body weight) and the longest average survival time in contrast to others or unencapsulated therapeutics [94]. Various examples of NPs used for the delivery of plant-based drugs confirm nanotechnology's effectiveness in treating malaria.

12.5 CONCLUSION

Plant bioactives such as alkaloids, terpenes, flavonoids, and other phytochemicals isolated from diverse plants (*Rhaphidophora decursiva, Aspidosperma olivaceum, Flindersia pimenteliana, Triphyophyllum peltatum, Ancistrocladus tanzaniensis, Melicope madagascariensis*, etc.) were proven effective against diverse strains of *Plasmodium*. These bioactives can minimize malarial infection with minimal or no side effects. Plant extracts produced from various sections of plants

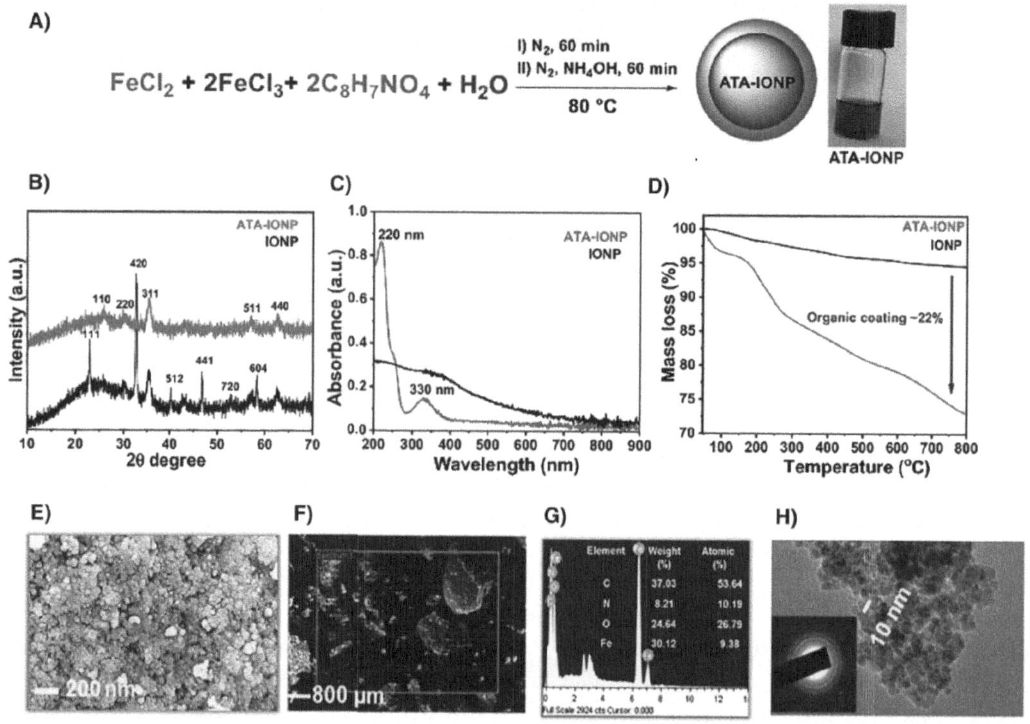

FIGURE 12.5 A) Synthetic scheme of 2-amino-terephthalic acid–coated iron oxide nanoparticles (ATA-IONPs), with digital images of their aqueous suspension (2.5 mg/mL); B) powder X-ray dispersion spectra, C) ultraviolet–visible spectra; D) Thermogravimetric analysis of coated ATA-IONPs and uncoated IONPs (control sample); E) Representative field-emission scanning electron microscope of ATA-IONPs, F–G) Energy-dispersive spectroscopy mapping of the respective region of ATA-IONPs and percentage atomic composition confirming the presence of organic ATA coating on IONPs in the selected area zone; and H) Transmission electron microscopy image showing the spherical shape morphology and the inset showing the selected area electron diffraction pattern exhibiting the crystalline domains. (From Kannan et al., 2019. *eBioMedicine*, vol. 45, pp. 261–277 [92]. With Permission.) Copyright (2019) Elsevier.

were screened for several antimalarial drugs with substantial efficacy against *Plasmodium* species. The effectiveness of plant bioactives has been found to be enhanced when loaded on nano-vehicles such as PLGA NPs, mesoporous silica NPs, LNPs, zein NPs, etc. These nano-based formulations have the ability to carry a high payload and release the bioactive in a sustainable manner. Other than these, AgNPs synthesized from plant extracts have proven effective against *Plasmodium* species. Thus plant-based nanomaterials or plant material encapsulated in NCs has shown great promise for malarial treatment.

REFERENCES

[1] D. S. Guttery, M. Zeeshan, D. J. P. Ferguson, A. A. Holder, and R. Tewari, *Annu. Rev. Microbiol.*, vol. 76, no. 1, 2022, doi: 10.1146/annurev-micro-041320-010046.

[2] S. Alven and B. Aderibigbe, *Molecules*, vol. 24, no. 19, 2019, doi: 10.3390/molecules24193601.

[3] N. P. Aditya, P. G. Vathsala, V. Vieira, R. S. R. Murthy, and E. B. Souto, *Adv. Colloid Interface Sci.*, vol. 201–202, pp. 1–17, 2013, doi: 10.1016/j.cis.2013.10.014.

[4] P. J. Guerin et al., *Lancet Infect. Dis.*, vol. 2, no. 9, pp. 564–573, 2002, doi: 10.1016/S1473-3099(02)00372-9.

[5] J. D. Mutoni, J. P. Coutelier, N. Rujeni, L. Mutesa, and P. D. Cani, *Microorganisms*, vol. 10, no. 4, pp. 1–10, 2022, doi: 10.3390/microorganisms10040721.
[6] A. Kantele and T. S. Jokiranta, *Clin. Infect. Dis.*, vol. 52, no. 11, pp. 1356–1362, 2011, doi: 10.1093/cid/cir180.
[7] H. Barabadi et al., *Nanomedicine Nanotechnology Biol. Med.*, vol. 18, pp. 221–233, 2019, doi: 10.1016/j.nano.2019.02.017.
[8] R. Tuteja, *FEBS J.*, vol. 274, no. 18, pp. 4670–4679, 2007, doi: 10.1111/j.1742-4658.2007.05997.x.
[9] N. Puttappa, R. S. Kumar, G. Kuppusamy, and A. Radhakrishnan, *Acta Trop.*, vol. 195, pp. 103–114, 2019, doi: 10.1016/j.actatropica.2019.04.020.
[10] L. Neves Borgheti-Cardoso et al., *J. Mater. Chem. B*, vol. 8, no. 41, pp. 9428–9448, 2020, doi: 10.1039/d0tb01398f.
[11] G. Z. Laporta et al., *Parasites and Vectors*, vol. 8, no. 1, pp. 1–13, 2015, doi: 10.1186/s13071-015-1038-4.
[12] I. C. Ezenyi and O. A. Salawu, *Curr. Top. Malar.*, 2016, doi: 10.5772/65658.
[13] P. F. Uzor, *Evidence-based Complement. Altern. Med.*, vol. 2020, 2020, doi: 10.1155/2020/8749083.
[14] J. Graßmann, *Vitam. Horm.*, vol. 72, no. 05, pp. 505–535, 2005, doi: 10.1016/S0083-6729(05)72015-X.
[15] A. Samanta, G. Das, and S. K. Das. *Int J Pharm Sci Tech*, vol. 6, no. 1, pp. 12–35, 2011.
[16] M. Rudrapal and D. Chetia, *Syst. Rev. Pharm.*, vol. 8, no. 1, pp. 13–18, 2016, doi: 10.5530/srp.2017.1.4.
[17] J. Xie, Z. Yang, C. Zhou, J. Zhu, R. J. Lee, and L. Teng, *Biotechnol. Adv.*, vol. 34, no. 4, pp. 343–353, 2016, doi: 10.1016/j.biotechadv.2016.04.002.
[18] M. Gogoi, *Biomed. Res. J.*, vol. 4, no. 2, p. 151, 2017, doi: 10.4103/2349-3666.240598.
[19] B. Yir-Erong, M. T. Bayor, I. Ayensu, S. Y. Gbedema, and J. Boateng, *J. Pharm.*, vol. 2018, pp. 1–12, 2018, doi: 10.1155/2018/1784645.
[20] M. Mosquera-Romero, L. Zuluaga-Idárraga, and A. Tobón-Castaño, *Malar. J.*, vol. 17, no. 1, pp. 1–9, 2018, doi: 10.1186/s12936-018-2591-z.
[21] S. Sinha, B. Medhi, and R. Sehgal, *Parasite*, vol. 21, 2014, doi: 10.1051/parasite/2014059.
[22] G. Kokwaro, *Malar. J.*, vol. 8, no. SUPPL. 1, pp. 1–6, 2009, doi: 10.1186/1475-2875-8-S1-S2.
[23] R. Kaur and H. Kaur, *J. Med. Plants Stud.*, vol. 5, no. 1, pp. 346–363, 2017 [Online]. Available: www.plantsjournal.com/archives/2017/vol5issue1/PartE/5-1-7-994.pdf.
[24] F. Mojab, *Avicenna J. Phytomed.*, vol. 2, no. 2, pp. 52–62, 2012, [Online]. Available: www.ncbi.nlm.nih.gov/pubmed/25050231%0Ahttp://www.pubmedcentral.nih.gov/articlerender.fcgi?artid=PMC4075661.
[25] A. Onguéné et al., *Malar. J.*, vol. 12, p. 449, 2013 [Online]. Available: www.malariajournal.com/content/12/1/449.
[26] J. Y. Li et al., *Arch. Pharm. (Weinheim).*, vol. 353, no. 11, 2020, doi: 10.1002/ardp.202000131.
[27] H. Zhang, S. Qiu, P. Tamez, G. T. Tan, Z. Aydogmus, and N. Van Hung, vol. 40, no. 03, pp. 221–224, 2002.
[28] T. P. Chierrito et al., *Malar. J.*, vol. 13, no. 1, pp. 1–10, 2014, doi: 10.1186/1475-2875-13-142.
[29] K. Kaur, M. Jain, T. Kaur, and R. Jain, *Bioorganic Med. Chem.*, vol. 17, no. 9, pp. 3229–3256, 2009, doi: 10.1016/j.bmc.2009.02.050.
[30] A. R. Carroll, S. J. Wild, S. Duffy, and V. M. Avery, *Tetrahedron Lett.*, vol. 53, no. 23, pp. 2873–2875, 2012, doi: 10.1016/j.tetlet.2012.03.126.
[31] G. Bringmann, K. Messer, B. Schwöbel, R. Brun, and L. Aké Assi, *Phytochemistry*, vol. 62, no. 3, pp. 345–349, 2003, doi: 10.1016/S0031-9422(02)00547-2.
[32] G. Bringmann, W. Saeb, M. Wohlfarth, K. Messer, and R. Brun, *Tetrahedron*, vol. 56, no. 32, pp. 5871–5875, 2000, doi: 10.1016/S0040-4020(00)00538-X.
[33] G. Bringmann et al., *Phytochemistry*, vol. 52, no. 2, pp. 321–332, 1999, doi: 10.1016/S0031-9422(99)00130-2.
[34] G. Bringmann et al., vol. 67, no. 5, pp. 5–10, 2004.
[35] X. F. Shang et al., *Med. Res. Rev.*, vol. 38, no. 5, pp. 1614–1660, 2018, doi: 10.1002/med.21492.
[36] V. E. Rasamison et al., *Nat. Products Bioprospect.*, vol. 6, no. 5, pp. 261–265, 2016, doi: 10.1007/s13659-016-0106-6.
[37] L. E. Cuca Suarez, M. E. Pattarroyo, J. M. Lozano, and F. Delle Monache, *Nat. Prod. Res.*, vol. 23, no. 4, pp. 370–374, 2009, doi: 10.1080/14786410802228439.

[38] X. Yang et al., *Planta Med.*, vol. 77, no. 14, pp. 1644–1647, 2011, doi: 10.1055/s-0030-1270963.
[39] K. K. Jean Duplex Wansi, H. Hussain, A. Tadjong Tcho, S. F. Kouam, S. Specht, S. Ramadan Sarite, and A. Hoerauf, *Phyther. Res.*, vol. 22, no. 4, pp. 544–549, 2008, doi: https://doi.org/10.1002/ptr.2894.
[40] J. Gershenzon and N. Dudareva, *Nat. Chem. Biol.*, vol. 3, no. 7, pp. 408–414, 2007, doi: 10.1038/nchembio.2007.5.
[41] E. Gonzalez-Burgos and M. P. Gomez-Serranillos, *Curr. Med. Chem.*, vol. 19, no. 31, pp. 5319–5341, 2012, doi: 10.2174/092986712803833335.
[42] I. A. Parshikov, A. I. Netrusov, and J. B. Sutherland, *Biotechnol. Adv.*, vol. 30, no. 6, pp. 1516–1523, 2012, doi: 10.1016/j.biotechadv.2012.03.010.
[43] S. S. Mahmoud and R. B. Croteau, *Trends Plant Sci.*, vol. 7, no. 8, pp. 366–373, 2002, doi: 10.1016/S1360-1385(02)02303-8.
[44] E. C. dos Santos et al., *PLoS One*, vol. 17, no. 5, p. e0268347, 2022, doi: 10.1371/journal.pone.0268347.
[45] M. Mojarrab, M. Rezaei, S. S. Miraghaee, G. Abdi, and Y. Shokoohinia, *J. Reports Pharm. Sci.*, vol. 3, no. 1, pp. 36–41, 2014.
[46] N. S. Lam, X. Long, X. Zhuan Su, and F. Lu, *Biomed. Pharmacother.*, vol. 130, p. 110624, 2020, doi: 10.1016/j.biopha.2020.110624.
[47] A. Ludwiczuk, K. Skalicka-Woźniak, and M. I. Georgiev, *Terpenoids*. 2017.
[48] F. Yu and R. Utsumi, *Cell. Mol. Life Sci.*, vol. 66, no. 18, pp. 3043–3052, 2009, doi: 10.1007/s00018-009-0066-7.
[49] M. Chadwick, H. Trewin, F. Gawthrop, and C. Wagstaff, *Int. J. Mol. Sci.*, vol. 14, no. 6, pp. 12780–12805, 2013, doi: 10.3390/ijms140612780.
[50] M. M. A. Ahmed et al., *Nat. Prod. Res.*, vol. 0, no. 0, pp. 1–9, 2022, doi: 10.1080/14786419.2022.2071268.
[51] J. H. Butler et al., *ACS Infect. Dis.*, vol. 6, no. 11, pp. 2994–3003, 2020, doi: 10.1021/acsinfecdis.0c00487.
[52] M. H. K. Tchuendem et al., *Phytochemistry*, vol. 52, no. 6, pp. 1095–1099, 1999, doi: 10.1016/S0031-9422(99)00344-1.
[53] R. C. S. Veneziani, S. R. Ambrósio, C. H. G. Martins, D. C. Lemes, and L. C. Oliveira, *Antibacterial Potential of Diterpenoids*, vol. 54, 2017.
[54] M. K. Dwivedi, S. Mishra, S. Sonter, and P. K. Singh, *Beni-Suef Univ. J. Basic Appl. Sci.*, vol. 10, no. 1, 2021, doi: 10.1186/s43088-021-00098-8.
[55] A. Endale, D. Bisrat, A. Animut, F. Bucar, and K. Asres, *Phyther. Res.*, vol. 27, no. 12, pp. 1805–1809, 2013, doi: 10.1002/ptr.4948.
[56] Y. A. Kim et al., *Bioorg. Chem.*, vol. 100, p. 103925, 2020, doi: 10.1016/j.bioorg.2020.103925.
[57] A. P. Shashank Kumara, *Sci. World J.*, 2013.
[58] R. Ortet et al., *J. Ethnopharmacol.*, vol. 138, no. 2, pp. 637–640, 2011, doi: 10.1016/j.jep.2011.09.039.
[59] A. M. Lehane and K. J. Saliba, *BMC Res. Notes*, vol. 1, pp. 1–5, 2008, doi: 10.1186/1756-0500-1-26.
[60] B. P. Raiguru, S. Mohapatra, S. Nayak, D. Sahal, M. Yadav, and B. N. Acharya, *Eur. J. Med. Chem. Reports*, vol. 4, no. November 2021, p. 100029, 2022, doi: 10.1016/j.ejmcr.2022.100029.
[61] S. G. Vincent and J. Joseph, *Mater. Today Proc.*, vol. 45, pp. 2159–2165, 2021, doi: 10.1016/j.matpr.2020.10.008.
[62] K. C. S. C. Liu, S. L. Yang, M. F. Roberts, B. C. Elford, and J. D. Phillipson, *Plant Cell Rep.*, vol. 11, no. 12, pp. 637–640, 1992, doi: 10.1007/BF00236389.
[63] D. Torres-Mendoza et al., *J. Nat. Prod.*, vol. 69, no. 5, pp. 826–828, 2006, doi: 10.1021/np050484i.
[64] Y. Liu, N. Murakami, H. Ji, P. Abreu, and S. Zhang, *Pharm. Biol.*, vol. 45, no. 4, pp. 278–281, 2007, doi: 10.1080/13880200701214748.
[65] D. Ganesh et al., *Parasitol. Res.*, vol. 110, no. 6, pp. 2289–2295, 2012, doi: 10.1007/s00436-011-2763-z.
[66] M. Rai, A. P. Ingle, P. Paralikar, I. Gupta, S. Medici, and C. A. Santos, *Int. J. Pharm.*, vol. 526, no. 1–2, pp. 254–270, 2017, doi: 10.1016/j.ijpharm.2017.04.042.
[67] N. S. Santos-Magalhães and V. C. F. Mosqueira, *Adv. Drug Deliv. Rev.*, vol. 62, no. 4–5, pp. 560–575, 2010, doi: 10.1016/j.addr.2009.11.024.
[68] J. Hemingway et al., *PLoS Biol.*, vol. 14, no. 3, pp. 1–14, 2016, doi: 10.1371/journal.pbio.1002380.
[69] M. J. Carnevale P, *Eur. PMC*, pp. 391–398, 1990.

[70] M. O. Ojemaye, S. O. Okoh, and A. I. Okoh, *Green Chem. Lett. Rev.*, vol. 14, no. 1, pp. 49–59, 2021, doi: 10.1080/17518253.2020.1861344.
[71] K. Chulasak, C. Punsawad, and P. Rattanakit, *Nanotechnol. Environ. Eng.*, vol. 7, no. 2, pp. 491–501, 2022, doi: 10.1007/s41204-022-00239-z.
[72] K. Okaiyeto, H. Hoppe, and A. I. Okoh, *J. Clust. Sci.*, vol. 32, no. 1, pp. 101–109, 2021, doi: 10.1007/s10876-020-01766-y.
[73] C. Panneerselvam et al., *Parasitol. Res.*, vol. 115, no. 3, pp. 997–1013, 2016, doi: 10.1007/s00436-015-4828-x.
[74] S. Arokiyaraj et al., *Environ. Sci. Pollut. Res.*, vol. 22, no. 13, pp. 9759–9765, 2015, doi: 10.1007/s11356-015-4148-9.
[75] D. Kumar, G. Kumar, R. Das, and V. Agrawal, *Process Saf. Environ. Prot.*, vol. 116, pp. 137–148, 2018, doi: 10.1016/j.psep.2018.02.001.
[76] D. Dinesh et al., *Parasitol. Res.*, vol. 114, no. 4, pp. 1519–1529, 2015, doi: 10.1007/s00436-015-4336-z.
[77] R. Bhuvaneswari, R. J. Xavier, and M. Arumugam, *J. King Saud Univ. Sci.*, vol. 28, no. 4, pp. 318–323, 2016, doi: 10.1016/j.jksus.2015.10.006.
[78] R. Ahmad, S. Srivastava, S. Ghosh, and S. K. Khare, *Colloids Surf. B Biointerfaces*, vol. 197, no. October 2020, p. 111389, 2021, doi: 10.1016/j.colsurfb.2020.111389.
[79] S. Krol et al., *Chem. Rev.*, vol. 113, no. 3, pp. 1877–1903, 2013, doi: 10.1021/cr200472g.
[80] H. Rashidzadeh, M. Salimi, S. Sadighian, K. Rostamizadeh, and A. Ramazani, *Curr. Drug Deliv.*, vol. 16, no. 10, pp. 923–930, 2019, doi: 10.2174/1567201816666191029121036.
[81] R. B. dos Santos et al., *Mater. Sci. Eng. C*, vol. 118, no. August 2020, p. 111356, 2021, doi: 10.1016/j.msec.2020.111356.
[82] B. Memvanga, R. Coco, and V. Préat, *J. Control. Release*, vol. 172, no. 3, pp. 904–913, 2013, doi: 10.1016/j.jconrel.2013.09.001.
[83] A. P. Nayak, W. Tiyaboonchai, S. Patankar, B. Madhusudhan, and E. B. Souto, *Colloids Surf. B Biointerfaces*, vol. 81, no. 1, pp. 263–273, 2010, doi: 10.1016/j.colsurfb.2010.07.020.
[84] N. P. Aditya, G. Chimote, K. Gunalan, R. Banerjee, S. Patankar, and B. Madhusudhan, *Exp. Parasitol.*, vol. 131, no. 3, pp. 292–299, 2012, doi: 10.1016/j.exppara.2012.04.010.
[85] P. P. Dandekar et al., *J. Pharm. Sci.*, vol. 99, no. 12, pp. 4992–5010, 2010, doi: https://doi.org/10.1002/jps.22191.
[86] F. Akhtar, M. M. A. Rizvi, and S. K. Kar, *Biotechnol. Adv.*, vol. 30, no. 1, pp. 310–320, 2012, doi: 10.1016/j.biotechadv.2011.05.009.
[87] Z. A. Busari et al., *Front. Pharmacol.*, vol. 8, no. SEP, pp. 1–9, 2017, doi: 10.3389/fphar.2017.00622.
[88] K. Martha Ariesaka, M. Taufiqurrahman, and M. Mirza Nuryady, *KnE Life Sci.*, vol. 2021, no. ISMoPHS 2020, pp. 391–399, 2021, doi: 10.18502/kls.v0i0.8898.
[89] A. Ramazani, M. Keramati, H. Malvandi, H. Danafar, and H. Kheiri Manjili, *Pharm. Dev. Technol.*, vol. 23, no. 9, pp. 911–920, 2018, doi: 10.1080/10837450.2017.1372781.
[90] P. Khanal, *Monatshefte fur Chemie*, vol. 152, no. 4, pp. 387–400, 2021, doi: 10.1007/s00706-021-02759-x.
[91] K. Dauda, Z. Busari, O. Morenikeji, and F. Afolayan, vol. 18, no. 11, pp. 977–985, 2017.
[92] D. Kannan et al., *EBioMedicine*, vol. 45, pp. 261–277, 2019, doi: 10.1016/j.ebiom.2019.06.026.
[93] Y. Boateng-Marfo, Y. Dong, W. K. Ng, and H. S. Lin, *Int. J. Mol. Sci.*, vol. 22, no. 3, pp. 1–22, 2021, doi: 10.3390/ijms22031141.
[94] S. A. Amolegbe et al., *Sci. Rep.*, vol. 8, no. 1, pp. 1–9, 2018, doi: 10.1038/s41598-018-21351-8.

13 Nanoparticles for the Delivery of Antiviral Phytotherapeutics

Sanjana Varma
(Biochemical Sciences Division, CSIR-National Chemical Laboratory, Pashan, Pune 411008, India, and Academy of Scientific and Innovative Research (AcSIR), Ghaziabad 201002, India, ss.varma@ncl.res.in)

Manali M. Bagade
(Biochemical Sciences Division, CSIR-National Chemical Laboratory, Pashan, Pune 411008, India, and Academy of Scientific and Innovative Research (AcSIR), Ghaziabad 201002, India, mm.bagade@ncl.res.in)

Dinesh Amalnerkar
(Nano Particles Technology Laboratory, School of Mechanical Engineering, Sungkyunkwan University, Suwon, Gyeonggi-do 16419, South Korea, and Department of Technology, Savitribai Phule Pune University, Pune 411007, India, dpa54@yahoo.co.in)

Manish D. Shinde
(Centre for Materials for Electronics Technology, Panchvati, Off Pashan Road, Pune 411008, India, ashman555@gmail.com)

**Bhushan P. Chaudhari*
(Biochemical Sciences Division, CSIR-National Chemical Laboratory, Pashan, Pune 411008, India, and Academy of Scientific and Innovative Research (AcSIR), Ghaziabad 201002, India, bp.chaudhari@ncl.res.in, Tel.: 91-20-25902226)

13.1 INTRODUCTION

Viral infections are one of the major factors contributing to public health problems globally, leading to significant mortality and causing adverse socioeconomic effects on the society (Cojocaru et al., 2020). Viruses are sub-microscopic intracellular parasites containing either DNA or RNA as the genetic material, surrounded by a protein coat known as capsid (Cojocaru et al., 2020; Milovanovic et al., 2020). Viruses use host machinery for their replication (Cojocaru et al., 2020). They are of several types, but, thus far, only approximately 5000 of them have been identified and characterized. However, the foremost infective viruses that cause most of the human illness and mortality are human papillomavirus (HPV), human immunodeficiency virus (HIV), herpes simplex virus (HSV), norovirus, coronaviruses, and hepatitis viruses (Delshadi et al., 2021). Viral infections lead to economic implications. Recently, during the COVID-19 pandemic, various socioeconomic consequences occurred globally, which led to the death of several people worldwide. Dengue is

DOI: 10.1201/9781003231721-13

highly prevalent in south-east Asia, and their prevalence rate is concerning in the current scenario. In 2018, approximately 1,64,000 dengue cases were reported globally. Over the past few years, Zika virus infection, which is usually endemic, was seen in various parts of the world. In India, the fatality rate resulting from Nipah virus has been in the range of 68%–100% since 2001. In 2018, Nipah virus caused 17 deaths out of a total of 19 reported cases in India. Influenza virus infection has been reported in many parts of the world and is considered as seasonal flu in various regions. During 2014–2016, among a total of 25,616 cases of Ebola virus in West Africa, 11,315 patients died (Chakravarty and Vora, 2021). Some of the recent viral outbreaks have claimed the lives of many individuals and imparted adverse economic effects on the society. A similar type of outbreak can increasingly occur in the future.

Viral infections account for approximately two million deaths each year (Delshadi et al., 2021). Thus, there is an immense need to develop effective antiviral therapy for viral infections. Vaccination is a very effective approach to combat against viruses, but currently, vaccines are available only for mumps, varicella (Ben-shabat et al., 2020), measles, and hepatitis A and B virus (Milovanovic et al., 2020). However, owing to the rapid mutation and replication feature of viruses, vaccine development has become a challenging and time-consuming task to undertake (Milovanovic et al., 2020). Synthetic antiviral drugs can treat only certain viral infections such as HIV, human cytomegalovirus (hCMV), herpes viruses (HSV), hepatitis viruses, varicella zoster virus (VZV), and influenza viruses (Ben-shabat et al., 2020). Some of the common synthetic antivirals are remdesivir, zidovudine, abacavir, zalcitabine, oseltamivir, stavudine, efavirenz, nelfinavir, and ritonavir, among others (Pradhan et al., 2021; Delshadi et al., 2021). These classic antiviral drugs have certain drawbacks in that some of them are ineffective *in vivo* but show significant results *in vitro* (Ben-shabat et al., 2020); these drugs are reported to have poor solubility in the aqueous medium, shorter half-life, and low bioavailability at the target site and are expensive. Owing to the efficacy and safety limitations of synthetic antivirals and given the urge to develop effective antiviral drugs to treat fatal viral infections, it is logical to shift to natural antiviral agents (Pradhan et al., 2021; Behl et al., 2021; Pour et al., 2019). In the case of viruses, rapid viral mutations and drug resistance are the reasons that always demand continuous studies for new antiviral drugs, and natural compounds are the appropriate candidates for antiviral therapy. Many types of phytochemicals have been studied for their efficacy as antiviral agents, and many of these phytochemicals are present in natural products, but they need to be evaluated for their therapeutic approaches (Pour et al., 2019). Plant-based antivirals have pleiotropic effects at various target sites, and additionally, their usage individually or in combination with other mainstream therapies augments their efficacy, making them a promising approach to treat viral infections. Different groups of antiviral phytochemicals act either directly, by inhibiting viral replication, or indirectly, by maintaining redox homeostasis (Baranwal et al., 2021). Different groups of phytochemicals with antiviral activity are alkaloids, polyphenols, proteins, lectins, terpenes, polysaccharides, and flavonoids (Behl et al., 2021).

Although the development of effective antiviral therapies is highly challenging because of frequent viral mutations and the use of host machinery by viruses for their replication, the limitations of pure phytochemicals, such as poor solubility, less bioavailability, low permeability through different membranes, and rapid metabolism, can be overcome with the use of a well-defined nano-drug delivery system (Ben-shabat et al., 2020; Delshadi et al., 2021).

Nanotechnology is emerging as the most promising area to circumvent the shortcomings of conventional antiviral therapy and to effectively cope with viral infections. The use of nanoparticles for the antiviral delivery of phytochemicals increases the efficacy and selectivity of drugs toward virus-infected cells, not the host cell (Chakravarty and Vora, 2021). A nanoparticle-based delivery system for plant-based antivirals improves its stability, bioavailability, solubility, and pharmacokinetics, leading to delayed metabolism (Ben-shabat et al., 2020) and aiding in target-specific delivery (Delshadi et al., 2021). These overall benefits of nano-drug delivery systems boost the antiviral activity of phytochemicals (Chakravarty and Vora, 2021; Delshadi et al., 2021).

Gradually, researchers are gaining increasing interest toward the use of nanoparticles as a drug delivery system owing to their small size, high surface area-to-volume ratio, easy tunability of structural and functional attributes, and quick internalization into the body (Delshadi et al., 2021; Milovanovic et al., 2020). Nanoparticles also aid in the controlled release of plant-based antivirals (Milovanovic et al., 2020). Furthermore, nano-based drug delivery systems are cost-effective, biocompatible, and feasible and are appropriate for the development of effective phytochemicals against viral infections (Gurunathan et al., 2020). Various types of nanoparticles that are potent phytotherapeutic delivery systems include carbon-based nanocarriers, cyclodextrin-based nanocarriers, micelles, liposomes, lipid polymeric nanocarriers, polymeric nanoparticles, nanoemulsions, inorganic metal-based nanoparticles, and dendrimers (Chakravarty and Vora, 2021; Maus et al., 2021).

13.2 VIRUSES

Viruses are a type of obligate intracellular parasite that depends on the host's biosynthetic machinery for their replication. Biological components that constitute a virus include nucleic acids (DNA or RNA) and capsid (a protein coat), and some viruses have an outer coating of lipid bilayer, called an envelope. The genetic material of viruses codes for different proteins required for their replication (Milovanovic et al., 2020). According to the International Committee on Taxonomy of Viruses, there are one phylum, two subphyla, six classes, ten orders, seven suborders, 89 families, 36 subfamilies, 387 genera, 59 subgenera, and 2202 species. Viruses are classified according to the Baltimore classification, which categorizes viruses based on the type of nucleic acid, its arrangement, and the mode of replication. There are seven classes of virus according to the Baltimore classification: (i) double-stranded DNA viruses, include herpes viruses, poxviruses, and adenoviruses; (ii) single-stranded DNA viruses, consist of a positive strand or a sense strand of DNA, for example, parvoviruses; (iii) double-stranded RNA viruses, for example, reoviruses; (iv) single-stranded RNA viruses, consist of a sense or a positive strand of RNA, for example, picornaviruses and togaviruses; (v) single-stranded RNA viruses, consist of an antisense or a negative strand, for example, rhabdoviruses and orthomyxoviruses; (vi) single-stranded reverse transcriptase RNA viruses involve DNA as an intermediate in the life cycle, and this class consists of viruses with a sense or a positive strand, for example, retroviruses; (vii) double-stranded reverse transcriptase DNA viruses involve RNA as an intermediate in the life cycle, and this class consists of hepadnaviruses (Cojocaru et al., 2020).

13.2.1 Viral Pathogenesis

Treatment of viral infections is challenging because most of them are subclinical. In the subclinical type of infection, wherein the host's immune response is generated, the viral replication is arrested at the initial point, and clinical symptoms do not appear or very few symptoms appear, but such types of infections have remarkable epidemiological importance because this can lead to the spreading of viral infections within and among populations (Maus et al., 2021; Chakravarty and Vora, 2021). Owing to the aforementioned reason, these subclinical viral infections cannot be diagnosed easily.

Viral pathogenesis is divided into six major steps, which are as follows: (i) attachment to the host cell surface—this step needs sufficient interaction between the viral spikes or proteins and the host cell surface receptors; (ii) penetration into the host cell by direct fusion of the viral membrane with the host cell membrane or by receptor-mediated endocytosis; (iii) uncoating of the viral genetic material (DNA or RNA) into the host cell's cytoplasm; (iv) viral replication, transcription, and translation, leading to the synthesis of viral proteins; (v) assembly of the virus, which involves the insertion of viral DNA or RNA into the naked capsid; (vi) release of the virus into the extracellular space, which is the final step, and it can be achieved through budding, exocytosis, cell lysis, or apoptosis (Milovanovic et al., 2020; Maus et al., 2021; Chakravarty and Vora, 2021).

Multiple factors can affect the ability of the virus to use and hijack the host cellular machine. First, the cells at the infection site should be accessible, and viral load should be adequate, that is, cells should be vulnerable to viral replication and the host defense mechanism must be futile, that is, the virus should have resistance to the host defense system. Further success of viral infections depends on the specific affinity between viral proteins and host cell receptors; viral structure; rate of viral multiplication, which decides the virulence; and physiological factors such as pH, moisture, and temperature. Apart from these factors, age, sex, race, and body weight of the host also affect virulence attainment (Maus et al., 2021; Chakravarty and Vora, 2021). These factors, which are responsible for the spread of viral infection, open up multiple possibilities to treat viral diseases through specific targeting (Maus et al., 2021).

Viruses destroy the host cell by seizing the cellular metabolic pathways that compromise the cell's energy. Moreover, the virus causes genetic mutations in the host cell genome by incorporating its viral genetic material into the host genome, leading to cell damage. Based on the replication rate of the virus, their load, virulence, genes responsible for virulence, and extent of spread of the viral infection, the course of infection is unraveled. The unpredictable behavior of the virus inside the host cell and the difficulty to estimate interactions of host defense systems with the viruses make the study of viral infections highly challenging (Chakravarty and Vora, 2021). Because of the volatile nature of viruses in the host, the development of effective antiviral therapy to treat viral diseases is a very difficult process to achieve and will require continuous exploration through research.

13.2.2 Challenges in Antiviral Therapy

The continuous appearance of novel viral infections and the occurrence of multidrug-resistant strains worldwide presents excessive challenges to the medical efficacy of antiviral therapies. Some of the major challenges with antiviral therapies are less bioavailability owing to poor solubility and permeability of the antivirals; therefore, treatment with a higher dose generates toxic effects. Furthermore, the short half-life of antivirals increases the need for frequent medication, which leads to deprived patient compliance and increases treatment expenses. Antiviral drugs can interact with the regularly prescribed drugs, leading to drug-drug interactions, and long-term antiviral medications can cause toxic effects, which may further hamper patients' health. Furthermore, prolonged exposure to antivirals increases the risk of developing drug resistance or drug tolerance, mainly in immunocompromised individuals. It is difficult for antiviral drugs to reach inaccessible anatomical areas such as the lymphatic system, central nervous system, and synovial fluid, which is the target site for many viruses, thereby decreasing the efficacy of antiviral therapy. The exclusive structure and function of different viruses pose another challenge in terms of developing a broad range of antiviral therapies. Furthermore, the selectiveness of antiviral drugs to target the virus rather than the host cell and the recognition of unique targets of the viral life cycle is one more major challenge in the development of effective antiviral therapies (Maus et al., 2021; Chakravarty and Vora, 2021). Many viral infections remain asymptomatic and dormant for a long period, which makes their detection and treatment extremely challenging to overcome (Chakravarty and Vora, 2021). Nanotechnology can provide promising solutions to the various challenges faced with currently available antiviral therapies. Worldwide, researchers are constantly putting in efforts to use nanotechnological approaches to upsurge the efficacy of traditional antiviral therapies (Maus et al., 2021).

13.3 DIFFERENT CLASSES OF PLANT-BASED ANTIVIRALS

The use of natural drugs for antiviral therapy is truly a valuable approach for combating fatal viral infections. European researchers first developed antiviral drugs after World War II, which led to the investigation of the antiviral potential of 288 plants and their phytocomponents against influenza A virus by a drug company in England. Based on the results of this examination, 12 plants displayed promising effects. To date, accumulating evidence suggests that viral infections can be cured through phytotherapies (Behl et al., 2021; Kumari and Chatterjee, 2021). Plant-based antiviral drugs are gaining importance because they have low toxicity, are inexpensive, and belong to natural origin, they have fewer side effects (Delshadi et al., 2021). Major bioactive phytochemicals that can fight against viral diseases are flavonoids, polyphenols, saponins, quinones, alkaloids, tannins, terpenes, proanthocyanidins, steroids, lignins, coumarins, polysaccharides, and thiosulfonates (Kapoor et al., 2017; Delshadi et al., 2021; S. Kumari and Chatterjee, 2021). Inhibition of viral replication, transcription, and translation; antioxidant activities; inhibition of the entry of viruses; and suppression of various steps of their life cycle are the prominent mechanisms of action demonstrated by the antiviral activities of phytochemicals (Delshadi et al., 2021; S. Kumari and Chatterjee, 2021). Several herbs and conventional plants with medicinal values have shown solid antiviral activity against hepatitis virus, H1N1 virus, and other such viruses (Kapoor et al., 2017). In the following sections, we present an overview of the various classes of plant-derived antiviral agents. Table 13.1 presents a summary of the phytochemicals with potent antiviral properties, and Figure 13.1 presents the general mechanism of action of the various classes of phytochemicals effective against viruses.

TABLE 13.1
Various Phytochemicals with Antiviral Properties

Phytochemicals	Group/Type	Active Against Virus	Refs
3-Hydroxy Caruilignan C	Lignan	HCV	(Wu et al., 2012)
5,7,4′-trihydroxy-8-methoxyflavone (f36)	Flavone	APR8 (mouse-adapted influenza virus)	(Nagai et al., 1995)
Andrographolide	Diterpenoid	Influenza virus, (DENV)	(Panraksa et al., 2017; Chen et al., 2009; PubChem ID: 5318517)
Apigenin	Flavonoid	Enterovirus-71, foot-and-mouth disease virus, African swine fever virus	(Hakobyan et al., 2016; Qian et al., 2015; Zhang et al., 2014)
Baicalin	Flavonoid	DENV, H1N1/H3N2	(Ding et al., 2014; Moghaddam et al., 2014)
Caffeoylquinic acid	Caffeate ester	RSV	(Li et al., 2005; PubChem ID: 10155076)
Chalcone compounds	Chalcones	Influenza A	(Dao et al., 2011)
Chalepin	Furanocoumarin	HCV	(Wahyuni et al., 2014; PubChem ID: 119066)
Compound 3 (from *Jatrophane*)	Diterpene esters	CHIKV	(Nothias-Scaglia et al., 2014)
Coumestrol	Isoflavonoid-like compound	HSV-1 and HSV-2	(Argenta et al., 2018)
Curcumin	Polyphenol	HPV-16, ZIKV, SARS-CoV-2, HSV-2, DENV, Rift Valley fever virus, HCV	(Bonfim et al., 2020; Pacho et al., 2021; Vitali et al., 2020; Jena et al., 2021; Nabila et al., 2020; Narayanan et al., 2012; Loutfy et al., 2020)
Catechin	Polyphenol	SARS-CoV-2, influenza virus	(Jena et al., 2021; Song et al., 2005)
Cyanidin-3-sambubiocide	Flavonoid	H1N1 influenza virus	(Kannan and Kolandaivel, 2018)
Dammarenolic acid	Triterpenoid	HIV-1	(Esimone et al., 2010)
Decanoylphorbol-13 acetate	Diterpene	CHIKV	(Corlay et al., 2014)
Delphinidin	Polyphenols	West Nile virus, DENV, and ZIKV	(Vázquez-Calvo et al., 2017)

(Continued)

TABLE 13.1 (Continued)
Various Phytochemicals with Antiviral Properties

Phytochemicals	Group/Type	Active Against Virus	Refs
Epigallocatechin-3-gallate	Polyphenol	ZIKV	(Carneiro et al., 2016)
Glycyrrhizic acid	Triterpene glycoside	EBV	(Lin, 2003; Baltina, 2003)
Honokiol	Lignan biphenol	DENV-2	(Fang et al., 2015)
Loliolide	Monoterpenoid lactone	HCV	(Chung et al., 2016)
Lignin–carbohydrate complex (PPS-2b)	Lignin–carbohydrate	HSV-1, HSV-2	(Yongwen Zhang et al., 2007)
Limonoids	Triterpenoid	DENV-2	(Cheng et al., 2014; Roy and Saraf, 2006)
Manassantin B	Lignans	EBV	(Cui et al., 2014)
Naringenin	Flavonoid	SARS-CoV-2, DENV, ZIKV	(Cataneo et al., 2019; Clementi et al., 2021; Frabasile et al., 2017)
Oxyresveratrol	Polyphenol	HSV-1, HSV-2	(Chen et al., 2016; Chuanasa et al., 2008)
Pseudane IX	Quinolone	HCV	(Wahyuni et al., 2014; Paulsen et al., 2020)
Quercetin	Flavonoid	HCV	(Bachmetov et al., 2012)
Quercetin 3-rhamnoside	Flavonoid	Anti-influenza A/WS/33 virus	(Choi et al., 2009)
Saikosaponins b2	Terpenoid	HCV	(Lin et al., 2015)
Samarangenin B	Polyphenol flavonoid	HSV-1	(Kuo et al., 2002)
Scopadulcic acid B	Diterpenoid	HSV-1	(Hayashi et al., 1988)
Sennoside A	Glycoside	HIV-1	(Esposito et al., 2016; PubChem ID: 5199)
Silibinin	Flavonolignan	HCV	(Ripoli et al., 2016)
Silvestrol	Benzofuran	Ebola	(Biedenkopf et al., 2017; Chen et al., 2015)
Spiroketalenol	Ether derivatives	HSV-1, HSV-2	(Álvarez et al., 2015)
Swerilactone A and B	Lactones	HBV	(Geng et al., 2009, 2010)
Tangeretin and nobiletin	Polymethoxylated flavones	RSV	(Xu et al., 2014)
Tetranortriterpenoid; 1-cinnamoyl-3; 11-dihydroxymeliacarpin	Triterpenoid	Vesicular stomatitis virus, HSV-1	(Barquero et al., 2006)
Xanthohumol	Flavonoid	HIV-1, SARS-CoV-2	(Lin et al., 2021; Wang et al., 2004; PubChem ID: 639665)

DENV: dengue virus; HCV: hepatitis C virus; RSV: respiratory syncytial virus; SARS-CoV-2: severe acute respiratory syndrome coronavirus disease 2; ZIKV: Zika virus; HIV: human immunodeficiency virus; HPV: human papilloma virus; EBV: Epstein-Barr virus: CHIKV: chikungunya virus

13.3.1 Flavonoids

Flavonoids are a class of potent and important antiviral agents, and many *in vivo* and *in vitro* studies have assessed the antiviral potential of flavonoids. Flavonoids are flower pigments that contain polyphenols of 15 carbon atoms. Flavonoids are further divided into subclasses, namely, anthocyanins, flavans, neoflavonoids, flavanones, flavones, flavanonols, flavonols, chalcones, and dihydrochalcones. The general antiviral mechanism of flavonoids involves the inhibition of

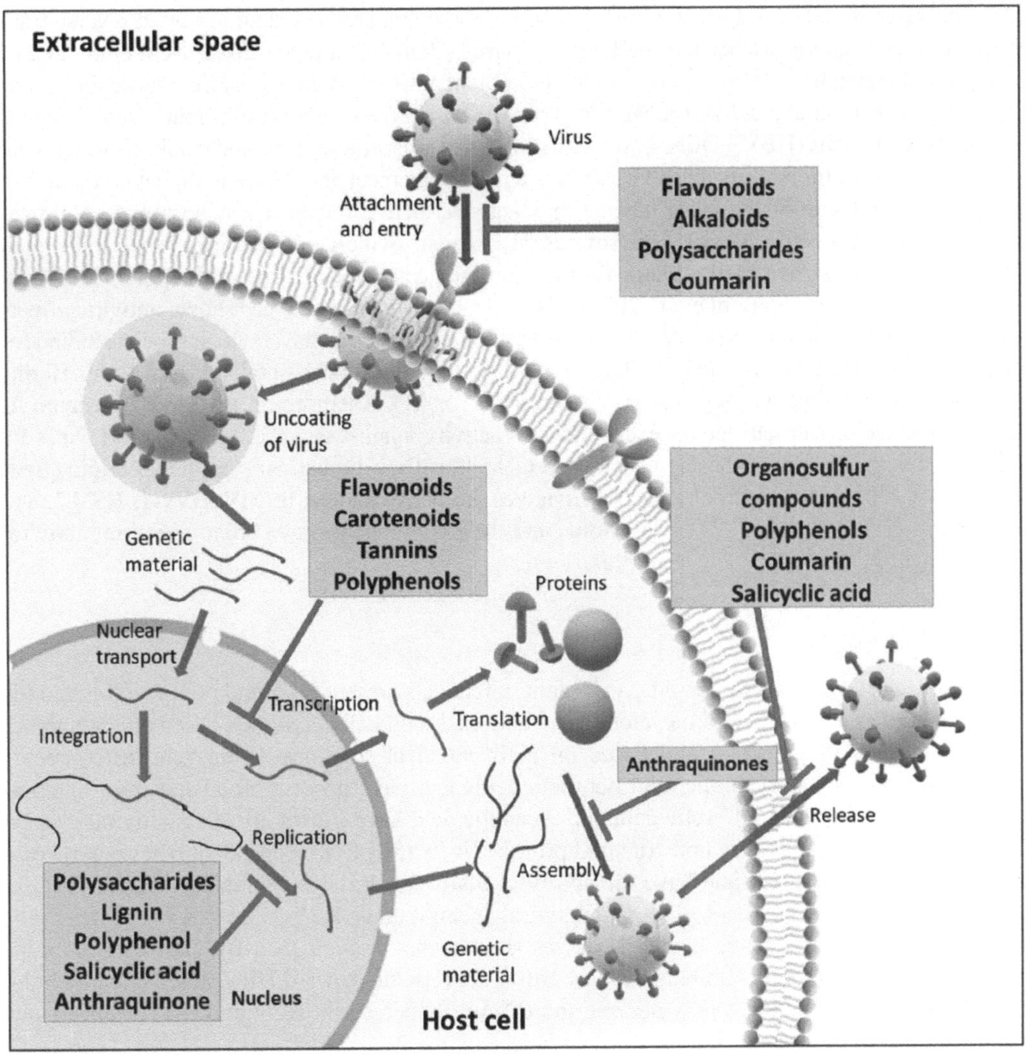

FIGURE 13.1 Overview of the general mechanism of action of different classes of phytochemicals against viruses.

various enzymes such as aldose reductase, xanthine oxidase, lipoxygenase, cyclooxygenase, and Ca^{2+}-ATPase (Behl et al., 2021). Flavonoids inhibit reverse transcriptase in some viruses (Behl et al., 2021; Pour et al., 2019). Some reports have mentioned that flavonoids possess antiviral potential, whereby they inhibit fusion and integration of viruses with the host genome (Pour et al., 2019). Anthocyanin exerts antiviral activity against influenza A and B virus, HSV-1, hepatitis A virus, coxsackievirus (CV-B1), adenovirus 36, BT2, T4 and simian rotavirus SA-11, avian influenza virus, and murine norovirus-1 (Behl et al., 2021; Pour et al., 2019). Chalcone was reported to have antiviral activity (Behl et al., 2021), and synthetically produced chalcone showed antiviral properties against hepatitis C virus (HCV) (Mateeva et al., 2017). Dihydrochalcone derivatives extracted from *Millettia leucantha* possess an effective antiviral potential for HSV (Hrutivorapongkul et al., 2003). Quercetin is one of the abundant phytochemicals in plants classified into flavonoids, and it is an aglycone of rutin. Quercetin shows antiviral activities against HSV, adenovirus, influenza virus, dengue virus (DENV) type 2, poliovirus, Epstein-Barr virus, Japanese encephalitis virus, and HCV

(Ben-shabat et al., 2020). Myricetin, quercetagetin, quercetin, and baicalein inhibited reverse transcriptase in HIV and Rauscher murine leukemia virus (Ben-shabat et al., 2020; Behl et al., 2021). Myricetin obtained from *Aronia melanocarpa* is a potent antiviral therapeutic effective against influenza viruses (Park et al., 2013). Baicalin also has antiviral effects against different viruses such as HIV, hepatitis B virus (HBV), DENV, Newcastle disease virus, enterovirus, and respiratory syncytial virus (RSV). Baicalin inhibits these viruses through various mechanisms (Ben-shabat et al., 2020). Apigenin, an aglycon of the class flavone, has antiviral activity against a wide range of viruses including HCV, influenza A virus, enterovirus 71, African swine fever virus, and foot-and-mouth disease virus (Ben-shabat et al., 2020). Furthermore, tea catechins are one of the potent antivirals for HIV and herpes virus (Behl et al., 2021). Naringin, a type of flavone, has shown antiviral effects against picornavirus and respiratory viruses. Abyssinone II, a type of flavanone, has been studied for its antiviral potential against HSV-1 (Behl et al., 2021). Isoflavanones obtained from *Bolus* Harms had evident antiviral activity against HIV (Behl et al., 2021). Kaempferol, apigenin, biochanin A, luteolin, baicalein, and naringenin exert antiviral activity against avian influenza H5N1 virus by inhibiting nucleoprotein synthesis, when tested in A549 cells of human lung epithelial origin (Ben-shabat et al., 2020). Kaempferol and its derivatives are active against hCMV, HSV-1, HSV-2, and coronaviruses (Behl et al., 2021). Flavonoids and their components have broad-spectrum antiviral potential against viruses causing multiple diseases.

13.3.2 Terpenoids

Terpenoids, also known as isoprenoids, are organic chemicals present in natural products. Terpenoids are considered as a derivative of mevalonic acid and are specifically classified based on the number of isoprene units they contain and based on their chemical structure. Pentacyclic triterpenoids mainly include ursolic, oleanolic, and betulinic acids and have possess broad-spectrum antiviral activity (Behl et al., 2021). Ursolic acid and oleanolic acid have shown effective antiviral properties against HCV by inhibiting nonstructural protein 5A (NS5B) RNA-dependent RNA polymerase activity (Kong et al., 2013) and have the ability to inhibit enterovirus 71 replication (Zhao et al., 2014). Borneol and isoborneaol are potent antiviral agents against HSV-1. Terpenoids possess antiviral activity against influenza A and B viruses (Ben-shabat et al., 2020). By inhibiting the transcription process, carotenoid and lutein show antiviral properties toward HBV. Another carotenoid, bacterioruberin, inhibits RNA polymerases and DNA polymerases in HCV and HBV, respectively, leading to the obstruction of viral replication (Behl et al., 2021).

13.3.3 Alkaloids

Alkaloids are commonly produced by many plant types but are primarily produced by flowering plants. These compounds are formed by plants through the amino acid synthesis pathways. Alkaloids are further classified according to their chemical structure. Alkaloids are mainly present in foods such as coffee, cocoa, tomatoes, and potatoes, among others (Behl et al., 2021). Alkaloids inhibit viral growth mainly by blocking their entry (Kapoor et al., 2017). Pericalline is more effective against polio type III virus among 36 alkaloids isolated from *Catharanthus lanceus*. It possesses antiviral potential against HIV (Behl et al., 2021). Berberine is one of the isoquinoline structures containing alkaloids and is an effective antiviral agent against severe acute respiratory syndrome coronavirus disease 2 (SARS-CoV-2), coronaviruses, RSV, herpes virus, and influenza virus (Behl et al., 2021). Hung et al. (2019) reported that berberine inhibits HCV infection by blocking viral entry and fusion. Furthermore, berberine is a potent antiviral agent against influenza H1N1 virus (Yan et. al., 2018). Berberine has demonstrated antiviral activity against hCMV by inhibiting viral replication; hence, it interferes with the transcription activator protein (immediate-to-early 2 [IE2]) protein function

(Luganini et al., 2019). Tomatidine is another steroidal alkaloid that has antiviral activity against DENV (Diosa-Toro et al., 2019) and chikungunya virus (CHIKV) (Troost et al., 2020). Some other alkaloids with great antiviral potential are periformyline, vincaleukoblastine, michellamines D/F, leurocristine, homonojirimycin, perivine, deoxymanojirimycin, castanospermine, acrimarine F, columbamine, australine, sesquiterpene, 5-hydroxynoracronycine, and palmitine (Thawabteh et al., 2019; Behl et al., 2021).

13.3.4 Organosulfur Compounds

This class includes glucosinolates, which undergo enzymatic or chemical hydrolysis to generate biologically active compounds, thiocyanates, and isothiocyanates. This class also includes allylic sulfur compounds, sulforaphane, and indoles. Vegetables such as broccoli, garlic, and onion are primary sources of these pungent-smelling organosulfur compounds. Some constituents of garlic, such as S-allyl cysteine and S-allyl-mercapto-cysteine, are biologically very active and has a less pungent odor. Allicin is one of the organosulfur compounds that can inhibit the spread of viral infections because this compound can cross the phospholipid bilayer and enter the host cell (Behl et al., 2021). Allyl alcohol, diallyl disulfide, and ajoene have the potential to act against HIV infection (Sharma, 2019).

13.3.5 Polysaccharides

Polysaccharides are composed of 200–2500 units of monosaccharides such as glucose, xylose, and galactose, and they have complex structures. Alginates, fucans, and galactans are important polysaccharides that possess antiviral properties. Carrageenan has antiviral activity against HIV and influenza virus (Behl et al., 2021). Polysaccharides derived from *Porphyridium* species have antiviral activity against VZV, HSV-1, and HSV-2 (Huleihel et al., 2001). Polysaccharides generally exert their antiviral activity by inhibiting viral replication and viral attachment to the cell (Kapoor et al., 2017).

13.3.6 Other Phenolic Compounds

The major components of this class that have an efficient antiviral property are lignans, hydroxyphenyl propenes, tannins, and curcumin. Lignan structure is highly complex, mainly composed of two or more phenylpropanoid units. Nordihydroguaiaretic acid, a lignan derived from *Larrea divaricata*, has antiviral potential against HIV, HPV, HSV-1, and HSV-2 (Behl et al., 2021). Niranthin is another kind of lignan that inhibits HBV by suppressing its replication process. Silymarin can inhibit HCV infection (Cui et al., 2020). Dibenzylbutyrolactones, substituted tetrahydrofurans, arylnaphthalenes, 2,6-diarylfurofurans, aryltetralins, and dibenzocyclooctenes are other potent antivirals belonging to the lignan group (Cui et al., 2020). 6-Gingerol, a type of hydroxyphenyl propene, is an antiviral phytochemical present in ginger (Behl et al., 2021; Rathinavel et al., 2020). 6-Gngerol is an effective antiviral therapy for COVID-19 (Rathinavel et al., 2020). It also exerts antiviral activity against human RSV (Behl et al., 2021).

Tannins are high-molecular-weight phenolic compounds. Tannins include catechin, epicatechin, ellagitannins, gallotannins chebulagic acid, and punicalagin. Ellagitannins obtained from *Phyllanthus urinaria* and *Phyllanthus myrtifolius* can inhibit Epstein-Barr virus (Behl et al., 2021). Ellagitannins extracted from *Tuberaria lignosa* have shown antiviral effects against HIV by suppressing reverse transcriptase (Wynn et al., 2004), although other reports suggest that ellagitannins inhibit protease and integrase in HIV-1 (Behl et al., 2021). Tannins also have antiviral activity against rotavirus, HSV, influenza A virus, and HIV (Behl et al., 2021).

Curcumin is another class of broad-spectrum phenolic antiviral. It possesses antiviral properties against a wide range of viruses such as influenza, HBV, HSV, HIV, and HCV (Delshadi et al., 2021). Curcumin reduces the infectivity of norovirus by 91% (Yang et al., 2016). Curcumin is also a potent antiviral against HIV because it can inhibit viral replication, multiplication, and infectivity (Prasad and Tyagi, 2015).

13.3.7 Essential Oils

Many herbs, plants, and spices contain nonpolar substances known as essential oils. These oils are generally used as flavoring agents in various foods. Essential oils also possess effective antiviral activity against several viruses. Essential oils contain many volatile compounds such as monoterpenoids (e.g., carvone, borneol, thymol, carvacrol, and citronellal), monoterpenes (e.g., P-cymene, limonene, sabinene, and α-pinene), and phenylpropanoids (e.g., safrole, cinnamaldehyde, eugenol, and vanillin). Essential oils can inhibit a wide range of viruses such as influenza, adenovirus type 3, poliovirus, Junin virus, norovirus, coxsackie virus B1, HSV (Delshadi et al., 2021), HIV, and yellow fever virus (Behl et al., 2021). Essential oil extracted from *Artemisia arborescens* L. showed antiviral activity against HSV-1(Sinico et al., 2005).

13.3.8 Other Antivirals

Resveratrol is categorized in the phenolic class of stilbenes. It is a potent antiviral against HIV, HCV, influenza virus, HSV (Ranjbar et al., 2020), SARS-CoV-2 (Ranjbar et al., 2020; Ramdani and Bachari, 2020), MERS-CoV (Ramdani and Bachari, 2020), simian virus (SA-11), and rotavirus (Gonçalves et al., 2005). Salicylic acid is an organic acid and a plant hormone. The antiviral potential of salicylic acid can be attributed to its ability to inhibit the replication and cell-to-cell movement of viruses (Behl et al., 2021). Anthraquinones, including emodin, aloe-emodine, and chrysophanol, are one of the important classes of phytochemical with antiviral properties. Chrysophanol isolated from *Dianella longifolia* is effective against poliovirus type 2 and 3 viruses and acts by inhibiting viral replication (Behl et al., 2021). Emodin is effective against SARS-related coronaviruses by blocking and inhibiting viral interaction with angiotensin-converting enzyme 2 receptors (Ho et al., 2007). Anthraquinones have antiviral property against SARS-CoV-2 by inhibiting viral protease (Behl et al., 2021). Coumarins are another class of compounds with antiviral properties. Some of the important coumarin derivatives that act as antiviral drugs include anilinocoumarins, 4-hydroxycoumarins, pyranocoumarins, 7-hydroxycoumarin analogs, arylcoumarins, 3-phenylcoumarins, coumarin-benzimidazole conjugates, coumestans, furanocoumarins, 4-phenylcoumarins, and toddacoumaquinone. Coumarins act by regulating cellular pathways and by inhibiting proteins essential for viral replication, entry, and spreading. Coumarin derivatives exert an inhibitory effect on influenza virus, coxsackievirus A16 (CVA16), HIV type 1, and enterovirus 71 (Mishra et al., 2020; Behl et al., 2021). Polyacetylenes are a class of phytochemicals that inhibit viral infections. They have shown promising results when tested against CMV herpes virus (Behl et al., 2021).

13.4 NANOTECHNOLOGY IN PHYTOTHERAPEUTIC DELIVERY

Nanotechnology is a branch of science that has the potential to overcome the limitations associated with the use of phytoantiviral drugs and can additionally intensify their therapeutic potential (Yang et al., 2019). Nanoparticles are particles that have only one major dimension, with their size being in the range of 1–100 nm; however, generally, any particle of size below 1000 nm is termed as a nanoparticle. The two major approaches used in the synthesis of nanoparticles are the top-down and the bottom-up approaches. In the first method, large bulk substances are broken by mechanical forces to form small particles, whereas in the second method, nanoparticles are formed from molecules by controlling physiochemical conditions and altering reagents required for nanoparticle synthesis. The size of the nanoparticles can be tuned by controlling many factors that are vital in their synthesis process (Delshadi et al., 2021).

Nanoparticles play an important role in effectively treating various infectious diseases (Yang et al., 2019). As discussed earlier, there are several challenges in antiviral therapy; however, a nanoparticle-based antiviral delivery system can be a major component to overcome the limitations associated with conventional antiviral therapies. Nanotechnology-based drug delivery systems are effective and useful for drug delivery because of the unique properties of nanoparticles, such as easy surface functionalization for target-specific delivery, high surface area-to-volume ratio, improved permeability and retention effect, and ability to easily cross various biological barriers owing to the small size (Delshadi et al., 2021; Pradhan et al., 2021). These attributes of nanoparticles help enhance the bioavailability of plant-based antivirals, improve the stability and solubility of natural antivirals, enable controlled and site-specific delivery of antivirals, increase the half-life of drugs, and improve the cellular uptake ability for antivirals. Furthermore, the programmable release of a drug at the target site using a nanotechnology-based drug delivery system helps maintain the optimum concentration of the drug and reduce the emergence of drug resistance. Nano-drug delivery systems also protect the drugs from degradation and reduce their toxicity (Yang et al., 2019; Pradhan et al., 2021; Lembo and Cavalli., 2010). Additionally, nano-based antiviral phytotherapeutic delivery approaches increase patient compliance and reduce treatment cost (Lembo and Cavalli., 2010). Nanotechnology paves the way for the preparation of novel effective phytotherapeutic medicine for treating fatal viral infections. In the current scenario, COVID-19 is the major viral pandemic, and nano-based phytotherapeutic delivery systems can be proven as effective approach to cure COVID-19. Table 13.2 presents information about various phytochemical-based nanoformulations prepared for treating viral infections.

Nanoparticle-based drug delivery system can be used to deliver drugs into the body through various routes depending on the site of action. The route of administration can be oral, nasal, intravenous, or others. As the nanoformulation reaches the site of action, the drug is transferred from extracellular fluids to intracellular fluids either actively or passively at the junctions. The targeting ability of nano-based drug carriers depends on the ligands attached to them, which can be antibodies, proteins (Z domains, DARPins, and fibronectin III), peptides (angiopep-2, epidermal growth factor, and cANF), organic molecules (carbohydrates, biotin, and folic acid). Furthermore, when reaching the target site, the drug carriers induce the generation of the anticipated immune response (Maus et al., 2021). The phytotherapeutic antiviral, which is released from the nanoparticle, inhibits viral replication by blocking either the viral receptors expressed on the surface, thus inhibiting the penetration and entry of the virus, or various steps of the viral life cycle, such as replication, transcription, translation, assembly of virus cell components, and viral release in the host cell (Delshadi et al., 2021). An overview of the general mechanism of action of nanoparticle-based antiviral phytotherapeutic delivery system against viruses is shown in Figure 13.2. In the following sections, we have discussed various types of potential nanoparticle-based drug delivery systems.

ORGANIC NANOCARRIERS

13.4.1 Lipid-Based Nanoformulations

Lipid-based drug delivery systems are preferred because they aid the delivery of both hydrophobic and hydrophilic drugs. Additionally, they have advantages such as easy permeability, protection of the drug from degradation, enhanced therapeutic potential of the drugs, increased stability, and targeted delivery of the drugs. However, they have some disadvantages such as leakage of the drug from the nanocarrier and poor solubility (Ahmad et al., 2021), but these limitations can be overcome while designing lipid-based drug delivery vehicles.

TABLE 13.2
Phytochemical-Based Nanoformulations to Treat Viral Infections

Phytochemical	Nanoformulations	Virus	Experimental Model	Size	Refs
Artemisia arborescens essential oil	Solid lipid nanoparticles encapsulating essential oil from *Artemisia arborescens*	Herpes simplex virus-1	African Green Monkey kidney cells (Vero)	219 nm (SLN1) and 223 nm (SLN 2)	(Lai et al., 2007)
Coumestrol	Coumestrol encapsulated in lipid nanoemulsion dispersed in hydroxyethylcellulose gel	HSV-1 and HSV-2	GMK AH1 cells, CCL81 Vero cell line	200–270 nm (varies according to formulation)	(Argenta et al., 2018)
Curcumin	Curcumin nanoemulsion (NE-Cur)	HPV-16	HPV-16-positive vulva carcinoma cells (A431)	195 nm	(Bonfim et al., 2020)
Curcumin	Curcumin nanoparticles	HSV-2	Female C57BL/6 mice	290–300 nm	(Vitali et al., 2020)
Curcumin	Self-nanoemulsion	DENV	A549 cells infected with DENV (four strains)	40.85 ± 0.919 nm	(Nabila et al., 2020)
Curcumin	Curcumin encapsulated in PLGA nanoparticles	Zika virus	-	-	(Pacho et al., 2021)
Curcumin	[mesoporous silica nanoparticles (MSN) with Eu+, GD+ combined with polymer PEGMA] MSN-EuGd-PEGMA loaded with curcumin	Zika virus	A549 (human alveolar basal epithelial adenocarcinoma cells) cells infected with Zika virus (African strain)	MSN-EuGd-PEGMA: radius 386.5 nm	(Lo et al., 2021)
Curcumin	Trimethoxysilane-based nanoparticles combined with curcumin	Rift Valley fever virus	HSAECs infected with MP-12 virus	-	(Narayanan et al., 2012)
Curcumin	Curcumin chitosan (CuCs) nanocomposite	Hepatitis C virus genotype 4a	Huh7 cells (human hepatoma cells)	29–39.5 nm	(Loutfy et al., 2020)
Olive leaf extract crude form	Self-emulsion	HSV-1	African Green Monkey kidney cells (Vero cell line)	13–19 nm	(Rania et al., 2016)
Silibinin	Silibinin entrapped in phytoliposomes	HCV	Huh7.5-derived Con1/FL-Neo S2204I cells permanently transfected with HCV	SP 165 ± 6 nm	(Ripoli et al., 2016)
Tannic acid	Tannic acid–modified silver nanoparticles	HSV-2	Female C57BL/6 mice	33 nm	(Orłowski et al., 2018)
Genistein	Genistein-loaded cationic nanoemulsions	HSV-1	Vero cell line (CCL81)	200–300 nm	(Argenta et al., 2016)
Curcumin	β-Cyclodextrin–functionalized graphene oxide composite	Respiratory syncytial virus	HEp-2 cells	127.7 nm	(Yang et al., 2017)
Curcumin	Silver nanoparticles (AgNPs)	Respiratory syncytial virus	Hep-2 cells	193.57 ± 14.84 nm	(Yang et al., 2016)

HCV: hepatitis C virus; HSAEC: human small airway epithelial cell; HSV: herpes simplex virus; DENV: dengue virus; PLGA: poly(lactic-co-glycolic acid)

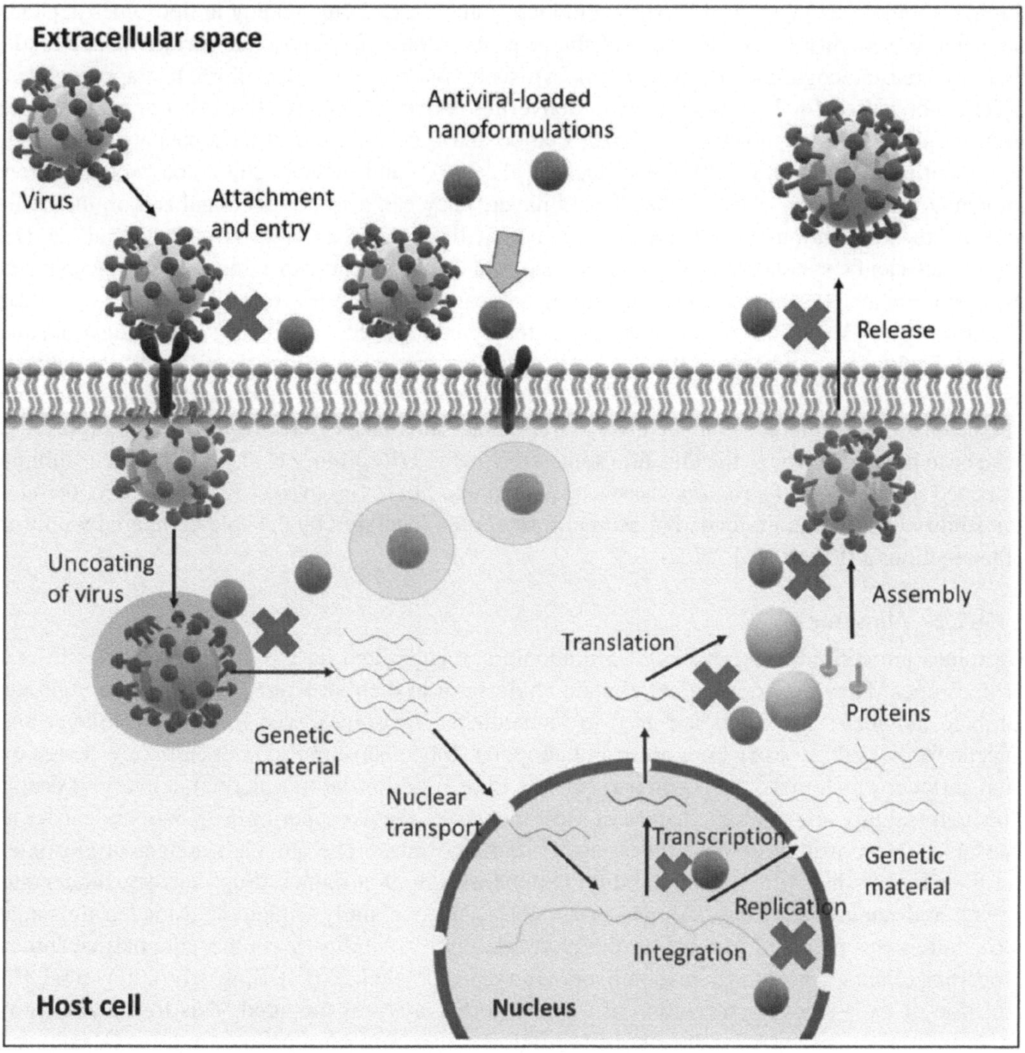

FIGURE 13.2 Overview of general mechanism of action of antiviral-loaded nanoformulations against viruses.

13.4.1.1 Liposomes

Liposomes are the most commonly used drug delivery system for antivirals (Delshadi et al., 2021). In 1965, liposomes were first commercialized for drug delivery. Liposomes are vesicular systems with spherical morphology and are made of at least one phospholipid bilayer (Milovanovic et al., 2020). Liposomes have a size range of approximately 15–1000 nm (Chakravarty and Vora, 2021; Maus et al., 2021). Depending on the presence or absence of the lipid bilayer, liposomes are divided into small unilamellar vesicles, large unilamellar vesicles, multilamellar vesicles, and multivesicular vesicles. As liposomes consist of a hydrophilic core and a hydrophobic lipid layer so that both hydrophilic and hydrophobic drugs can be loaded (Milovanovic et al., 2020; Pradhan et al., 2021). Various kinds of lipids used for liposome formation are phosphatidylserine, 1-stearoyl-2-linoleoyl-sn-glycerol-3-[phospho-L-serine], phosphatidylcholine, diotadecyl dimethyl ammonium bromide/chloride (DODAB/C), dioleoyl trimethylammonium propane, and cholesterol. Cholesterol present in the phospholipid layer increases membrane stability and decreases membrane permeability. Apart from this, it also imparts flexibility to the liposomes, as a result of which they can pass through small-sized anatomical barriers (Pradhan et al., 2021). Drug loading in liposomes depends on membrane fluidity, liposome size, and charge of its surface. The liposomal surface can be modified to increase recognition ability and endocytosis by macrophages. Therefore, this approach can be efficiently used for HIV treatment by delivering antivirals against HIV because HIV resides in macrophages (Milovanovic et al., 2020; Lembo and Cavalli., 2010). Liposomes are biodegradable, nontoxic (Singh et al., 2017; Delshadi et al., 2021), and biocompatible and produce a low immunogenic response owing to their lipid bilayer. They can also be functionalized on their surface by attaching various ligands for the site-specific delivery of antivirals (Delshadi et al., 2021). Liposomes can be synthesized by various methods such as solvent dispersion method, mechanical dispersion methods such as microfluidization, sonication, homogenization (Delshadi et al., 2021; Chakravarty and Vora, 2021), membrane contactor technology, dual asymmetric centrifugation, and others (Chakravarty and Vora, 2021). Wei et al. (2014) prepared baicalin and tested its delivery using an antiviral phytochemical-loaded liposomal formulation, which increased the bioavailability of baicalin. This formulation can be used to treat viral infections. Quercetin-loaded nanoliposomes have been prepared through the thin-film sonication method (Rodriguez et al., 2019). When silibinin extracted from *Silybum marianum* was encapsulated into liposomes to treat HCV infection, the bioavailability of the silibinin-loaded liposomal formulation increased by 2.4-fold compared with that of pure silibinin (Ripoli et al., 2016).

13.4.1.2 Niosomes

Niosomes are colloidal aggregates of a nonionic surfactant and have a structure similar to that of liposomes. They may or may not contain cholesterol in their structure. Owing to the nontoxic attribute, nonionic surfactants are used in biomedicine. Niosomes exist in the lipid bilayer and micelle forms and can carry both polar and nonpolar drugs. Niosomes do not carry any charge on their surface, which make them efficient carriers to be used for the transdermal delivery of drugs. The high stability and nontoxic nature of niosomes make them a superior drug delivery carrier to liposomes. Moreover, niosomes are less costly than liposomes. The other advantages of niosomes are their longer shelf-life, higher stability, better function of sustained drug delivery, biodegradability, and nonimmunogenic (Ahmad et al., 2021). In one study, ellagic acid-loaded niosomal formulation was prepared to increase the bioavailability, solubility, and other biopharmaceutical properties of ellagic acid for dermal delivery (Junyaprasert et al., 2012). Similarly, a niosomal formulation of the alkaloid-rich fraction of *Fumaria officinalis* was prepared. This formulation was used for treating various inflammatory conditions, diabetes, and neuropathic condition (Raafat and Zahaby, 2020). In this way, niosomes can be used to deliver different phytochemicals with antiviral activity.

13.4.1.3 Solid Lipid Nanoparticles

Solid lipid nanoparticles (SLPs) are colloidal drug delivery systems comprising solid lipid matrices. Their size ranges from 10 to 100 nm (Chakravarty and Vora, 2021). SLPs are an alternative to conventional nanocarriers because they have the ability to overcome limitations associated with conventional systems (Lembo and Cavalli, 2010; Pradhan et al., 2021). The solid lipids used for the preparation of SLPs are partial glycerides, triglycerides, fatty acids, waxes, and steroids. Hydrophobic drugs can be loaded in the central core of SLPs, but the efficiency of drug entrapment depends on the type of fatty acid used to prepare SLPs. Some of the fatty acids used for SLP synthesis are stearic acid, tetradecanoic acid, and palmitic acid; however, because of the presence of long chains, SLPs prepared with stearic fatty acid displays great drug entrapment efficiency (Pradhan et al., 2021). The attributes of SLPs that make them a potent drug nanocarrier are biocompatibility, low toxicity, great drug release profile, easy functionalization for targeted drug delivery, high stability, and safety (Pradhan et al., 2021; Chakravarty and Vora, 2021). One of the unique properties of SLPs is that they can be sterilized in the autoclave (Ahmad et al., 2021). The techniques used for the preparation of SLPs are supercritical assisted injection in a liquid antisolvent, high-pressure homogenization, solvent emulsification, ultrasonication, spray-drying, fluid extraction of emulsions, and emulsification–diffusion. The synthesis of SLPs is scalable and cost-effective. SLPs were explored for the delivery of various synthetic antivirals (Chakravarty and Vora, 2021). Curcumin was loaded in SLPs to increase its oral bioavailability and blood–brain distribution. The formulation was found to be stable and demonstrated controlled drug release (Ramalingam and Ko, 2015). Naringenin-loaded spherical SLPs of size 98 nm have shown increased bioavailability of water-insoluble naringenin in Sprague Dawley rats (Ji et al., 2016). In one study, an essential oil-loaded SLP formulation of *Artemisia arborescens* was effective against HSV in *in vitro* study (Lai et al., 2007). In a similar manner, SLPs can be used for antiviral phytotherapy owing to their commendable characteristics.

Nanostructured lipid carriers (NLCs) are composed of liquid lipids. They are considered as second-generation SLNs (Chakravarty and Vora, 2021). The lipids of NLCs are partially solidified and have irregular crystallinity so that they can accumulate more drugs (Delshadi et al., 2021). Compared with SLPs, NLCs have good stability, a well-controlled drug release profile, and a good ability to load the drug. Additionally, they can be altered for site-specific drug delivery (Chakravarty and Vora, 2021).

13.4.1.4 Micelles

Micelles are another efficient antiviral drug delivery system that are amphiphilic and have colloidal structures, with a size range of 5–100 nm. Micelles comprise amphiphilic molecules that exist separately in the aqueous solution at lower concentrations, but as the concentration increases, aggregation of the molecules occurs. The critical micelle concentration is important for micelle formation because, at this concentration, the amphiphile will aggregate, leading to the synthesis of micelles. The aggregation of molecules occurs at certain concentration intervals only. Another important factor is critical micellization temperature; at this temperature, the monomer will aggregate to form micelles, but below this temperature, the molecules remain separated. Further micelle formation can be achieved only after removing the hydrophobic fragments of micellar molecules from an aqueous solvent and reconstructing hydrogen bonds in water. Micelles have the ability to carry both hydrophilic and lipophilic drugs (Lembo and Cavalli, 2010; Milovanovic et al., 2020). Micelles formed of low-molecular-weight amphiphiles have good structures. Some of the low-molecular-weight amphiphiles used for micelle formation are sodium dodecyl sulfate, sodium deoxycholate, and poly(oxyethylene) 20 sorbitan monooleate (Ahmad et al., 2021). Quercetin-loaded mixed micellar formulation was developed by Lu et. al. (2018), which improved the solubility and bioavailability of the polyphenol quercetin.

13.4.2 Polymer-Based Nanoformulations

Many types of polymers are available, which may be present in natural products or synthetically synthesized to form polymeric nanoformulations. Natural polymers include polysaccharides such as chitosan, alginate, agarose, and dextran and proteins such as lectin, gelatin, and albumin. These polymers are hydrophilic as compared to synthetic polymers, which are hydrophobic. Some of the synthetic polymers are poly(ε-caprolactone) (PECL), polylactic acid, poly(isobutylcyanoacrylate), polyhexylcyanoacrylate, polylactide-co-glycolic acid (PLGA), polymethacrylate, and polybutylcyanoacrylate. Polymers' surfaces are altered to avoid nonspecific interaction with serum components and to modify pharmacokinetic factors of the drug through the uptake of phagocytosis. Many polymeric nanoparticles have been prepared in such a manner that drug release can be stimulated by varying different environmental factors. Polymeric nanoparticles not only increase drug bioavailability but also reduce toxic effects of the drug at nontarget sites (Chakravarty and Vora, 2021).

13.4.2.1 Polymeric Micelles

Polymeric micelle is the type of micelle comprising amphiphilic block copolymers. Some of the widely used core-forming polymers are polyesters, poloxamers, polycaprolactone (PCL), and poly(L-amino acids) (Patra et al., 2018; Chakravarty and Vora, 2021). These micelles enhance the bioavailability and stability of the drug. Synthesis of polymeric micelles can be carried out by directly suspending polymers in the solvent and subsequently dialyzing them or by adding a solvent, leading to aggregation of one block (Patra et al., 2018). Polymeric micelles are formed at the critical micellization concentration and the critical micellization temperature. Their size ranges from 10 to 100 nm, and they can carry both hydrophilic and hydrophobic antiviral phytotherapeutic agents (Chakravarty and Vora, 2021). Polymeric micelles have more therapeutic potential than conventional micelles. Micelles show a slower rate of dissociation, thereby enhancing drug retention time and concentrating a large amount of drug at the target site (Chakravarty and Vora, 2021; Singh et al., 2017). Viruses enter the cell with the aid of lectin receptors on the host cell. Hence, micelles made of polyethylene glycol (PEG)–polylactide copolymer along with surface modification with galactose can be prepared to target viruses because galactose interacts with lectins (Milovanovic et al., 2020). Kheiri et al. (2017) prepared curcumin-loaded mono methoxy-PEG-PECL copolymeric micelles of size 81 nm. The therapeutic efficacy of curcumin increased after loading. This approach can be used for antiviral therapy, as curcumin has antiviral potential.

13.4.2.2 Dendrimers

Dendrimers are polymeric nanoparticles comprising a hyperbranched, three-dimensional structure. They contain a central core, and branches originating from it are made of recurring monomeric units termed as dendrons, and the outer region comprises multiple functions mainly for binding ligands (Chakravarty and Vora, 2021; Milovanovic et al., 2020; Singh et al., 2017). Dendrimers are attractive carriers for antiviral delivery because of their size being less than 100 nm (Milovanovic et al., 2020), central cargo property, controlled release of the drug, and effective translocation from various anatomical barriers (Pradhan et al., 2021). Dendrimers are mainly useful for drugs with lower solubility and have a hydrophobic nature. They deliver the drug by simply releasing it owing to a change in physiological conditions or by breaking the covalent bond between the drug and the dendrimer (Ahmad et al., 2021). Dendrimer synthesis is a stepwise approach that includes either convergent or divergent branching methods. Some of the examples of dendrimers are poly(glycerol), poly(glycerol-co-succinic acid), polyamidoamine (PAMAM), poly(L-lysine), melamine, poly(2,2-bis(hydroxymethyl)propionic acid), and poly(propylene imine) (Pradhan et al., 2021). In one study, polyamidoamine dendrimers of a different generation, that is, G0–G3, were used as a carrier for the phytochemical quercetin. The size of different formulations was in the range of 34.39–100.3 nm. This delivery system improved the solubility of quercetin (Madaan et al., 2016).

13.4.2.3 Nanogels

Nanogels are also known as hydrogels. They are made of a cross-linked polymer net with 3D structures, which impart them the unique structure consisting of water (95%) and the unique feature of being swelled up without being dissolved into the aqueous solvent. Their size is in the range of 10–1000 nm. Nanogels can be prepared from natural products, synthetic sources, or a combination of both polymers. The attributes of nanogels that render them as valuable nanocarriers are their great drug-loading capacity, ability to protect the drug from degradation, stability, biocompatibility, and hydrophilicity. Additionally, nanogels show a stimulus-responsive behavior, wherein they respond to various external and internal stimuli such as temperature, enzyme concentration, pH, light, redox environment, and magnetic field. Nanogels are prepared by chemical cross-linking of the polymer, emulsion polymerization, and physical cross-linking of the polymer (Maus et al., 2021; Lembo and Cavalli, 2010). In one study, chitosan-based nanogels were prepared to load myricetin. The size of the myricetin-loaded nanogel formulation was in the range of 100–300 nm (Yao et al., 2016). Similarly, nanogels can be used as a carrier for the delivery of various phytotherapeutics.

13.4.2.4 Cyclodextrin-Based Nanodelivery System

Cyclodextrins are cyclic oligosaccharides derived from starch and have the shape of truncated cones owing to the glucopyranose monomers arranged in a chair conformation. The alpha, beta, and gamma cyclodextrins are the predominant components, each containing 6, 7, and 8 glucopyranose units, respectively (Lembo and Cavalli, 2010). These oligosaccharides are amphiphilic in nature with a hydrophobic inner area and a hydrophilic surface owing to the presence of the -OH group. The cyclodextrin nanodelivery system enhances the stability, solubility, and bioavailability of drugs. Beta cyclodextrins are more commonly used than alpha and gamma cyclodextrins (Chakravarty and Vora, 2021). Shulman et al. (2011) complexed naringenin, a potent antiviral effective against HCV with hydroxypropoyl-β-cyclodextrin. This delivery system increased the solubility and bioavailability of naringenin Another β-cyclodextrin–based formulation of baicalin was prepared by Li et al. (2017). Here also, the dissolution rate and solubility of baicalin were enhanced, attributable to β-cyclodextrin.

13.4.2.5 Other Polymeric Nanoparticles

Polymeric nanoparticles are potent nanocarriers made of either synthetic or natural polymers. The methods used for the preparation of polymeric nanoparticles are solvent evaporation, nanoprecipitation, ionic gelation, salting out, emulsion–diffusion evaporation, dialysis, polymer dispersion, coacervation, phase inversion temperature, supercritical fluid technology, controlled interfacial polymerization, and spray-drying. The major benefits of using polymeric nanoparticles, which makes them a potent carrier for antiviral delivery, are enhanced cellular uptake, low toxicity, controlled drug release, target-specific delivery, and protection of the drug from degradation. Polymeric nanoparticles are majorly explored for their increasing efficacy of antiretroviral drugs because of their high pertinency in targeting macrophages and monocytes, which are the primary pool for viral diffusion during HIV infection, thereby being efficient in treating HIV infection. Nanocapsules are a unique type of polymeric nanoparticles wherein the drug is loaded in the core. Their size ranges from 50 to 300 nm. Nanospheres are nanoparticles made of polymer, of size 10–200 nm, wherein the drug is adsorbed onto their surface or implanted into the matrix (Chakravarty and Vora, 2021). Chitosan nanoparticles are one of the polymeric nanospheres with great potential for the site-specific delivery of plant-based antivirals and synthetic compounds (Delshadi et al., 2021). Curcumin was loaded on chitosan nanoparticles to treat HCV infection. The nanoformulation inhibited HCV genotype 4a (HCV-4a) replication and entry, thereby establishing itself the potent phytotherapeutic to treat HCV infection. The size of the nanoformulation was 29–39.5 nm, and the zeta potential was 33 mV (Loutfy et al., 2020). Another nanoformulation of curcumin with PLGA nanoparticles was prepared to treat Zika virus infection. The formulation inhibited viral translation and RNA synthesis (Pacho et al., 2021).

Protein-based nanoparticles also have the potential for use in antiviral drug delivery (Delshadi et al., 2021). Protein-based nanoparticles can be synthesized by methods such as electrospraying, coacervation/desolvation, complex coacervation, and emulsion/solvent extraction. There are two types of proteins: water-insoluble proteins including gliadin and zein, and water-soluble proteins including human and bovine serum albumin. These nanoparticles can be easily metabolized, functionalized, and decomposed (Patra et al., 2018), which make them a good carrier for targeted drug delivery application.

13.4.3 Nanoemulsions

Nanoemulsions are colloidal systems consisting of emulsions of size less than 200 nm. They aid in protecting and encapsulating antivirals with great efficacy. Nanoemulsions are thermodynamically stable and consist of water, oil, surfactants, and cosurfactants. The two common types of nanoemulsions are water-in-oil (W/O) and oil-in-water (O/W) types. Water-soluble antivirals are encapsulated into W/O nanoemulsions and oil-soluble in O/W nanoemulsions. There are also other types of nanoemulsions such as water-in-oil-in-water (W/O/W) and oil-in-water-in-oil (O/W/O) (Delshadi et al., 2021). Nanoemulsions can be synthesized by using various approaches such as sonication, high-pressure homogenization, microfluidization, emulsion phase inversion, spontaneous emulsification, or phase inversion temperature (Delshadi et al., 2021; Chakravarty and Vora, 2021). Nanoemulsions are potent drug delivery vehicles because they increase solubility, absorption, and bioavailability of drugs and can be used for targeted drug delivery and helps in sustained release (Chakravarty and Vora, 2021). Cationic nanoemulsions were used to encapsulate genistein by spontaneous emulsification methods. Further *in vitro* study proved that the formulation was effective against HSV-1, strain 29R (Argenta et al., 2016). In one study, curcumin-loaded O/W nanoemulsions were effective for treating lesions caused by HPV infection. The size of the nanoformulations is 195 ± 3.1 nm (Bonfim et al., 2020). Self-nanoemulsifying drug delivery systems (SNEDDS) are thermostable lipid-based monotropic systems prepared using the spontaneous emulsification method. To prepare SNEDDS, oil or lipid is emulsified with water by using surfactant, solvents, co-surfactants, and co-solvents under gentle stirring. SNEDDS have good penetration and bioavailability. It is a useful delivery system for lipophilic drugs (Chakravarty and Vora, 2021). Naringenin-loaded SNEDDS increased the solubility and bioavailability of the drug (Khan et al., 2015). Another formulation of SNEDDS with myricetin increased its solubility and permeability (Qian et al., 2017). SNEDDS can be used for phytotherapeutic delivery of the antiviral.

INORGANIC NANOPARTICLES

13.4.4 Metal Nanoparticles

The properties of metal nanoparticles, such as surface plasmon resonance (SPR), biocompatibility, and easy surface functionalization, make them ideal for use in various biomedical applications such as targeted drug delivery, bioimaging, hyperthermia, and biosensing. Surfaces of metal nanoparticles can be easily modified with drugs, ligands, antibodies, and other molecules for various biomedical applications, mainly for targeted drug delivery. Some of the widely studied metal nanoparticles are gold, copper, silver, and iron nanoparticles (Patra et al., 2018). The size of metal nanoparticles is in the range of 1–100 nm. They are synthesized using either the top-down or bottom-up approach. The size, shape, drug encapsulating efficiency, and drug release profile of metal nanoparticles can be altered by altering the reaction conditions (Singh et al., 2017). However, drug delivery-related studies using metal nanoparticles need more investigations owing to nanotoxicity issues (Patra et al., 2018).

13.4.4.1 Silver Nanoparticles

The inherent antimicrobial activity of silver nanoparticles against different bacteria, viruses, and fungi makes them a very potent metal nanoparticle. The mechanism involved in antimicrobial function is that silver nanoparticles interact with the electron transport chain and various respiratory chain enzymes, viral proteins, and with the genetic material of the microorganisms, which causes cellular dysfunction, leading to cell death (Pradhan et al., 2021; Singh et al., 2017; Chakravarty and Vora, 2021). The properties of silver nanoparticles that are responsible for their use in therapeutics are their chemical stability, catalytic properties, stability, high conductivity, and localized SPR (Singh et al., 2017; Chakravarty and Vora, 2021). Silver nanoparticles have efficient activity against a broad range of viruses such as HBV, HIV-1, HSV, RSV, and retrovirus owing to their vast surface area and their small size, which helps in rapid dissolution (Pradhan et al., 2021; Chakravarty and Vora, 2021). As silver nanoparticles can attack multiple targets of microorganisms, the chances of resistance development are less (Chakravarty and Vora, 2021). Silver nanoparticles of diameter less than 10 nm have effective antiviral properties toward HIV-1, as proposed in various studies (Pradhan et al., 2021). Curcumin-modified silver nanoparticles have shown effective antiviral activity against RSV (Yang et al., 2016). In one study, tannic acid-modified silver nanoparticles of size 33 nm were used for HSV-2 infection. *In vitro* and *in vivo* studies have shown that modified silver nanoparticles inhibit viral entry and their spread into the host cells (Szymańska et al., 2018).

13.4.4.2 Gold Nanoparticles

Gold nanoparticles (AuNPs) have antimicrobial activity against various microorganisms owing to their functional and structural properties (Pradhan et al., 2021). AuNPs are extensively studied metal nanoparticles because of their ease of synthesis and functionalization, biocompatibility, very good conductivity (Singh et al., 2017), and unique SPR (Chakravarty and Vora, 2021). They show unique chemical, optical, biological, and physical properties due to which they are gaining interest in pharmaceutical industries (Pradhan et al., 2021). Gold nanoparticles exhibit antiviral potential against a wide range of viruses such as influenza virus, HBC, HIV-1, and HSV. Literature evidence suggests that the antiviral activity of AuNPs is dependent on the size and the interaction between the virus cell wall and gold nanoparticles, for example, AuNPs of size 7–70 nm are effective against influenza virus A/47/1/FM(H1N1) in Madin-Darby canine kidney cell line. AuNPs of size of up to 30 nm have significant antiviral potential. The different shapes and sizes of AuNPs influence the immune response and their cellular uptake by the host cells (Pradhan et al., 2021). AuNPs of size 7.86 nm synthesized using gallic acid were effective in treating HSV infection by inhibiting viral attachment and entry (Halder et al., 2018). AuNPs can be used as a potential carrier of various phytochemicals to treat viral infections. They can also be used as the delivery vehicle for phytochemicals that can treat COVID-19.

13.4.5 Metal Oxide Nanoparticles

Metal oxide nanoparticles can be used as an efficient drug delivery vehicle to treat viral infection. They can be used for the site-specific delivery of the drug by modifying their surface (Maus et al., 2021). These nanoparticles are effective against a wide range of viruses such as influenza (H3N2 and H1N1), HIV-1, vesicular stomatitis virus, DENV type-2, HBV, HSV, and foot-and-mouth disease virus (Chakravarty and Vora, 2021). Copper oxide nanoparticles have the ability to inhibit viruses and other microorganisms owing to their excellent catalytic properties. Enhanced oxidative stress of Cu^{2+} ions makes them a potent antiviral against HIV-1, HSV, HAV, and influenza virus. Copper oxide nanoparticles have toxicity issues (Pradhan et al., 2021). Although these nanoparticles have antiviral potential, more studies need to be undertaken regarding their use as a drug delivery carrier for antivirals. Zinc oxide nanoparticles are potent antimicrobial agents owing to their high surface area-to-volume ratio, and their robust electrostatic interactions between zinc oxide nanoparticles and cell walls of the microorganisms lead to the production of reactive oxygen species, which consequently leads to cell death. Their properties such as brilliant biocompatibility and great reducing ability encourage researchers to use these nanoparticles in various biomedical applications (Pradhan et al., 2021).

13.4.6 Carbon-Based Nanoparticles

Carbon-based nanoparticles include carbon nanotubes (CNTs), graphene, and fullerenes. These nanoparticles are a potent vehicle for antiviral delivery owing to less toxicity and ease in altering their properties according to the application (Maus et al., 2021). Carbon-based nanoparticles can be used for antiviral phytotherapeutic delivery.

13.4.6.1 Carbon Nanotubes

CNTs are the allotropic form of carbon. As the name suggests, they are hollow cylindrical tubes made by rolling graphene sheets. The ends of nanotubes are capped with fullerene sheets. The two most common types of CNTs are single-walled CNTs and multiple-walled CNTs (MWCNT). CNTs can be used as a drug nanocarrier owing to their characteristics such as high aspect ratio; high surface area, which increases drug encapsulating ability; good electrical and thermal conductivity; lightweight; good cell penetration property; and great mechanical strength (Ahmad et al., 2021; Chakravarty and Vora, 2021). Some of the methods used for the synthesis of CNTs are laser-ablation technique, electric arc-discharge techniques, and chemical vapor deposition technique (Chakravarty and Vora, 2021). The inner hollow part of CNTs contains drugs, and the outer surface can be functionalized for targeted drug delivery. CNTs are commonly used for delivering anticancer drugs. The use of CNTs have some drawbacks such as their toxicity and less solubility; however, these limitations can be overcome by functionalizing CNTs (Ahmad et al., 2021; Chakravarty and Vora, 2021). CNTs have shown antiviral potential against RSV and can be used as a candidate for vaccine delivery (Chakravarty and Vora, 2021). In one study, it was found that silver and copper MWCNTs functionalized with *Terminalia arjuna*'s bark extract inhibited bacteria and fungi. The nanomaterial did not show any toxicity for normal epithelial cell lines (Yallappa et al., 2016).

13.4.6.2 Graphene

Graphene is a carbon-based nanomaterial that shows efficient antiviral potential. It consists of 2D sheets of sp2-hybridized carbon atoms that are organized in a hexagonal pattern. Graphenes can be used as graphene oxide (GO) for various biomedical applications. GO nanomaterials are amphiphilic in nature; hence, they can be used for loading both hydrophilic and lipophilic drugs. The functional groups consisting of oxygen on its surface help in GO functionalization; it also has less toxicity, biocompatibility, and better dispersibility. These functional groups help in attachment with nucleic acids (DNA and RNA) and proteins (Chakravarty and Vora, 2021). GO can inhibit viruses through various mechanisms owing to its ability to form electrostatic interactions with the virus cell membrane with its negatively charged surface, easy penetration because of their sharp edges, and photocatalytic activity for photodegradation of viruses (Maus et al., 2021). The advantages of graphene-based nanomaterials are their excellent mechanical strength, high surface area, and good drug-loading capacity. All these attributes make them a suitable agent for antiviral therapeutics (Chakravarty and Vora, 2021). In one study, functionalized GO was loaded with curcumin to treat RSV infection. The formulation inhibited virus entry and directly inactivated RSV. The GO-based formulation was biocompatible with less toxicity (Yang et al., 2017).

13.4.6.3 Fullerenes

Fullerenes are carbon-based nanostructures that are discovered earlier than other carbon-based nanocarriers. It has gained significant attention in antiviral investigations. Fullerenes are cage-like hollow spheres consisting of carbon atoms; the most common fullerene is Buckminster fullerene (C60) with 60 carbon atoms. The great symmetry of fullerene makes them ideal agents for drug delivery, and their antimicrobial activity provides them with huge scope for derivatization; they also have free radical scavenging properties and less toxicity. Fullerenes are effective against HIV infection because they can block the active site of HIV protease, thereby inhibiting the enzyme (Chakravarty and Vora, 2021; Maus et al., 2021). Fullerenes can inhibit viruses by generating singlet oxygen particles and they function as photosensitizers. The drawback associated with fullerenes is their less solubility. However, fullerenes play a promising role in treating various viral infections caused by HCV, influenza virus (H1N1), HSV, RSV, DENV, zika virus, and HCV (Maus et al., 2021).

Despite having certain brilliant properties, the use of inorganic nanoparticles for drug delivery applications is limited because of their toxicity issue. More research in this direction may lead to their use in targeted drug delivery applications. Figure 13.3 presents an overview of the different

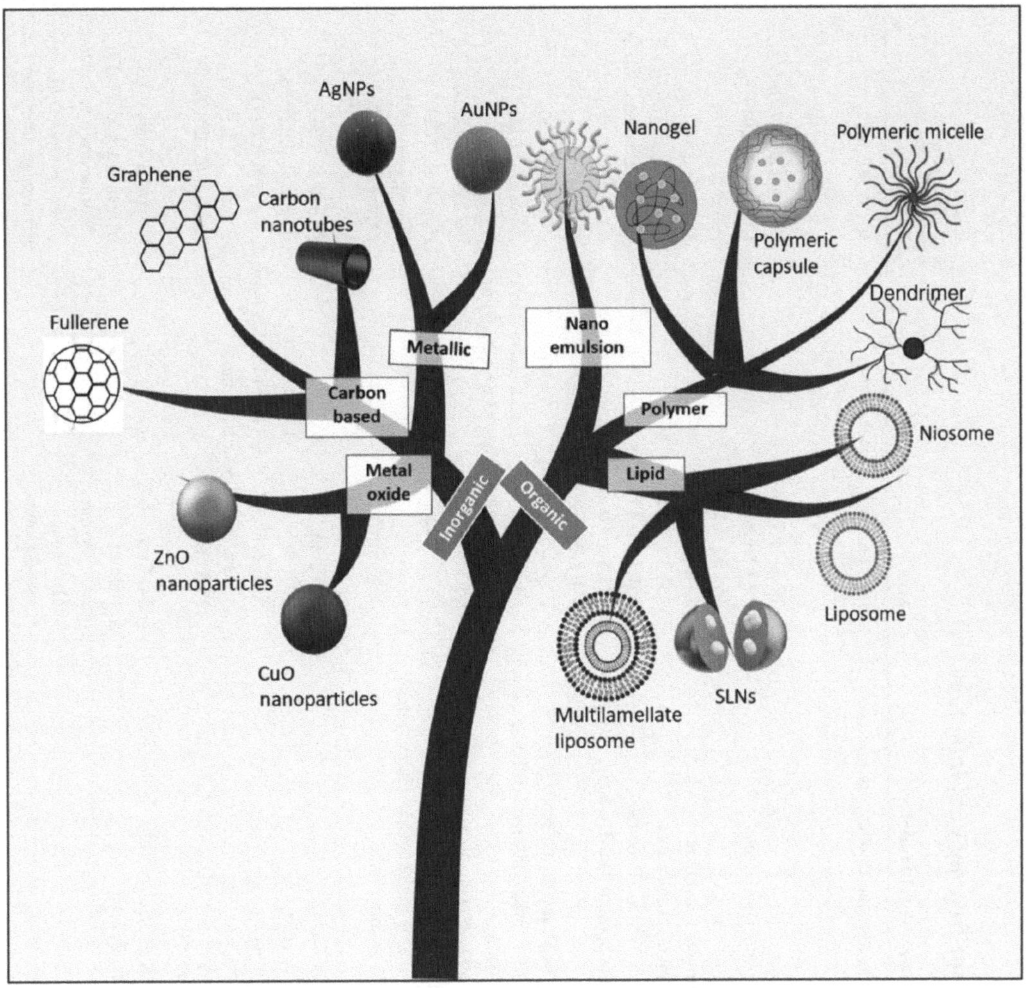

FIGURE 13.3 Different types of nanocarriers for delivering antiviral agents.

types of nanocarriers that can be used for drug delivery applications. Table 13.3 provides details about the potential nanoformulations that can be used to treat viral infections.

13.5 SURFACE MODIFICATION AND CHARACTERIZATION OF NANOCARRIERS

The size of nanocarriers itself enhances their property as a drug delivery vehicle over traditional drug carriers. Surface modification further enhances the properties of nanocarriers and helps in overcoming their limitations (Maus et al., 2021). Usually, nanocarriers are modified by either coating them or adding charges to their surface. Surface coating enables nanoparticles to maintain their unique properties and guarantees that the drug remains uniformly dispersed. The surface is the major deciding factor for the behavior of nanocarriers, and charges, regardless of positive, negative, or neutral, rule properties such as absorption, compatibility, toxicity, and elimination rate of nanoparticles. The charges can be modified with agents such as polymers or lipids. To function as an efficient drug delivery vehicle, different nanocarriers require a different charge. For example, polymeric nanoparticles and nanoemulsions favor a positive charge so that they can effortlessly penetrate

TABLE 13.3
Potent Phytochemical-Based Nanoformulations Used for Treating Viral Infections

Nanoparticles	Phytochemical	Experimental Model	Size	Refs
Methoxy Poly(ethylene glycol)-B-poly (L-lactic acid) micelle	Quercetin	Male Sprague Dawley rats	88.5 nm	(Lv et al., 2017)
Polymeric nanoparticle carrier	Myricetin	-	12.5–46.7 nm	(Sims et al., 2020)
Exosomes	Curcumin	Sprague Dawley rats	93 ± 6 nm	(Aqil et al., 2017)
Micelles	Quercetin	Wistar rats	16.20 ± 2.12 nm	(Lu et al., 2018)
Solid lipid nanoparticles	Naringenin	Sprague Dawley rats	98 nm	(Ji et al., 2016)
Lactoferrin nanoparticles	Oleanolic acid	Sprague Dawley rats	202.2 ± 8.3 nm	(Xia et al., 2017)
Poly(N-butylcyanoacrylate) nanoparticles	Quercetin	Male Wistar Albino rats	161.1 ± 0.44 nm and 166.6 ± 0.33 nm	(Bagad and Khan, 2015)
Nanosuspensions	Andrographolide	Male Sprague Dawley rats	244.6 ± 3.0 nm	(Qiao et al., 2017)
Poly-D,L-lactide nanoparticles	Quercetin	-	130 ± 30 nm	(Kumari et al., 2010)
Nanoemulsion	Naringenin	Male Albino Wistar rats	Below 50 nm (34.5 ± 2.91 nm)	(Khan et al., 2015)
Nanosuspension	Oleanolic acid	A549 human non–small cell lung cancer cell line	Varies according to formulation: average size for 16 batches 93.67 ± 53.53 nm	(Li et al., 2014)
Micelle	Myricetin	Glioblastoma cells, Sprague Dawley rats	96.3 nm	(Wang et al., 2016)
Solid lipid nanoparticles	Baicalin	Rabbit	91.42 ± 1.02 nm	(Liu et al., 2011)
PEGylated lipid nanoparticles	Baicalin	Acute MI (myocardial ischemia) rats	83.9 nm	(Zhang et al., 2016)
Lactobionic acid–modified chitosan lactate polymer nanoparticles	Baicalin	Swiss Albino mice	490 nm	(Ahmed et al., 2020)
Nanosuspension	Apigenin	HepG2 cells	159.2 nm	(Wu et al., 2017)
Magnetic $Fe_2O_3/Fe_3O_4@mSiO_2$ nanocomposite	Apigenin	A549 cells	-	(Liu et al., 2021)
PLGA-DMSA nanoparticles	Apigenin	A549 cells	APG-NPs: 92.18 ± 2.1 and DMSA conj. APG-NPs: 93.41 ± 2.4 nm	(Sen et al., 2021)
Polymeric micelle–poly (ethylene glycol)-poly(D,L-lactide) copolymer	Chalcone	MCF-7 and MDA-MB-231 cells	20.8 nm	(Kim et al., 2019)
Gold nanoparticles	Chalcone	HEK293-treated cells	2–12 nm	(Omolaja et al., 2021)
Lignin nanoparticles	Curcumin	Caco-2 cells	104 nm	(Alqahtani et al., 2019)
Lignin nanoparticles	Folic acid	-	200–500 nm	(Pourmoazzen et al., 2020)
Zinc oxide-quercetin nanocomposite	Quercetin	MCF-7	21–39 nm	(Sathishkumar et al., 2021)
B-Lactoglobulin-based nanoparticles	Quercetin	-	180–300 nm	(Chavoshpour-Natanzi and Sahihi, 2019)

Small extracellular vesicles	Delphinidin	Human aortic endothelial cells	111.7 ± 1.8 nm	(Barkallah et al., 2021)
Oligomerized (-)-epigallocatechin-3-O-gallate	Lycopene	Male C57BL/6 mice	152 ± 32 nm	(Li et al., 2017)
Pectin nanoparticles	Honokiol	HepG2 cell	640.50 ± 11.97 nm	(Zhang et al., 2015)
Polymeric nanoparticles	Honokiol	HNE-1 cells	20.77 ± 7.56 nm	(Yang et al., 2017)
Poly(ethylene glycol)–poly(E-caprolactone)–poly(ethylene glycol) micelles	Honokiol	HEK293, A549 cells	58–61 nm (redissolved lyophilized nanoparticles)	(Gong et al., 2011)
PLGA nanoparticles	Oleanolic acid and ursolic acid	Male New Zealand Albino rabbits	<225 nm	(Alvarado et al., 2015)
Silver nanoparticles	Sennoside A	*Staphylococcus aureus* and *Escherichia coli*: two reference yeast strains (*Candida albicans* and *Candida parapsilosis*)	12 nm	(Al-Ghamdi et al., 2020)
PLGA nanoparticles	Xanthohumol	Human corneal epithelial cells (HCE-T)	~200 nm	(Ghosh et al., 2021)
PLGA nanoparticles	Xanthohumol	B16F10, malignant cutaneous melanoma, and RAW 264.7, macrophagic, mouse cell lines	312 ± 41 nm	(Fonseca et al., 2021)

MI: myocardial infarction; PLGA: poly(lactic-co-glycolic acid); DMSA: meso-2,3-dimercaptosuccinic acid; NP: nanoparticle negatively charged skin and have better cellular uptake along with high drug-loading capacity. In contrast to this, either neutral or negative charges are preferred for liposomes and dendrimers on their surface, as they prevent phagocytosis, do not cause toxicity, and lead to a longer retention time in plasma (Chakravarty and Vora, 2021; Maus et al., 2021). Likewise, silver nanoparticles and CNTs favor a positive charge, whereas SLPs and micelles prefer a negative charge (Chakravarty and Vora, 2021).

Chemical treatment is another way to modify nanoparticles as drug delivery systems. Using chemical treatment, the dispersion stability of nanoparticles can be improved. The adsorption of silane coupling agents on nanoparticles is one of the examples of chemical treatment. Silane coupling is formed by its absorption on the hydrophilic outer surface along with the interaction with -OH groups. This method increases the compatibility between polymeric surfaces and nanoparticles (Maus et al., 2021).

Another approach in surface modification is grafting and ligand exchange. The surface topology of nanocarriers and the functionality of the groups present on the surface of nanoparticles can be altered by grafting synthetic polymers on their surface. Polymer grafting can be achieved by grafting the complete polymer either directly or in a stepwise manner, which is preferred by adding a single anchoring group and, subsequently, building the polymer on nanoparticles' surface. Polymers used for grafting are polyacrylamide (PAAM), polystyrene, and methyl methacrylate (Maus et al., 2021). Ligand exchange helps to add an appropriate ligand on the surfaces of the nanoparticles, which supports binding and other functions (Maus et al., 2021). Surface modification of nanocarriers is an important tool for the application of nanophytotherapeutics against viral diseases. This facilitates the cellular uptake of nanoformulation by crossing various biological barriers and improves the stability of nanoformulations. Subsequently, these attributes lead to enhancement of the antiviral potential of loaded compounds, which improves the pharmacodynamics and pharmacokinetics of antivirals and reduces the chances of drug resistance and toxicity by allowing controlled release of drugs (Fouad, 2021). In this way, surface modification of nanocarriers is necessary to overcome limitations such as toxicity and solubility, among others and further assures the effectiveness of the prepared nanoformulation at the site of action.

There are various techniques by which nanocarriers and nanoformulations can be characterized to study various aspects such as their morphology, size, surface charge, surface groups, their nature, and their arrangement to confirm drug loading, analyze surface modification, and for many other purposes. Overall characterization is essential to study the physicochemical properties of nanoparticles and alter them according to the application for which it has to be used. These techniques include ultraviolet-visible spectroscopy, to study their stability; dynamic light scattering, to study particle size distribution and zeta potential; transmission and scanning electron microscopy, to study the size, shape, and surface of the nanoparticles; and X-ray diffraction, to study the crystallinity. Some other techniques are Fourier transform infrared spectroscopy, atomic force microscopy, energy-dispersive X-ray analysis, field-emission scanning electron microscopy, X-ray photoelectron spectroscopy, localized SPR, and electrophoresis laser Doppler (Fouad, 2021).

13.6 ROUTES OF DELIVERY AND POTENTIAL BIOLOGICAL BARRIERS

Various routes for the delivery of antiviral drugs are oral, transdermal, intravenous, and ocular. In the oral delivery route, the drug is absorbed and transported into systemic circulation by passing first through the gastrointestinal tract (GIT) and then through the small intestine. The bioavailability of the antiviral through this route depends on its accessibility, stability, and absorption. Many plant-based antiviral compounds have low solubility; hence, their solubility and accessibility are increased in the GIT by ingesting them with lipids. Lipid-based nanoparticles can help deliver antivirals by this route. Nanoparticles also help protect the antiviral drug from harsh conditions in the GIT and enhance their bioavailability. In transdermal delivery, antivirals are administered through the skin. Delivery through this route provides easy application and increases bioaccessibility. Many synthetic antiviral drugs are delivered by this route. The prerequisite for transdermal delivery is that antivirals should fuse with the lipids of the skin and penetrate deep through the tissues present beneath. However, it is difficult for free antiviral compounds to withstand the aforementioned requirement, necessitating the need for a nanoparticle-based drug delivery system, which enhances the bioavailability of antivirals by enhancing their absorbance and penetration. Another factor that needs attention regarding the delivery of antivirals by this route is that the active plasma level of the drug should be achieved, as well as its absorption in the systemic circulation. The delivery of antiviral compounds in the vein by injection is the intravenous route of delivery. This is the best route in emergencies because it bypasses various barriers involved in absorption. The limitation associated with this route is that drugs can undergo opsonization and, subsequently, phagocytic uptake, but this limitation can be solved using the nanoparticle delivery system. Moreover, this route is costly and injection-related concerns may arise. Neutral nanoparticle-based delivery systems are preferred for this route because they have a longer circulation time. In contrast to this, cationic nanoparticles are removed from the circulation rapidly through opsonization when administered intravenously (Delshadi et al., 2021). These are the commonly used delivery routes for antiviral compounds. However, other routes can be explored for these compounds, such as ocular delivery.

The biological barriers protect the internal systems from foreign substances and act as the first line of defense. These barriers provide selective access to nutrients and therapeutic agents. Intracellular drug delivery can be achieved by overcoming biological barriers such as the skin, mucus, blood–brain barrier (BBB), and cell membrane. Mucus is the protective layer present almost everywhere in the body. It contains mainly water (95%), as well as cholesterol, proteins, fibers, lipids, and salts. According to its position, and its amount, thickness, and pH differ, although it is produced continuously. The mucosal barrier can be overcome for drug delivery by mucoadhesion, which involves interactions of different groups on nanocarriers with the mucus components through the formation of different bonds or mucopenetration, mainly by interacting with mucus and chemically shifting its features or by enhancing the hydrophilicity to penetrate the mucus. Skin, the largest organ of the body, functions to maintain hydroelectrolytic balance and temperature in the body. The skin has two layers: the dermis and epidermis. Functionalization of nanocarriers with different compatible functional groups helps overcome the skin barrier to deliver antiviral compounds. Nanotechnological approaches enhance the absorption and penetration of drugs delivered through the skin. The selectively permeable lipid bilayer of the cell membrane separates the extracellular components from the cytoplasm. Nano-based delivery systems can pass through the cell membrane by simple diffusion, receptor-mediated endocytosis, pinocytosis, or nonspecific phagocytosis (Cojocaru et al., 2020) depending on the nanocarrier type used for delivery. The BBB restricts the entry of any foreign substance into the brain. The BBB separates the brain parenchyma from systemic circulation. It is a layer of selectively permeable tightly packed endothelial cells (Delshadi et al., 2021). Small lipophilic substances can pass through the BBB by diffusion; another mechanism of transport through the BBB is active efflux transport. Although the BBB is a complex and difficult-to-access barrier, nanotechnological approaches can help deliver antiviral compounds across this barrier, owing to the small size of nanoparticles, and their surface can be functionalized specifically (Cojocaru et al., 2020). Various nanocarriers have been investigated to deliver drugs across the BBB. Nanotechnology-based drug delivery vehicles are gradually gaining increasing importance for the preparation of nanoformulations to treat viral infections and several other diseases owing to their excellent advantages in terms of crossing various anatomical barriers and drug deliver at the target site, which enhances drug efficacy.

13.7 FUTURE NANOSYSTEMS FOR ANTIVIRAL PHYTOTHERAPEUTICS

Different types of novel nano-delivery systems, including nanotraps, nanobubbles, nanorobots, nanodiamonds, and nanofibers, which have the potential to be used in antiviral phytotherapy in the future are discussed in the following sections.

13.7.1 Nanotraps

As the name suggests, nanotraps function to selectively trap viral proteins, virions, and the target virus and exclude other high-molecular-weight proteins. The trapping is executed using electrostatic and hydrophobic interactions. Nanotraps are nanoparticles in the form of homogeneous hydrogels wherein the core is made of charged, high-affinity aromatic compounds and the sieving shell surrounding the core is made by polymerization. However, they can be physically and chemically reformed for various applications. Nanotraps are highly stable at different temperatures and storage environments. The sieve shell captures the virus and retains it in the inner core, but it is not necessary that the virus undergoes degradation inside the core. Rather, nanotraps have the ability to protect the protein and trapped particles from the action of proteases. In the future, nanotraps can be used for the diagnosis of HIV infection (Chakravarty and Vora, 2021; Maus et al., 2021). Some of the compounds used for designing the core of nanotraps include cibacron blue, acrylic acid, pigment red, and Acid Black 48. The feature of molecular sieving through the sieve shell helps in sensitive and selective diagnosis using nanotraps (Chakravarty and Vora, 2021). Nanotraps have a great potential for their application in diagnostics in the future owing to their unique features.

13.7.2 Nanodiamonds

Nanodiamonds are a type of carbon-based nanocarriers. The core of the nanodiamonds comprises sp3 stable carbon; however, the surface is faceted and can coordinate with the water molecules present in the surrounding. Thus, nanodiamonds can efficiently deliver water-insoluble antiviral agents/drugs. Nanodiamonds can be prepared by detonation in a high-temperature and high-pressure environment. The useful features of nanodiamonds are their biocompatibility, nontoxicity, structural stability, longer circulation time, high drug-loading capacity, increased drug solubility and stability. Nanodiamonds can be functionalized with functional groups or ligands for targeted drug delivery, and their surface can be modified with growth factors for use in localized bone formation and tissue engineering (Chakravarty and Vora, 2021; Maus et al., 2021). Nanodiamonds have been explored for the delivery of hormones, vitamins, small interference RNAs, anticancer drugs, synthetic antiviral drugs, and proteins (Chakravarty and Vora, 2021).

13.7.3 Nanobubbles

Nanobubbles are echogenic nanosystems that contain oxygen in their core. The drugs can be loaded into the core, and combining drug loading with the ultrasound technique helps in targeted drug delivery. The outer shell of nanobubbles consists of polymers or lipids along with the surfactant, whereas the inner core comprises gases such as decafluoropentane or perfluoropentane. The outer shell provides nanobubbles with a suitable half-life and stability and models their acoustic parameters, which aid in site-specific drug release. Owing to their echogenic nature and minimal invasiveness, ultrasound-guided target-specific drug delivery can be carried out. This also leads to increased stability and residence time of the nanoformulation. Nanobubbles can also be used for theranostics by modifying their surface. Nanobubbles have been studied for anticancer drug and antibiotic delivery. Further, owing to their unique structure, their application can be investigated for antiviral therapy with the advantage of visualizing the site of infection and the target drug by simultaneous ultrasound imaging. Nanobubbles can also be explored for ultrasonic and photoacoustic imaging (Chakravarty and Vora, 2021; Maus et al., 2021).

13.7.4 Nanofibers

Nanofibers are one of the potential drug delivery vehicles in the future owing to their good mechanical strength, large surface area, scope of surface modification, and better porosity owing to the presence of small pores. Nanofibers are prepared using the electrospinning technique in which polymers such as chitosan, cellulite acetate, polyethylene oxide, and polyvinyl alcohol are spun. The drug can be exclusively integrated into nanofibers either in its original form or in the nanoparticulate form. One limitation associated with drug delivery based on nanofibers is the initial burst release, which can be avoided by using core-sheath designs (Chakravarty and Vora, 2021; Maus et al., 2021). Nanofibers prepared from soy protein were investigated against the influenza virus (Chakravarty and Vora, 2021). In one of the studies, the tunable release of different synthetic hydrophilic anti-HIV drugs was achieved by using nanofibers prepared from PLGA and PCL. By tuning the ratio of these two polymers, sustained drug release can be controlled (Carson et al., 2016). Similarly, nanofibers can be explored for antiviral phytotherapeutic delivery.

13.7.5 Nanorobots

Nanorobots are nanosized machines with multiple functions. They comprise of organic or inorganic materials and contain biomimetic materials to achieve various functions with abundant accuracy. The nanorobotic system is made of sensors, power sources, pumps, actuators, structural support, onboard computers, a payload compartment for drug loading, and a miniature camera to visualize its movement in the bloodstream. Nanorobots can be used to treat and diagnose various diseases and disorders with minimal invasion. Multiple functions of nanorobots, such as sensing, actuating, propulsion, self-replicating, signaling, and delivering, contribute to the targeted drug delivery at a specific site. Nanorobots require further exploration for delivering antiviral therapeutics, which can be achieved in the future by incorporating nanobiosensors, nanochips, nanotubes, and nanocontainers in the nanorobotic system (Chakravarty and Vora, 2021; Maus et al., 2021). Nanorobots are fascinating carriers for drug delivery.

13.8 PATENTS

Patents pertaining to the nanoformulation used for the delivery of phytochemicals for antiviral treatment and potent phytochemicals that can be used in nanoformulations for antiviral therapy are listed in Table 13.4. Antiviral curcumin-containing carbon quantum dots have been used against enterovirus. The spherical nanoparticles were monodispersed with a size range from 5.4 to 7 nm (US20200138896) (Huang et al., 2020). Lignosulfonate is effective against HIV-1; hence, it has the potential to be used in nanoformulations for treating HIV infection (CN102091091) (Zhiwei et al., 2011). In another study, oxyethyl group-substituted silybin derivatives were effective against HSV-1; these derivatives can be delivered as nanoformulations (CN101474172) (Kexin et al., 2011).

13.9 CONCLUSIONS AND FUTURE PROSPECTS

This chapter expands knowledge about several antiviral phytotherapeutics and the potential nanosystems available for their delivery. The scarcity of antiviral therapies against fatal viral diseases encourages researchers to scrutinize this area with utmost enthusiasm. Only limited studies have been conducted in this area, which subsequently gives rise to many unanswered questions that need to be taken care of, to gain detailed insights into the nanoparticle-based drug delivery systems of phytochemicals for the treatment of several viral infections.

Natural antiviral compounds are preferred over synthetic compounds owing to less toxicity and more efficacy. A variety of plant-based compounds with antiviral potential have been discovered but have not been replicated into the well-versed biomedicine, owing to the limitations of antivirals, such as low bioavailability, insolubility, poor stability, and many more as discussed in this chapter. Nanotechnological approaches can help overcome these drawbacks of antiviral compounds, and the numerous advantages of nanocarriers make them superior to conventional drug delivery systems.

TABLE 13.4
Patents on Phytochemical-Based Nanocarrier Systems with Antiviral Property and Phytochemicals with Potential to be Used with Nanocarriers Against Viral Infections

Patent Number	Formulation	Virus	Refs
	Patents on Phytochemical-Based Nanoformulation for Viral Infections		
US20200138896	Curcumin carbon quantum dots or nanoparticles	Enterovirus	(Huang et al., 2020)
CN1498630	*Stellaria media* acid A formulation of nanometer size	Adenovirus-3 of the influenza virus, parainfluenza virus, rhinovirus, echovirus, herpesvirus	(Gengxin et al., 2004)
	Patents on Phytochemicals with Antiviral Properties That Have the Potential to be Used as Nanoformulations		
CN103239466	Punicalin	Hepatitis C virus	(Xulin et al., 2013)
CN101474172	Oxyethyl group–substituted silybin	Herpes simplex virus-1	(Kexin et al., 2011)
CN101293830	1-Oxygen–[3-aryl replacement-acrylyl] quinic acid	Hepatitis B virus	(Xiaokun et al., 2011)
CN102091091	Lignosulfonate	HIV-1	(Zhiwei et al., 2011)
CN101543491	M-chlorobenzoyl–substituted dehydrosilybin	Herpes simplex virus-1	(Xiaokun et al., 2009)

To develop plant component-based nanoformulations against viral infections, initially certain subareas such as the mechanism of action of various phytochemicals against viruses, pharmacokinetics and pharmacodynamics of phytochemicals, types of nanocarrier compatible with the specific phytochemical, its complete *in vivo* behavior, and the targeted virus behavior inside the host needed to be explored. Furthermore, several types of organic nanocarriers have been examined in detail for antiviral phytotherapeutic delivery as compared to inorganic nanocarriers. Along with the conventional nanocarriers, other novel nanocarriers such as nanobubbles, nanodiamonds, nanorobots, nanotraps, and nanofibers deserve the attention of researchers for the treatment of viral diseases. More studies are required to understand the nanotoxicity issues in the body. Nanovaccine is a new field that involves the use of nanoparticles for developing viral vaccines. Antiviral phytochemicals can be nanotized to use directly as biomedicine for viral infections. This can help in the treatment of various brain-related viral infections by crossing the BBB. Quantum dots are nanocarriers that have efficacy to be used along with antiviral agents for brain-related viral diseases owing to their small size so that they can cross the BBB.

In the case of viruses, the development of one type of phytochemical-based nanoformulation is not enough because new strains are emerging each day. Viruses are difficult-to-treat microorganisms owing to their rapid mutations, leading to the development of drug resistance and dependency on the host cell; hence, continuous research is needed to treat viral infections. Like the COVID-19 pandemic, there are possibilities of more such outbreaks in the future; hence, it is important to prepare for the same to efficiently save more lives worldwide, which will be economically beneficial.

REFERENCES

Ahmad, R., Srivastava, S., Ghosh, S., and Khare, S. K. 2021. *Colloids and Surfaces B: Biointerfaces* 197, no. 111389 (January):1–13. https://doi.org/10.1016/j.colsurfb.2020.111389.

Ahmed, I. S., Rashed, H. M., Fayez, H., Farouk, F., and Shamma, R. N. 2020. *Pharmaceutics* 12, no. 2 (January): 1–21. https://doi.org/10.3390/pharmaceutics12020107.

Al-Ghamdi, A. D., Zaheer, Z., and Aazam, E. S. 2020. *Saudi Pharmaceutical Journal* 28, no. 8 (August): 1035–1048. https://doi.org/10.1016/j.jsps.2020.07.003.

Alqahtani, M. S., Alqahtani, A., Al-Thabit, A., Roni, M., and Syed, R. 2019. *Journal of Materials Chemistry B* 7, no. 28 (June): 4461–4473. https://doi.org/10.1039/c9tb00594c.

Alvarado, H. L., Abrego, G., Garduño-Ramirez, M. L., Clares, B., Calpena, A. C., and García, M. L. 2015. *Nanomedicine: Nanotechnology, Biology, and Medicine* 11, no. 3 (April): 521–530. https://doi.org/10.1016/j.nano.2015.01.004.

Álvarez, Á. L., Habtemariam, S., Moneim, A. E. A., Melón, S., Dalton, K. P., and Parra, F. 2015. *Antiviral Research* 119: 8–18. https://doi.org/10.1016/j.antiviral.2015.04.004.

Aqil, F., Munagala, R., Jeyabalan, J., Agrawal, A. K., and Gupta, R. 2017. *The AAPS Journal* 19, no. 6 (October): 1691–1702. https://doi.org/10.1208/s12248-017-0154-9.

Argenta, D. F., Bidone, J., Misturini, F. D., Koester, L. S., Bassani, V. L., Simões, C. M. O., and Teixeira, H. F. 2016. *Journal of Nanoscience and Nanotechnology* 16, no. 2 (February): 1282–1290. https://doi.org/10.1166/jnn.2016.11676.

Argenta, D. F., Bidone, J., Koester, L. S., Bassani, V. L., Simões, C. M. O., and Teixeira, H. F. 2018. *AAPS PharmSciTech* 19, no. 1 (January): 192–200. https://doi.org/10.1208/s12249-017-0828-8.

Bachmetov, L., Gal-Tanamy, M., Shapira, A., Vorobeychik, M., Giterman-Galam, T., Sathiyamoorthy, P., Golan-Goldhirsh, A., Benhar, I., Tur-Kaspa, R., and Zemel, R. 2012. *Journal of Viral Hepatitis* 19, no. 2 (February): e81–e88. https://doi.org/10.1111/j.1365-2893.2011.01507.x.

Bagad, M., and Khan, Z. A. 2015. *International Journal of Nanomedicine* 10: 3921–3935. https://doi.org/10.2147/IJN.S80706.

Baltina, L. A. 2003. *Current Medicinal Chemistry* 10, no. 2 (January): 155–171. https://doi.org/10.2174/0929867033368538.

Baranwal, M., Gupta, Y., Dey, P., and Majaw, S. 2021. *Phytotherapy Research* 35, no. 11 (November): 6148–6169. https://doi.org/10.1002/ptr.7222.

Barkallah, M., Nzoughet-kouassi, J., Simard, G., Thoulouze, L., Marze, S., Ropers, Marie-Hélène., and Andriantsitohaina, R. 2021. *Nutrients* 13, no. 12 (December): 1–11. https://doi.org/10.3390/nu13124378.

Barquero, A. A., Michelini, F. M., and Alché, L. E. 2006. *Biochemical and Biophysical Research Communications* 344, no. 3 (June): 955–962. https://doi.org/10.1016/j.bbrc.2006.03.226.

Behl, T., Rocchetti, G., Chadha, S., Zengin, G., Bungau, S., Kumar, A., et al. 2021. *Pharmaceuticals* 14, no. 4 (April): 1–46. https://doi.org/10.3390/ph14040381.

Ben-shabat, S., Yarmolinsky, L., Porat, D., and Dahan, A. 2020. *Drug Delivery and Translational Research* 10: 354–367. https://doi.org/10.1007/s13346-019-00691-6.

Biedenkopf, N., Lange-Grünweller, K., Schulte, F. W., Weißer, A., Müller, C., Becker, D., Becker, S., Hartmann, R. K., and Grünweller, A. 2017. *Antiviral Research* 137: 76–81. https://doi.org/10.1016/j.antiviral.2016.11.011.

Bonfim, C. M. do, Monteleoni, L. F., Calmon, M. de F., Cândido, N. M., Provazzi, P. J. S., Lino, V. de S., Rabachini, T., et al. 2020. *Artificial Cells, Nanomedicine, and Biotechnology* 48, no. 1 (February): 515–524. https://doi.org/10.1080/21691401.2020.1725023.

Carneiro, B. M., Batista, M. N., Braga, A. C. S., Nogueira, M. L., and Rahal, P. 2016. *Virology* 496: 215–218. https://doi.org/10.1016/j.virol.2016.06.012.

Carson, D., Jiang, Y., and Woodrow, K. A. 2016. *Pharmaceutical Research* 33, no. 1 (January): 125–136. https://doi.org/10.1007/s11095-015-1769-0.

Cataneo, A. H. D., Kuczera, D., Koishi, A. C., Zanluca, C., Silveira, G. F., de Arruda, T. B., et al. 2019. *Scientific Reports* 9, no. 1 (November): 1–15. https://doi.org/10.1038/s41598-019-52626-3.

Chakravarty, M., and Vora, A. 2021. *Drug Delivery and Translational Research* 11:748–787. https://doi.org/10.1007/s13346-020-00818-0.

Chavoshpour-Natanzi, Z., and Sahihi, M. 2019. *Food Hydrocolloids*, 96: 493–502. https://doi.org/10.1016/j.foodhyd.2019.05.051.

Chen, J.-X., Xue, H.-J., Ye, W.-C., Fang, B.-H., Liu, Y.-H., Yuan, S.-H., Yu, P., and Wang, Y.-Q. 2009. *Biological and Pharmaceutical Bulletin* 32, no. 8 (August): 1385–1391. https://doi.org/10.1248/bpb.32.1385.

Chen, W. L., Burdette, J. E., and Swanson, S. M. 2015. *Planta Medica* 81, no. 11 (July): PW1. https://doi.org/10.1055/s-0035-1556429.

Chen, W., Yeo, S. C. M., Elhennawy, M. G. A. A., and Lin, H.-S. 2016. *Journal of Functional Foods* 22: 122–131. https://doi.org/10.1016/j.jff.2016.01.020.

Cheng, Y.-B., Chien, Y.-T., Lee, J.-C., Tseng, C.-K., Wang, H.-C., Lo, I.-W., et al. 2014. *Journal of Natural Products* 77, no. 11 (November): 2367–2374. https://doi.org/10.1021/np5002829.

Choi, H. J., Song, J. H., Park, K. S., and Kwon, D. H. 2009. *European Journal of Pharmaceutical Sciences* 37, no. 3–4 (June), 329–333. https://doi.org/10.1016/j.ejps.2009.03.002.

Chuanasa, T., Phromjai, J., Lipipun, V., Likhitwitayawuid, K., Suzuki, M., Pramyothin, P., Hattori, M., and Shiraki, K. 2008. *Antiviral Research* 80, no. 1 (October), 62–70. https://doi.org/10.1016/j.antiviral.2008.05.002.

Chung, C.-Y., Liu, C.-H., Burnouf, T., Wang, G.-H., Chang, S.-P., Jassey, A., et al. 2016. *Antiviral Research* 130: 58–68. https://doi.org/10.1016/j.antiviral.2016.03.012.

Clementi, N., Scagnolari, C., D'Amore, A., Palombi, F., Criscuolo, E., Frasca, F., et al. 2021. *Pharmacological Research* 163:1–4. https://doi.org/10.1016/j.phrs.2020.105255.

Cojocaru, F., Botezat, D., Gardikiotis, I., Uritu, C., Dodi, G., Trandafir, L., et al. 2020. *Pharmaceutics* 12, no. 2 (February): 1–34. https://doi.org/10.3390/pharmaceutics12020171.

Corlay, N., Delang, L., Girard-Valenciennes, E., Neyts, J., Clerc, P., Smadja, J., et al. 2014. *Fitoterapia* 97: 87–91. https://doi.org/10.1016/j.fitote.2014.05.015.

Cui, H., Xu, B., Wu, T., Xu, J., Yuan, Y., and Gu, Q. 2014. *Journal of Natural Products* 77, no. 1 (January): 100–110. https://doi.org/10.1021/np400757k.

Cui, Q., Du, R., Liu, M., and Rong, L. 2020. *Molecules* 25, no. 1 (January): 1–17. https://doi.org/10.3390/molecules25010183.

Dao, T. T., Nguyen, P. H., Lee, H. S., Kim, E., Park, J., Lim, S. Il, and Oh, W. K. 2011. *Bioorganic and Medicinal Chemistry Letters* 21, no. 1 (January): 294–298. https://doi.org/10.1016/j.bmcl.2010.11.016.

Delshadi, R., Bahrami, A., Julian, D., Moore, M. D., and Williams, L. 2021. *Journal of Controlled Release* 331: 30–44. https://doi.org/10.1016/j.jconrel.2021.01.017.

Ding, Y., Dou, J., Teng, Z., Yu, J., Wang, T., Lu, N., Wang, H., and Zhou, C. 2014. *Archives of Virology* 159, no. 12 (December): 3269–3278. https://doi.org/10.1007/s00705-014-2192-2.

Diosa-Toro, M., Troost, B., van de Pol, D., Heberle, A. M., Urcuqui-Inchima, S., Thedieck, K., and Smit, J. M. 2019. *Antiviral Research* 161: 90–99. https://doi.org/10.1016/j.antiviral.2018.11.011.

Esimone, C. O., Eck, G., Nworu, C. S., Hoffmann, D., Überla, K., and Proksch, P. 2010. *Phytomedicine* 17, no. 7 (June): 540–547. https://doi.org/10.1016/j.phymed.2009.10.015.

Esposito, F., Carli, I., Del Vecchio, C., Xu, L., Corona, A., Grandi, N., et al. 2016. *Phytomedicine* 23, no. 12 (November): 1383–1391. https://doi.org/10.1016/j.phymed.2016.08.001.

Fang, C.-Y., Chen, S.-J., Wu, H.-N., Ping, Y.-H., Lin, C.-Y., Shiuan, D., et al. 2015. *Viruses* 7, no. 9 (September): 4894–4910. https://doi.org/10.3390/v7092852.

Fonseca, M., Macedo, A. S., Costa Lima, S. A., Reis, S., Soares, R., and Fonte, P. 2021. *Materials* 14, no. 21 (October): 1–16. https://doi.org/10.3390/ma14216421.

Fouad, G. I. 2021. *Bulletin of the National Research Centre* 45: 1–22. https://doi.org/10.1186/s42269-021-00487-0.

Frabasile, S., Koishi, A. C., Kuczera, D., Silveira, G. F., Verri, W. A., Dos Santos, C. N. D., and Bordignon, J. 2017. *Scientific Reports* 7, no. 1 (February): 1–10. https://doi.org/10.1038/srep41864.

Geng, C.-A., Jiang, Z.-Y., Ma, Y.-B., Luo, J., Zhang, X.-M., Wang, H.-L., et al. 2009. *Organic Letters* 11, no. 18 (August): 4120–4123. https://doi.org/10.1021/ol901592f.

Geng, C.-A., Zhang, X.-M., Ma, Y.-B., Jiang, Z.-Y., Luo, J., Zhou, J., Wang, H.-L., and Chen, J.-J. 2010. *Tetrahedron Letters* 51, no. 18 (May): 2483–2485. https://doi.org/10.1016/j.tetlet.2010.02.156.

Gengxin, Z., Yi, Z., and Zelin, L. 2004. Method for preparing broad-spectrum antiviral drug of nano total acid of chickweed as well as usage and chickweed acid A. CN1498630, China. https://patents.google.com/patent/CN1498630A/en?oq=CN1498630.

Ghosh, A. K., Thapa, R., Hariani, H. N., Volyanyuk, M., Ogle, S. D., Orloff, K. A., et al. 2021. *Pharmaceutics* 13, no. 9 (September): 1–18. https://doi.org/10.3390/pharmaceutics13091362.

Gonçalves, J. L. S., Lopes, R. C., Oliveira, D. B., Costa, S. S., Miranda, M. M. F. S., Romanos, M. T. V, Santos, N. S. O., and Wigg, M. D. 2005. *Journal of Ethnopharmacology* 99, no. 3 (July): 403–407. https://doi.org/10.1016/j.jep.2005.01.032.

Gong, C. Y., Wang, Y. J., Wang, X. H., Wei, X. W., Wu, Q. J., Wang, B. L., et al. 2011. *Journal of Nanoparticle Research* 13, no. 2 (February): 721–731. https://doi.org/10.1007/s11051-010-0071-7.

Gurunathan, S., Qasim, M., Choi, Y., Do, J. T., Park, C., Hong, K., Kim, J., and Song, H. 2020. *Nanomaterials* 10, no. 9 (September): 1–29. https://doi.org/10.3390/nano10091645.

Hakobyan, A., Arabyan, E., Avetisyan, A., Abroyan, L., Hakobyan, L., and Zakaryan, H. 2016. *Archives of Virology* 161, no. 12 (December): 3445–3453. https://doi.org/10.1007/s00705-016-3061-y.

Halder, A., Das, S., Ojha, D., Chattopadhyay, D., and Mukherjee, A. 2018. *Materials Science and Engineering C* 89: 413–421. https://doi.org/10.1016/j.msec.2018.04.005.

Hayashi, K., Niwayama, S., Hayashi, T., Nago, R., Ochiai, H., and Morita, N. 1988. *Antiviral Research* 9, no. 6 (September): 345–354. https://doi.org/10.1016/0166-3542(88)90036-8.

Ho, T. Y., Wu, S. L., Chen, J. C., Li, C. C., and Hsiang, C. Y. 2007. *Antiviral Research* 74, no. 2 (May): 92–101. https://doi.org/10.1016/j.antiviral.2006.04.014.

Hrutivorapongkul, A. P., Ipipun, V. L., Uangrungsi, N. R., Irtikara, K. K., Ishikawa, K. N., Aruyama, S. M., Atanabe, T. W., and Shikawa, T. I. 2003. *Chemical and Pharmaceutical Bulletin* 51, no. 2 (February): 187–190. https://doi.org/10.1248/cpb.51.187.

Huleihel, M., Ishanu, V., Tal, J., and Arad, S. 2001. *Journal of Applied Phycology* 13, no. 2 (April): 127–134. https://doi.org/10.1023/A:1011178225912.

Huang, C-C., Lin, C-J., Chen, S-Y., Chang, L., and Lin, H-J. 2020. Curcumin carbon quantum dots and use thereof. US20200138896, United States. https://patents.google.com/patent/US20200138896A1/en?oq=US20200138896.

Hung, T., Jassey, A., Liu, C., Lin, C., Lin, C., Wong, S. H., Wang, J. Y., et al. 2019. *Phytomedicine* 52: 62–69. https://doi.org/10.1016/j.phymed.2018.09.025.

Jena, A. B., Kanungo, N., Nayak, V., Chainy, G. B. N., and Dandapat, J. 2021. *Scientific Reports* 11, no. 1 (January): 1–14. https://doi.org/10.1038/s41598-021-81462-7.

Ji, P., Yu, T., Liu, Y., Jiang, J., Xu, J., Zhao, Y., et al. 2016. *Drug Design, Development and Therapy* 10: 911–925. https://doi.org/10.2147/DDDT.S97738.

Junyaprasert, V. B., Singhsa, P., Suksiriworapong, J., and Chantasart, D. 2012. *International Journal of Pharmaceutics*, 423, no. 2 (February): 303–311. https://doi.org/10.1016/j.ijpharm.2011.11.032.

Kannan, S., and Kolandaivel, P. 2018. *Journal of Biomolecular Structure and Dynamics* 36, no. 16 (December): 4255–4269. https://doi.org/10.1080/07391102.2017.1413422.

Kapoor, R., Sharma, B., and Kanwar, S. S. 2017. *Biochemistry and Physiology* 6, no. 2 (January): 1–7. https://doi.org/10.4172/2168-9652.1000220.

Khan, A. W., Kotta, S., Ansari, S. H., Sharma, R. K., and Ali, J. 2015. *Drug Delivery* 22, no. 4 (May): 552–561. https://doi.org/10.3109/10717544.2013.878003.

Kheiri, H., Ghasemi, P., Malvandi, H., Sajjad, M., and Attari, E. 2017. *European Journal of Pharmaceutics and Biopharmaceutics* 116: 17–30. https://doi.org/10.1016/j.ejpb.2016.10.003.

Kim, Y. J., Lee, K. P., Lee, D. Y., Kim, Y. T., Koh, D., Lim, Y., and Yoon, M. S. 2019. *Macromolecular Research* 27, no. 1 (January): 48–54. https://doi.org/10.1007/s13233-019-7002-y.

Kong, L., Li, S., Liao, Q., Zhang, Y., Sun, R., Zhu, X., et al. 2013. *Antiviral Research* 98, no. 1 (April): 44–53. https://doi.org/10.1016/j.antiviral.2013.02.003.

Kumari, A., Yadav, S. K., Pakade, Y. B., Singh, B., and Yadav, S. C. 2010. *Colloids and Surfaces B: Biointerfaces* 80, no. 2 (October): 184–192. https://doi.org/10.1016/j.colsurfb.2010.06.002.

Kumari, S., and Chatterjee, K. 2021. *APL Bioengineering* 5, no. 1 (March): 1–14 https://doi.org/10.1063/5.0029486.

Kexin, H., Hang, S., Jingxu, G., Yanguang, W., Yingqiu, P., Jinhan, X., Yunwei, Z., Yu, Z., Jia, Q., and Xiaoku, L. 2011. Pharmacy use of oxyethyl group substituted silybin for inhibiting herpes simplex virus. CN101474172, China. https://patents.google.com/patent/CN101474172B/en.

Kuo, Y.-C., Lin, L.-C., Tsai, W.-J., Chou, C.-J., Kung, S.-H., and Ho, Y.-H. 2002. *Antimicrobial Agents and Chemotherapy* 46, no. 9 (September): 2854–2864. https://doi.org/10.1128/AAC.46.9.2854-2864.2002.

Lai, F., Sinico, C., De Logu, A., Zaru, M., Müller, R. H., and Fadda, A. M. 2007. *International Journal of Nanomedicine* 2, no. 3 (September): 419–425.

Lembo, D. and Cavalli, R. 2010. *Antiviral Chemistry and Chemotherapy* 21, no. 2 (December): 53–70. https://doi.org/10.3851/IMP1684.

Li, J., Jiang, Q., Deng, P., Chen, Q., Yu, M., Shang, J., and Li, W. 2017. *Journal of Pharmacy and Pharmacology* 69, no. 6 (June): 663–674. https://doi.org/10.1111/jphp.12708.

Li, Weikun, Yalcin, M., Lin, Q., Ardawi, M. S. M., and Mousa, S. A. 2017. *Journal of Controlled Release* 248: 117–124. https://doi.org/10.1016/j.jconrel.2017.01.009.

Li, Wenji, Ng, K. Y., and Heng, P. W. S. 2014 *Biological and Pharmaceutical Bulletin* 37, no. 6 (June): 926–937. https://doi.org/10.1248/bpb.b13-00864.

Li, Y., But, P. P. H., and Ooi, V. E. C. 2005. *Antiviral Research* 68, no. 1 (October): 1–9. https://doi.org/10.1016/j.antiviral.2005.06.004.

Lin, J.-C. 2003. *Antiviral Research* 59, no. 1 (June): 41–47. https://doi.org/10.1016/S0166-3542(03)00030-5.

Lin, L.-T., Chung, C.-Y., Hsu, W.-C., Chang, S.-P., Hung, T.-C., Shields, J., et al. 2015. *Journal of Hepatology* 62, no. 3 (March), 541–548. https://doi.org/10.1016/j.jhep.2014.10.040.

Lin, Y., Zang, R., Ma, Y., Wang, Z., Li, L., Ding, S., et al. 2021. *International Journal of Molecular Sciences* 22, no. 22 (January): 1–10. https://doi.org/10.3390/ijms222212134.

Liu, R., Rong, G., Liu, Y., Huang, W., He, D., and Lu, R. 2021. *Materials Science and Engineering: C* 120: 111719. https://doi.org/10.1016/j.msec.2020.111719.

Liu, Z., Zhang, X., Wu, H., Li, J., Shu, L., Liu, R., Li, L., and Li, N. 2011. *Drug Development and Industrial Pharmacy* 37, no. 4 (April), 475–481. https://doi.org/10.3109/03639045.2010.522193.

Lo, T. H., Wu, Z. Y., Chen, S. Y., Meng, F. Y., Chou, P. T., Wang, C. M., and Lin, H. M. 2021. *Microporous and Mesoporous Materials* 314: 1–8. https://doi.org/10.1016/j.micromeso.2021.110886.

Loutfy S.A., Elberry M.H., Farroh K.Y., Mohamed H.T., Mohamed A.A., Mohamed E.B., et al. 2020. *International Journal of Nanomedicine* 15: 2699–2715. https://doi.org/10.2147/IJN.S241702.

Lu, Z., Bu, C., Hu, W., Zhang, H., Liu, M., Lu, M., and Zhai, G. 2018. *Bioscience, Biotechnology and Biochemistry* 82, no. 2 (February): 238–246. https://doi.org/10.1080/09168451.2017.1419852.

Luganini, A., Mercorelli, B., Messa, L., Palù, G., Gribaudo, G., and Loregian, A. 2019. *Antiviral Research* 164: 52–60. https://doi.org/10.1016/j.antiviral.2019.02.006.

Lv, L., Liu, C., Li, Z., Song, F., Li, G., and Huang, X. 2017. *BioMed Research International* 2017: 1-7. https://doi.org/10.1155/2017/1750895.

Madaan, K., Lather, V., and Pandita, D. 2016. *Drug Delivery* 23, no. 1 (January): 254–262. https://doi.org/10.3109/10717544.2014.910564.

Mateeva, N., Eyunni, S. V. K., Redda, K. K., Ononuju, U., Hansberry, T. D., Aikens, C., and Nag, A. 2017. *Bioorganic and Medicinal Chemistry Letters* 27, no. 11 (June), 2350–2356. https://doi.org/10.1016/j.bmcl.2017.04.034.

Maus, A., Strait, L., and Zhu, D. 2021. *Engineered Regeneration* 2: 31–46. https://doi.org/10.1016/j.engreg.2021.03.001.

Milovanovic, M., Arsenijevic, A., Milovanovic, J., Kanjevac, T. and Arsenijevic, N., 2017. In: *Antimicrobial Nanoarchitectonics*, A.M. Grumezescu, ed. 383–410. Elsevier.

Mishra, S., Pandey, A., and Manvati, S. 2020. *Heliyon* 6, no. 1 (January): 1–7. https://doi.org/10.1016/j.heliyon.2020.e03217.

Moghaddam, E., Teoh, B.-T., Sam, S.-S., Lani, R., Hassandarvish, P., Chik, Z., et al. 2014. *Scientific Reports* 4: 1–8. https://doi.org/10.1038/srep05452.

Nabila, N., Suada, N. K., Denis, D., Yohan, B., Adi, A. C., Veterini, A. S., et al. 2020. *Pharmaceutical Nanotechnology* 8, no. 1 (February): 54–62. https://doi.org/10.2174/2211738507666191210163408.

Nagai, T., Moriguchi, R., Suzuki, Y., Tomimori, T., and Yamada, H. 1995. *Antiviral Research* 26, no. 1 (January): 11–25. https://doi.org/10.1016/0166-3542(94)00062-D.

Narayanan, A., Kehn-Hall, K., Senina, S., Lundberg, L., Van Duyne, R., Guendel, I., et al. 2012. *Journal of Biological Chemistry* 287, no. 40 (September): 33198–33214. https://doi.org/10.1074/jbc.M112.356535.

Nothias-Scaglia, L.-F., Retailleau, P., Paolini, J., Pannecouque, C., Neyts, J., Dumontet, V., et al. 2014. *Journal of Natural Products* 77, no. 6 (June): 1505–1512. https://doi.org/10.1021/np500271u.

Omolaja, A. A., Pearce, B., Omoruyia, S. I., Badmus, J. A., Ismail, E., Marnewick, J., et al. 2021. *Surfaces and Interfaces* 25: 1–12. https://doi.org/10.1016/j.surfin.2021.101251.

Orłowski, P., Kowalczyk, A., Tomaszewska, E., Ranoszek-Soliwoda, K., Węgrzyn, A., Grzesiak, J., et al. 2018. *Viruses* 10, no. 10 (October): 1–15. https://doi.org/10.3390/v10100524.

Pacho, M. N., Pugni, E. N., Díaz Sierra, J. B., Morell, M. L., Sepúlveda, C. S., Damonte, E. B., et al. 2021. *Journal of Pharmacy and Pharmacology* 73, no. 3 (March), 357–365. https://doi.org/10.1093/jpp/rgaa045.

Panraksa, P., Ramphan, S., Khongwichit, S., and Smith, D. R. 2017. *Antiviral Research* 139, 69–78. https://doi.org/10.1016/j.antiviral.2016.12.014.

Park, S., Il, J., Lee, I., Lee, S., Hwang, M., Bae, J., et al. 2013. *Biochemical and Biophysical Research Communications* 440, no. 1 (October): 14–19. https://doi.org/10.1016/j.bbrc.2013.08.090.

Patra, J. K., Das, G., Fraceto, L. F., Vangelie, E., Campos, R., Rodriguez, P., et al. 2018. *Journal of Nanobiotechnology* 16: 1–33. https://doi.org/10.1186/s12951-018-0392-8.

Paulsen, S. S., Isbrandt, T., Kirkegaard, M., Buijs, Y., Strube, M. L., Sonnenschein, E. C., Larsen, T. O., and Gram, L. 2020. *Scientific Reports* 10, no. 1 (December), 1–13. https://doi.org/10.1038/s41598-020-78439-3.
Pour, P. M., Fakhri, S., Asgary, S., Farzaei, M. H., and Echeverría, J. 2019. *Frontiers in Pharmacology* 10: 1–23. https://doi.org/10.3389/fphar.2019.01207.
Pourmoazzen, Z., Sadeghifar, H., Yang, G., and Lucia, L. 2020. *Colloids and Surfaces B: Biointerfaces* 186: 1–6. https://doi.org/10.1016/j.colsurfb.2019.110685.
Pradhan, D., Biswasroy, P., Goyal, A., Ghosh, G., and Rath, G. 2021. *AAPS PharmSciTech* 22: 1–19. https://doi.org/10.1208/s12249-020-01908-5.
Prasad, S., and Tyagi, A. K. 2015. *Food and Function* 6, no. 11 (November): 3412–3419. https://doi.org/10.1039/c5fo00485c.
Qian, J., Meng, H., Xin, L., Xia, M., Shen, H., Li, G., and Xie, Y. 2017. *Colloids and Surfaces B: Biointerfaces* 160, no. 1 (December): 101–109. https://doi.org/10.1016/j.colsurfb.2017.09.020.
Qian, S., Fan, W., Qian, P., Zhang, D., Wei, Y., Chen, H., and Li, X. 2015. *Viruses*, 7, no. 4 (April), 1613–1626. https://doi.org/10.3390/v7041613.
Qiao, H., Chen, L., Rui, T., Wang, J., Chen, T., Fu, T., Li, J., and Di, L. 2017. *International Journal of Nanomedicine* 12: 1033–1046. https://doi.org/10.2147/IJN.S120887.
Raafat, K. M., and Zahaby, S. A. El. 2020. *Chinese Medicine* 15, 1–22. https://doi.org/10.1186/s13020-020-00321-1.
Ramalingam, P., and Ko, Y. T. 2015. *Pharmaceutical Research* 32: 389–402. https://doi.org/10.1007/s11095-014-1469-1.
Ramdani, L. H., and Bachari, K. 2020. *Acta Virologica* 64, no. 3 (January): 276–280. https://doi.org/10.4149/av.
Rania, A. K., Alaa, E. D. S. H., Mostafa, A. A., Rania, H. F., and Nariman, A. E. 2016. *African Journal of Microbiology Research* 10, no. 22 (June): 820–828. https://doi.org/10.5897/ajmr2016.7972.
Ranjbar, A., Jamshidi, M., and Torabi, S. 2020. *European Review for Medical and Pharmacological Sciences* 24: 7834–7844.
Rathinavel, T., Palanisamy, M., Palanisamy, S., Subramanian, A., and Thangaswamy, S. 2020. *International Journal of Advanced Science and Engineering* 6, no. 4 (May): 1482–1489. https://doi.org/10.29294/ijase.6.4.2020.1482-1489.
Ripoli, M., Angelico, R., Sacco, P., Ceglie, A., and Mangia, A. 2016. *Journal of Biomedical Nanotechnology* 12, no. 4 (April): 770–780. https://doi.org/10.1166/jbn.2016.2161.
Rodriguez, E. B., Almeda, R. A., Vidallon, M. L. P., and Reyes, C. T. 2019. *Journal of the Science of Food and Agriculture* 99, no. 4 (March): 1980–1989. https://doi.org/10.1002/jsfa.9396.
Roy, A., and Saraf, S. 2006. *Biological and Pharmaceutical Bulletin* 29, no. 2 (February): 191–201.
Sathishkumar, P., Li, Z., Govindan, R., Jayakumar, R., Wang, C., and Long Gu, F. 2021. *Applied Surface Science* 536: 1–14. https://doi.org/10.1016/j.apsusc.2020.147741.
Sen, R., Ganguly, S., Ganguly, S., Debnath, M. C., Chakraborty, S., Mukherjee, B., and Chattopadhyay, D. 2021. *Molecular Pharmaceutics* 18, no. 5 (March), 1920–1938. https://doi.org/10.1021/acs.molpharmaceut.0c00977.
Sharma, N. 2019. *International Journal of Research in Pharmaceutical Sciences* 10, no. 4 (November), 3578–3586. https://doi.org/10.26452/ijrps.v10i4.1738.
Shulman, M., Cohen, M., Soto-gutierrez, A., Yagi, H., Wang, H., Lee-parsons, C. W., et al. 2011. *PLoS One* 6, no. 4 (April): 1–8. https://doi.org/10.1371/journal.pone.0018033.
Sims, Jr. K. R., He, B., Koo, H., and Benoit, D. S. W. 2020. *ACS Omega* 5, no. 22 (May): 12649–12659. https://doi.org/10.1021/acsomega.9b04101.
Singh, L., Kruger, H. G., Maguire, G. E. M., Govender, T., and Parboosing R. 2017. *Therapeutic Advances in Infectious Disease* 4, no. 4 (July): 105–131. https://doi.org/10.1177/2049936117713593.
Sinico, C., Logu, A. De, Lai, F., Valenti, D., Manconi, M., Loy, G., Bonsignore, L., and Maria, A. 2005. European Journal of Pharmaceutics and Biopharmaceutics 59, no. 1 (January): 161–168. https://doi.org/10.1016/j.ejpb.2004.06.005.
Song, J.-M., Lee, K.-H., and Seong, B.-L. 2005. *Antiviral Research* 68, no. 2 (November), 66–74. https://doi.org/10.1016/j.antiviral.2005.06.010.
Szymańska, E., Orłowski, P., Winnicka, K., Tomaszewska, E., Bąska, P., Celichowski, G., et al. 2018. *International Journal of Molecular Sciences* 19, no. 2 (February): 1–21. https://doi.org/10.3390/ijms19020387.
Thawabteh, A., Juma, S., Bader, M., Karaman, D., Scrano, L., Bufo, S. A., and Karaman, R. 2019. *Toxins* 11, no. 656 (November): 1–28. https://doi.org/10.3390/toxins11110656.

Troost, B., Mulder, L. M., Diosa-Toro, M., van de Pol, D., Rodenhuis-Zybert, I. A., and Smit, J. M. 2020. *Scientific Reports* 10: 1–12. https://doi.org/10.1038/s41598-020-63397-7.

Vázquez-Calvo, Á., Jiménez de Oya, N., Martín-Acebes, M. A., Garcia-Moruno, E., and Saiz, J.-C. 2017. *Frontiers in Microbiology* 8: 1–8. https://doi.org/10.3389/fmicb.2017.01314.

Vitali, D., Bagri, P., Wessels, J. M., Arora, M., Ganugula, R., Parikh, A., et al. 2020. *International Journal of Molecular Sciences* 21, no. 1 (January): 1–19. https://doi.org/10.3390/ijms21010337.

Wahyuni, T. S., Widyawaruyanti, A., Lusida, M. I., Fuad, A., Fuchino, H., Kawahara, N., et al. 2014. *Fitoterapia* 99: 276–283. https://doi.org/10.1016/j.fitote.2014.10.011.

Wang, G., Wang, J. J., Li, F., and To, S. S. T. 2016. *Journal of Pharmaceutical Sciences* 105, no. 4 (April): 1535–1543. https://doi.org/10.1016/j.xphs.2016.01.016.

Wang, Q., Ding, Z.-H., Liu, J.-K., and Zheng, Y.-T. 2004. *Antiviral Research* 64, no. 3 (December):189–194. https://doi.org/10.1016/j.antiviral.2004.08.005.

Wei, Y., Guo, J., Zheng, X., Wu, J., Zhou, Y., Yu, Y., et al. 2014. *International Journal of Nanomedicine* 9: 3623–3630. https://doi.org/10.2147/IJN.S66312.

Wu, S., Lin, C., Chuang, Y., Chang, F., Tseng, C., Wu, Y., and Lee, J. 2012. *Journal of Viral Hepatitis* 19, no. 5 (May), 364–370. https://doi.org/10.1111/j.1365-2893.2011.01558.x.

Wu, W., Zu, Y., Wang, L., Wang, L., Wang, H., Li, Y., Wu, M., Zhao, X., and Fu, Y. 2017. *Drug Delivery* 24, no. 1 (January): 1713–1720. https://doi.org/10.1080/10717544.2017.1399302.

Wynn, G. H., Zapor, M. J., Smith, B. H., Wortmann, G., Oesterheld, J. R., Armstrong, S. C., and Cozza, K. L. 2004. *Psychosomatics* 45, no. 3 (May), 262–270. https://doi.org/10.1176/appi.psy.45.3.262.

Xia, X., Liu, H., Lv, H., Zhang, J., Zhou, J., and Zhao, Z. 2017. *Drug Design, Development and Therapy* 11: 1417–1427. https://doi.org/10.2147/DDDT.S133997.

Xiaokun, L., Feng, W., Kexin, H., Haibo, L., Jingxu, G., Xiumei, W., Su, Z., Yu, Z., and Jia, Q. 2009. Application of m-chlorobenzoyl substituted dehydrosilybin for preparing antiviral drugs. CN101543491, China. https://patents.google.com/patent/CN101543491A/en.

Xiaokun, L., Lijuan, Z., Haibo, L., Jianyong, D., Xiumei, W., Kexin, H., and Jia, Q. 2011. 1-oxygen-[3-aryl substituted-alkene propionyl]quinic acid compounds and uses. CN101293830, China. https://patents.google.com/patent/CN101293830A/en?oq=CN101293830.

Xu, J.-J., Wu, X., Li, M.-M., Li, G.-Q., Yang, Y.-T., Luo, H.-J., et al. 2014. *Journal of Agricultural and Food Chemistry* 62, no. 10 (March): 2182–2189. https://doi.org/10.1021/jf404310y.

Xulin, C., Wenyan, G., Yang, W., and Jiao, L. Q. 2013. Application of punicalin in preparation of medicine for treating or preventing hepatitis C virus infection. CN103239466, China. https://patents.google.com/patent/CN103239466B/en?oq=CN103239466.

Yallappa, S., Manjanna, J., Dhananjaya, B. L., Vishwanatha, U., Ravishankar, B., Gururaj, H., Niranjana, P., and Hungund, B. S. 2016. *Nano-Micro Letters* 8, no. 2 (April): 120–130. https://doi.org/10.1007/s40820-015-0066-0.

Yan, Y.Q., Fu, Y.J., Wu, S., Qin, H.Q., Zhen, X., Song, B.M., et al. 2018. *Phytotherapy Research* 32, no. 12 (December): 2560–2567. https://doi.org/10.1002/ptr.6196.

Yang, B., Ni, X., Chen, L., Zhang, H., Ren, P., Feng, Y., Chen, Y., Fu, S., and Wu, J. 2017. *Drug Delivery* 24, no. 1 (January): 660–669. https://doi.org/10.1080/10717544.2017.1303854.

Yang, G., Chen, S., and Zhang, J. 2019. *Frontiers in Pharmacology* 10: 1–17. https://doi.org/10.3389/fphar.2019.00751.

Yang, M., Lee, G., Si, J., Lee, S. J., You, H. J., and Ko, G. P. 2016. *Molecules* 21, no. 10 (October): 1–14. https://doi.org/10.3390/molecules21101401.

Yang, X. X., Li, C.M. and Huang, C.Z., 2016. *Nanoscale* 8, no. 5 (January): 3040–3048. https://doi.org/10.1039/C5NR07918G.

Yang, X. X., Li, C. M., Li, Y. F., Wang, J., and Huang, C. Z. 2017. *Nanoscale* 9, no. 41 (September): 16086–16092. https://doi.org/10.1039/c7nr06520e.

Yao, Y., Xia, M., Wang, H., Li, G., Shen, H., Ji, G., Meng, Q. and Xie, Y., 2016. *European Journal of Pharmaceutical Sciences* 91: 144–153. https://doi.org/10.1016/j.ejps.2016.06.014.

Zhang, S., Wang, J., and Pan, J. 2016. *Drug Delivery*, 23, no. 9 (November): 3696–3703. https://doi.org/10.1080/10717544.2016.1223218.

Zhang, W., Qiao, H., Lv, Y., Wang, J., Chen, X., Hou, Y., Tan, R., and Li, E. 2014. *PLoS One* 9, no. 10 (October): 1–9. https://doi.org/10.1371/journal.pone.0110429.

Zhang, Yongwen, But, P. P.-H., Ooi, V. E.-C., Xu, H.-X., Delaney, G. D., Lee, S. H. S., and Lee, S. F. 2007. *Antiviral Research* 75, no. 3 (September): 242–249. https://doi.org/10.1016/j.antiviral.2007.03.010.

Zhang, Yuxia, Chen, T., Yuan, P., Tian, R., Hu, W., Tang, Y., Jia, Y., and Zhang, L. 2015. *Carbohydrate Polymers* 133: 31–38. https://doi.org/10.1016/j.carbpol.2015.06.102.

Zhao, C., Xu, J., Zhang, Y., Zhao, L., and Feng, B. 2014. *Chemical and Pharmaceutical Bulletin* 62, no. 8 (August): 764–771. https://doi.org/10.1248/cpb.c14-00088.

Zhiwei, W., Ying, C., Min, Q., Qin, W., Hongyong, S., Yibin, S., and Haiyan, J. 2011. Application of lignosulfonate and pharmaceutical composition thereof, CN102091091, China. https://patents.google.com/patent/CN102091091A/en.

National Center for Biotechnology Information (2021). PubChem Compound Summary for CID 5318517, Andrographolide. Retrieved December 29, 2021 from https://pubchem.ncbi.nlm.nih.gov/compound/Andrographolide.

National Center for Biotechnology Information (2021). PubChem Compound Summary for CID 10155076, 1-Caffeoylquinic acid. Retrieved December 29, 2021 from https://pubchem.ncbi.nlm.nih.gov/compound/1-Caffeoylquinic-acid.

National Center for Biotechnology Information (2021). PubChem Compound Summary for CID 119066, Chalepin. Retrieved December 29, 2021 from https://pubchem.ncbi.nlm.nih.gov/compound/Chalepin.

National Center for Biotechnology Information (2021). PubChem Compound Summary for CID 5199, Sennosides. Retrieved December 29, 2021 from https://pubchem.ncbi.nlm.nih.gov/compound/Sennosides.

National Center for Biotechnology Information (2021). PubChem Compound Summary for CID 639665, Xanthohumol. Retrieved December 29, 2021 from https://pubchem.ncbi.nlm.nih.gov/compound/Xanthohumol.

14 Benefits and Pitfalls of Metal Nanotechnology

Tan Kin Fai
(UCSI University, Faculty of Pharmaceutical Sciences, Department of Pharmaceutical Technology, Malaysia, kinfai0101@gmail.com)

*Palanirajan Vijayaraj Kumar**
(UCSI University, Faculty of Pharmaceutical Sciences, Department of Pharmaceutical Technology, Malaysia, vijayarajkumar_p@yahoo.com)

14.1 INTRODUCTION

Nanotechnology is an integrative field of study and is applied in many sectors such as agricultural, biomedical, industrial, and pharmaceutical sectors to develop various nanofertilizers, nanofabrics, drug carriers, antimicrobial agents, biosensors, and so on (Karupannan, Dowlath, and Arunachalam 2020). Nanotechnology commonly involves the use of particles with a size of at least 1–100 nm (Lyddy 2009) The prefix "*nano*" in the term "nanotechnology" indicates a billionth (1×10^{-9}) (Poole Jr and Owens 2003). The idea of "nano" was first presented by Richard Zsigmondy, the 1925 Nobel Prize Laureate in chemistry. He used this term for characterizing particle size while examining the gold colloid under a microscope (Hulla, Sahu, and Hayes 2015). The idea was further explored by Richard Feynman, the 1965 Nobel Prize Laureate in physics during the 1959 American Physical Society meeting by introducing the concept of manipulating matter at the atomic level (Feynman 1959). Thus far, various revolutionary advancements have been made.

Several types of nanoparticles are available, categorized depending on their physical and chemical characteristics, such as carbon-based nanoparticles, metal nanoparticles, ceramic nanoparticles, semiconductor nanoparticles, polymeric nanoparticles, and lipid-based nanoparticles (Figure 14.1). Among all the types of nanoparticles, metal nanoparticles have shown an outstanding performance owing to their unique optical property, known as surface plasmon resonance (SPR) (Khan, Saeed, and Khan 2019, Humes and Sabo-Attwood 2020).

In terms of application, phytonanotechnology, a branch of nanotechnology, has gained attention. Phytonanotechnology generally involves the application of nanotechnology in the plant system (Karupannan, Dowlath, and Arunachalam 2020, Li and Yan 2020). However, it involves the utilization of plant-based material, known as phytochemicals, in the synthesis of nanoparticles (Elemike et al. 2020, Deshmukh and Kim 2020). This chapter discusses the advantages and disadvantages of metal nanoparticles and the application of phytochemicals in overcoming the disadvantages of metal nanoparticles by utilizing its benefits, as well as the potential problems faced when using phytochemicals in the synthesis of metal nanoparticles.

14.2 METAL NANOPARTICLES

In this decade, despite the presence of various types of nanoparticles, metal nanoparticles have shown an outstanding performance when compared with other nanoparticles and have been extensively utilized in various applications. Their involvement has been extended from a small theoretical idea to become

DOI: 10.1201/9781003231721-14

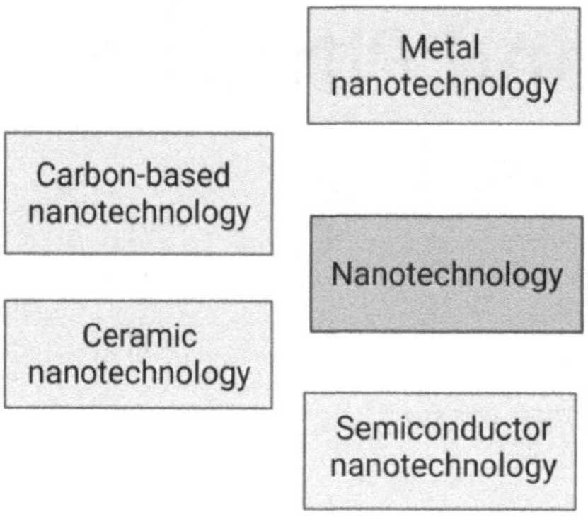

FIGURE 14.1 Various types of nanoparticles.

one of the main materials to be studied and applied for industrial, pharmaceutical, and biomedical purposes (Fai and Kumar 2021, Alarcon, Griffith, and Udekwu 2015, Rubilar et al. 2013, Khan, Al Mamun, and Ara 2021, Saldan et al. 2015, Jeevanandam et al. 2018). For example, in Raman spectroscopy (Stamplecoskie et al. 2011), as a drug carrier (Oh et al. 2008), antibacterial agent (Yin et al. 2020), anticancer agent (Chen et al. 2020, Riley and Day 2017), catalytic agent (Hajfathalian et al. 2015, Pizzutilo et al. 2017, Ojha et al. 2017, Wang et al. 2008, Zhou et al. 2006), and biosensor (Pizzutilo et al. 2017, Sun et al. 2020, Pourreza et al. 2015).

Because of the nano-size and unique nature of the metal nanoparticles, they possess a unique property known as SPR (Pelton, Aizpurua, and Bryant 2008, Turkevich, Stevenson, and Hillier 1951). The projection of light (electromagnetic wave) induces a synergistic vibration of electrons within the metal nanoparticles, leading to the SPR phenomenon (Lu et al. 2009). SPR enables the confirmation of the successful formation of the metal nanoparticles and their application as a photothermal agent and in surface-enhanced Raman spectroscopy (Papastefanaki et al. 2015, Yuan et al. 2012, Itoh et al. 2006). Moreover, the nanosize of the metal nanoparticles, when compared with their bulk, also provides a high surface area-to-volume for catalytic action, which makes them an excellent catalytic agent (Mody et al. 2010, Ojha et al. 2017, Tran et al. 2018).

14.3 PHYTOSYNTHESIS OF METAL NANOPARTICLES

14.3.1 Chemical and Physical Methods for the Synthesis of Metal Nanoparticles

Several varieties of metal nanoparticles exist; for example, silver nanoparticles (AgNPs), palladium nanoparticles, copper nanoparticles, gold nanoparticles, and so on. However, all of them are merely the same in terms of the preparation method, with the only difference being the different types of metal precursors used.

Different techniques are used in the preparation of metal nanoparticles, such as photochemical reduction, microwave, laser ablation, electrochemical method, micro-emulsion, gamma ray radiation, chemical reduction, and so on (Zhang et al. 2016, Herizchi et al. 2016, Liu et al. 2021, Jeyaraj et al. 2019). The two main methods used for the synthesis of metal nanoparticles are chemical and physical methods (Figure 14.2) (Zhang et al. 2016, Herizchi et al. 2016, Liu et al. 2021, Jeyaraj et al. 2019). The chemical method generally requires three important ingredients: a metal precursor, a reducing agent, and a stabilizing or capping agent (Zhang et al. 2016). Sometimes, if a dual functional agent such as sodium citrate is being used in the synthesis process, then only one of the three agents is required, as the dual functional agent has both reducing and stabilizing capability (Wuithschick et al. 2015). When a reducing agent is used, the metal precursor is first reduced to form a metal atom and then stabilized by the capping agent. The process is repeated until

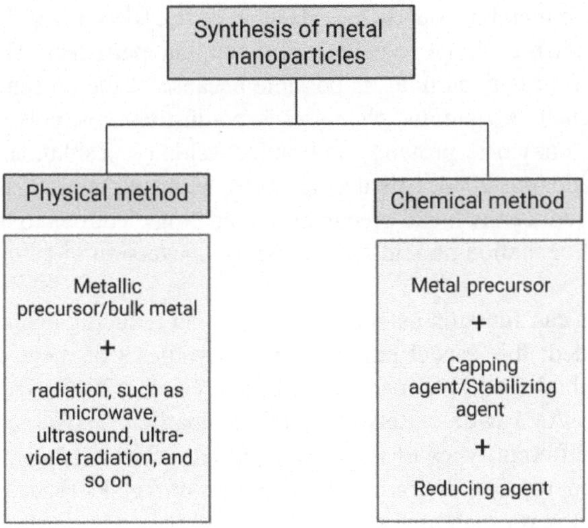

FIGURE 14.2 Main components required in the synthesis of metal nanoparticles through physical and chemical methods.

a stable nanoscale particle is formed (Zhang et al. 2016). Even though the chemical method owns the advantages of low cost, ease of production, and high yield, the capping and reducing agents used are generally toxic or hazardous to the human body (Duan, Wang, and Li 2015). The physical method generally involves the use of certain types of radiation, such as microwave, ultrasound, and ultraviolet radiation, to reduce the metal precursor/bulk metal or a laser to ablate the bulk metal to form metal nanoparticles (Liu et al. 2021). In contrast to the chemical method, the physical method does not require hazardous capping and reducing agents, but it produces a low yield, which requires high-energy consumption and an extreme environmental condition such as high pressure and temperature (Zhang et al. 2016, Liu et al. 2021). Therefore, the disadvantages of the physical and chemical methods have limited the application of metal nanoparticles.

14.3.2 Green Synthesis of Metal Nanoparticles

To overcome the shortcomings of the physical and chemical methods while retaining their advantages such as high yield, ease of production, and so on, another method for the biological synthesis of metal nanoparticles has been explored, known as the "green synthesis method." The biological method involves the use of microalgae, fungi, bacteria, and plants as stabilizing and reducing agents to form metal nanoparticles; therefore, this method is advantageous in terms of being more ecological friendly, having less toxicity, and requiring less extreme synthesis conditions than those required in the chemical and physical methods (Pourreza et al. 2015, Nayak et al. 2011, Goodsell 2004, GnanaJobitha, Annadurai, and Kannan 2012, Qian et al. 2019, Patil and Kim 2017, 2018). Additionally, this method is a nonpathogenic, nontoxic, inexpensive, and one-step procedure that allows large-scale preparation to proceed with the advantages of the additional medicinal value from the green materials used (Rajeshkumar, Bharath, and Geetha 2019, Khatua et al. 2020, Patil and Kim 2017, 2018).

Compared with the use of microalgae, fungi, and bacteria, the use of plants is easier, as it is time consuming to maintain a microbial culture, not to mention the cost involved in it (Farooqui et al. 2010, GnanaJobitha, Annadurai, and Kannan 2012, Patil et al. 2019), not to mention that the use of microorganisms involves extra effort to isolate, identify, and maintain them in a highly specific environment (Manivasagan, Nam, and Oh 2016). Furthermore, when compared with microorganism-based preparation of metal nanoparticles, plant-based preparation provides a higher reduction rate and produces more stable metal nanoparticles (Hasnain et al. 2019). Plant-based synthesis of metal nanoparticles is also known as "phytosynthesis of metal nanoparticles." The use of plant parts in the phytosynthesis of metal nanoparticles is possible because of the presence of different types of components in plants, such as vitamins, phenolics, saponins, tannins, polysaccharides, terpenoids, flavonoids, coenzymes, enzymes, proteins, and amino acids (Rajeshkumar, Bharath, and Geetha 2019, Kulkarni and Muddapur 2014, Srikar et al. 2016, Mohanta et al. 2016, Khatua et al. 2020). These components can reduce the metal precursor by donating electrons to the metal ion during the reaction, such as dehydrogenation of acids and alcohols, conversion of keto to enol, and so on (Jha et al. 2009).

In most cases, plants can function as both stabilizing and reducing agents, with zero additional agent needed to be added; this aspect has made phytosynthesis of metal nanoparticles an alternative eco-friendly method (Rajeshkumar, Bharath, and Geetha 2019, Sathishkumar et al. 2009, Ankamwar et al. 2005). As a wide variety of metal nanoparticles exists, different plant parts can be used for preparing different types of metal nanoparticles. For example, angiosperm plants are proven to be the most appropriate option for the synthesis of AgNPs (Rai, Yadav, and Gade 2009). However, different plant parts such as flowers, leaves, callus, seeds, fruits, stems, roots, and bark, can been used to produce a variety of metal nanoparticles of different geometrical shapes and sizes (Rai, Yadav, and Gade 2009).

14.3.2.1 Steps Involved in the Phytosynthesis of Metal Nanoparticles

As reported previously (Rajeshkumar, Bharath, and Geetha 2019), with reference to the phytosynthesis of AgNPs, the general steps (Figure 14.3) involved in the phytosynthesis of metal nanoparticles are as follows: Different plant parts are first collected and then rinsed with a suitable detergent, after which they are rinsed for a few times with ultrapure water to ensure the complete removal of the detergent. The plant parts are then dried at room temperature. To extract the essential components from the plant parts, a suitable solvent is needed for the immersion of the plant parts at a suitable temperature. In the absence of a suitable solvent, the plant parts can be heated in ultrapure water at 60–80 °C for 15 min. The final solution is then filtered to remove any sediment from the plant part, and the filtrate is stored in a refrigerator until further use. To induce the formation of the metal nanoparticles, the filtrate is mixed with the metal precursor under stirring at room temperature. The color change will indicate the successful formation of the metal nanoparticles.

14.4 APPLICATION OF PLANT-BASED METAL NANOPARTICLES

As previously mentioned, many types of metal nanoparticles exist, such as ruthenium nanoparticles, platinum nanoparticles, gold nanoparticles, AgNPs, and so on. It is difficult to list out all of their applications, as they are many fields are involved. However, the two most common types of plant-derived metal nanoparticles are gold nanoparticles and AgNPs. Therefore, their predominant applications are discussed in the following sections.

14.4.1 Plant-Derived Silver Nanoparticles in Antibacterial Treatment

Plant-derived AgNPs are used in various applications such as anticancer agents (Jalilian et al. 2020, Panja et al. 2016, Taghavizadeh Yazdi et al. 2020), insecticides (Ghramh et al. 2020), antioxidants (Sriranjani et al. 2016), and so on (Table 14.1). However, because of the antibacterial activity of silver (Clement and Jarrett 1994), phytosynthesized AgNPs are most commonly used as an antibacterial agent, as discussed below. For example, AgNPs synthesized from *Elettaria cardamomum* (GnanaJobitha, Annadurai, and Kannan 2012), *Setcreasea pallida* (MubarakAli et al. 2011), and *Saraca asoca* (Fatema et al. 2019) exhibited strong antibacterial property. Almost all of the

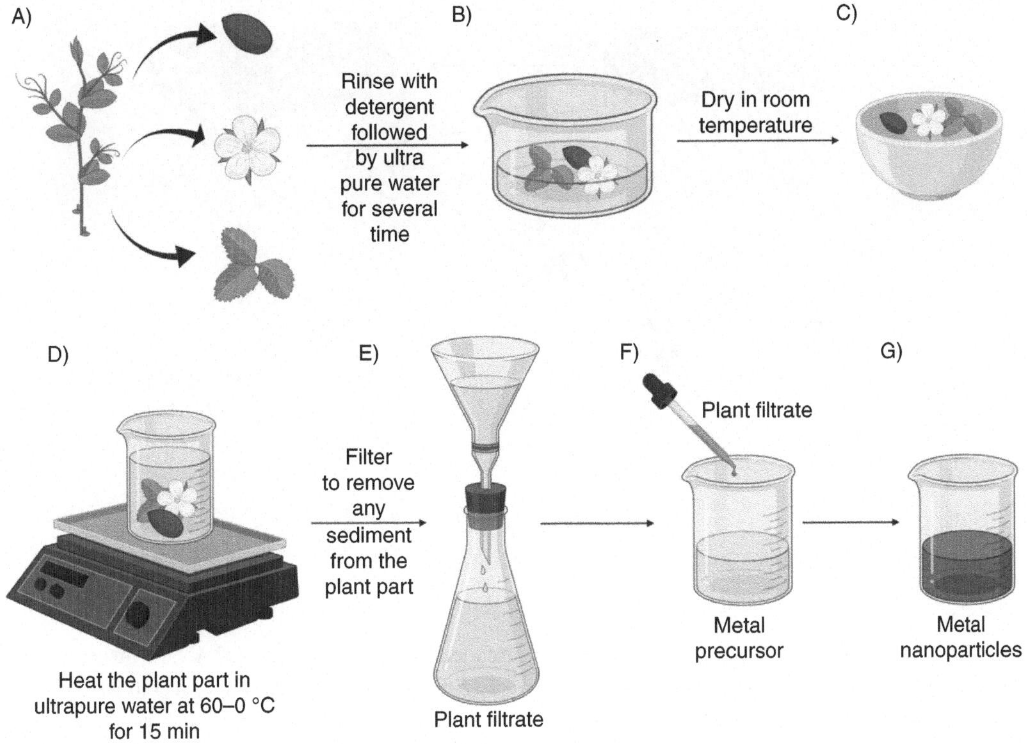

FIGURE 14.3 Schematic diagram of the steps involved in the phytosynthesis of metal nanoparticles. A) Collection of different parts of the plant. B) Rinsing of plant parts with a suitable detergent and ultrapure water. C) Drying of plant parts at room temperature. D) Heating of the plant parts in ultra-pure water at 60–80 °C for 15 minutes. E) Filtration to remove the plant part. F) Adding of the plant filtrate to the metal precursor. G) Formation of the metal nanoparticles.

plant-based AgNPs exhibited certain antibacterial property owing to the presence of Ag, but the exact mechanism of antibacterial property exhibited by AgNPs remains only partially understood. Three primary mechanisms have been proposed in the antibacterial activity of AgNPs (Yun'an Qing et al. 2018): contact killing (Azizian-Shermeh, Einali, and Ghasemi 2017), reactive oxygen species (ROS) generation, and silver ion-mediated killing (Suman et al. 2013).

Contact killing occurs when the AgNPs attach onto the cell wall of the bacterium and penetrate into the cell. After penetration, the AgNPs conjugate with the protein on the bacterial membrane, thus destroying the membrane and inducing leakage of the bacterial contents, finally killing the bacterium (Figure 14.4) (Seong and Lee 2017). Additionally, the AgNPs can enter into the cytoplasm of the bacterium, further damaging the intracellular structure and proteins (You et al. 2012). AgNPs with a smaller size exhibited a higher antibacterial activity than those with a larger size owing to the higher surface area and easier penetration of smaller AgNPs (Khalandi et al. 2017). Therefore, the antibacterial property of the AgNPs can be regarded as being size-dependent.

Additionally, AgNPs exhibit antibacterial property by generating ROS (Gomaa 2017, Kędziora et al. 2021, Das et al. 2017). AgNPs have the ability to produce ROS to disrupt the integrity of biomolecules such as ATP, lipids, and DNA and cause an apoptosis-like response (Korshed et al. 2016, Hwang et al. 2012). Generally, the generated ROS can be removed by antioxidant compounds such as glutathione. However, AgNPs can cause disturbance in the synthesis of the antioxidant enzymes, thus indirectly increasing the ROS production in the bacteria (Ramalingam, Parandhaman, and Das 2016, Quinteros et al. 2016).

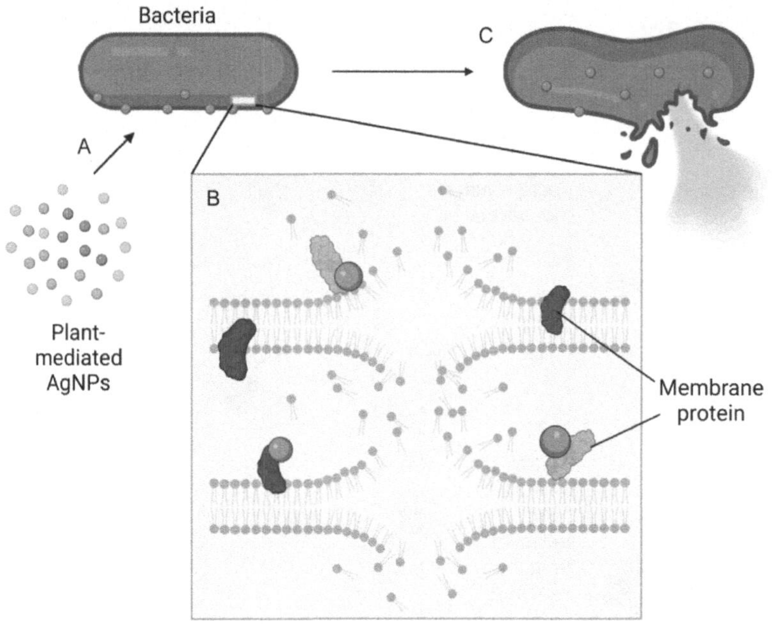

FIGURE 14.4 Schematic diagram of the contact killing mechanism of plant-derived AgNPs. A) Binding of the plant-derived AgNPs to the bacterial cell. B) Binding of the plant-derived AgNPs to membrane proteins, causing the destruction of the bacterial membrane structure. C) Leakage of the bacterial content and further penetration of the AgNPs into the cytoplasm of the bacterial cell.

The AgNPs can exhibit antibacterial property through the release of Ag^+ (Kim et al. 2016, Jin et al. 2017). In this antibacterial mechanism, the positive charge on the Ag ion is very important for its antibacterial property. The surface area of AgNPs is the main factor for the antibacterial property, as AgNPs with a high surface area can release a higher concentration of Ag+ or has a higher release rate of Ag+ when compared with Ag+ with a smaller size (Jin et al. 2017, Xiu et al. 2012). The Ag+ can conjugate with the sulfhydryl groups of biological components crucial for ATP production, thus inactivating them (Klueh et al. 2000). Moreover, Ag+ can also limit the activity of the respiratory chain enzyme and cell division and increase oxidative stress (Yun'an Qing et al. 2018). The electrostatic interaction between the positive charge on Ag+ with the negative charge on the bacterial membrane can further destroy the cell wall by forming a pit on it (Hamouda et al. 2001, Dibrov et al. 2002, Dragieva et al. 1999, Raut et al. 2010, Sondi and Salopek-Sondi 2004). Therefore, because of the release of Ag^+ ions into the environment, AgNPs are an ideal bactericidal agent (Nel et al. 2009).

According to Hasnain et al. (Hasnain et al. 2019), the inhibitory zone in the bacterial culture caused by plant-derived AgNPs is larger than that caused by antibacterial agents such as minocycline. Generally, plant-derived AgNPs exhibit bactericidal activity toward both gram-positive and gram-negative bacteria. However, some studies have shown that they have a higher bactericidal activity toward gram-positive bacteria than toward gram-negative bacteria (Dada et al. 2019), whereas some studies showed the converse finding: that is, AgNPs have a higher antibacterial activity toward gram-negative bacteria than toward gram-positive bacteria (Jalilian et al. 2020), and some even showed zero inhibitory activity toward gram-negative bacteria (Fatema et al. 2019). These variations in bactericidal activity are due to the different plant parts used in the phytosynthesis of AgNPs. Every plant part can produce unique AgNPs, which is one of the advantages and the disadvantages of using plant-derived AgNPs as bactericidal agents. The lesser activity of AgNPs against gram-positive bacteria can be explained by the solid peptidoglycan layer surrounding the cell, which can prevent the

TABLE 14.1
Application of Plant-Derived Silver Nanoparticles

Type of Metal Nanoparticle	Plant Species	Plant Part	Application	Refs.
Silver nanoparticles	*Elettaria cardamomum*	Seed	Antimicrobial agent	(GnanaJobitha, Annadurai, and Kannan 2012)
	Setcreasea pallida	Leaf	Antibacterial agent	(Hasnain et al. 2019)
	Mentha piperita	Leaf	Antibacterial agent	(MubarakAli et al. 2011)
	Citrus sinensis	Peel	Antibacterial agent	(Kaviya et al. 2011)
	Ocimum gratissimum	Leaf	Antibacterial agent	(Das et al. 2017)
	Saraca asoca	Leaf	Antibacterial agent	(Fatema et al. 2019)
	Acalypha wilkesiana	Leaf	Antibacterial agent	(Dada et al. 2019)
	Cassia fistula	Leaf	Antioxidant, antibacterial, and anticancer agent	(Mohanta et al. 2016)
	Wedelia chinensis	Leaf	Antioxidant, antibacterial, and anticancer agent	(Das et al. 2018)
	Indigofera tinctoria	Leaf	Antioxidant, antibacterial, anticancer, and catalytic agent	(Vijayan, Joseph, and Mathew 2018)
	Prunus japonica	Leaf	Antibacterial and antioxidant agent	(Saravanakumar et al. 2017)
	Morinda citrifolia	Root	Anticancer agent	(Suman et al. 2013)
	Maclura pomifera	Leaf	Antibacterial	(Azizian-Shermeh, Einali, and Ghasemi 2017)
	Rauvolfia serpentina	Leaf	Antimicrobial, larvicidal, anticancer agent	(Panja et al. 2016)
	Ruta graveolens	Leaf	Anticancer and insecticidal agent	(Ghramh et al. 2020)
	Ruta graveolens	Shoot	Antimicrobial and anticancer agent	(Taghavizadeh Yazdi et al. 2020)
	Clerodendrum phlomidis	Leaf	Antioxidant and anticancer agent	(Sriranjani et al. 2016)
	Acalypha indica	Leaf	Anticancer agent	(Krishnaraj et al. 2014)
	Mukia maderaspatanus	Leaf	Anticancer agent	(Devi and Sathishkumar 2017)

uptake of AgNPs (Das et al. 2018, Saravanakumar et al. 2017). However, the reason for the lesser activity of AgNPs against gram-negative bacteria requires further study. Moreover, plant-derived AgNPs exhibit a high antibacterial activity toward multidrug-resistant bacteria, which making them an even higher potential antibacterial agent requiring further study (Das et al. 2017, Kaviya et al. 2011, MubarakAli et al. 2011, Jalilian et al. 2020). Overall, plant-derived AgNPs have shown a very promising potential as an antibacterial agent. However, more studies must be conducted to validate their potential and accuracy.

14.4.2 Plant-Derived Gold Nanoparticles for Anticancer Treatment

Similar to AgNPs, plant-derived gold nanoparticles (AuNPs) were used in a wide variety of applications. For example, they have been used as a catalytic agent (Latha et al. 2018, Unal et al. 2020), larvicide (Balasubramani, Ramkumar, Krishnaveni, Sowmiya, et al. 2015), antimicrobial agent (Folorunso et al. 2019, Shittu, Stephen, and Kure 2017, Kumar, Kala, and Prakash 2019), antioxidant agent (Ponnanikajamideen et al. 2019), and so on (Table 14.2). However, one of the most pronounced applications of plant-derived AuNPs is their use as an anticancer agent. In anticancer treatment, plant-derived AuNPs exhibited excellent activity and potential. For example, AuNPs prepared from *Pongamia pinnata* (Khatua et al. 2020) exhibited an IC_{50} value of 200 μg/mL in cervical cancer, and AuNPs prepared from *Alternanthera sessilis* (Qian et al. 2019) exhibited a pronounced cell death in HeLa cells when administered at a higher amount ranging from 10 to 15 mg/mL and so on. Even though plant-derived AuNPs showed a high potential in anticancer treatment, their cytotoxicity might vary according to the plant part used, which is one of the disadvantages of plant-derived AuNPs. However, all plant-derived AuNPs exhibited a dose-dependent cytotoxicity in anticancer treatment (Khatua et al. 2020, Qian et al. 2019, Patil et al. 2019). According to Khatua et al. (Khatua et al. 2020) and Patil et al. (Patil et al. 2019), plant-derived AuNPs exhibited nontoxicity toward human cell lines such as human embryonic kidney cell line.

A few mechanisms have been proposed with regard to the cytotoxicity of plant-derived AuNPs. One of the most promising mechanisms is the generation of ROS (Figure 14.5) (Khatua et al. 2020). First, plant-mediated AuNPs were taken up by cancer cells through receptor-mediated endocytosis (Kim et al. 2006). After uptake into the cell, plant-mediated AuNPs increase ROS levels, which, in turn, alter the mitochondrial membrane potential. This can lead to changes in the permeability of the mitochondrial membrane and the collapse of the mitochondria (Marchi et al. 2012). Changes in the mitochondrial membrane potential is often related to cell death, as the loss of mitochondrial membrane function can cause the release of apoptotic proteins such as cytochrome c and apoptosis-inducing factor, leading to programmed cell death or apoptosis (Elmore 2007). A different study has been proposed by Qian et al. (Qian et al. 2019), who showed an increase in the number of apoptosis cells after treatment with plant-derived AuNPs.

Moreover, plant-derived AuNPs induce apoptosis by increasing apoptotic gene expression in cancer cells (Qian et al. 2019). Two main types of apoptosis pathways exist: intrinsic and extrinsic pathways. The intrinsic pathway is controlled by Bcl-2-mediated caspase-3 activation and leads to apoptosis (Pistritto et al. 2016). However, during treatment with plant-derived AuNPs, an increase in Bax expression and a decrease in Bid and Bcl-2 expression were observed in cancer cells were observed, indirectly leading to the increase in the activity of caspase 3 and, finally, apoptosis (Qian et al. 2019). Another study showed that plant-derived AuNPs induced apoptosis through intrinsic apoptotic pathway in tumors (Baharara et al. 2016).

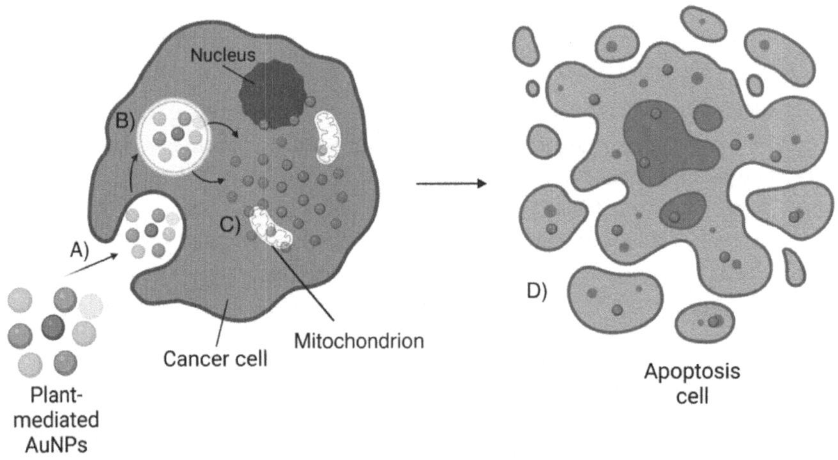

FIGURE 14.5 Schematic diagram showing the cytotoxicity mechanism of the plant-derived AuNPs. A) Internalization of the plant-derived AuNPs into the cancer cell through endocytosis. B) Release of plant-derived AuNPs into the cytoplasm of the bacterial cell. C) Increasing levels of reactive oxygen species in the cytoplasm of the bacterial cell, causing disruption of the mitochondrial membrane potential. D) Apoptosis of the cancer cell.

TABLE 14.2
Application of Gold Nanoparticles

Type of Metal Nanoparticle	Plant Species	Plant Part	Application	Refs.
Gold nanoparticle	*Paullinia pinnata*	Leaf	Anticancer agent	(Khatua et al. 2020)
	Lonicera japonica	Flower	Anticancer agent	(Patil et al. 2019)
	Alternanthera sessilis	Leaf	Anticancer agent	(Qian et al. 2019)
	Zataria multiflora	Leaf	Anticancer agent	(Baharara et al. 2016)
	Sasa borealis	Leaf	Anticancer agent	(Patil et al. 2018)
	Acalypha indica	Leaf	Anticancer agent	(Krishnaraj et al. 2014)
	Antigonon leptopus	Leaf	Antioxidant and anticancer agent	(Balasubramani, Ramkumar, Krishnaveni, Pazhanimuthu, et al. 2015)
	Podophyllum hexandrum	Leaf		(Jeyaraj et al. 2014)
	Justicia adhatoda	Leaf	Catalytic agent	(Latha et al. 2018)
	Chloroxylon swietenia	Leaf	Larvicide	(Balasubramani, Ramkumar, Krishnaveni, Sowmiya, et al. 2015)
	Andrographis paniculata	Leaf	Cytotoxicity agent	(Kumari et al. 2019)
	Elaeis guineensis	Leaf	Anticancer agent	(Ahmad, Irfan, and Bhattacharjee 2016)
	Camellia sinensis	Leaf	Anticancer agent	(Ahmeda, Zangeneh, and Zangeneh 2020)
	Chamaecostus cuspidatus	Leaf	Antioxidant agent	(Ponnanikajamideen et al. 2019)
	Brassica oleracea	Leaf	Catalytic agent	(Unal et al. 2020)
	Catharanthus roseus	Leaf	Antibacterial agent	(Shittu, Stephen, and Kure 2017)
	Mukia maderaspatanus	Leaf	Anticancer agent	(Devi and Sathishkumar 2017)
	Croton caudatus	Leaf	Antimicrobial and anticancer agent	(Kumar, Kala, and Prakash 2019)

In cell cycle analysis, the number of cells in the G1, G2, and M phases decreased but the number of cells in the S phase increased after treatment with plant-derived AuNPs when compared with untreated cells (Khatua et al. 2020). Therefore, it was assumed that plant-derived AuNPs induced an S-phase arrest during their cytotoxicity activity. However, another study demonstrated that plant-derived AuNPs induced arrest of the G2/M phase growth due to DNA damage and caspase-mediated apoptotic death of cancer cells (Jeyaraj et al. 2014). Therefore, the difference in the mechanism might be due to the different plant part used in the synthesis of AuNPs.

Morphological alteration in cancer cells, with shrinkage of the cytoplasm, was observed during treatment with plant-derived AuNPs. Moreover, many cells detached and began floating, and dead cells with fragmented nucleus were also observed (Khatua et al. 2020). In *in vitro* wound healing or cell migration assay, plant-derived AuNPs exhibited an excellent effect in inhibiting the migration of cancer cells (Khatua et al. 2020). Even though plant-mediated AuNPs have some shortcomings such as inconsistency with the plant part used, they still proved to be a very strong candidate as an anticancer agent.

ACKNOWLEDGMENT

This project is supported by Fundamental Research Grant Scheme (FRGS) (reference code: FRGS/1/2021/ SKK0/UCSI/02/5), Ministry of Higher Education (MOHE), and Research Excellence and Innovation Grant (project code: REIG-FPS-2020-052) under Centre of Excellence in Research, Value Innovation and Entrepreneurship (CERVIE), UCSI University, Malaysia.

REFERENCES

Ahmad, Tausif, Muhammad Irfan, and Sekhar Bhattacharjee. 2016. *Procedia Engineering* 148:1396–1401.

Ahmeda, Ahmad, Akram Zangeneh, and Mohammad Mahdi Zangeneh. 2020. *Applied Organometallic Chemistry* 34(3):e5290.

Alarcon, Emilio I, May Griffith, and Klas I Udekwu. 2015. *Springer International Publishing.*

Ankamwar, Balaprasad, Chinmay Damle, Absar Ahmad, and Murali Sastry. 2005. *Journal of Nanoscience and Nanotechnology* 5(10):1665–1671.

Azizian-Shermeh, Omid, Alireza Einali, and Ali Ghasemi. 2017. *Advanced Powder Technology* 28 (12):3164–3171.

Baharara, Javad, Tayebe Ramezani, Adeleh Divsalar, Marzieh Mousavi, and Arefeh Seyedarabi. 2016. *Avicenna Journal of Medical Biotechnology* 8(2):75.

Balasubramani, Govindasamy, Rajendiran Ramkumar, Narayanaswamy Krishnaveni, Annamalai Pazhanimuthu, Thillainathan Natarajan, Rajamani Sowmiya, and Pachiappan Perumal. 2015. *Journal of Trace Elements in Medicine and Biology* 30:83–89.

Balasubramani, Govindasamy, Rajendiran Ramkumar, Narayanaswamy Krishnaveni, Rajamani Sowmiya, Paramasivam Deepak, Dhayalan Arul, and Pachiappan Perumal. 2015. *Journal of Photochemistry and Photobiology B: Biology* 148:1–8.

Chen, Jihua, Yinchu Ma, Wei Du, Tianyue Dai, Yanfang Wang, Wei Jiang, Yifei Wan, Yucai Wang, Gaolin Liang, and Guangfeng Wang. 2020. *Advanced Functional Materials* 30(50):2001566.

Clement, Julia L, and Penelope S Jarrett. 1994. *Metal-Based Drugs* 1(5–6):467–482.

Dada, Adewumi Oluwasogo, Folahan Amoo Adekola, Fehintoluwa Elizabeth Dada, Adunola Tabitha Adelani-Akande, Micheal Oluwasesan Bello, Chidiogo Rita Okonkwo, Adejumoke Abosede Inyinbor, Abimbola Peter Oluyori, Adeniyi Olayanju, and Kolawole Oluseyi Ajanaku. 2019. *Heliyon* 5(10):e02517.

Das, Balaram, Sandeep Kumar Dash, Debasish Mandal, Totan Ghosh, Sourav Chattopadhyay, Satyajit Tripathy, Sabyasachi Das, Sankar Kumar Dey, Debasis Das, and Somenath Roy. 2017. *Arabian Journal of Chemistry* 10(6):862–876.

Das, Merina Paul, Jeyanthi Rebecca Livingstone, Pandiyarasan Veluswamy, and Jayabrata Das. 2018. *Journal of Food and Drug Analysis* 26(2):917–925.

Deshmukh, Aarti R., and Beom Soo Kim. 2020. In *Phytonanotechnology*, edited by N. Thajuddin and Silvy Mathew, 97–121. Elsevier.

Devi, Guruviah Karthiga, and Kannaiyan Sathishkumar. 2017. *IET Nanobiotechnology* 11(2):143–151.

Dibrov, Pavel, Judith Dzioba, Khoosheh K Gosink, and Claudia C Häse. 2002. *Antimicrobial Agents and Chemotherapy* 46(8):2668–2670.

Dragieva, I, S Stoeva, P Stoimenov, E Pavlikianov, and K Klabunde. 1999. *Nanostructured Materials* 12(1–4):267–270.

Duan, Haohong, Dingsheng Wang, and Yadong Li. 2015. *Chemical Society Reviews* 44(16):5778–5792.

Elemike, Elias E., Kenneth A. Ibe, Justina I. Mbonu, and Damian C. Onwudiwe. 2020. In *Phytonanotechnology*, edited by N. Thajuddin and Silvy Mathew, 71–96. Elsevier.

Elmore, Susan. 2007. *Toxicologic Pathology* 35(4):495–516.

Fai, Tan Kin, and Palanirajan Vijayaraj Kumar. 2021. *Current Pharmaceutical Design.*

Farooqui, Md Arshad, Prakash Singh Chauhan, Praveen Krishnamoorthy, and Jameel Shaik. 2010. *Digest Journal of Nanomaterials and Biostructures* 5(1):43–49.

Fatema, Samreen, Mahendra Shirsat, Mazahar Farooqui, and Mohd Arif Pathan. 2019. *International Journal of Nanodimension* 10(2):163–168.

Feynman, Richard P. 1959. *Engineering and Science* 23(5).

Folorunso, Aderonke, Sunday Akintelu, Abel Kolawole Oyebamiji, Samuel Ajayi, Babawale Abiola, Ibrahim Abdusalam, and Adetoun Morakinyo. 2019. *Journal of Nanostructure in Chemistry* 9(2):111–117.

Ghramh, Hamed A, Essam H Ibrahim, Mona Kilnay, Zubair Ahmad, Sadeq K Alhag, Khalid Ali Khan, Ramadan Taha, and Fawziah M Asiri. 2020. *Bioinorganic Chemistry and Applications*

Gnana Jobitha, Gnana Dhas, Gurusamy Annadurai, and Chellapandian Kannan. 2012. *International Journal of Pharmaceutical Sciences and Research* 3 (3):323–330.

Gomaa, Eman Zakaria. 2017. *The Journal of General and Applied Microbiology* 63(1):36–43.

Goodsell, David S. 2004. *Bionanotechnology: Lessons from Nature.* John Wiley & Sons.

Hajfathalian, Maryam, Kyle D Gilroy, Ali Yaghoubzade, Aarthi Sundar, Teng Tan, Robert A Hughes, and Svetlana Neretina. 2015. *The Journal of Physical Chemistry C* 119(30):17308–17315.

Hamouda, Tarek, Andrzej Myc, Brian Donovan, Amy Y Shih, Jon D Reuter, and James R Baker. 2001. *Microbiological Research* 156(1):1–7.

Hasnain, M Saquib, Md Noushad Javed, Md Sabir Alam, Poonam Rishishwar, Sanjay Rishishwar, Sadath Ali, Amit Kumar Nayak, and Sarwar Beg. 2019. *Materials Science and Engineering: C* 99:1105–1114.

Herizchi, Roya, Elham Abbasi, Morteza Milani, and Abolfazl Akbarzadeh. 2016. *Artificial Cells, Nanomedicine, and Biotechnology* 44(2):596–602.

Hulla, JE, SC Sahu, and AW Hayes. 2015. *Human & Experimental Toxicology* 34(12):1318–1321.

Humes, Sara T, and Tara Sabo-Attwood. 2020. In: *Information Resources in Toxicology (Fifth Edition)*, edited by Philip Wexler, 425–436. Academic Press.

Hwang, In-sok, Ji Hong Hwang, Hyemin Choi, Keuk-Jun Kim, and Dong Gun Lee. 2012. *Journal of Medical Microbiology* 61(12):1719–1726.

Itoh, Tamitake, Vasudevanpillai Biju, Mitsuru Ishikawa, Yasuo Kikkawa, Kazuhiro Hashimoto, Akifumi Ikehata, and Yukihiro Ozaki. 2006. *The Journal of Chemical Physics* 124(13):134708.

Jalilian, Fereshteh, Azam Chahardoli, Komail Sadrjavadi, Ali Fattahi, and Yalda Shokoohinia. 2020. *Advanced Powder Technology* 31(3):1323–1332.

Jeevanandam, Jaison, Ahmed Barhoum, Yen S Chan, Alain Dufresne, and Michael K Danquah. 2018. *Beilstein Journal of Nanotechnology* 9(1):1050–1074.

Jeyaraj, M, R Arun, G Sathishkumar, D MubarakAli, M Rajesh, G Sivanandhan, G Kapildev, M Manickavasagam, N Thajuddin, and A Ganapathi. 2014. *Materials Research Bulletin* 52:15–24.

Jeyaraj, Muniyandi, Sangiliyandi Gurunathan, Muhammad Qasim, Min-Hee Kang, and Jin-Hoi Kim. 2019. *Nanomaterials* 9(12):1719.

Jha, Anal K, K Prasad, Kamlesh Prasad, and AR Kulkarni. 2009. *Colloids and Surfaces B: Biointerfaces* 73(2):219–223.

Jin, Jian-Cheng, Xiao-Juan Wu, Juan Xu, Bei-Bei Wang, Feng-Lei Jiang, and Yi Liu. 2017. *Biomaterials Science* 5(2):247–257.

Karupannan, Sathish Kumar, Mohammed Junaid Hussain Dowlath, and Kantha D Arunachalam. 2020. In: *Phytonanotechnology*, 303–322. Elsevier.

Kaviya, S, J Santhanalakshmi, B Viswanathan, J Muthumary, and K Srinivasan. 2011. *Spectrochimica Acta Part A: Molecular and Biomolecular Spectroscopy* 79(3):594–598.

Kędziora, Anna, Robert Wieczorek, Mateusz Speruda, Iva Matolínová, Tomasz M Goszczyński, Ireneusz Litwin, Vladimír Matolín, and Gabriela Bugla-Płoskońska. 2021. *Frontiers in Microbiology* 12:1707.

Khalandi, Bahar, Nahideh Asadi, Morteza Milani, Soodabeh Davaran, Azam Jafari Najaf Abadi, Elham Abasi, and Abolfazl Akbarzadeh. 2017. *Drug Research* 11(02):70–76.

Khan, Ibrahim, Khalid Saeed, and Idrees Khan. 2019. *Arabian Journal of Chemistry* 12(7):908–931. https://doi.org/10.1016/j.arabjc.2017.05.011.

Khan, Md Abu Rayhan, Muhammad Shamim Al Mamun, and Mosummath Hosna Ara. 2021. *Microchemical Journal* 171:106840.

Khatua, Ashapurna, Abhinav Prasad, Eepsita Priyadarshini, Amiya Kumar Patel, Aparupa Naik, Muthupandian Saravanan, Hamed Barabadi, Bernard Paul, Rajamani Paulraj, and Ramovatar Meena. 2020. *Journal of Cluster Science* 31(6):1329–1340.

Kim, Jaeyun, Sungjin Park, Ji Eun Lee, Seung Min Jin, Jung Hee Lee, In Su Lee, Ilseung Yang, Jun-Sung Kim, Seong Keun Kim, and Myung-Haing Cho. 2006. *Angewandte Chemie* 118(46):7918–7922.

Kim, Taeho, Gary B Braun, Zhi-gang She, Sazid Hussain, Erkki Ruoslahti, and Michael J Sailor. 2016. *ACS Applied Materials & Interfaces* 8(44):30449–30457.

Klueh, U, V Wagner, S Kelly, A Johnson, and JD Bryers. 2000. *Journal of Biomedical Materials Research: An Official Journal of The Society for Biomaterials, The Japanese Society for Biomaterials, and The Australian Society for Biomaterials and the Korean Society for Biomaterials* 53(6):621–631.

Korshed, Peri, Lin Li, Zhu Liu, and Tao Wang. 2016. *PLoS One* 11(8):e0160078.

Krishnaraj, C, P Muthukumaran, R Ramachandran, MD Balakumaran, and PT Kalaichelvan. 2014. *Biotechnology Reports* 4:42–49.

Kulkarni, Narendra, and Uday Muddapur. 2014. *Journal of Nanotechnology* 2014.

Kumar, P Vijaya, S Mary Jelastin Kala, and KS Prakash. 2019. *Materials Letters* 236:19–22.

Kumari, Shalini, Pritam Kumar Panda, Nandini Pramanik, Suresh K Verma, and MA Mallick. 2019. *Environmental Nanotechnology, Monitoring & Management* 11:100201.

Latha, D, S Sampurnam, C Arulvasu, P Prabu, K Govindaraju, and V Narayanan. 2018. *Materials Today: Proceedings* 5(2):8968–8972.

Li, Chengjun, and Bing Yan. 2020. *Environmental Science: Nano* 7(10):2863–2874.

Liu, Xin-Yu, Jing-Quan Wang, Charles R Ashby Jr, Leli Zeng, Ying-Fang Fan, and Zhe-Sheng Chen. 2021. *Drug Discovery Today* 26(5):1284–1292.

Lu, Xianmao, Matthew Rycenga, Sara E Skrabalak, Benjamin Wiley, and Younan Xia. 2009. *Annual Review of Physical Chemistry* 60:167–192.

Lyddy, Ruth. 2009. In *Information Resources in Toxicology (Fourth Edition)*, edited by Philip Wexler, Steve G. Gilbert, Pertti J. Hakkinen and Asish Mohapatra, 321–328. San Diego: Academic Press.

Manivasagan, Panchanathan, Seung Yun Nam, and Junghwan Oh. 2016. *Critical Reviews in Microbiology* 42 (6):1007–1019.

Marchi, Saverio, Carlotta Giorgi, Jan M Suski, Chiara Agnoletto, Angela Bononi, Massimo Bonora, Elena De Marchi, Sonia Missiroli, Simone Patergnani, and Federica Poletti. 2012. *Journal of Signal Transduction* 2012.

Mody, Vicky V, Rodney Siwale, Ajay Singh, and Hardik R Mody. 2010. *Journal of Pharmacy and Bioallied Sciences* 2(4):282.

Mohanta, Yugal Kishore, Sujogya Kumar Panda, Kunal Biswas, Abiral Tamang, Jaya Bandyopadhyay, Debashis De, Dambarudhar Mohanta, and Akshaya Kumar Bastia. 2016. *IET Nanobiotechnology* 10(6):438–444.

MubarakAli, D, N Thajuddin, K Jeganathan, and M Gunasekaran. 2011. *Colloids and Surfaces B: Biointerfaces* 85(2):360–365.

Nayak, Rati Ranjan, Nilotpala Pradhan, Debadhyan Behera, Kshyama Madhusikta Pradhan, Srabani Mishra, Lala Behari Sukla, and Barada Kanta Mishra. 2011. *Journal of Nanoparticle Research* 13(8):3129–3137.

Nel, Andre E, Lutz Mädler, Darrell Velegol, Tian Xia, Eric Hoek, Ponisseril Somasundaran, Fred Klaessig, Vince Castranova, and Mike Thompson. 2009. *Nature Materials* 8(7):543–557.

Oh, Keun Sang, Ree Sun Kim, Jinho Lee, Dongmin Kim, Sun Hang Cho, and Soon Hong Yuk. 2008. *Journal of Applied Polymer Science* 108(5):3239–3244.

Ojha, Nisha Kant, Grigory V Zyryanov, Adinath Majee, Valery N Charushin, Oleg N Chupakhin, and Sougata Santra. 2017. *Coordination Chemistry Reviews* 353:1–57.

Panja, Sudipta, Indranil Chaudhuri, Kalyani Khanra, and Nandan Bhattacharyya. 2016. *Asian Pacific Journal of Tropical Disease* 6(7):549–556.

Papastefanaki, Florentia, Igor Jakovcevski, Nafsika Poulia, Nevena Djogo, Florian Schulz, Tamara Martinovic, Darko Ciric, Gabrielle Loers, Tobias Vossmeyer, and Horst Weller. 2015. *Molecular Therapy* 23(6):993–1002.

Patil, Maheshkumar Prakash, Enkhnaran Bayaraa, Paul Subedi, Lei Lottice Anne Piad, Naresh Hiralal Tarte, and Gun-Do Kim. 2019. *Journal of Drug Delivery Science and Technology* 51:83–90.

Patil, Maheshkumar Prakash, Xing Jin, Natasha Chantal Simeon, Jaymee Palma, Daniel Kim, Daniel Ngabire, Nan-Hee Kim, Naresh Hiralal Tarte, and Gun-Do Kim. 2018. *Artificial Cells, Nanomedicine, and Biotechnology* 46(1):82–88.

Patil, Maheshkumar Prakash, and Gun-Do Kim. 2017. *Applied Microbiology and Biotechnology* 101(1):79–92.

Patil, Maheshkumar Prakash, and Gun-Do Kim. 2018. *Colloids and Surfaces B: Biointerfaces* 172:487–495.

Pelton, Matthew, Javier Aizpurua, and Garnett Bryant. 2008. *Laser & Photonics Reviews* 2(3):136–159.

Pistritto, Giuseppa, Daniela Trisciuoglio, Claudia Ceci, Alessia Garufi, and Gabriella D'Orazi. 2016. *Aging (Albany NY)* 8(4):603.

Pizzutilo, Enrico, Simon J Freakley, Serhiy Cherevko, Sriram Venkatesan, Graham J Hutchings, Christian H Liebscher, Gerhard Dehm, and Karl JJ Mayrhofer. 2017. *ACS Catalysis* 7(9):5699–5705.

Ponnanikajamideen, MohamedIbrahim, Shanmugam Rajeshkumar, Mahendran Vanaja, and Gurusamy Annadurai. 2019. *Canadian Journal of Diabetes* 43(2):82–89. e6.

Poole Jr, Charles P, and Frank J Owens. 2003. *Introduction to Nanotechnology.* John Wiley & Sons.

Pourreza, Nahid, Hamed Golmohammadi, Tina Naghdi, and Hossein Yousefi. 2015. *Biosensors and Bioelectronics* 74:353–359.

Qian, Linhua, Weihua Su, Yan Wang, Minyan Dang, Wenzhi Zhang, and Changlin Wang. 2019. *Artificial Cells, Nanomedicine, and Biotechnology* 47(1):1173–1180.

Quinteros, MA, V Cano Aristizábal, Pablo Roberto Dalmasso, María Gabriela Paraje, and Paulina Laura Páez. 2016. *Toxicology In Vitro* 36:216–223.

Rai, Mahendra, Alka Yadav, and Aniket Gade. 2009. *Biotechnology Advances* 27(1):76–83.

Rajeshkumar, S, LV Bharath, and R Geetha. 2019. In: *Green synthesis, Characterization and Applications of Nanoparticles*, 429–444. Elsevier.

Ramalingam, Baskaran, Thanusu Parandhaman, and Sujoy K Das. 2016. *ACS Applied Materials & Interfaces* 8(7):4963–4976.

Raut, Rajesh W, Niranjan S Kolekar, Jaya R Lakkakula, Vijay D Mendhulkar, and Sahebrao B Kashid. 2010. *Nano-Micro Letters* 2(2):106–113.

Riley, Rachel S, and Emily S Day. 2017. *Wiley Interdisciplinary Reviews: Nanomedicine and Nanobiotechnology* 9(4):e1449.

Rubilar, Olga, Mahendra Rai, Gonzalo Tortella, Maria Cristina Diez, Amedea B Seabra, and Nelson Durán. 2013. *Biotechnology Letters* 35(9):1365–1375.

Saldan, Ivan, Yuriy Semenyuk, Iryna Marchuk, and Oleksandr Reshetnyak. 2015. *Journal of Materials Science* 50(6):2337–2354.

Saravanakumar, Arthanari, Mei Mei Peng, Mani Ganesh, Jayabalan Jayaprakash, Murugan Mohankumar, and Hyun Tae Jang. 2017. *Artificial Cells, Nanomedicine, and Biotechnology* 45(6):1165–1171.

Sathishkumar, M, K Sneha, SW Won, C-W Cho, S Kim, and Y-S Yun. 2009. *Colloids and Surfaces B: Biointerfaces* 73(2):332–338.

Seong, Minju, and Dong Gun Lee. 2017. *Current Microbiology* 74(6):661–670.

Shittu, OK, DI Stephen, and AH Kure. 2017. *African Journal of Biomedical Research* 20(2):195–202.

Sondi, Ivan, and Branka Salopek-Sondi. 2004. *Journal of Colloid and Interface Science* 275(1):177–182.

Srikar, Sista Kameswara, Deen Dayal Giri, Dan Bahadur Pal, Pradeep Kumar Mishra, and Siddh Nath Upadhyay. 2016. *Green and Sustainable Chemistry* 6(1):34–56.

Sriranjani, R, B Srinithya, Vadivel Vellingiri, Pemaiah Brindha, Savarimuthu Philip Anthony, Aravind Sivasubramanian, and Meenakshi Sundaram Muthuraman. 2016. *Journal of Molecular Liquids* 220:926–930.

Stamplecoskie, Kevin G, Juan C Scaiano, Vidhu S Tiwari, and Hanan Anis. 2011. *The Journal of Physical Chemistry C* 115(5):1403–1409.

Suman, TY, SR Radhika Rajasree, A Kanchana, and S Beena Elizabeth. 2013. *Colloids and Surfaces B: Biointerfaces* 106:74–78.

Sun, Jingwei, Yuexiang Lu, Liuying He, Jiawei Pang, Fengyi Yang, and Yueying Liu. 2020. *TrAC Trends in Analytical Chemistry* 122:115754.

Taghavizadeh Yazdi, Mohammad Ehsan, Majid Darroudi, Mohammad Sadegh Amiri, Hasan Ali Hosseini, Fahimeh Nourbakhsh, Mohammad Mashreghi, Malihe Farjadi, Seyed Mousa Mousavi Kouhi, and Seyed Hadi Mousavi. 2020. *Micro & Nano Letters* 15(14):1046–1050.

Tran, Tien D, Mai TT Nguyen, Hoang V Le, Duc N Nguyen, Quang Duc Truong, and Phong D Tran. 2018. *Chemical Communications* 54(27):3363–3366.

Turkevich, John, Peter Cooper Stevenson, and James Hillier. 1951. *Discussions of the Faraday Society* 11:55–75.

Unal, Ilay Sema, Ayse Demirbas, Irem Onal, Nilay Ildiz, and Ismail Ocsoy. 2020. *Journal of Photochemistry and Photobiology B: Biology* 204:111800.

Vijayan, Remya, Siby Joseph, and Beena Mathew. 2018. *Artificial Cells, Nanomedicine, and Biotechnology* 46(4):861–871.

Wang, Chao, Hideo Daimon, Taigo Onodera, Tetsunori Koda, and Shouheng Sun. 2008. *Angewandte Chemie* 120(19):3644–3647.

Wuithschick, Maria, Alexander Birnbaum, Steffen Witte, Michael Sztucki, Ulla Vainio, Nicola Pinna, Klaus Rademann, Franziska Emmerling, Ralph Kraehnert, and Jörg Polte. 2015. *ACS Nano* 9(7):7052–7071.

Xiu, Zong-ming, Qing-bo Zhang, Hema L Puppala, Vicki L Colvin, and Pedro JJ Alvarez. 2012. *Nano Letters* 12(8):4271–4275.

Yin, Iris Xiaoxue, Jing Zhang, Irene Shuping Zhao, May Lei Mei, Quanli Li, and Chun Hung Chu. 2020. *International Journal of Nanomedicine* 15:2555.

You, Chuangang, Chunmao Han, Xingang Wang, Yurong Zheng, Qiyin Li, Xinlei Hu, and Huafeng Sun. 2012. *Molecular Biology Reports* 39(9):9193–9201.

Yuan, Hsiangkuo, Christopher G Khoury, Christy M Wilson, Gerald A Grant, Adam J Bennett, and Tuan Vo-Dinh. 2012. *Nanomedicine: Nanotechnology, Biology and Medicine* 8(8):1355–1363.

Yun'an Qing, Lin Cheng, Ruiyan Li, Guancong Liu, Yanbo Zhang, Xiongfeng Tang, Jincheng Wang, He Liu, and Yanguo Qin. 2018. *International Journal of Nanomedicine* 13:3311.

Zhang, Xi-Feng, Zhi-Guo Liu, Wei Shen, and Sangiliyandi Gurunathan. 2016. *International Journal of Molecular Sciences* 17(9):1534.

Zhou, Wei Ping, Adam Lewera, Robert Larsen, Rich I Masel, Paul S Bagus, and Andrzej Wieckowski. 2006. *The Journal of Physical Chemistry B* 110(27):13393–13398.

15 Conclusion and Future Perspectives of Phytonanotechnology

Neha Parveen[1], Vidhu Aeri[2], Waleed H. Almalki[3], Amirhossein Sahebkar[4,5,6], and Prashant Kesharwani[1,7]*

[1]Department of Pharmaceutics, School of Pharmaceutical Education and Research, Jamia Hamdard University, New Delhi 110062, India
[2]Department of Pharmacognosy and Phytochemistry, School of Pharmaceutical Education and Research, Jamia Hamdard University, New Delhi 110062, India
[3]Department of Pharmacology and Toxicology, Faculty of Medicine, Umm Al-Qura University, Makkah, Saudi Arabia
[4]Applied Biomedical Research Center, Mashhad University of Medical Sciences, Mashhad, Iran
[5]Biotechnology Research Center, Pharmaceutical Technology Institute, Mashhad University of Medical Sciences, Mashhad, Iran
[6]School of Medicine, The University of Western Australia, Perth, Australia
[7]University Institute of Pharma Sciences, Chandigarh University, Mohali, Punjab, India

15.1 INTRODUCTION

Phytonanotechnology is a newly emerging and interesting field of medical scientific research. "Phytonanotechnology" is a term that refers to the use of nanotechnology for plant production practices and plant sciences.[1] Nanoparticles are a remarkable class of nano-sized substances with distinctive attributes and broad-spectrum applications in numerous sectors.[2–6] Nanomaterials are regarded as a linkage between atomic structures and bulk material size. Because of their favorable surface-to-mass ratio and small size, inorganic nanoparticles possess unique properties.[7–13] Based on their shape, size, and distribution, nanoparticles exhibit novel or enhanced abilities.[14–17] Nowadays, plant extracts have been used as capping and reducing agents for the synthesis of nanoparticles; this method may be preferable to microorganism-based synthesis, as it does not require the complex process of cultivating and maintaining microbial cultures. Research on plant-derived nanoparticle synthesis has revealed that biopolymers, microorganisms, and plants are widely used as biological agents, with very few findings for animal origin. The synthesis process involves the use of plant materials such as seeds, fruits, flowers, roots, bark, leaves, and other parts. For the synthesis of nanoparticles, a variety of plants with potent properties are used as sources.[18]

Conventional techniques for synthesizing nanoparticles through chemical and physical methods are harmful when used in biomedical applications and are not eco-friendly.[19] Several synthetic processes including biochemical, chemical, and physical methods have been used to synthesize nanoparticles.[20]

The use of harmful stabilizing and reducing agents such as toxic solvents, N,N-dimethylformamide, and sodium borohydride is a common reason why chemical-based synthesis processes are prohibited.

DOI: 10.1201/9781003231721-15

Green synthesis is becoming increasingly important in modern science, with more attention being paid to substances such as microorganisms, soluble starch, chitosan, and glucose. Because of the ability to maintain an aseptic environment throughout the procedure, the biochemical production of nanoparticles from plant extracts as capping and reducing agents has drawn particular attention.[21] Moreover, because of the versatility and eco-friendliness, synthesis of plant-derived nanoparticles has attracted attention in contemporary nanotechnology and science. Therefore, medicinal herbs with a history of successful therapeutic usage are frequently used in the shape- and size-controlled production of nanoparticles.

Silver nanoparticles, which have been extensively explored, are crucial components in the synthesis of plant-derived nanoparticles. Different types of biological, chemical, and physical methods can be used to synthesize these nanoparticles.[22–24] Plant-mediated reduction resulting from phytochemicals is the primary technique under consideration. Carboxylic acids, amides, aldehydes, ketones, flavones, and terpenoids are the primary phytochemicals implicated for the plant-mediated reduction. Quinones, organic acids, and flavones are water-soluble phytochemicals that facilitate instantaneous reduction of the ions. The extensive use of silver nanoparticles in the medical industry is attributed to their anti-inflammatory and antibacterial properties.

Because of the improved electrical, optical, and catalytic abilities, nanoparticles, particularly silver nanoparticles, have achieved widespread application in therapeutics, antimicrobial agents, diagnostics, detection, optoelectronics, and catalysis.[25] Amin et al. investigated the green synthesis of silver nanoparticles from berry extract of *Solanum xanthocarpum* L. The authors identified urease inhibitory and antibacterial effects against *Helicobacter pylori* and compared these effects to those of $AgNO_3$ and four drugs commonly used to treat *Helicobacter pylori* infection, namely, tetracycline, metronidazole, clarithromycin, and amoxicillin. The nanoparticles used in this study were spherical, monodisperse, and nearly 10 nm in size.[26] Additionally, silver nanoparticles synthesized from *Tephrosia tinctoria* have shown antidiabetic activity by significantly increasing the rate of glucose absorption, inhibiting carbohydrate digestion enzymes, and scavenging free radicals.[27] Nanoparticles are synthesized from a variety of plant parts including leaves.[28]

Plant-derived biomolecules mediate the conversion of silver ions from silver nitrate to silver nanoparticles. Various determination methods such as X-ray diffraction, ultraviolet (UV)-visible spectroscopy, and transmission electron microscopy can be used to assess nanoparticle synthesis. Silver nanoparticles can be easily biosynthesized because of the extracellular and quick reduction reaction. The distinctive features of nanoparticles, such as shape, size, and other size-dependent characteristics, may be crucial for their application in cosmetics, agriculture, forensic sciences, food chemistry, optics, biomedicine, and other uses in the healthcare industry.

Phytonanotechnology addresses the issues using a nanotechnological approach by developing nano-pesticides and herbicides, nano-fertilizers, nutrient enrichment, food packaging, and storage. Phytonanotechnology offers new approaches and possibilities for the stable, simple, quick, eco-friendly, and economical synthesis of different nanoparticles. Some benefits include the simplicity of nanoparticle synthesis from water, the world's most versatile solvent, as well as their biocompatibility and medical utility. Gold and silver nanoparticles have been prepared successfully from the roots and leaves of medicinal herbs.[29] Nanoparticles have been synthesized from fruit and stem extracts, in addition to extracts of other plant parts.[30] The synthesis processes differ between various plant species, and in-depth research has revealed various processes for producing nanoparticles from plants.

Nanotechnology has slowly started gaining application in agriculture when compared with that in other industries including medicine, energy, electronics, and materials. The successful use of nanomaterials has sparked considerable attention in agro-nanotechnology, which makes it feasible for the immediate application of pesticide molecules for treating plant diseases. To lower production prices and boost yields, sustainable crop protection involves replacing chemicals with locally sourced biological inputs and expertise. The foundation of sustainable agriculture is a blend of historical wisdom and modern scientific discoveries. This approach supports economic feasibility, agricultural output, and reduction in environmental deterioration.[31] In-depth research focusing on agricultural uses and plant-based synthesis of nanoparticles can facilitate the development of this approach. The use of nanoparticles in agriculture can decrease the need for plant protection products, reduce the loss of nutrients during fertilization, and boost yields by improved nutritional management.[32] Compared with chemical and physical methods for nanoparticle synthesis, the use of phytonanotechnology has several advantages including scalability, biocompatibility, environmental friendliness, affordability, and the potential for therapeutic uses of the nanoparticles produced (Figure 15.1).

15.2 SYNTHESIS OF PHYTONANOPARTICLES

Disease-free, healthy, and fresh plant parts are chosen for the preparation of nanoparticles from plant extracts, and these parts are frequently rinsed with Milli-Q deionized water eliminate any surface impurities. The plant parts are then boiled for 5 min, after which they are decanted. This produces the crude extract, which should be filtered using Whatman No. 1 filter paper. The filtrate is then used to extract 5 mL of plant material, which is mixed with 500 mL of solution. The resulting mixture is allowed to stand until the colorless solution changes to a reddish-brown solution, signifying nanoparticle synthesis. Generally, this procedure is carried out at room temperature. The synthesized particles are then centrifuged for cleaning. The colloids are cleaned with deionized water approximately three times to eliminate any unbound ions. The colloids are then lyophilized and kept in ambient storage in screw-capped containers for subsequent characterization and use. Several publications have described the production of nanoparticles from plant extracts (Figure 15.1).

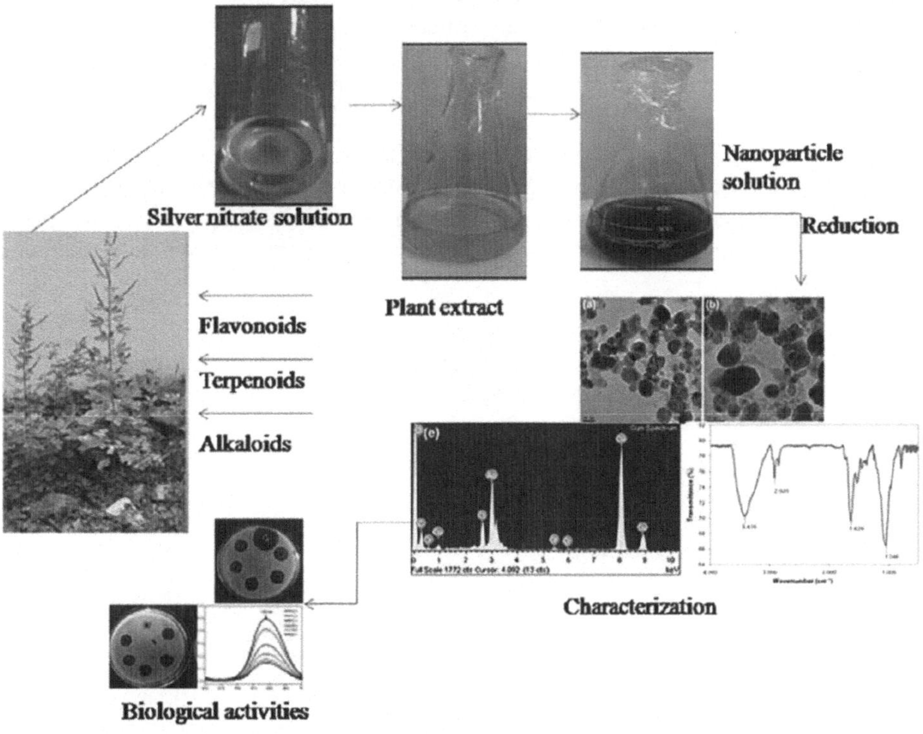

FIGURE 15.1 Diagrammatic representation of the synthesis and bioassay of nanoparticles using plant extracts. (From Mathew 2020.[33] With Permission.)

15.3 PROPERTIES AND CHARACTERIZATION OF PHYTONANOPARTICLES

Nanoparticles exhibit desirable features such as high stability and the ability to quickly change their surface properties (Figure 15.2).[34–39] The development stability of metal nanoparticles can be determined by calculating the optical characteristics of the biosynthesized nanoparticles, which is an essential and widely used method. Metal nanoparticles exhibit a significant electromagnetic wave absorption in the visible region owing to surface plasmon resonance (SPR). Because of the collective oscillations of conduction electrons, the SPR effect develops when nanoparticles are exposed to the visible region. Characterization of green-synthesized nanoparticles is carried out using energy-dispersive X-ray spectroscopy analysis and transmission electron microscopy.

Raman scattering (Raman spectroscopy) is used to characterize nanoparticles by identifying the vibrational signals of various chemical entities that are bonded to the interfaces of nanoparticles during manufacturing and obtain information about the surface appearance of the nanoparticles. Fourier-transform infrared spectroscopy, X-ray photon spectroscopy, atomic force microscopy, transmission electron microscopy (TEM), scanning electron microscopy (SEM), energy-dispersive X-ray spectroscopy, and inductively coupled plasma atomic emission spectroscopy were used to characterize and synthesize gold nanoparticles using *Diospyros kaki* and *Magnolia kobus* leaf extracts. TEM and SEM analyses have revealed the synthesis of nanoparticles in a variety of plate (hexagons, pentagons, and triangles) and spherical shapes (Figure 15.3).[40]

15.4 PLANT–PHYTONANOPARTICLE INTERACTIONS AND TRANSLOCATION

The accumulation, translocation, and absorption of nanoparticles in plants depend on the delivery method, stability, functionalization, physiological property, growth environment, age, and plant species. Because of their small pore size, nanoparticles pass through the cell walls of plants by attaching to carrier proteins and forming complexes with trichome bases in leaves, stomata, membrane transporters, or root exudates.[40] The nanoparticles eventually reach the plasma membrane. To improve nanoparticle uptake, functionalized nanoparticles should increase the pore size or induce the formation of a new cell wall.[41] Following entry into the cell, the nanoparticles pass from one cell to another through the plasmodesmata.[42] Only few publications have assessed the direct translocation, interaction, and uptake of nanoparticles such as virus nanoparticles, iron oxide nanoparticles, titanium dioxide nanoparticles, gold nanoparticles, quantum dots, fullerene nanoparticles, carbon nanotubes, silica nanoparticles, and mesoporous silica nanoparticles (Figure 15.4). The plant cells undergo the development of new cuticles and cell walls, bigger pores, and structural alterations, all

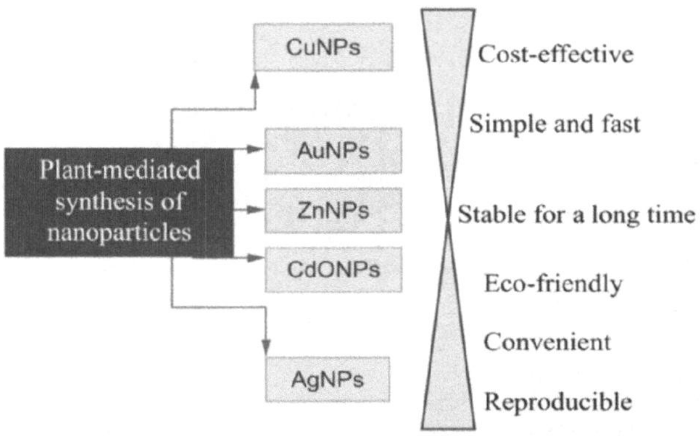

FIGURE 15.2 Diagram displaying the properties and characteristics of the as-synthesized plant-derived nanoparticles. (From Mathew 2020.[33] With Permission.)

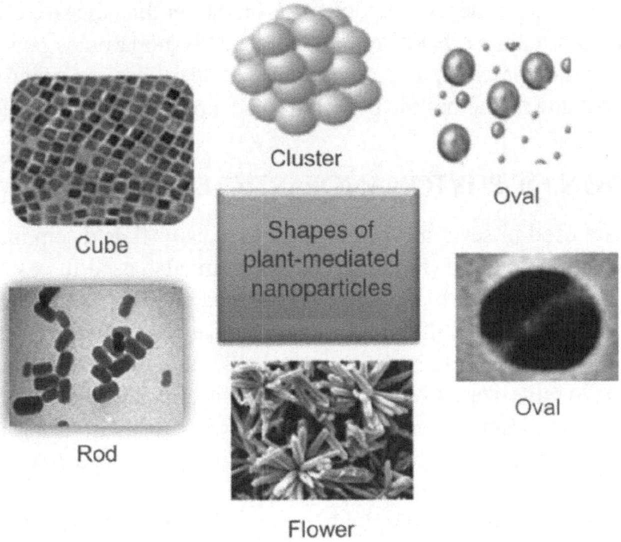

FIGURE 15.3 Diagrammatic representation of the different potential morphology of plant-derived nanoparticles. (From Mathew 2020.[33] With Permission.)

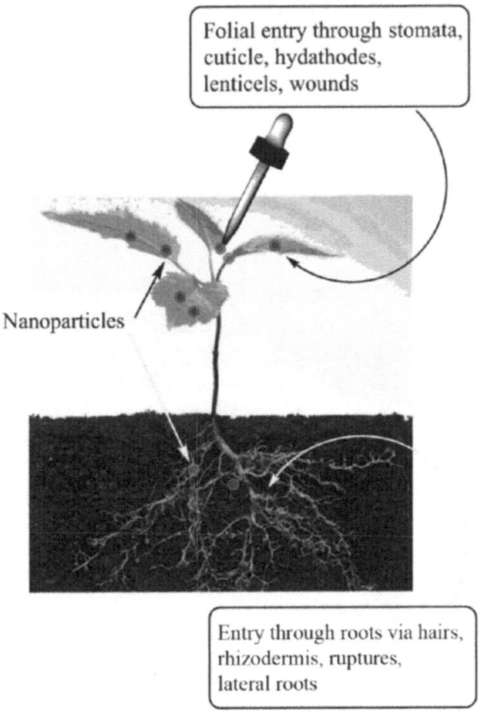

FIGURE 15.4 Plant cell system's potential interaction points for nanoparticles. (From Khan et al., 2019.[45] With Permission.)

of which are due to the effect of the nanoparticles that are larger than the size exclusion limit. Larger nanoparticles can therefore have a better internalization.[43] Nanoparticles penetrate leaves through the epidermis and stomata. For example, cerium oxide nanoparticles can penetrate *Zea mays* leaves; however, it remains unknown whether the penetration occurred through the epidermis or stomata.[44]

15.5 APPLICATION OF PHYTONANOPARTICLES

Phytonanoparticles are used in several types of industries as well as disciplines including, but not limited to, medicine and agriculture (Figure 15.5). They are also used in cosmetic and biomedical applications owing to their increased bioactivity when compared with physicochemical nanoparticles. For instance, compared with chemically synthesized nanoparticles, phytonanoparticles demonstrate greater biocompatibility with Chinese hamster ovary cells, cardiomyoblasts, and endothelial cells.[46]

Silver nanoparticles synthesized through phytosynthesis are mainly used in the biomedical field because of their exceptional antibacterial properties. Owing to their strong antibacterial properties, metal nanoparticles such as zinc, silver, and others are used in catheters, food packaging, and wound dressings. Titanium- and zinc-based nanoparticles are mostly used in aesthetic applications owing to their UV-blocking ability. One of the notable uses of phytonanoparticles is in the production of sensors that can detect numerous analytes; these sensors have been used in various fields such as environmental, diagnostic, and agricultural sectors. Additionally, phytonanoparticles are recommended for use in cell labelling, drug delivery, and gene transfer, providing significant biological and medical uses. Metal oxide and metal nanomaterials have several extremely intriguing uses that have not already been completely investigated, including magnetically induced, photoimaging, and photothermal-induced drug delivery.

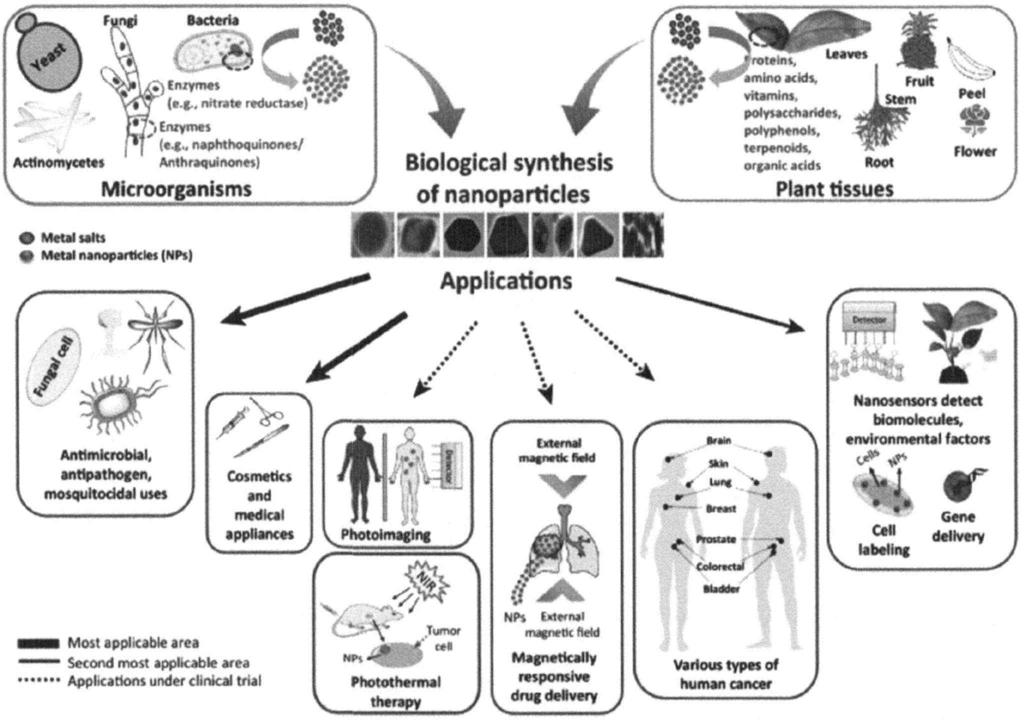

FIGURE 15.5 Applications of metal nanoparticles in environmental and biomedical sectors and their biological synthesis. (From Singh et al., 2016.[29] With Permission.)

15.5.1 Agricultural Applications of Phytonanoparticles

Phytonanoparticles are an eco-friendly approach and are used in agriculture for stress endurance, plant growth, and plant disease control.[47] Silver nanoparticles synthesized from plants have demonstrated encouraging antifungal activity against wheat disease, caused by *Bipolaris sorokiniana*.[48] Furthermore, silver nanoparticles synthesized from *Serratia* sp. inhibited the genes involved in melanin production by *Bipolaris sorokiniana*.[49] Biosynthesized nanoparticles are susceptible to several different phytopathogens, indicating their ability to treat a variety of disorders in agricultural plants.[50] Biosynthesized TiO_2 nanoparticles enhance both rhizosphere microbial community and *Vigna radiata* growth, and both effects are very promising.[51] Additionally, the viability and productivity of cluster bean plants were improved by the use of biosynthesized MgO nanoparticles. The use of such biosynthesized nanoparticles improved shoot–root development and total chlorophyll pigmentation in plants. Acid phosphatase, alkaline phosphatase, and phytase activities in the cluster bean rhizosphere were improved by biosynthesized ZnO nanoparticles.[51] Nanoparticles produced through phytosynthesis are extremely interesting for preventing disease epidemics and encouraging plant growth because of their fertilizing and antimicrobial properties. These studies highlight the possibility of different phytonanoparticles in numerous agricultural applications.

15.5.2 Environmental Applications of Phytonanoparticles

Environmental contamination with harmful substances due to human activity remains a remarkable concern. Water bodies and polluted soils are decontaminated through a variety of techniques and technology. The main metal pollutants include toxic metals such as nickel, mercury, chromium, barium, lead, arsenic, and cadmium. The main organic pollutants include herbicides, pesticides, plastics, organic solvents, and petroleum. A variety of chemical and physical methods are used for environmental cleanup; however, they are not secure, eco-friendly, or economical. Therefore, it is crucial to develop new methods that use eco-friendly techniques to eliminate military trash, petroleum sludge, and organic pollutants as well as metallic residues.

15.5.2.1 Removal of Heavy Metals

Lead and other heavy metals are major pollutants in water and soil. To remove heavy metals from water and soil, $FeCl_3$ can be reduced by the application of *Sargassum acinarium* and *Padina pavonica* to create Fe_3O_4 nanoparticles. To eliminate Pb, the resulting Fe_3O_4 nanoparticles were loaded with calcium alginate particles. After 75 min, the Fe_3O_4 nanoparticle-loaded alginate beads prepared using *Padina pavonica* outperformed those prepared using *Sargassum acinarium* in terms of their ability to remove 91% of total Pb content.

15.5.2.2 Water Purification

It is difficult to practically ascertain the accessibility and quality of drinking water in all nations worldwide. There is an increasing demand to discover new technology to improve water management. Nanotechnology has considerable promise for the development of cutting-edge industrial and domestic water management systems. As for the use of water filters, increasing interest is being paid to surface modification or coating of nanoparticles with antimicrobial compounds to stop the development of pathogenic microorganisms. Silver nanoparticles produced through phytosynthesis have increased antibacterial activity and can be used in membrane filters in water purification systems. The World Health Organization has approved the use of silver nanoparticles for this purpose.[52] Antimicrobial coatings should have strong antibacterial activity, simple production process, and low toxicity. Silver nanoparticle coatings on the surface have mostly been used as antibacterial coatings.[53]

15.5.2.3 Waste Water Management

Wastewater treatment entails the elimination of toxic pollutants from sewage or wastewater, followed by converting the wastewater into an effluent that can be directly reused for domestic use or sent back to the water loop with little environmental effects. In addition to biomass, the main pollutants in wastewater are inorganic–organic compounds and bacteria. Zinc oxide, titanium dioxide, and silver nanoparticles have significant therapeutic potential for wastewater management. Phytosynthesized nanoparticles are very promising and have possible benefits over physically or chemically produced nanoparticles. In addition to having strong bactericidal properties against several bacteria, gold and silver nanoparticles prepared using the aqueous extract of burdock root showed outstanding catalytic activity in the breakdown of contaminants in wastewater.[54] One important step in the treatment of wastewater is the reductive conversion of nitroaromatic chemicals. Metals such as palladium are excellent catalytic materials that can be utilized to eliminate these chemicals. As a potent catalyst for reducing nitroaromatics, palladium nanoparticles produced from *Pantoea* sp. has been investigated.[55] When nitroaromatic substances such as 2,6-dichloro-4-nitroaniline, 4-nitroaniline, 4-nitrophenol, and 2-nitroaniline were reduced, palladium nanoparticles demonstrated better catalytic activity.

15.5.3 Cosmetic Application

Nanotechnology has tremendous opportunities for application in the cosmeceutical sector, as nanoparticles can rapidly penetrate through the skin and reach the desired tissue. The use of nanomaterials can frequently enhance the characteristics of various formulations and products in the cosmetics industry. Additionally, nanoparticles can provide more stability and better sensory properties to cosmetic formulations. For rendering the cosmetic product with a good feel and spreading ability, titanium dioxide and zinc oxide nanoparticles are used, as they provide a superior shielding effect from the harmful UV rays. Photosynthesized silver nanoparticles can be used to prepare antimicrobial cream formulations.[56] Cream base containing silver nanoparticles prepared from *Withania somnifera* extract was investigated for their antibacterial properties against *Candida albicans*, *Proteus vulgaris*, *Staphylococcus aureus*, *Pseudomonas aeruginosa*, and *Escherichia coli*. The results indicated that creams containing silver nanoparticles have a greater ability to suppress these bacteria. Similarly, shampoos containing silver nanoparticles prepared from *Solanum trilobatum* Linn. extract exerted antimicrobial activity against numerous fungi and bacteria.[57] Shampoo containing phytonanoparticles demonstrated antidandruff properties against fungi that cause dandruff such as *Pityrosporum ovale* and *Pityrosporum folliculitis*.

When a person often closes and opens cosmetic containers or bottles, there is a considerable potential that microorganisms will enter into the products. Preservatives must therefore be added to cosmetic goods to stop microbiological infection. The antibacterial characteristics of phenoxyethanol and parabens are typically utilized in cosmetic products; however, these products can briefly cause skin irritation and enhance the skin's UV susceptibility.[58] Silver nanoparticles are effective against microorganisms, which make these nanoparticles as safe substitutes for these preservatives in cosmetics. Thus, preservatives prepared from photosynthesized silver nanoparticles can be added to toothpastes and shampoos. Although there is only little information about the use of photosynthesized nanoparticles in cosmetics, they hold great opportunities as active components in cosmetics.

15.5.4 Biomedical Application

Numerous biological applications such as biosensors, cancer therapy, antimicrobial dressings, and image contrast agents make considerable use of different kinds of nanoparticles. Therefore, environmentally and biologically safe nanoparticles prepared from plants and their metabolites have a variety of uses in biomedical and medicinal products.[59]

15.5.4.1 Wound Healing Agents

Numerous biopolymers including gelatin,[60] chitosan,[61] polyvinyl alcohol,[62] polylactic acid,[63] and polycaprolactone[64] are frequently used in wound-dressing purposes. A reliable method for creating porous membranes with microbial resistance qualities for wound-dressing purposes is electrospinning.[65] Silver nanoparticles prepared from *Piper nigrum* leaf residue were added to electrospun polycaprolactone membranes to confer bactericidal characteristics to them.[66] To create an antimicrobial wound dressing, nanomaterials prepared from *Biophytum sensitivum* were embedded into nano-micro dual-porous calcium pectinate nanostructures.[67] When compared with standard wound dressings, biosilver nanoparticle treatment lowered the time required for fibroblast differentiation into myofibroblasts and shortened the inflammatory process.

15.5.4.2 Antibacterial Agents

Phytonanoparticles possess better antibacterial effects than their empty counterparts because these nanoparticles have phytochemicals. The natural antibacterial activity of silver nanoparticles is improved because of the phytochemicals derived from plants.[68] Intercellular components of the fungus and the *Candida* species' membranes are damaged by phytosynthesized silver nanoparticles.[53] Silver phytonanoparticles prepared from the leaf extract of *Garcinia mangostana* were incredibly efficient against a variety of drug-resistant diseases.[69] Silver nanoparticles prepared from *Artemisia nilagirica* demonstrated sensitivity to various bacteria.[70]

The findings of various studies suggested that the use of phytonanoparticles as antiviral drugs, such as silver nanoparticles, is a promising strategy that offers additional therapeutic options against deadly viral diseases. In general, therapeutic agents that are less expensive, safer, and newer to combat many viral infections can be developed through plant-mediated production of metal or metal oxide nanoparticles, with little to no cytotoxicity on mammalian cells.

15.5.4.3 Anticancer Agents

Several types of nanomaterials are widely used in cancer treatment. Pro-apoptotic pathways controlled by the production of reactive oxygen species (ROS) can be activated by nanoparticles. *Oregano vulgare* plant extract was used to prepare silver nanoparticles, and they showed a concentration-dependent inhibitory effect against a lung tumor cell line.[71] Malignant cells exhibited apoptosis when treated with gold nanoparticles at a comparatively low concentration.[72] Nanoparticles prepared from plants have the potential to improve cancer therapy without compromising the survivability of healthy cells. When compared with other chemotherapy medications, green-synthesized silver nanoparticles prepared from *Morinda citrifolia* showed better cytotoxic activity in HeLa cell lines.[73] Silver nanoparticles produced biochemically from the yeast *Cryptococcus laurentii* were visible. When these nanoparticles were treated at low concentrations, they effectively reduced the growth of breast carcinoma cells and caused membrane permeability. *In vitro* tests with Dalton lymphoma cell lines revealed that silver nanoparticles prepared from *Syzygium cumini* fruit residue exerted anticancer[74] and antioxidant properties.[75]

15.5.5 Drug Delivery Application

Numerous kinds of nanoparticles are used as drug delivery systems for pharmaceuticals.[76] Nanotechnology-based new drug delivery methods have been used in gene therapy and the treatment of human ailments such as microbial infections, diabetes, and cancer. These therapies have the advantage that the medication affects only cancer cells, thus improving the safety profile of the drug by reducing side effects on healthy cells. The main benefit of nanodrug delivery systems is the ability to combine various kinds of medicines and deliver them to target sites with the aid of nanoparticles and through a variety of techniques. Evidently, green synthesis methods produce nanoparticles that are safer for drug delivery purposes than those made through chemical synthesis. Only fewer studies have utilized the phytosynthesized gold nanoparticles for drug delivery purposes, despite the fact that chemically synthesized gold nanoparticles have been substantially explored. Vancomycin-conjugated gold nanoparticles prepared from *Trichoderma viride* effectively prevented the development of vancomycin-resistant *Staphylococcus aureus* at a low dosage.[77] It is also feasible to deliver anticancer medicines to cancerous cells by utilizing biosynthesized silver and gold nanoparticles. Doxorubicin-conjugated nanoparticle-based drug delivery methods significantly reduced the proliferation of cancer cells as compared with methods using free medication.[78] Green-synthesized gold nanoparticles prepared from *Peltophorum pterocarpum* is an effective drug transport strategy for DOX. Figure 15.6 provides a schematic representation of the detrimental effects of nanoparticles on plants. Additionally, melanoma tumor mouse models and *in vivo* and *in vitro* anticancer experiments involving DOX-loaded drug delivery methods were conducted on B16F10 and A549 cancer cells. When compared with the nanoconjugated form, free DOX uptake and release were slower, and the biosynthesized gold nanoparticle–PP-DOX conjugates had higher tumor-suppressive potential.

15.6 CHALLENGES IN PHYTONANOTECHNOLOGY

Because of their small size, nanoparticles are quickly absorbed by plants and exert a variety of effects including expression of genes, physiological and biological attributes, and altered morphology.[79]

CuO nanoparticles improved the phytohormone indole-3-acetic acid in *Pisum sativum*. The buildup of ZnO nanoparticles in the phloem and xylem slowed down the photosynthetic effect and wheat seedling growth. Ascorbate-glutathione cycle enzymes such as glutathione reductase, ascorbate peroxidase, and monodehydroascorbate reductase and the related metabolites ascorbate and glutathione were inhibited by ZnO nanoparticles, which increased lipid peroxidation and hydrogen peroxidation.[80] Silver nanoparticles reduced ascorbate and glutathione levels, thus causing damage to the anatomical integrity of the leaf and root. *Pisum sativum* photosynthetic pigments, chlorophyll fluorescence, and seedling growth were reduced because of increased deposition of silver in the pea seedlings' shoots and roots.[81] In *Arabidopsis thaliana*, the large quantity of silver nanoparticles upregulated the expression of oxidative stress genes and upregulated the expression of hormonal and pathogen stimuli genes, all of which considerably reduced plant growth.[82] Single-walled carbon nanotubes suppress the growth of *Zea mays* root hairs.[83]

Soyabean, corn, and carrot root elongation are inhibited by aluminum nanoparticles. In tomato and tobacco tissues, multiwalled carbon nanotubes altered gene expression. Exposure of fullerene and ZnO nanoparticles in *Arabidopsis thaliana* resulted in a decrease in the phytotoxic effect and biomass, along with the downregulated and upregulated expression of genes.[84,85] According to Van Aken, nanoparticles have an effect on the expression of crucial genes involved in the photosynthetic system.[86]

In *Vicia faba*, TiO_2 nanoparticles reduced genotoxicity, activity of glutathione reductase and ascorbate peroxidase, micronuclei and chromosomal abnormalities, and mitotic index, especially after silver nanoparticle treatment.[87] When single-walled nanoparticles were treated to the protoplast of *Arabidopsis thaliana* and *Oryza sativa*, the plants showed cell death, ROS production, and DNA damage.[88] Soya beans exposed to CeO_2 and ZnO caused homologous recombination, addition or deletion, DNA damage, and mutation.[89] In *Lycopersicon esculentum* and *Spinacia oleracea*, stress modified the photosystem II quantum yield and light absorption (Figure 15.6).[90]

The quantum yields a photosynthetic process are impacted by silver and gold nanoparticles.[92] In agriculture, nanoparticles and nanomaterials play crucial roles.[93] Food products and groundwater are contaminated when the nanoparticles are present in excess concentrations in the ecosystem.[94] The shape and size of the nanoparticles facilitates their penetration into the surface and persistence in environmental systems.[95] Permeated nanoparticles accumulate inside the cells, leading to intracellular alterations and production of reactive nitrogen species and ROS.[96] Various nanoparticles

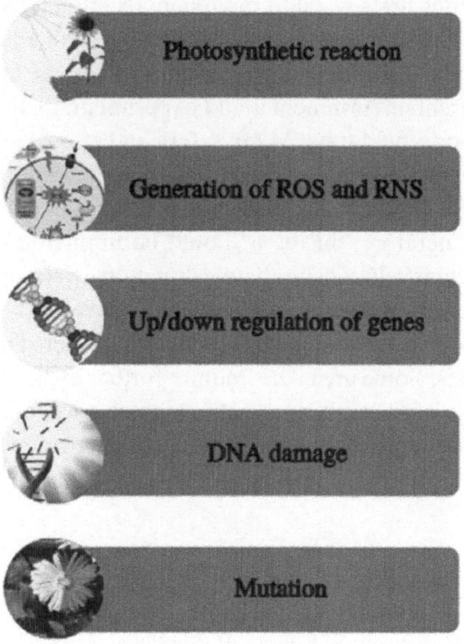

FIGURE 15.6 Detrimental effects of nanoparticles on plants. (From Karupannan et al., 2020.[91] With Permission.)

and nanomaterials are thought to be hazardous to people and other living things.[97] Particle shape, size, abiotic variables, and other such factors have an effect on the properties of nanoparticles. ROS production, physical limitations, or chemical and physical characteristics are examples of how the indirect effects of nanoparticles manifest.[98]

15.7 SUMMARY AND OUTLOOK

An effective and environmentally acceptable way of creating different nanoparticles with different compositions is through phytosynthesis. When phytonanoparticles are formed, phytochemicals (biochemicals present in the plant extract) can function as both stabilizing and reducing agents. The main advantages of using phytonanoparticles are their bioactivity and biocompatibility because they contain biomolecules from plants. Because of their outstanding qualities and environmentally friendly manufacturing process, phytonanoparticles hold great promise in a variety of industries including biomedical applications, bioremediation, cosmetics, and agriculture. Customizing the optical, electrical, and electronic characteristics of nanoparticles by adjusting their size and shape is a significant challenge in nanotechnology. Utilizing phytosynthesized nanoparticles with a variety of surface modifications, shapes/sizes, and chemical compositions may effectively be a cutting-edge, environmentally friendly, and practical method for reducing the use of harmful chemicals in conventional practices.

In several fields, including environmental cleanup, medicines, agriculture, and others, phytonanotechnology has a potential for future use. To manipulate the genetic data in plants, phytonanotechnology is used to create tools for the transport and delivery of agrochemicals and bioactive substances.

Phytosynthesized nanoparticles offer long-term product safety and are more economical, time- and labor-efficient, and environmentally friendly. Future research can assist in debunking the myths in this subject to transfer the technology to on-field uses and develop techniques to discover their potential in numerous other industries because phytonanotechnology still has several unknown and unexplored regions and little laboratory-to-field work. The growth and development of plants have benefited from several laboratory studies. The long-term effects of the usage of the nanoparticles must be assessed under different environmental and experimental situations before moving further with a large-scale procedure or a field type. Many safety and ethical concerns have been raised: for instance, it is mostly unclear how nanoparticles are absorbed, transported, and finally eliminated.

In conclusion, phytonanotechnology has a potential prospects in the treatment of various diseases and drug delivery systems. Therefore, this area should be improved for further field applications. Clarification of the biomolecular relationship between nanoparticles and the control of the expression of genes is required. The sub-molecular- and molecular-level activities of nanomaterials require further research. Drug transport in the animal tissue requires further study in the future, similar to the agrochemical delivery in plants. Some areas that require further exploration in phytonanotechnology include the continuous release of therapeutic agents, targeted particular release, the capacity to halt the discharge of drugs within the animal body, and nontoxic degradation or removal of the carrier nanoparticle.

REFERENCES

1 P. Wang, E. Lombi, F. J. Zhao and P. M. Kopittke, *Trends Plant Sci.*, 2016, **21**, 699–712.

2 A. Singh, M. Handa, M. Ruwali, S. J. Flora, R. Shukla and P. Kesharwani, *Process Biochem.*, 2021, **109**, 198–206.

3 V. Gawande, H. Choudhury and P. Kesharwani. In: *Dendrimer-Based Nanotherapeutics*, Elsevier, 2021, pp. 75–94.

4 D. Sahoo, R. Bandaru, S. K. Samal, R. Naik, P. Kumar, P. Kesharwani and R. Dandela, *Theory Appl. Nonparenteral Nanomedicines*, 2021, 181–207.

5 R. Thakare, G. Kaul, M. Shukla, P. Kesharwani, N. Srinivas, A. Dasgupta and S. Chopra, *Drug Discov. Target. Drug-Resistant Bact.*, 2020, 105–138.

6 A. Sheikh, S. Md and P. Kesharwani, *Biomed. Pharmacother.*, 2022, **146**, 112530.

7 M. Fatima, A. Sheikh, N. Hasan, A. Sahebkar, Y. Riadi and P. Kesharwani, *Eur. Polym. J.*, 2022, **170**, 111156.

8 V. Singh, S. Md, N. A. Alhakamy and P. Kesharwani, *Eur. Polym. J.*, 2022, **162**, 110883.

9 F. Zeeshan, T. Madheswaran, J. Panneerselvam, R. Taliyan and P. Kesharwani, *J. Pharm. Sci.*, DOI:10.1016/j.xphs.2021.05.001.

10 R. Kumar Bandaru, S. R. Rout, G. Kenguva, B. Gorain, N. A. Alhakamy, P. Kesharwani and R. Dandela, *Front. Pharmacol.*, DOI:10.3389/FPHAR.2021.780582.

11 S. A. Kumbhar, B. Gorain, H. Choudhury and P. Kesharwani. In: *Dendrimer-Based Nanotherapeutics*, Elsevier, 2021, pp. 143–162.

12 B. Gorain, H. Choudhury, S. Patro Sisinthy and P. Kesharwani, *Nanotechnology-Based Approaches for Tuberculosis Treatment*, 2020, 175–191.

13 R. Shah, T. Soni, U. Shah, B. N. Suhagia, M. N. Patel, T. Patel, G. A. Gabr, B. Gorain and P. Kesharwani, *J. Biomater. Sci. Polym. Ed.*, DOI:10.1080/09205063.2020.1870378.

14 S. Gorantla, G. Wadhwa, S. Jain, S. Sankar, K. Nuwal, A. Mahmood, S. K. Dubey, R. Taliyan, P. Kesharwani and G. Singhvi, *Drug Deliv. Transl. Res.*, DOI:10.1007/S13346-021-01097-Z.

15 S. Kumar Dubey, R. Pradhan, S. Hejmady, G. Singhvi, H. Choudhury, B. Gorain and P. Kesharwani, *Int. J. Pharm.*, 2021, **600**, 120499.

16 V. Srivastava, A. Singh, G. K. Jain, F. J. Ahmad, R. Shukla and P. Kesharwani, *Process Biochem.*, 2021, **109**, 130–142.

17 A. Singh, S. K. Maharana, R. Shukla and P. Kesharwani, *Process Biochem.*, 2021, **110**, 181–194.

18 G. Nian Xu, X. Liang Qiao, X. Lin Qiu and J. Guo Chen, *Colloids Surf. A Physicochem. Eng. Asp.*, 2008, **320**, 222–226.

19 H. Wang, X. Qiao, J. Chen and S. Ding, *Colloids Surf. A Physicochem. Eng. Asp.*, 2005, **256**, 111–115.

20 J. Huang, Q. Li, D. Sun, Y. Lu, Y. Su, X. Yang, H. Wang, Y. Wang, W. Shao, N. He, J. Hong and C. Chen, *Nanotechnology*, 2007, **18**, 105104.

21 S. Basavaraja, S. D. Balaji, A. Lagashetty, A. H. Rajasab and A. Venkataraman, *Mater. Res. Bull.*, 2008, **43**, 1164–1170.

22 V. K. Sharma, R. A. Yngard and Y. Lin, *Adv. Colloid Interface Sci.*, 2009, **145**, 83–96.

23 R. A. Bapat, T. V. Chaubal, C. P. Joshi, P. R. Bapat, H. Choudhury, M. Pandey, B. Gorain and P. Kesharwani, *Mater. Sci. Eng. C*, 2018, **91**, 881–898.

24 H. Choudhury, M. Pandey, Y. Q. Lim, C. Y. Low, C. T. Lee, T. C. L. Marilyn, H. S. Loh, Y. P. Lim, C. F. Lee, S. K. Bhattamishra, P. Kesharwani and B. Gorain, *Mater. Sci. Eng. C*, 2020, 112.

25 D. Cruz, P. L. Falé, A. Mourato, P. D. Vaz, M. Luisa Serralheiro and A. R. L. Lino, *Colloids Surfaces B Biointerfaces*, 2010, **81**, 67–73.

26 M. Amin, F. Anwar, M. R. S. A. Janjua, M. A. Iqbal and U. Rashid, *Int. J. Mol. Sci.* 2012, **13**, 9923–9941.

27 K. Rajaram, D. C. Aiswarya and P. Sureshkumar, *Mater. Lett.*, 2015, **138**, 251–254.

28 D. Philip, *Spectrochim. Acta Part A Mol. Biomol. Spectrosc.*, 2011, **78**, 327–331.

29 P. Singh, Y. J. Kim, D. Zhang and D. C. Yang, *Trends Biotechnol.*, 2016, **34**, 588–599.

30 K. Murugan, G. Benelli, C. Panneerselvam, J. Subramaniam, T. Jeyalalitha, D. Dinesh, M. Nicoletti, J. S. Hwang, U. Suresh and P. Madhiyazhagan, *Exp. Parasitol.*, 2015, **153**, 129–138.

31 C. A. Francis, http://dx.doi.org/10.1300/J064v01n01_08, 2008, **1**, 97–106.

32 A. Gogos, K. Knauer and T. D. Bucheli, *J. Agric. Food Chem.*, 2012, **60**, 9781–9792.

33 S. Mathew, *Phytonanotechnology Challenges Prospect.*, 2020, 1–20.

34 R. T. Tom, V. Suryanarayanan, P. G. Reddy, S. Baskaran and T. Pradeep, *Langmuir*, 2004, **20**, 1909–1914.

35 H. Choudhury, M. Pandey, T. H. Yin, T. Kaur, G. W. Jia, S. Q. L. Tan, H. Weijie, E. K. S. Yang, C. G. Keat, S. K. Bhattamishra, P. Kesharwani, S. Md, N. Molugulu, M. R. Pichika and B. Gorain, *Mater. Sci. Eng. C*, 2019, 101, 596–613.

36 P. Kesharwani, V. Gajbhiye, R. K. Tekade and N. K. Jain, 2011, 12, 1478–1497.

37 D. Luong, P. Kesharwani, B. A. Killinger, A. Moszczynska, F. H. Sarkar, S. Padhye, A. K. Rishi and A. K. Iyer, *J. Colloid Interface Sci.*, DOI:10.1016/j.jcis.2016.08.061.

38 B. Gorain, H. Choudhury, A. B. Nair, S. K. Dubey and P. Kesharwani, *Drug Discov. Today*, 2020, **25**, 1174–1188.

39 H. Choudhury, B. Gorain, M. Pandey, R. K. Khurana and P. Kesharwani, *Int. J. Pharm.*, 2019, **565**, 509–522.

40 J. Y. Song, H. K. Jang and B. S. Kim, *Process Biochem.*, 2009, **44**, 1133–1138.

41 J. Kurepa, T. Paunesku, S. Vogt, H. Arora, B. M. Rabatic, J. Lu, M. B. Wanzer, G. E. Woloschak and J. A. Smalle, *Nano Lett.*, 2010, **10**, 2296–2302.

42 C. M. Rico, S. Majumdar, M. Duarte-Gardea, J. R. Peralta-Videa and J. L. Gardea-Torresdey, *J. Agric. Food Chem.*, 2011, **59**, 3485–3498.

43 H. Wang, X. Kou, Z. Pei, J. Q. Xiao, X. Shan and B. Xing, *Nanotoxicology*, 2011, **5**, 30–42.

44 K. Birbaum, R. Brogioli, M. Schellenberg, E. Martinoia, W. J. Stark, D. Günther and L. K. Limbach, *Environ. Sci. Technol.*, 2010, **44**, 8718–8723.

45 M. A. Khan, T. Khan, Z. ur R. Mashwani, M. S. Riaz, N. Ullah, H. Ali and A. Nadhman, *Compr. Anal. Chem.*, 2019, **84**, 23–54.

46 S. Mukherjee, D. Chowdhury, R. Kotcherlakota, S. Patra, V. B, M. P. Bhadra, B. Sreedhar and C. R. Patra, *Theranostics*, DOI:10.7150/thno.7819.

47 V. Mishra, R. K. Mishra, A. Dikshit and A. C. Pandey, *Emerg. Technol. Manag. Crop Stress Toler. Biol. Tech.*, 2014, **1**, 159–180.

48 S. Mishra, B. R. Singh, A. Singh, C. Keswani, A. H. Naqvi and H. B. Singh, *PLoS One*, 2014, **9**, e97881.

49 S. Mishra and H. B. Singh, *Microbiol. Res.*, 2015, **172**, 16–18.

50 V. Kumar and S. K. Yadav, *J. Chem. Technol. Biotechnol.*, 2009, **84**, 151–157.

51 R. Raliya, I. Rathore and J. C. Tarafdar, *J. Bionanoscience*, 2013, **7**, 590–596.

52 V. V Makarov, M. B. B., A. J. Love, Л. А., O. V. Sinitsyna, С. О. В., S. S. Makarova, М. С. С., I. V. Yaminsky, Я. И. В., M. E. Taliansky, Т. М. Э., N. O. Kalinina and К. Н. О., *Acta Naturae*, 2014, **6**, 35–44.

53 P. Logeswari, S. Silambarasan and J. Abraham, *Sci. Iran.*, 2013, **20**, 1049–1054.

54 T. T. N. Nguyen, T. T. Vo, B. N. H. Nguyen, D. T. Nguyen, V. S. Dang, C. H. Dang and T. D. Nguyen, *Environ. Sci. Pollut. Res.* 2018, **25**, 34247–34261.
55 W. Liu, Y. Yu, J. Du and C. Jing, *J. Hazard. Mater.*, 2019, **366**, 338–345.
56 J. Pardeike, A. Hommoss and R. H. Müller, *Int. J. Pharm.*, 2009, **366**, 170–184.
57 G. Pant, N. Nayak and R. Gyana Prasuna, *Appl. Nanosci.*, 2013, **3**, 431–439.
58 A. Herman, *Curr. Microbiol.* 2018, **76**, 744–754.
59 R. Augustine and A. Hasan, *J. Drug Deliv. Sci. Technol.*, 2020, **56**, 101516.
60 R. Shi, H. Geng, M. Gong, J. Ye, C. Wu, X. Hu and L. Zhang, *J. Colloid Interface Sci.*, 2018, **509**, 275–284.
61 A. A. Zahid, R. Ahmed, S. Raza ur Rehman, R. Augustine, M. Tariq and A. Hasan, *Int. J. Biol. Macromol.*, 2019, **136**, 901–910.
62 R. Ahmed, M. Tariq, I. Ali, R. Asghar, P. Noorunnisa Khanam, R. Augustine and A. Hasan, *Int. J. Biol. Macromol.*, 2018, **120**, 385–393.
63 X. Xu, W. Zhong, S. Zhou, A. Trajtman and M. Alfa, *J. Appl. Polym. Sci.*, 2010, **118**, 588–595.
64 B. Joseph, R. Augustine, N. Kalarikkal, S. Thomas, B. Seantier and Y. Grohens, *Mater. Today Commun.*, 2019, **19**, 319–335.
65 K. M. Zadeh, A. S. Luyt, L. Zarif, R. Augustine, A. Hasan, M. Messori, M. K. Hassan and H. C. Yalcin, *Emergent Mater.*, 2019, **2**, 141–151.
66 R. Augustine, N. Kalarikkal and S. Thomas, *Appl. Nanosci.*, 2016, **6**, 337–344.
67 R. Augustine, A. Augustine, N. Kalarikkal and S. Thomas, *Prog. Biomater.*, 2016, **5**, 223–235.
68 R. Augustine, N. Kalarikkal and S. Thomas, *Appl. Nanosci.*, 2014, **4**, 809–818.
69 R. Veerasamy, T. Z. Xin, S. Gunasagaran, T. F. W. Xiang, E. F. C. Yang, N. Jeyakumar and S. A. Dhanaraj, *J. Saudi Chem. Soc.*, 2011, **15**, 113–120.
70 M. Vijayakumar, K. Priya, F. T. Nancy, A. Noorlidah and A. B. A. Ahmed, *Ind. Crops Prod.*, 2013, **41**, 235–240.
71 R. Sankar, A. Karthik, A. Prabu, S. Karthik, K. S. Shivashangari and V. Ravikumar, *Colloids Surf. B Biointerfaces*, 2013, **108**, 80–84.
72 C. Dipankar and S. Murugan, *Colloids Surf. B Biointerfaces*, 2012, **98**, 112–119.
73 T. Y. Suman, S. R. Radhika Rajasree, A. Kanchana and S. B. Elizabeth, *Colloids Surf. B Biointerfaces*, 2013, **106**, 74–78.
74 R. Augustine, D. L. Ashkenazi, R. S. Arzi, V. Zlobin, R. Shofti and A. Sosnik, *Acta Biomater.*, 2018, **74**, 344–359.
75 A. K. Mittal, J. Bhaumik, S. Kumar and U. C. Banerjee, *J. Colloid Interface Sci.*, 2014, **415**, 39–47.
76 R. Augustine, A. P. Mathew and A. Sosnik, *Appl. Mater. Today*, 2017, **7**, 91–103.
77 A. Mohammed Fayaz, M. Girilal, S. A. Mahdy, S. S. Somsundar, R. Venkatesan and P. T. Kalaichelvan, *Process Biochem.*, 2011, **46**, 636–641.
78 S. Patra, S. Mukherjee, A. K. Barui, A. Ganguly, B. Sreedhar and C. R. Patra, *Mater. Sci. Eng. C*, DOI:10.1016/j.msec.2015.04.048.
79 D. K. Tripathi, Shweta, S. Singh, S. Singh, R. Pandey, V. P. Singh, N. C. Sharma, S. M. Prasad, N. K. Dubey and D. K. Chauhan, *Plant Physiol. Biochem.*, 2017, **110**, 2–12.
80 D. K. Tripathi, R. K. Mishra, S. Singh, S. Singh, K. Vishwakarma, S. Sharma, V. P. Singh, P. K. Singh, S. M. Prasad, N. K. Dubey, A. C. Pandey, S. Sahi and D. K. Chauhan, *Front. Plant Sci.*, 2017, **8**, 1.
81 D. K. Tripathi, S. Singh, S. Singh, P. K. Srivastava, V. P. Singh, S. Singh, S. M. Prasad, P. K. Singh, N. K. Dubey, A. C. Pandey and D. K. Chauhan, *Plant Physiol. Biochem.*, 2017, **110**, 167–177.
82 R. Kaveh, Y. S. Li, S. Ranjbar, R. Tehrani, C. L. Brueck and B. Van Aken, *Environ. Sci. Technol.*, 2013, **47**, 10637–10644.
83 S. Yan, L. Zhao, H. Li, Q. Zhang, J. Tan, M. Huang, S. He and L. Li, *J. Hazard. Mater.*, 2013, **246–247**, 110–118.
84 H. Chu, H. J. Kim, J. Su Kim, M. S. Kim, B. D. Yoon, H. J. Park and C. Y. Kim, *Radiat. Phys. Chem.*, 2012, **81**, 180–184.
85 H. Qian, X. Peng, X. Han, J. Ren, L. Sun and Z. Fu, *J. Environ. Sci.*, 2013, **25**, 1947–1956.
86 B. Van Aken, *Curr. Opin. Biotechnol.*, 2015, **33**, 206–219.

87 A. S. Foltête, J. F. Masfaraud, E. Bigorgne, J. Nahmani, P. Chaurand, C. Botta, J. Labille, J. Rose, J. F. Férard and S. Cotelle, *Environ. Pollut.*, 2011, **159**, 2515–2522.
88 C. X. Shen, Q. F. Zhang, J. Li, F. C. Bi and N. Yao, *Am. J. Bot.*, 2010, **97**, 1602–1609.
89 M. L. López-Moreno, G. De La Rosa, J. A. Hernández-Viezcas, J. R. Peralta-Videa and J. L. Gardea-Torresdey, *J. Agric. Food Chem.*, 2010, **58**, 3689–3693.
90 C. M. Rico, J. R. Peralta-Videa and J. L. Gardea-Torresdey, *Nanotechnol. Plant Sci. Nanoparticles Their Impact Plants*, 2015, 1–17.
91 S. K. Karupannan, M. J. H. Dowlath and K. D. Arunachalam, *Phytonanotechnology Challenges Prospect.*, 2020, 303–322.
92 A. O. Govorov and I. Carmeli, *Nano Lett.*, 2007, **7**, 620–625.
93 S. Berkner, K. Schwirn and D. Voelker, *Environ. Toxicol. Chem.*, 2016, **35**, 780–787.
94 J. Jampilek, K. Zaruba, M. Oravec, M. Kunes, P. Babula, P. Ulbrich, I. Brezaniova, R. Opatrilova, J. Triska and P. Suchy, *Biomed Res. Int.*, DOI:10.1155/2015/812673.
95 M. Auffan, J. Rose, J. Y. Bottero, G. V. Lowry, J. P. Jolivet and M. R. Wiesner, *Nat. Nanotechnol.*, 2009.
96 J. Vestel, D. J. Caldwell, L. Constantine, V. J. D'Aco, T. Davidson, D. G. Dolan, S. P. Millard, R. Murray-Smith, N. J. Parke, J. J. Ryan, J. O. Straub and P. Wilson, *Environ. Toxicol. Chem.*, 2016, **35**, 1201–1212.
97 E. Navarro, A. Baun, R. Behra, N. B. Hartmann, J. Filser, A. J. Miao, A. Quigg, P. H. Santschi and L. Sigg, *Ecotoxicol.* 2008, **17**, 372–386.
98 R. D. Handy, R. Owen and E. Valsami-Jones, *Ecotoxicol.* 2008, **17**, 315–325.